In All Likelihood

In All Likelihood

Statistical Modelling and Inference Using Likelihood

Yudi Pawitan
Department of Medical Epidemiology and Biostatistics
Karolinska Institutet
Stockholm, Sweden
yudi.pawitan@ki.se

CLARENDON PRESS · OXFORD
2001

OXFORD
UNIVERSITY PRESS

Great Clarendon Street, Oxford OX2 6DP

Oxford University Press is a department of the University of Oxford.
It furthers the University's objective of excellence in research, scholarship,
and education by publishing worldwide in

Oxford New York

Auckland Cape Town Dar es Salaam Hong Kong Karachi
Kuala Lumpur Madrid Melbourne Mexico City Nairobi
New Delhi Shanghai Taipei Toronto

With offices in

Argentina Austria Brazil Chile Czech Republic France Greece
Guatemala Hungary Italy Japan South Korea Poland Portugal
Singapore Switzerland Thailand Turkey Ukraine Vietnam

Oxford is a registered trade mark of Oxford University Press
in the UK and in certain other countries

Published in the United States
by Oxford University Press Inc., New York

First published 2001

British Library Cataloguing in Publication Data
Data available

Library of Congress Cataloging in Publication Data

ISBN 978 0 19 850765 9

10 9 8 7

Typeset by Yudi Pawitan.
Printed in Great Britain on acid-free paper by the
MPG Books Group, Bodmin and King's Lynn

Preface

Likelihood is the central concept in statistical modelling and inference. *In All Likelihood* covers the essential aspects of likelihood-based modelling as well as likelihood's fundamental role in inference. The title is a gentle reminder of the original meaning of 'likelihood' as a measure of uncertainty, a Fisherian view that tends to be forgotten under the weight of likelihood's more technical role.

Fisher coined the term 'likelihood' in 1921 to distinguish the method of maximum likelihood from the Bayesian or inverse probability argument. In the early days its application was fairly limited; few statistical techniques from the 1920s to 1950s could be called 'likelihood-based'. To see why, let us consider what we mean by 'statistical activities':

- *planning:* making decisions about the study design or sampling protocol, what measurements to take, stratification, sample size, etc.
- *describing:* summarizing the bulk of data in few quantities, finding or revealing meaningful patterns or trends, etc.
- *modelling:* developing mathematical models with few parameters to represent the patterns, or to explain the variability in terms of relationship between variables.
- *inference:* assessing whether we are seeing a real or spurious pattern or relationship, which typically involves an evaluation of the uncertainty in the parameter estimates.
- *model checking:* assessing whether the model is sensible for the data. The most common form of model checking is residual analysis.

A lot of early statistical works was focused on the first two activities, for which likelihood thinking does not make much contribution. Often the activity moved directly from description to inference with little modelling in between. Also, the early modelling scene was dominated by the normal-based linear models, so statisticians could survive with least-squares, and t tests or F tests (or rank tests if the data misbehaved).

The emergence of likelihood-based modelling had to wait for both the advent of computing power and the arrival of more challenging data analysis problems. These problems typically involve nonnormal outcome data, with possible complexities in their collection such as censoring, repeated measures, etc. In these applications, modelling is important to impose structure or achieve simplification. This is where the likelihood becomes indispensable.

Plan of the book

The chapters in this book can be categorized loosely according to

- modelling: Chapters 4, 6, 11, 14, 17, 18;
- inference: Chapters 2, 3, 5, 7, 10, 13, 15, 16.

The inference chapters describe the anatomy of likelihood, while the modelling chapters show its physiology or functioning. The other chapters are historical (Chapter 1) or technical support (Chapters 8, 9, 12).

There is no need to proceed sequentially. Traditionally, likelihood inference requires the large sample theory covered in Chapter 9, so some instructors might feel more comfortable to see the theory developed first. *Some sections are starred* to indicate that they can be skipped on first reading, or they are optional as teaching material, or they involve ideas from future sections. In the last case, the section is there more for organizational reasons, so some 'nonlinear' reading might be required.

There is much more material here than can be covered in two semesters. In about 50 lectures to beginning graduate students I covered a selection from Chapters 2 to 6, 8 to 11, 13 and 14. Chapter 1 is mostly for reading; I use the first lecture to discuss the nature of statistical problems and the different schools of statistics. Chapter 7 is also left as reading material. Chapter 12 is usually covered in a separate statistical computing course. Ideally Chapter 15 is covered together with Chapters 13 and 14, while the last three chapters also form a unit on mixed models. So, for a more leisurely pace, Chapters 13 to 14 can be removed from the list above, and covered separately in a more advanced modelling course that covers Chapters 13 to 18.

Prerequisites

This book is intended for senior students of statistics, which include advanced undergraduate or beginning graduate students. Students taking this course should already have

- two semesters of introductory applied statistics. They should be familiar with common statistical procedures such as z, t, and χ^2 tests, P-value, simple linear regression, least-squares principle and analysis of variance.
- two semesters of introduction to probability and theory of statistics. They should be familiar with standard probability models such as the binomial, negative binomial, Poisson, normal, exponential, gamma, etc.; with the concepts of conditional expectation, Bayes theorem, transformation of random variables; with rudimentary concepts of estimation, such as bias and the method of moments; and with the central limit theorem.

- two semesters of calculus, including partial derivatives, and some matrix algebra.
- some familiarity with a flexible statistical software package such as `Splus` or `R`. Ideally this is learned in conjunction with the applied statistics course above.

The mathematical content of the book is kept relatively low (relative to what is possible). I have tried to present the whole spectrum of likelihood ideas from both applied and theoretical perspectives, both showing the depth of the ideas. To make these accessible I am relying (most of the time) on a nontechnical approach, using heuristic arguments and encouraging intuitive understanding. What is intuitive for me, however, may not be so for the reader, so sometimes the reader needs to balance the personal words with the impersonal mathematics.

Computations and examples

Likelihood-based methods are inherently computational, so computing is an essential part of the course. Inability to compute impedes our thought processes, which in turn will hamper our understanding and willingness to experiment. For this purpose it is worth learning a statistical software package. However, not all packages are created equal; different packages have different strengths and weaknesses. In choosing a software package for this course, bear in mind that here we are not trying to perform routine statistical analyses, but to learn and understand what is behind them, so graphics and programming flexibility are paramount.

All the examples in this book can be programmed and displayed quite naturally using `R` or `Splus`. `R` is *free* statistical programming software developed by a dedicated group of statisticians; it can be downloaded from `http://cran.r-project.org`.

Most educators tell us that understanding is best achieved through direct experience, in effect letting the knowledge pass through the fingers rather than the ears and the eyes only. Students can get such an experience from verifying or recreating the examples, solving the exercises, asking questions that require further computations, and, best still, trying out the methodology with their own data. To help, I have put all the `R` programs I used for the examples in `http://www.meb.ki.se/~yudpaw`.

Acknowledgements

This book is an outgrowth of the lecture notes for a mathematical statistics course in University College Dublin and University College Cork. I am grateful to the students and staff who attended the course, in particular Phil Boland, John Connolly, Gabrielle Kelly and Dave Williams, and to University College Dublin for funding the sabbatical leave that allowed me to transform the notes into tidy paragraphs. Jim Lindsey was generous with comments and encouragement. Kathleen Loughran and Áine Allen were of enormous help in hunting down various errors; I would kindly ask the reader to please let me know if they spot any remaining errors. It is impossible to express enough gratitude to my parents. During the sabbatical leave my mother took a special interest in the book, checking every morning whether it was finished. In her own way she made it possible for me to concentrate fully on writing. And finally, *buíochas le mo bhean chéile* Marie Reilly.

Y.P.
Cork
April, 2001

Notes on the 2003 corrected edition

Since the book was published in 2001 I received many kind words as well as corrections, which I took as indications that the book is being read and that the effort of writing it had been worthwhile. Special thanks go to John Nelder and Hiroshi Okamura for such a careful reading of the book. I am also grateful to Pat Altham, Harry Southworth, Aji Hamim Wigena, Jin Peng and various other people for comments and for informing me of errors. The current website for R programs, datasets and list of errors is moved to `http://www.mep.ki.se/~yudpaw`.

Y.P.
Stockholm
June, 2003

Notes on the 2008 corrected edition

I have corrected around 15 errors found since 2003 (these are listed in my website); the good news is that – assuming people are still reading this book – the rate of errors found has reduced dramatically. The current website for R programs, datasets and list of errors is moved to `http://www.meb.ki.se/~yudpaw`.

Y.P.
Stockholm
September, 2008

Contents

1
Introduction

Statistical modelling and inference have grown, above all else, to deal with variation and uncertainty. This may sound like an ambitious undertaking, since anyone going through life, even quietly, realizes ubiquitous uncertainties. It is not obvious that we can say something rigorous, scientific or even just sensible in the face of uncertainty.

Different schools of thought in statistics have emerged in reaction to uncertainty. In the Bayesian world all uncertainties can be modelled and processed through the standard rules of probability. Frequentism is more sceptical as it limits the type of uncertainty that can be studied statistically. Our focus is on the likelihood or Fisherian school, which offers a Bayesian–frequentist compromise. The purpose of this chapter is to discuss the background and motivation of these approaches to statistics.

1.1 Prototype of statistical problems

Consider the simplest nontrivial statistical problem, involving only *two* values. Recent studies show a significant number of drivers talk on their mobile phones while driving. Has there been an impact on accident rates? Suppose the number of traffic deaths increases from 170 last year to 190 this year. Numerically 190 is greater than 170, but it is not clear if the increase is 'real'. Suppose instead the number this year is 174, then in this case we feel intuitively that the change is not 'real'. If the number is 300 we feel more confident that it is a 'real' increase (although it is a totally different matter whether the increase can be attributed to mobile-phone use; see Redelmeier and Tibshirani (1997) for a report on the risk of car collision among drivers while using mobile phones).

Let us say that a change is 'significant' if we sense that it is a 'real' change. At the intuitive level, what is this sense of significance? It definitely responds to a numerical stimulus since we 'feel' 174 is different from 300. At which point do we change from being uncertain to being more confident? There is nothing in the basic laws of arithmetic or calculus that can supply us with a numerical answer to this problem. And for sure the answer cannot be found in the totality of the data itself (the two values in this case).

Uncertainty is pervasive in problems that deal with the real world, but statistics is the only branch of science that puts systematic effort into dealing with uncertainty. Statistics is suited to problems with inherent uncertainty due to limited information; it does not aim to remove uncertainty, but in many cases it merely quantifies it; uncertainty can remain even after an analysis is finished.

Aspirin data example

In a landmark study of the preventive benefits of low-dose aspirin for healthy individuals (Steering Committee of the Physicians' Health Study Research Group 1989), a total of 22,071 healthy physicians were randomized to either aspirin or placebo groups, and were followed for an average of 5 years. The number of heart attacks and strokes during follow-up are shown in Table 1.1.

Group	Heart attacks	Strokes	Total
Aspirin	139	119	11,037
Placebo	239	98	11,034
Total	378	217	22,071

Table 1.1: *The number of heart attacks and strokes during follow-up in the Physicians' Health Study.*

The main medical question is statistical: is aspirin beneficial? Obviously, there were fewer heart attacks in the aspirin group, 139 versus 239, but we face the same question: is the evidence strong enough so we can answer the question with confidence? The side effects, as measured by the number of strokes, were greater in the aspirin group, although 119 versus 98 are not as convincing as the benefit.

Suppose we express the benefit of aspirin as a relative risk of

$$\frac{139/11,037}{239/11,034} = 0.58.$$

A relative risk of one indicates aspirin is not beneficial, while a value much less than one indicates a benefit. Is 0.58 'far enough' from one? Answering such a question requires a *stochastic model* that describes the data we observe. In this example, we may model the number of heart attacks in the aspirin group as binomial with probability θ_1 and those in the placebo group as binomial with probability θ_2. Then the true relative risk is $\theta \equiv \theta_1/\theta_2$.

Let us denote the observed relative risk by $\widehat{\theta} = 0.58$. No uncertainty is associated with this number, so it fails to address the statistical nature of the original question. Does the trial contain information that $\widehat{\theta}$ is truly 'much' less than one? Now suppose the study is 10 times larger, so, assuming similar event rates, we observed 1390 versus 2390 heart attacks. Then $\widehat{\theta} = 0.58$ as before, but intuitively the information is now stronger. So, the data must have contained some measure of precision about $\widehat{\theta}$, from which we can assess our confidence that it is far from one.

We can now state the basic problem of statistical inference: *how do we go from observed data to statements about the parameter of interest θ?*

1.2 Statistical problems and their models

Stochastic element

In a statistical problem there is an obvious *stochastic* or random element, which is not treated by the basic laws of arithmetic. In the traffic example, we intuitively accept that there are various contingencies or random events contributing to the number of deaths; in fact, we would be surprised if the two numbers were exactly the same. Thus statistical methods need stochastic models to deal with this aspect of the problem. The development of models and methods is the deductive or mathematical aspect of statistics.

While the mathematical manipulation of models is typically precise and potentially free from arguments, the choice of the model itself is, however, uncertain. This is important to keep in mind since the validity of most statistical analysis is conditional on the model being correct. It is a trade-off: we need some model to proceed with an analysis, especially with sparse data, but a wrong model can lead to a wrong conclusion.

Inductive process

Statistical problems are *inductive*: they deal with questions that arise as consequences of observing specific facts. The facts are usually the outcome of an experiment or a study. The questions are typically more general than the observations themselves; they ask for something not directly observed, but somehow logically contained in the observed data. We say we 'infer' something from the data. In the traffic deaths example, we want to compare the underlying accident/death rates after accounting for various contingencies that create randomness.

For deductive problems like mathematics, sometimes only parts of the available information are needed to establish a new theorem. In an inductive problem every piece of the data should be accounted for in reaching the main conclusion; ignoring parts of the data is generally not acceptable. An inductive problem that has some parallels with statistical inference is a court trial to establish the guilt or the innocence of a defendant. The witness's oath to tell 'the truth, the whole truth, and nothing but the truth' embodies the requirements of the inductive process.

In deductive problems the truth quality of the new theorem is the same as the quality of the 'data' (axioms, definitions and previous theorems) used in establishing it. In contrast, the degree of certainty in an inductive conclusion is typically stronger than the degree in the data constituent, and the truth quality of the conclusion improves as we use more and more data.

However, a single new item of information can destroy a carefully crafted conclusion; this aspect of inductive inference is ideal for mystery novels or courtroom dramas, but it can be a bane for practising statisticians.

Suppose we want to estimate the number of BSE- (Bovine Spongiform Encephalopathy, or 'mad-cow') infected cattle that entered the food chain in Ireland. This is not a trivial problem, but based on the observed number of BSE cases and some assumptions about the disease, we can estimate the number of infected animals slaughtered prior to showing symptoms. New but last-minute information on exported cattle might invalidate a current estimate; further information that exported animals have a different age distribution from the animals for domestic consumption will also change the estimate.

Statistics plays an important role in science because all scientific endeavours are inductive, although many scientific questions are deterministic rather than stochastic. The emergence of statistical science is partly the result of the effort to make the inductive process rigorous. However, Lipton (1993), a philosopher of science, warns that

> inductive inference is about weighing evidence and judging likelihood, not definite proof.

The inductive process is inherently underdetermined: the input does not guarantee a unique solution, implying that even a correct induction is fallible.

Empirical or mechanistic models

The models used to deal with statistical problems can be either empirical or mechanistic. The latter is limited to applications where there is detailed knowledge regarding the underlying processes. For example, Newtonian laws in physics or Mendelian laws in genetics are mechanistic models. Here the exact relationships between the different quantities under observation are proposed mostly by some subject matter consideration rather than by looking at the data. A mechanistic model describes an underlying mechanism that explains the observed data.

Models in the applied sciences, such as medicine, epidemiology, psychology, climatology or agriculture, tend to be empirical. The analytical unit such as a human being or an area of land is usually too complex to be described by a scientific formula. If we model the number of deaths in the traffic example as having a Poisson distribution, we barely explain why we observe 170 rather than 100 deaths. Empirical models can be specified just by looking at the data without much subject matter consideration (this of course does not mean it is acceptable for a statistician to work on a desert island). The main requirement of an empirical model is that it explains the variability, rather than the underlying mechanism, in the observed data.

The separation between these two types of models is obviously not complete. There will be grey areas where some empirical evidence is used to help develop a mechanistic model, or a model may be composed of partly mechanistic and partly empirical submodels. The charge on the electron, for example, is an empirical quantity, but the (average) behaviour of electrons is mechanistically modelled by the quantum theory.

In the 19th and early 20th centuries most experiments were performed in the basic sciences; hence scientific models then were mostly mechanistic. The rise of empirical modelling was a liberating influence. Now experiments can be performed in most applied sciences, or even 'worse': data can be collected from observational studies rather than controlled experiments. Most of the general models in statistics, such as classes of distributions and linear or nonlinear regression models, are empirical models. Thus the rise of statistical modelling coincides with empirical modelling.

While empirical models are widely applicable, we must recognize their limitations; see Example 1.1. A mechanistic model is more satisfying than an empirical model, but a current empirical model may be a future mechanistic model. In some areas of statistical applications, there may never be a mechanistic model; for example, there will never be a mechanistic model for the number of traffic accidents. The compromise is an empirical model with as much subject matter input as possible.

The role of models from a statistical point of view is discussed further in Lehmann (1990) and Cox (1990).

Example 1.1: A classic example of an empirical model is the 18th century Bode's geometric law of progression of the planetary distance d_k from the Sun. Good (1969) and Efron (1971) provided a statistical evaluation of the 'reality' of this law, which specifies

$$d_k = 4 + 3 \times 2^k,$$

where $k = -\infty, 0, 1, \dots$ and d_k is scaled so that $d_1 = 10$ for Earth. With some 'jiggling' the law fitted very well for the known planets at the time it was proposed (planets as far as Saturn can be seen by the naked eye). To get a better fit, Jupiter was shifted up to position $k = 4$, leaving a missing spot at $k = 3$ between Mars and Jupiter. After the law was proposed there was a search for the 'missing planet'. Uranus at $k = 6$ was discovered first at the predicted distance, hence strengthening the confidence in the law. The missing planet was never found; there is, however, a band of asteroids at approximately the predicted distance.

Planet	k	Bode's law	Observed distance	Fourth-degree polynomial
Mercury	$-\infty$	4	4.0	4.1
Venus	0	7	7.2	6.7
Earth	1	10	10	10.2
Mars	2	16	15.3	16.0
?	3	28	?	26.9
Jupiter	4	52	51.9	50.0
Saturn	5	100	95.5	97.0
Uranus (1781)	6	196	191.4	186.5
Neptune (1846)	7	388	300.0	312.8
Pluto (1930)	8	772	394.6	388.2

Even though the formula fits the data well (up to Uranus; see Figure 1.1), the question remains: is this a 'real' physical law? As it happened, the law did not fit Neptune or Pluto. A better fit to the data is given by a fourth-degree polynomial, but now it is clear that we cannot attach much mechanistic value to the model. □

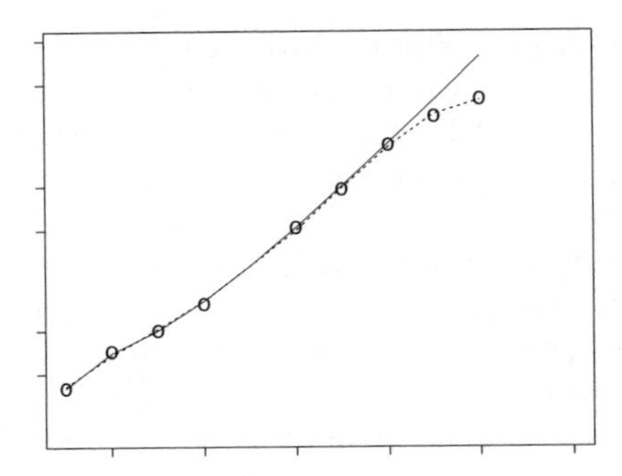

Figure 1.1: *Empirical model of planetary distances in terms of the order number from the Sun: Bode's law (solid) and a fourth-degree polynomial fit (dotted).*

1.3 Statistical uncertainty: inevitable controversies

> As far as the laws of mathematics refer to reality, they are not certain; and as far as they are certain they do not refer to reality. – Albert Einstein (1879–1955)

The characteristics discussed in the previous section, especially for empirical problems, militate to make statistical problems appear vague. Here it is useful to recognize two types of statistical uncertainty:

(i) *stochastic uncertainty:* this includes the uncertainty about a fixed parameter and a random outcome. This uncertainty is relatively easy to handle. Uncertainty about a fixed parameter, in principle, can always be reduced by performing a larger experiment. Many concepts in statistical inference deal with this uncertainty: sampling distribution, variability, confidence level, P-value, etc.

(ii) *inductive uncertainty:* owing to incomplete information, this uncertainty is more difficult to deal with, since we may be unable to quantify or control it.

Mathematically, we can view stochastic uncertainty as being conditional on an assumed model. Mathematics within the model can be precise and potentially within the control of the statistician. However, the choice of model itself carries an inductive uncertainty, which may be less precise and potentially beyond the control of the statistician.

The contrast between these two uncertainties is magnified when we are analysing a large dataset. Now the stochastic uncertainty becomes less important, while the inductive uncertainty is still very much there: Have we chosen the right class of models? Can we generalize what we find in the data? Have we considered and measured all the relevant variables? Are we asking the right questions? Given a set of data, depending on the way it was collected, there is usually an uncertainty about its variable definitions or meaning, wording and ordering of questions, representativeness of the sample, etc.

While it is possible to deal with stochastic uncertainty in an axiomatic way, it is doubtful that inductive uncertainty would ever yield to such an effort. It is important to recognize that, in statistical data analysis, inductive uncertainty is typically present in addition to the stochastic nature of the data itself. Due to the inductive process and the empirical nature of statistical problems, controversy is sometimes inevitable.

The traffic deaths example illustrates how controversies arise. If the number of deaths increases from 170 to 300, it would seem like a 'real' change and it would not be controversial to claim that the accident rate has increased, i.e. the uncertainty is small. But what if further scrutiny reveals one major traffic accident involving 25 cars and a large number of deaths, or an accident involving a bus where 40 people died? At this point we start thinking that, probably, a better way to look at the problem is by considering the number of accidents rather than deaths. Perhaps most accidents this year happened in the winter, whereas before they were distributed over the year. Possibly the number of younger drivers has increased, creating the need to split the data by age group. Splitting the data by years of driving experience may make more sense, but such a definition is only meaningful for drivers, while the death count also includes passengers and pedestrians!

This inductive process, which is very much a scientific process, raises two problems: one is that it tends to increase the stochastic uncertainty, since, by splitting the original observations into smaller explanatory groups, we are bound to compare smaller sets of numbers. The other problem is deciding where to stop in finding an explanation. There is no formal or precise answer to this question, so statisticians or scientists would have to deal with it on a case-by-case basis, often resorting to a judgement call. The closest guideline is to stop at a point where we have a reasonable control of stichastic uncertainty, deferring any decision on other factors of interest where too much uncertainty exists. Statisticians will have different experience, expertise, insight and prejudice, so from the same set of observations they might arrive at different conclusions. Beware! This is where we might find 'lies, damned lies and statistics'.

Pedagogic aspect

It is easier to learn, teach or describe methods that deal with stochastic uncertainty, and these have some chance of being mastered in a traditional academic or classroom setting. The unavoidable limitation of statistical texts is that they tend to concentrate on such methods. The joy and the pain of data analysis come as a reaction to uncertainties, so this discussion is not merely pedantic. Some might argue that the vagueness is part of the problem rather than of statistics, but even if we view it as such, the consequent difficulty in empirical model building and model selection is very much part of statistics and a statistician's life. This discussion also contains a warning that statisticians cannot work in a vacuum, since most of the relevant factors that create inductive uncertainties in a problem are subject matter specific.

1.4 The emergence of statistics

> It is impossible to calculate accurately events which are determined by chance. – Thucydides (*c.* 400BC)

There were two strands in the emergence of statistics. One was the development of the theory of probability, which had its original motivation in the calculation of expectation or uncertainties in gambling problems by Pascal (1623–1662) and Fermat (1601–1665). The theory was later developed on the mathematical side by Huygens (1629–1695), the Bernoulli brothers, in particular James Bernoulli (1654–1705), de Moivre (1667–1754) and Laplace (1749–1827), and on the logical side by Bayes (1701–1761), Boole (1815–1864) and Venn (1834–1923).

The growth of probability theory was an important milestone in the history of science. Fisher liked to comment that it was unknown to the Greek and the Islamic mathematicians (Thucydides was a historian); Persi Diaconis once declared that our brain is not wired to solve probability problems. With probability theory, for the first time since the birth of mathematics, we can make rigorous statements about uncertain events. The theory, however, is mostly deductive, which makes it a true branch of mathematics. Probability statements are evaluated as consequences of axioms or assumptions rather than specific observations. Statistics as the child of probability theory was born with the paper of Bayes in 1763 and was brought to maturity by Laplace.

The second strand in the emergence of statistics was an almost parallel development in the theory of errors. The main emphasis was not on the calculation of probabilities or uncertainties, but on summarizing observational data from astronomy or surveying. Gauss (1777–1855) was the main contributor in this area, notably with the principle of *least squares* as a general method of estimation. The important ingredient of this second line of development was the data-rich environment. In this connection Fisher noted the special role of Galton (1822-1911) in the birth of modern statistics towards the end of the 19th century. A compulsive data gatherer, Galton had a passionate conviction in the power of quantitative and statistical methods to deal with 'variable phenomena'.

Further progress in statistics continues to depend on data-rich environments. This was first supplied by experiments in agriculture and biometry, where Fisher was very much involved. Later applications include: industrial quality control, the military, engineering, psychology, business, medicine and health sciences. Other influences are found in data gathering and analysis for public or economic policies.

Bayesians and frequentists

The Bayesian and frequentist schools of statistics grew in response to problems of uncertainty, in particular to the way probability was viewed. The early writers in the 18th and 19th centuries considered it both a (subjective) degree of belief and (objective) long-run frequency. The 20th century brought a strong dichotomy. The frequentists limit probability to mean only a long-run frequency, while for the Bayesians it can carry the subjective notion of uncertainty.

This Bayesian–frequentist divide represents the fundamental tension between the need to say something relevant on a specific instance/dataset and the sense of objectivity in long-run frequencies. If we toss a coin, we have a sense of uncertainty about its outcome: we say the probability of heads is 0.5. Now, think about the *specific* next toss: can we say that our sense of uncertainty is 0.5, or is the number 0.5 meaningful only as a long-term average? Bayesians would accept both interpretations as being equally valid, but a true frequentist allows only the latter.

Since the two schools of thought generate different practical methodologies, the distinction is real and important. These disagreements do not hinder statistical applications, but they do indicate that the foundation of statistics is not settled. This tension also provides statistics with a fruitful dialectical process, at times injecting passion and emotion into a potentially dry subject. (Statisticians are probably unique among scientists with constant ponderings of the foundation of their subject; physicists are not expected to do that, though Einstein did argue with the quantum physicists about the role of quantum mechanics as the foundation of physics.)

Inverse probability: the Bayesians

The first modern method to assimilate observed data for quantitative inductive reasoning was published (posthumously) in 1763 by Bayes with his *Essay towards Solving a Problem in the Doctrine of Chances.* He used an inverse probability, via the now-standard Bayes theorem, to estimate a binomial probability. The simplest form of the Bayes theorem for two events A and B is

$$P(A|B) = \frac{P(AB)}{P(B)} = \frac{P(B|A)P(A)}{P(B|A)P(A) + P(B|\bar{A})P(\bar{A})}. \tag{1.1}$$

Suppose the unknown binomial probability is θ and the observed number of successes in n independent trials is x. Then, in modern notation, Bayes's solution is

$$f(\theta|x) = \frac{f(x,\theta)}{f(x)} = \frac{f(x|\theta)f(\theta)}{\int f(x|\theta)f(\theta)d\theta}, \tag{1.2}$$

where $f(\theta|x)$ is the conditional density of θ given x, $f(\theta)$ is the so-called prior density of θ and $f(x)$ is the marginal probability of x. (Note that we have used the symbol $f(\cdot)$ as a generic function, much like the way we use $P(\cdot)$ for probability. The named argument(s) of the function determines what the function is. Thus, $f(\theta, x)$ is the joint density of θ and x, $f(x|\theta)$ is the conditional density of x given θ, etc.)

Leaving aside the problem of specifying $f(\theta)$, Bayes had accomplished a giant step: he had put the problem of inductive inference (i.e. learning from data x) within the clean deductive steps of mathematics. Alas, 'the problem of specifying $f(\theta)$' a priori is an equally giant point of controversy up to the present day.

There is nothing controversial about the Bayes theorem (1.1), but (1.2) is a different matter. Both A and B in (1.1) are random events, while in the Bayesian use of (1.2) only x needs to be a random outcome; in a typical binomial experiment θ is an unknown fixed parameter. Bayes was well aware of this problem, which he overcame by considering that θ was generated in an *auxiliary physical experiment* – throwing a ball on a level square table – such that θ is expected to be uniform in the interval $(0, 1)$. Specifically, in this case we have $f(\theta) = 1$ and

$$f(\theta|x) = \frac{\theta^x(1-\theta)^{n-x}}{\int_0^1 u^x(1-u)^{n-x}du}. \tag{1.3}$$

Fisher was very respectful of Bayes's seeming apprehension about using an axiomatic prior; in fact, he used Bayes's auxiliary experiment to indicate that Bayes was not a Bayesian in the modern sense. If θ is a random variable then there is nothing 'Bayesian' in the use of the Bayes theorem. Frequentists do use Bayes theorem in applications that call for it.

Bayes did, however, write a *Scholium* (literally, a 'dissertation'; see Stigler 1982) immediately after his proposition:

> ... the same rule [i.e. formula (1.3) above] is a proper one to be used in the case of an event concerning the probability of which we absolutely know nothing antecedently to any trial made concerning it.

In effect, he accepted the irresistible temptation to say that if we know nothing about θ then it is equally probable to be between zero and one. More significantly, he accepted that the uniform prior density, which now can be purely axiomatic, can be processed with the objective binomial probability to produce a posterior probability. So, after all, Bayes was a Bayesian, albeit a reluctant one. (In hindsight, probability was then the only available concept of uncertainty, so Bayes did not have any choice.)

Bayes's paper went largely unnoticed until Pearson (1920). It was Laplace, who, after independently discovering Bayes theorem, developed Bayesian statistics as we understand it today. Boole's works on the probability theory (e.g. *Laws of Thought*, published in 1854), which discussed Bayes theorem in the 'problem of causes', clearly mentioned Laplace as the main reference. Laplace's *Théorie Analytique des Probabilités* was first published in 1812 and became the standard reference for the rest of the century. Laplace used the flat or uniform prior for all estimation problems, presented or justified as a reasonable expression of ignorance. The principle of inverse probability, hence Bayesian statistics, was an integral part of the teaching of probability until the end of the 19th century. Fisher (1936) commented that that was how he learned inverse probability in school and 'for some years saw no reason to question its validity'.

Statistical works by Gauss and others in the 19th and early 20th centuries were largely Bayesian with the use of inverse probability arguments. Even Fisher, who later became one of the strongest critics of axiomatic Bayesianism, in his 1912 paper 'On an absolute criterion for fitting frequency curves', erroneously called his maximum likelihood the 'most probable set of values', suggesting inverse probability rather than likelihood, although it was already clear he had distinguished these two concepts.

Repeated sampling principle: the frequentists

A dominant section of statistics today views probability formally as a long-run frequency based on repeated experiments. This is the basis of the frequentist ideas and methods, where the truth of a mathematical model must be validated through an objective measure based on externally observable quantities. This feels natural, but as Shafer (1990) identified, 'the rise of frequentism' in probability came only in the mid-19th century from the writings of empiricist philosophers such as John Stuart Mill. Population counting and classification was also a factor in the empirical meaning of probability when it was used for modelling.

The *repeated sampling principle* specifies that procedures should be evaluated on the basis of repeat experimentation under the same conditions. The sampling distribution theory, which expresses the possible outcomes from the repeated experiments, is central to the frequentist methodology. Many concepts in use today, such as bias, variability and standard error of a statistic, P-value, type I error probability and power of a test, or confidence level, are based on the repeated sampling principle. The dominance of these concepts in applied statistics today proves the practical power of frequentist methods. Neyman (1894–1981) and Wald (1902–1950) were the most influential exponents of the frequentist philosophy. Fisher contributed enormously to the frequentist methodology, but did not subscribe fully to the philosophy.

True frequentism states that measures of uncertainties are to be interpreted only in a repeated sampling sense. In areas of statistical application, such as medical laboratory science or industrial quality control, where procedures are naturally repeated many times, the frequentist measures are very relevant.

The problem arises as the requirement of repeat experimentation is allowed to be hypothetical. There are many areas of science where experiments are unlikely to be repeated, for example in archaeology, economics, geology, astronomy, medicine, etc. A reliance on repeated sampling ideas can lead to logical paradoxes that appear in common rather than esoteric procedures.

Extreme frequentism among practical statisticians is probably quite rare. An extremist will insist that an observed 95% confidence interval, say $1.3 < \theta < 7.1$, either covers the parameter or it does not, we do not know which, and there is no way to express the uncertainty; the 95% applies only to the procedure, not to the particular interval. That is in fact the orthodox interpretation of the confidence interval. It neglects the evidence contained in a particular interval/dataset, because measures of uncertainty are only interpreted in hypothetical repetitions.

Most scientists would probably interpret the confidence interval intuitively in a subjective/Bayesian way: there is a 95% probability the interval contains the true parameter, i.e. the value 95% has some evidential attachment to the observed interval.

Bayesians versus frequentists

> A great truth is a truth whose opposite is also a great truth. – Thomas Mann (1875–1955)

In Bayesian computations one starts by explicitly postulating that a parameter θ has a distribution with prior density $f(\theta)$; for example, in a problem to estimate a probability θ, one might assume it is uniformly distributed on (0,1). The distinguishing attitude here is that, since θ does not have to be a random outcome of an experiment, this prior can be specified axiomatically, based on thinking alone. This is the methodological starting point that separates the Bayesians from the frequentists, as the latter cannot accept that a parameter can have a distribution, since such a distribution does not have an external reality. Bayesians would say there is an uncertainty about θ and insist any uncertainty be expressed probabilistically. The distribution of θ is interpreted in a subjective way as a degree of belief.

Once one accepts the prior $f(\theta)$ for θ and agrees it can be treated as a regular density, the way to proceed is purely deductive and (internally) consistent. Assuming that, given θ, our data x follows a statistical model $p_\theta(x) = f(x|\theta)$, then the information about θ contained in the data is given by the *posterior* density, using the Bayes theorem as in (1.2),

$$f(\theta|x) = \frac{f(x|\theta)f(\theta)}{f(x)}.$$

In Bayesian thinking there is no operational difference between a prior density $f(\theta)$, which measures belief, and $f(x|\theta)$, which measures an observable quantity. These two things are conceptually equal as measures of uncertainty, and they can be mixed using the Bayes theorem.

The posterior density $f(\theta|x)$, in principle, captures from the data all the information that is relevant for θ. Hence, it is an update of the prior $f(\theta)$. In a sequence of experiments it is clear that the current posterior can function as a future prior, so the Bayesian method has a natural way of accumulating information.

When forced, most frequentists would probably admit that a degree of belief does exist subjectively. The disagreement is not that a parameter can assume a density, since frequentists could also think of $f(\theta)$ as a prior likelihood (the likelihood of the parameter before we have any data). Two genuine concerns exist:

(i) the practical problem of choosing an appropriate prior. Leaving aside the problem of subjective interpretation, there is an ongoing controversy on how we should pick $f(\theta)$. Several early writers such as Boole (1854, pages 384, 392) and Venn (1876) had criticized the arbitrariness in the axiomatic choice of $f(\theta)$; Fisher was also explicit in his rejection of any axiomatic prior, although *he did not rule out* that some applications, such as genetics, may have physically meaningful $f(\theta)$. Modern Bayesians seem to converge toward the so-called 'objective priors' (e.g. Gatsonis *et al.* 1997), but there are many shades of Bayesianism (Berger 2000).

(ii) the 'rules of engagement' regarding a subjective degree of belief. There is nothing really debatable about how one feels, and there is nothing wrong in thinking of probability in a subjective way. However, one's formal action based on such feeling is open to genuine disagreement. Treating a subjective probability density like a regular density function means, for example, that it can be integrated out, and it needs a Jacobian term when transformed to a different scale. The latter creates a lack of invariance in the choice of prior: seeming ignorance in one scale becomes information in another scale (see Section 2.8).

Efron (1998) compares the psychological differences between the two schools of thought. A comparative study highlights the strengths and weaknesses of each approach. The strength of the Bayesian school is its unified approach to all problems of uncertainty. Such unity provides clarity, especially in complex problems, though it does not mean Bayesian solutions are practical. In fact, until recently Bayesians could not solve complex problems because of computational difficulties (Efron 1986a). While, bound by fewer rules, the strength of a frequentist solution is usually its practicality.

Example 1.2: A new eye drug was tested against an old one on 10 subjects. The two drugs were randomly assigned to both eyes of each person. In all cases the new drug performed better than the old drug. The P-value from the observed data is $2^{-10} = 0.001$, showing that what we observe is not likely due to chance alone, or that it is very likely the new drug is better than the old one. □

Such simplicity is difficult to beat. Given that a physical randomization was actually used, very little extra assumption is needed to produce a valid conclusion. And the final conclusion, that the new drug is better than the old one, might be all we need to know from the experiment. The achieved simplicity is a reward of focus: we are only interested in knowing if chance alone could have produced the observed data. In real studies, of course, we might want to know more about the biological mechanism or possible side effects, which might involve more complicated measurements.

The advent of cheap computer power and Monte Carlo techniques (e.g. Gilks *et al.* 1995) have largely dismantled the Bayesian computational wall. Complex problems are now routinely solved using the Bayesian methodology. In fact, being pragmatic, one can separate the Bayesian numerical methods from the underlying philosophy, and use them as a means of obtaining likelihood functions. This is a recent trend, for example, in molecular genetics. In Section 10.6 we will see that the Bayesian and likelihood computations have close numerical connections.

Luckily, in large-sample problems, frequentist and Bayesian computations tend to produce similar numerical results, since in this case the data dominate the prior density and the level of uncertainty is small. In small- to medium-sized samples, the two approaches may not coincide, though in real data analysis the difference is usually of smaller order of magnitude than the inductive uncertainty in the data and in the model selection.

The following 'exchange paradox', discussed in detail by Christensen and Utts (1992), illustrates how our handling of uncertainty affects our logical thinking. To grasp the story quickly, or to entertain others with it, replace x by 100.

Example 1.3: A swami puts an unknown amount of money in one envelope and twice that amount in another. He asks you to pick one envelope at random, open it and then decide if you would exchange it with the other envelope. You pick one (randomly), open it and see the outcome $X = x$ dollars. You reason that, suppose Y is the content of the other envelope, then Y is either $x/2$ or $2x$ with probability 0.5; if you exchange it you are going to get $(x/2+2x)/2 = 5x/4$, which is bigger than your current x. 'With a gleam in your eye', you would exchange the envelope, wouldn't you?

The reasoning holds for any value of x, which means that you actually *do not need to open the envelope* in the first place, and you would still want to exchange it! Furthermore, when you get the second envelope, the same reasoning applies again, so you should exchange it back. A discussion of the Bayesian and frequentist aspects of this paradox is left as an exercise. □

1.5 Fisher and the third way

The likelihood approach offers a distinct 'third way', a Bayesian-frequentist compromise. We might call it Fisherian as it owes most of its conceptual development to Fisher (1890–1962). Fisher was clearly against the use of the axiomatic prior probability fundamental to the Bayesians, but he was equally emphatic in his rejection of long-run frequency as the only way to interpret probability. Fisher was a frequentist in his insistence that statistical inference should be objectively verifiable; however, his advocacy of likelihood inference in cases where probability-based inference is not available puts him closer to the Bayesian school.

In a stimulating paper on Fisher's legacies, Efron (1998) created a statistical triangle with Fisherian, Bayesian and frequentist nodes. He then placed various statistical techniques within the triangle to indicate their flavour.

Fisher's effort for an objective inference without any use of prior probability led him to the idea of *fiducial probability* (Fisher 1930, 1934). This concept prompted the confidence interval procedure (Neyman 1935). It appears that Fisher never managed to convince others what fiducial probability was, despite his insistence that, conceptually, it is 'entirely identical with the classical probability of the early writers' (Fisher 1973, page 54). In some models the fiducial probability coincides with the usual frequentist/long-run-frequency probability. The problems occur in more complex models where exact probability statements are not possible.

From his last book *Statistical Methods and Scientific Inference* (1973, in particular Chapter III) it is clear that Fisher settled with the idea that

- whenever possible to get exact results we should base inference on probability statements, otherwise it should be based on the likelihood;
- the likelihood can be interpreted subjectively as a rational degree of belief, but it is weaker than probability, since it does not allow an external verification, and
- in large samples there is a strengthening of likelihood statements where it becomes possible to attach some probabilistic properties ('asymptotic approach to a higher status' – Fisher 1973, page 78).

These seem to summarize the Fisherian view. (While Fisher's probability was fiducial probability, let us take him at his own words that it is 'entirely identical with the classical probability'.) About 40 years elapsed between the explicit definition of the likelihood for the purpose of estimation and Fisher's final judgement about likelihood inference. The distinguishing view is that *inference is possible directly from the likelihood function*; this is neither Bayesian nor frequentist, and in fact both schools would reject such a view as they allow only probability-based inference.

These Fisherian views also differ from the so-called 'pure likelihood view' that considers the likelihood as the sole carrier of uncertainty in statistical inference (e.g. Royall 1997, although he would call it 'evidence' rather than 'uncertainty'). Fisher recognized two 'well-defined levels of logical status' for uncertainty about parameters, one supplied by probability and the other by likelihood. A likelihood-based inference is used to 'analyze, summarize and communicate statistical evidence of types too weak to supply true probability statements' (Fisher 1973, page 75). Furthermore, when available, a probability statement must allow for an external verification (a verification by observable quantities), so it is clear that frequentist consideration is also an important aspect of the Fisherian view.

Fisher's requirement for an exact probability inference is more stringent than the so-called 'exact inference' in statistics today (Fisher 1973, pages 69–70). His prototype of an exact probability-based inference is the confidence interval for the normal mean (even though the term 'confidence interval' is Neyman's). The statement

$$P(\overline{x} - 1.96\sigma/\sqrt{n} < \mu < \overline{x} + 1.96\sigma/\sqrt{n}) = 0.95$$

is unambiguous and exactly/objectively verifiable; it is an ideal form of inference. However, the so-called 'exact 95% confidence interval' for the binomial proportion (see Section 5.8) in fact does not have exactly 95% coverage probability, so logically it is of lower status than the exact interval for the normal model. It is for this situation the likelihood is indicated.

For Fisher, both likelihood and probability are measures of uncertainty, but they are on a different footing. This is a non-Bayesian view, since for Bayesians all uncertainty is measured with probability. The subjective element in the interpretation of likelihood, however, is akin to a Bayesian/nonfrequentist attitude. It is worth noting that, when backed up with large-sample theory to supply probability statements, the mechanics and numerical results of likelihood inference are generally acceptable to frequentist statisticians. So, in their psychology, Fisherians are braver than the frequentists in saying that inference is possible from the likelihood function alone, but not as brave as the Bayesians to admit an axiomatic prior into the argument.

Legacies

By 1920 the field of statistics must have been a confusing place. Yates (1990) wrote that it was the age of correlation and coefficients of all kinds. To assess association in 2×2 tables there were the coefficient of association, coefficient of mean square contingency, coefficient of tetrachoric correlation, equiprobable tetrachoric correlation, and coefficient of colligation, but the idea of estimating the association and its test of significance were mixed up. There were many techniques available, such as the least squares principle, the method of moments, the inverse probability method, the χ^2 test, the normal distribution, Pearson's system of curves, the central limit theorem, etc., but there was no firm logical foundation.

The level of confusion is typified by the title of Edgeworth's paper in 1908 and Pearson's editorial in *Biometrika* in 1913: '*On the probable errors of frequency constants*', which in modern terminology would be 'the standard error of fixed parameters'. There was simply no logical distinction or available terms for a parameter and its estimate. On the mathematical side, the χ^2 test of association for the 2×2 tables had 3 degrees of freedom!

A more serious source of theoretical confusion seems to be the implicit use of inverse probability arguments in many early statistical works, no doubt the influence of Laplace. The role of the prior distribution in inverse probability arguments was never seriously questioned until early 20th century. When explicitly stated, the arbitrariness of the prior specification was probably a stumbling block to a proper appreciation of statistical questions as objective questions. Boole (1854) wrote in the *Laws of Thoughts* (Chapter XX, page 384) that such arbitrariness

> seems to imply, that definite solution is impossible, and to mark the point where inquiry ought to stop.

Boole discussed the inverse probability method at length and identified its weakness, but did not see any alternative; he considered the question of inductive inference as

> second to none other in the Theory of Probabilities in importance, [I hope it] will receive the careful attention which it deserves.

In his works on the theory of errors, Gauss was also aware of the problem, but he got around it by justifying his method of estimation in terms of the least-squares principle; this principle is still central in most standard introductions to regression models, which is unfortunate, since (i) in itself it is devoid of inferential content and (ii) it is not natural for general probability models, so it creates an unnecessary conceptual gap with the far richer class of generalized linear models.

Fisher answered Boole's challenge by clearly identifying the likelihood as the key inferential quantity that is free of subjective prior probabilities. He stressed that if, prior to the data, we know absolutely nothing about a parameter (recall Bayes's *Scholium* in Section 1.4) then all of the information from the data is in the likelihood. In the same subjective way the Bayesians interpret probability, the likelihood provides a 'rational degree of belief' or an 'order of preferences' on possible parameter values; the fundamental difference is that *the likelihood does not obey probability laws.* So probability and likelihood are different concepts available to deal with different levels of uncertainty.

There were earlier writers, such as Daniel Bernoulli or Venn, who had used or mentioned the idea of maximum likelihood in rudimentary forms (see Edwards 1992, Appendix 2). It usually appeared under the name of 'most probable value', indicating the influence of inverse probability argument. Even Fisher in 1912 used that name, even though it was clear from the discussion he had likelihood in mind. The confusion was only cleared in 1921 when Fisher invented the term 'likelihood'.

In a series of the most influential papers in statistics Fisher (in particular in 1922 and 1925) introduced order into the chaos by identifying and naming the fundamental concepts such as 'parameter', 'statistic', 'variance', 'sufficiency', 'consistency', 'information', and 'estimation','maximum likelihood estimate', 'efficiency' and 'optimality'. He was the first to use Greek letters for unknown parameters and Latin letters for the estimates. He set up the agenda for statistical research by identifying and formulating the important questions.

He 'fixed' the degree of freedom of the χ^2 test for the 2×2 tables in 1922. He recognized the paper by 'Student' in 1908 on the t-test, which was ignored by the large-sample-based statistical world at the time, as a milestone in the history of statistics: it was the first exact test. He emphasized the importance of inference based on exact distribution and identified 'the problem of distribution' as a respectable branch of theoretical statistics. Fisher was unsurpassed in this area, being the first to derive the exact distribution of the t and F statistics, as well as that of the sample correlation and multiple correlation coefficient.

Fisher's influence went beyond the foundation of statistics and the likelihood methods. His *Statistical Methods for Research Workers*, first published in 1925, brought the new ideas to generations of practical research workers. Fisher practically invented the field of experimental design, introducing the fundamental ideas of randomization, replication, blocking, factorial experiments, etc., and its analysis of variance. His *Design of Experiments*, first published in 1935, emphasized the importance of carefully collected data to simplify subsequent analysis and to arrive at unambiguous conclusions. He contributed significantly to areas of sampling distribution theory, regression analysis, extreme value theory, nonparametric and multivariate analysis. In a careful study of Fisher's legacy, Savage (1976) commented that it would be a lot faster to list areas in statistics where Fisher did *not* contribute fundamentally, for example sequential analysis and time series modelling.

Outside statistics, many geneticists consider Fisher as the most important evolutionary biologist after Darwin. In 1930 Fisher was the first to provide a key synthesis of Mendelian genetics and Darwin's theory of evolution, thus giving a quantitative basis for the latter. Fisher was never a professor of statistics: he was Galton Professor of Eugenics at University College London, then Balfour Professor of Genetics at Cambridge University.

For a statistician, his writings can be inspirational as they are full of conviction on the fundamental role and contributions of statistical methods in science and in 'refinement of human reasoning'. Fisher (1952) believed that

> Statistical Science was the peculiar aspect of human progress which gave to the twentieth century its special character. ... it is to the statistician that the present age turns for what is most essential in all its more important activities.

The 'important activities' include the experimental programmes, the observational surveys, the quality control engineering, etc. He identified the crucial contribution of statistical ideas to the fundamental scientific advances of the 19th century such as in Lyell's *Principles of Geology* and Darwin's theory of evolution.

It is an unfortunate turn of history that Fisher's articles and books are no longer standard reading in the study of statistics. Fisher was often criticized for being obscure or hard to read. Savage (1976), however, reported that his statistical mentors, which included Milton Friedman and W. Allen Wallis, gave the advice: 'To become a statistician, practice statistics and mull Fisher over with patience, respect and scepticism'. Savage closed his 1970 Fisher Memorial Lecture with 'I do hope that you won't let a week go by without reading a little bit of Fisher'.

Fisher's publications were collected in the five-volume *Collected Papers of R.A. Fisher*, edited by Bennett and Cornish (1974). His biography, entitled *R.A. Fisher, The Life of a Scientist*, was published by his daughter Joan Fisher Box in 1978. Other notable biographies, memoirs or reviews of his works were written by Barnard (1963), Bartlett (1965), Yates and Mather (1963), Kendall (1963), Neyman (1961, 1967), Pearson (1974) and Savage (1976). Recent articles include Aldrich (1997), Efron (1998) and Hald (1999). Edwards's (1992) book on likelihood was largely influenced by Fisher and the Appendices contain useful accounts of the history of likelihood and Fisher's key contributions. Fienberg and Hinkley (1980) contains a wide-ranging discussion of Fisher's papers and his impact on statistics.

1.6 Exercises

Exercise 1.1: Discuss the stochastic and inductive uncertainty in the following statements:

(a) A study shows that children of mothers who smoke have lower IQs than those of non-smoking mothers.

(b) A report by Interpol in 1994 shows a rate of (about) 55 crimes per 1000 people in the USA, compared to 100 in the UK and 125 in Sweden. ('Small' note: the newspaper that published the report later published a letter by an official from the local Swedish Embassy saying that, in Sweden, if a swindler defrauds 1000 people the case would be recorded as 1000 crimes.)

(c) Life expectancy in Indonesia is currently 64 years for women and 60 years for men. (To which generation do these numbers apply?)

(d) The current unemployment rate in Ireland is 4.7%. (What does 'unemployed' mean?)

(e) The total fertility rate for women in Kenya is 4.1 babies.

(f) The population of Cairo is around 16 million people. (Varies by a few million between night and day.)

(g) The national clinical trial of aspirin, conducted on about 22,000 healthy male physicians, established the benefit of taking aspirin. (To what population does the result apply?)

Exercise 1.2: What is wrong with the reasoning in the exchange paradox in Example 1.3? Discuss the Bayesian and frequentist aspects of the paradox, first assuming the 'game' is only played once, then assuming it is played repeatedly.

2

Elements of likelihood inference

2.1 Classical definition

The purpose of the likelihood function is to convey information about unknown quantities. The 'information' is incomplete, and the function will express the degree of incompleteness. Unknown quantities in statistical problems may be *fixed* parameters, with the associated *estimation* problem, or unobserved *random* values; in a real *prediction* problem the two unknowns can be easily mixed. We will consider the extended definition that solves the prediction problem in Section 16.2.

Recall first the standard mode of deductive mathematical thinking: given a probabilistic model and parameter values we derive a description of data. In a deductive mode we derive the consequences of certain assumptions. For example, if we perform a binomial experiment with parameters $n = 10$ and $\theta = 1/3$, and denote X to be the number of successes, then $P_\theta(X = 0) = 0.0282$, etc. This means if we repeat the experiment 10,000 times, we expect around 282 of them would yield no successes.

Now suppose we toss a coin 10 times and observe $X = 8$ heads. Based on this information alone, what is the probability of heads θ? (That is, assuming we know absolutely nothing about it prior to the experiment.) Information about θ is not complete, so there will be some uncertainty. Now, θ cannot be zero and is very unlikely to be very small. We can say this, since, *deductively* we know $P_\theta(X = 8)$ is zero or very tiny. In contrast, $\theta = 0.6$ or $\theta = 0.7$ is likely, since $P_\theta(X = 8) = 0.1209$ or 0.2335. We have thus found a deductive way of comparing different θ's: compare the probability of the observed data under different values of θ. As a function of the unknown parameter

$$L(\theta) = P_\theta(X = 8)$$

is called the *likelihood* function: see Figure 2.1. The plot shows θ is unlikely to be less than 0.5 or to be greater than 0.95, but is more likely to be in between. Given the data alone (and no other information) we should prefer values between 0.5 and 0.95 over values outside this interval.

In a simple and deductive way we have found *a numerical quantity to express the order of preferences on* θ. Of course we still do not know exactly where θ is, but we have captured the information provided in the data by showing where θ is likely to fall. The uncertainty in the data is inherent, and that is what is conveyed in the likelihood function.

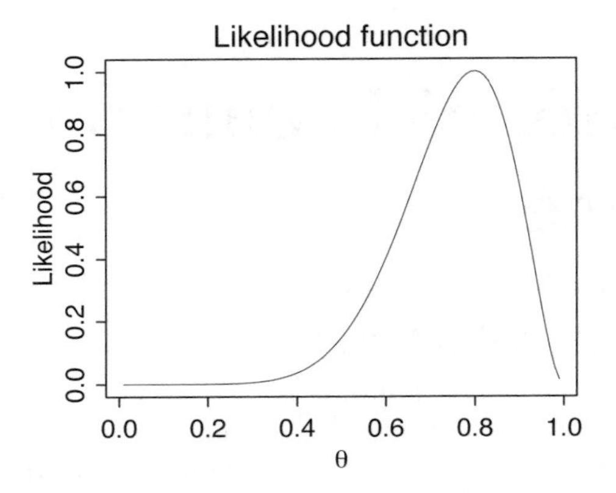

Figure 2.1: *Likelihood function of the success probability θ in a binomial experiment with $n = 10$ and $x = 8$. The function is normalized to have unit maximum.*

The likelihood provides us with a measure of relative preferences for various parameter values. Given a model, the likelihood $L(\theta)$ is an exact and objective quantity, hence a measure of 'rational belief'; it is objective in the sense that it exists outside any subjective preference. This is an important fact about the likelihood function as it implies that *quantities that we compute from the function are also exact and objective.* In practice, what we do or how we act given the information from the likelihood are another matter.

There is a tendency in classical teaching to focus immediately on the maximum of the likelihood and disregard the function itself. That is not a fruitful thought process regarding what we want to learn about θ from the data. Barnard *et al.* (1962) were emphatic that one should try the habit of sketching the likelihood functions for some time to realize how helpful they are. It is the *entire* likelihood function that is the carrier of information on θ, not its maximizer. In the above example the likelihood is maximized at 0.8, but there is a range of values of θ which are almost equally likely. In Section 2.5 we will examine in more detail the role of the maximum likelihood estimate.

Definition 2.1 *Assuming a statistical model parameterized by a fixed and unknown θ, the likelihood $L(\theta)$ is the probability of the observed data x considered as a function of θ.*

The generic data x include any set of observations we might get from an experiment of *any complexity*: a range of values rather than exact measurements, a vector of values, a matrix, an array of matrices, a time series or a

2D image. The generic parameter θ can also be as complex as the model requires; in particular it can be a vector of values. In the future chapters we will embark on a grand tour to show the richness of the likelihood world from a simple toy model to very complex studies.

Discrete models

There is no ambiguity about the probability of the observed data in the discrete models, since it is a well-defined nonzero quantity. For the binomial example above, the likelihood function is

$$\begin{aligned} L(\theta) &= P_\theta(X = x) \\ &= \binom{n}{x} \theta^x (1-\theta)^{n-x}. \end{aligned}$$

Figure 2.2 shows four likelihood functions computed from four binomial experiments with $n = 10$ and $x = 0, 2, 5, 10$. Interpretation of the functions is immediate. For example, when $x = 0$ the likelihood is concentrated near zero, indicating strong evidence that θ is very likely to be near zero.

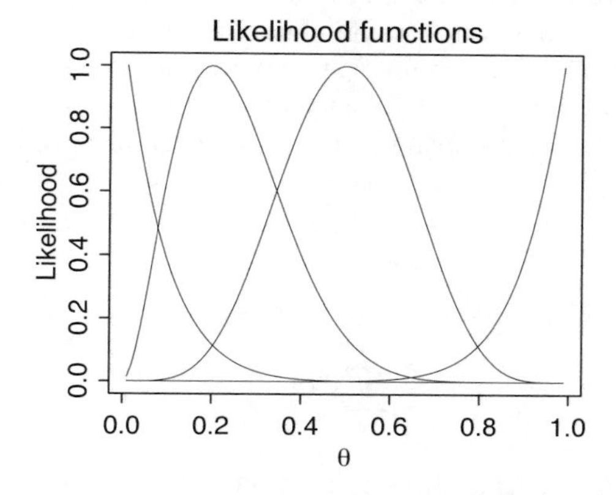

Figure 2.2: *Likelihood functions of the success probability θ in four binomial experiments with $n = 10$ and $x = 0, 2, 5, 10$. The functions are normalized to have unit maximum.*

Continuous models

A slight technical issue arises when dealing with continuous outcomes, since theoretically the probability of any point value x is zero. We can resolve this problem by admitting that in real life there is only a finite precision: observing x is short for observing $x \in (x - \epsilon/2, x + \epsilon/2)$, where ϵ is the precision limit. If ϵ is small enough, on observing x the likelihood for θ is

$$\begin{aligned} L(\theta) &= P_\theta\{X \in (x - \epsilon/2, x + \epsilon/2)\} \\ &= \int_{x-\epsilon/2}^{x+\epsilon/2} p_\theta(x)dx \approx \epsilon\ p_\theta(x). \qquad (2.1) \end{aligned}$$

For the purpose of comparing θ within the model $p_\theta(x)$ the likelihood is only meaningful up to an arbitrary constant (see Section 2.4), so we can ignore ϵ. Hence, *in all continuous models where the outcome x is observed with good precision we will simply use the density function $p_\theta(x)$ to compute the likelihood.*

In many applications, continuous outcomes are not measured with good precision. It can also happen that the outcome has been categorized into a few classes, so the data involve a range of values. In clinical studies observations are commonly censored: the lifetime of a subject is only known to be greater than a certain point. In these cases the simple approximation (2.1) does not apply, and either the integral must be evaluated exactly or other more appropriate approximations used.

Mathematical convention

The treatment of discrete or continuous random variables differs little in probability theory, and the term 'probability density' or 'probability' will be applied to cover both discrete and continuous models. In most cases the likelihood is the probability density seen as a function of the parameter. The original definition is important when the data are both discrete and continuous, or when we are comparing separate models.

When we say we 'integrate' a density this means (i) the usual integration when we are dealing with continuous random variables:

$$\int h(x)dx$$

or (ii) summation when dealing with discrete ones

$$\sum_x h(x)$$

where $h(x)$ is some density function. When we discuss a particular example, we might use either an integration or a summation, depending on the context, but it should be understood that the idea under consideration usually covers both the continuous and discrete cases.

2.2 Examples

Example 2.1: Suppose 100 seeds were planted and it is known only that $x \leq 10$ seeds germinated. The exact number of germinating seeds is unknown. Then the information about θ is given by the likelihood function

$$\begin{aligned} L(\theta) &= P(X \leq 10) \\ &= \sum_{x=0}^{10} \binom{100}{x} \theta^x (1-\theta)^{n-x}. \end{aligned}$$

Figure 2.3(a) compares this likelihood with the likelihood based on $x = 5$. □

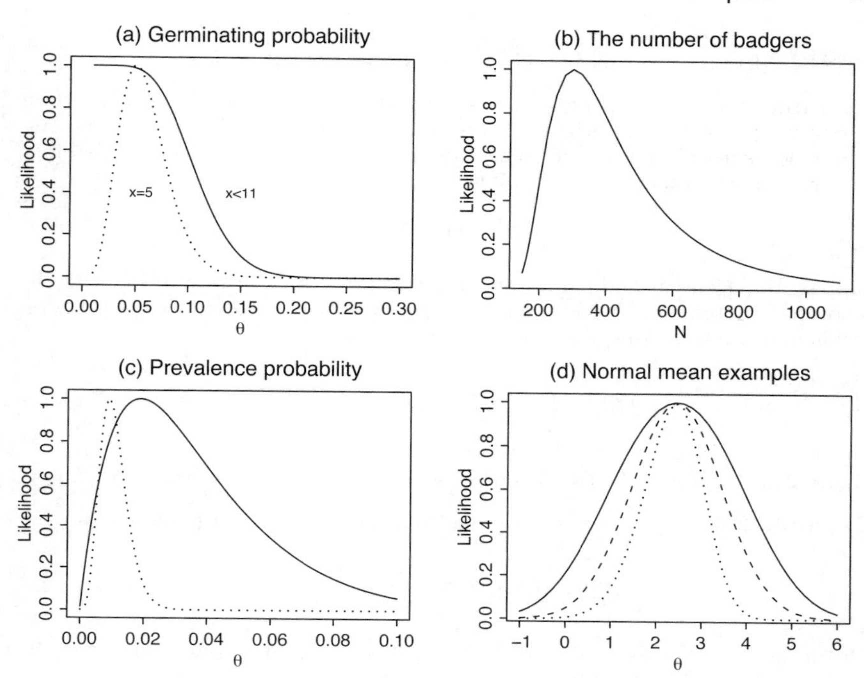

Figure 2.3: *(a) Likelihood functions from two binomial experiments:* $n = 100$ *and* $x < 11$, *and* $n = 100$ *and* $x = 5$. *(b) The likelihood of the number of badgers. (c) The likelihood of the prevalence of a certain genotype. (d) The likelihood of the normal mean based on observing* $0.9 < x < 4$ *(solid line),* $x = 2.45$ *(dashed line), and the maximum* $x_{(5)} = 3.5$ *(dotted line). All likelihoods are set to have unit maximum.*

Example 2.2: A useful technique for counting a population is to mark a subset of the population, then take a random sample from the mixture of the marked and unmarked individuals. This capture–recapture technique is used, for example, to count the number of wild animals. In census applications a post-enumeration survey is conducted and one considers the previously counted individuals as 'marked' and the new ones as 'unmarked'; the proportion of new individuals in the survey would provide an estimate of the undercount during the census. To estimate the number of people who attend a large rally one can first distribute colourful hats, then later on take a random sample from the crowd.

As a specific example, to estimate the number of badgers (N) in a certain region, the Department of Agriculture tags $N_1 = 25$ of them. Later on it captures $n = 60$ badgers, and finds $n_2 = 55$ untagged and $n_1 = 5$ tagged ones. Assuming the badgers were caught at random, the likelihood of N can be computed based on the hypergeometric probability:

$$L(N) = P(n_1 = 5) = \frac{\binom{25}{5}\binom{N-25}{55}}{\binom{N}{60}}.$$

Figure 2.3(b) shows the likelihood function for a range of N. □

Example 2.3: A team of geneticists is investigating the prevalence of a certain rare genotype, which makes its first appearance on the 53rd subject analysed. Assuming the subjects are independent, the likelihood of the prevalence probability θ is given by the geometric probability

$$L(\theta) = (1-\theta)^{52}\theta.$$

Suppose the scientists had planned to stop when they found five subjects with the genotype of interest, at which point they analysed 552 subjects. The likelihood of θ is now given by a negative binomial probability

$$L(\theta) = \binom{552-1}{5-1}\theta^5(1-\theta)^{552-5}.$$

Figure 2.3(c) shows these likelihoods in solid and dotted lines, respectively. □

Example 2.4: Suppose x is a sample from $N(\theta, 1)$; the likelihood of θ is

$$L(\theta) = \phi(x-\theta) \equiv \frac{1}{\sqrt{2\pi}} e^{-\frac{1}{2}(x-\theta)^2}.$$

The dashed curve in Figure 2.3(d) is the likelihood based on observing $x = 2.45$.

Suppose it is known only that $0.9 < x < 4$; then the likelihood of θ is

$$L(\theta) = P(0.9 < X < 4) = \Phi(4-\theta) - \Phi(0.9-\theta),$$

where $\Phi(z)$ is the standard normal distribution function. The likelihood is shown in solid line in Figure 2.3(d).

Suppose $x_1, \ldots, x_n$ are an identically and independently distributed (iid) sample from $N(\theta, 1)$, and only the maximum $x_{(n)}$ is reported, while the others are missing. The distribution function of $x_{(n)}$ is

$$\begin{aligned} F(t) &= P(X_{(n)} \le t) \\ &= P(X_i \le t, \text{ for each } i) \\ &= \{\Phi(t-\theta)\}^n. \end{aligned}$$

So, the likelihood based on observing $x_{(n)}$ is

$$L(\theta) = p_\theta(x_{(n)}) = n\{\Phi(x_{(n)}-\theta)\}^{n-1}\phi(x_{(n)}-\theta).$$

Figure 2.3(d) shows this likelihood as a dotted line for n=5 and $x_{(n)} = 3.5$.

There is a general heuristic to deal with order statistics for an iid sample from continuous density $p_\theta(x)$. Assume a finite precision ϵ, and partition the real line into a regular grid of width ϵ. Taking an iid sample $x_1, \ldots, x_n$ is like performing a multinomial experiment: throw n balls to cells with probability $p(x)\epsilon$ and record where they land. For example, the probability of the order statistics $x_{(1)}, \ldots, x_{(n)}$ is approximately

$$n!\epsilon^n \prod_i p_\theta(x_{(i)}).$$

Knowing only the maximum $x_{(n)}$, the multinomial argument yields immediately the likelihood given above. If only $x_{(1)}$ and $x_{(n)}$ are given, the likelihood of θ is

$$L(\theta) = \frac{n(n-1)}{2}\epsilon^2 p_\theta(x_{(1)})p_\theta(x_{(n)})\{F_\theta(x_{(n)}) - F_\theta(x_{(1)})\}^{n-2},$$

where $F_\theta(x)$ is the underlying distribution function. □

Example 2.5: Let us now solve the exchange paradox described in Example 1.3 using the likelihood. Treat the amounts of money as the unknown parameters θ and 2θ. On seeing the amount x in the envelope, the *likelihood* of θ is

$$\begin{aligned} L(\theta = x) &= P(X = x|\theta = x) \\ &= P(X = \theta|\theta = x) = 0.5, \end{aligned}$$

and

$$\begin{aligned} L(\theta = x/2) &= P(X = x|\theta = x/2) \\ &= P(X = 2\theta|\theta = x/2) = 0.5, \end{aligned}$$

which means the data x cannot tell us any preference over two possible values of θ. The (unknown) amount in the other envelope is either $x/2$ (if $\theta = x/2$) or $2x$ (if $\theta = x$) with equal likelihood, *not probability*, and we have to stop there. The likelihood analysis provides that, given x alone and no other information, there is no rational way for preferring one envelope over the other. The paradox is avoided as we cannot take an average using the likelihood values as weights. □

2.3 Combining likelihoods

The likelihood definition immediately provides a simple rule for combining likelihoods from different datasets. If x_1 and x_2 are independent datasets with probabilities $p_{1,\theta}(x_1)$ and $p_{2,\theta}(x_2)$ that share a common parameter θ, then the likelihood from the combined data is

$$\begin{aligned} L(\theta) &= p_{1,\theta}(x_1)p_{2,\theta}(x_2) \\ &= L_1(\theta)L_2(\theta), \end{aligned} \tag{2.2}$$

where $L_1(\theta)$ and $L_2(\theta)$ are the likelihoods from the individual datasets. In log scale this property is a simple additive property

$$\log L(\theta) = \log L_1(\theta) + \log L_2(\theta),$$

giving a very convenient formula for combining information from independent experiments: simply add the log-likelihoods. For analytical and computational purposes it is usually more convenient to work in the log-likelihood scale. It turns out that most of the (frequentist) properties of the likelihood function are associated with the log-likelihood and quantities derived from it.

The simplest case occurs if x_1 and x_2 are an iid sample from the same density $p_\theta(x)$, so

$$L(\theta) = p_\theta(x_1)p_\theta(x_2),$$

or $\log L(\theta) = \log p_\theta(x_1) + \log p_\theta(x_2)$. So, if $x_1, \ldots, x_n$ are an iid sample from $p_\theta(x)$ we have

$$L(\theta) = \prod_{i=1} p_\theta(x_i),$$

or $\log L(\theta) = \sum_{i=1}^n \log p_\theta(x_i)$.

Example 2.6: Let $x_1, \ldots, x_n$ be an iid sample from $N(\theta, \sigma^2)$ with known σ^2. The contribution of x_i to the likelihood is

$$L_i(\theta) = \frac{1}{\sqrt{2\pi\sigma^2}} \exp\left\{-\frac{(x_i - \theta)^2}{2\sigma^2}\right\},$$

and the total log-likelihood is

$$\begin{aligned} \log L(\theta) &= \sum_{i=1}^{n} \log L_i(\theta) \\ &= -\frac{n}{2}\log(2\pi\sigma^2) - \frac{1}{2\sigma^2}\sum_{i=1}^{n}(x_i - \theta)^2. \ \square \end{aligned}$$

Example 2.7: Suppose we have two independent samples taken from $N(\theta, 1)$. From the first sample it is reported that the sample size is $n_1 = 5$, and the maximum $x_{(5)} = 3.5$. The second sample has size $n_2 = 3$, and only the sample mean $\overline{y} = 4$ is reported. From Example 2.4 we have

$$L_1(\theta) = 5\{\Phi(x_{(5)} - \theta)\}^4 \phi(x_{(5)} - \theta),$$

and, since $\overline{y}$ is $N(\theta, 1/3)$,

$$L_2(\theta) = \frac{1}{\sqrt{2\pi/3}} \exp\left\{-\frac{3}{2}(\overline{y} - \theta)^2\right\}.$$

The log-likelihood from the combined data is

$$\log L(\theta) = \log L_1(\theta) + \log L_2(\theta).$$

These likelihoods are shown in Figure 2.4. $\square$

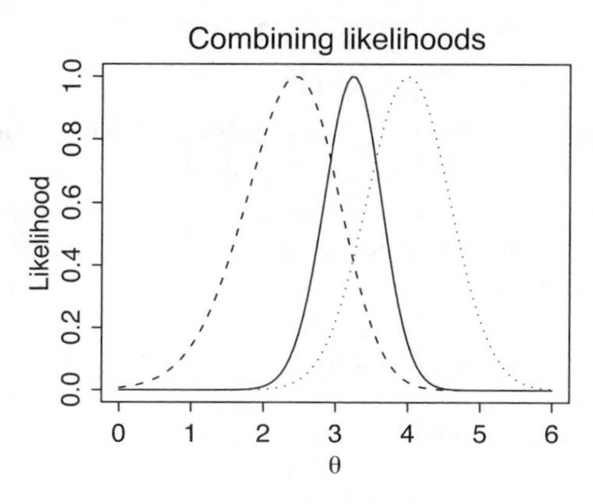

Figure 2.4: *Likelihood based on the maximum $x_{(5)}$ of the first sample (dashed line), on the sample mean $\overline{y} = 4$ of the second sample (dotted line), and on the combined data (solid line).*

Connection with Bayesian approach*

Recall that in Bayesian computation we begin with a prior $f(\theta)$ and compute the posterior

$$\begin{aligned} f(\theta|x) &= \text{constant} \times f(\theta) f(x|\theta) \\ &= \text{constant} \times f(\theta) L(\theta), \end{aligned} \qquad (2.3)$$

where, to follow Bayesian thinking, we use $f(x|\theta) \equiv p_\theta(x)$. Comparing (2.3) with (2.2) we see that the Bayesian method achieves the same effect as the likelihood method: it combines the information from the prior and the current likelihood by a simple multiplication.

If we treat the prior $f(\theta)$ as a 'prior likelihood' then the posterior is a combined likelihood. If *we know absolutely nothing about θ* prior to observing $X = x$ (recall Bayes's *Scholium* in Section 1.4), the *prior likelihood* is $f(\theta) \equiv 1$, and the likelihood function expresses the current information on θ after observing x. Using a uniform prior and scaling the functions to integrate to one, the posterior density and the likelihood functions would be the same.

2.4 Likelihood ratio

How should we compare the likelihood of different values of a parameter, say $L(\theta_1)$ versus $L(\theta_2)$? Suppose y is a one-to-one transformation of the observed data x; if x is continuous,

$$p_\theta(y) = p_\theta(x(y)) \left| \frac{\partial x}{\partial y} \right|,$$

so the likelihood based on the new data y is

$$L(\theta; y) = L(\theta; x) \left| \frac{\partial x}{\partial y} \right|.$$

Obviously x and y should carry the same information about θ, so to compare θ_1 and θ_2 only the likelihood ratio is relevant since it is invariant with respect to the transformation:

$$\frac{L(\theta_2; y)}{L(\theta_1; y)} = \frac{L(\theta_2; x)}{L(\theta_1; x)}.$$

Since only the ratio is important, within a model $p_\theta(x)$, the likelihood function is only meaningful up to a multiplicative constant. This means, for example, in setting up the likelihood we can ignore terms not involving the parameter. Proportional likelihoods are equivalent as far as evidence about θ is concerned and we sometimes refer to them as being the same likelihood. To make it unique, especially for plotting, it is customary to normalize the likelihood function to have unit maximum, i.e. we divide the function by its maximum. From now on if we report a likelihood value as a percentage it is understood to be a normalized value. Alternatively, we can set the log-likelihood to have zero maximum.

It may be tempting to normalize the likelihood so that it integrates to one, but there are reasons for not doing that. In particular there will be an invariance problem when we deal with parameter transformation; see Section 2.8.

Example 2.8: Suppose x is a sample from the binomial(n, θ), where n is known. We have, ignoring irrelevant terms,

$$L(\theta) = \theta^x (1-\theta)^{n-x},$$

or $\log L(\theta) = x \log \theta + (n-x)\log(1-\theta)$. □

It is stated earlier that the likelihood gives us a measure of rational belief or relative preferences. How do we interpret the actual values of the likelihood function or likelihood ratio? In the binomial example with $n = 10$ and outcome $x = 8$, how should we react to the statement

$$\frac{L(\theta = 0.8)}{L(\theta = 0.3)} \approx 209 \equiv N?$$

Is there a way to calibrate this numerical value with something objective? The answer is yes, but for the moment we will try to answer it more subjectively with an analogy.

Imagine taking a card at random from a deck of N well-shuffled cards and consider the following two hypotheses:

H_0: the deck contains N different cards labelled as 1 to N.
H_2: the deck contains N similar cards labelled as, say, 2.

Suppose we obtain a card with label 2; the likelihood ratio of the two hypotheses is

$$\frac{L(H_2)}{L(H_0)} = N;$$

that is, H_2 is $N = 209$ times more likely than H_0. That is how we can gauge our 'rational belief' about $\theta = 0.8$ versus $\theta = 0.3$ based on observing $x = 8$. Interpretations like this, unfortunately, cannot withstand a careful theoretical scrutiny (Section 2.6), which is why we call it only a subjective interpretation.

2.5 Maximum and curvature of likelihood

Fisher (1922) introduced likelihood in the context of estimation via the method of maximum likelihood, but in his later years he did not think of it as simply a device to produce parameter estimates. The likelihood is a tool for an objective reasoning with data, especially for dealing with the uncertainty due to the limited amount of information contained in the data. It is the entire likelihood function that captures all the information in the data about a certain parameter, not just its maximizer.

The obvious role of the maximum likelihood estimate (MLE) is to provide a point estimate for a parameter of interest; the purpose of having a point estimate is determined by the application area. In cases where a model parameter has a physical meaning, it is reasonable to ask what is the best estimate given by the data; the uncertainty is in a way a nuisance, not part of the scientific question. The MLE is usually a sensible answer. Another important role is for simplifying a multiparameter likelihood through a profile likelihood (Section 3.4): nuisance parameters are replaced by the MLEs.

We should view the MLE as a device to simplify the presentation of the likelihood function, especially in a real data analysis situation; a number is a lot simpler than a function. Imagine the standard task of describing the characteristics of a study population: it is still possible for our mind to absorb, communicate, compare and reason with 10 or even 20 sample means or proportions, but it would be futile to keep referring to 20 likelihood functions.

Generally, a single number is not enough to represent a function; the MLE is rarely enough to represent a likelihood function. If the log-likelihood is well approximated by a quadratic function, then we need at least two quantities to represent it: the location of its maximum and the curvature at the maximum. In this case we call the likelihood function 'regular'. When our sample becomes large the likelihood function generally does become regular; the large-sample theory in Chapter 9 establishes this practical fact.

To repeat this crucial requirement, *regular problems are those where we can approximate the log-likelihood around the MLE by a quadratic function*; for such cases we will also say that the likelihood function is regular. (Not to be pedantic, when we say 'a likelihood function has a good quadratic approximation', we mean the log-likelihood does.) This approximation is the port of entry for calculus into the likelihood world. For simplicity we will start with a scalar parameter; the multiparameter case is discussed in Section 3.3. First we define the score function $S(\theta)$ as the first derivative of the log-likelihood:

$$S(\theta) \equiv \frac{\partial}{\partial\theta} \log L(\theta).$$

Hence the MLE $\widehat{\theta}$ is the solution of the *score equation*

$$S(\theta) = 0.$$

At the maximum, the second derivative of the log-likelihood is negative, so we define the curvature at $\widehat{\theta}$ as $I(\widehat{\theta})$, where

$$I(\theta) \equiv -\frac{\partial^2}{\partial\theta^2} \log L(\theta).$$

A large curvature $I(\widehat{\theta})$ is associated with a tight or strong peak, intuitively indicating less uncertainty about θ. In likelihood theory $I(\widehat{\theta})$ is a key quantity called the *observed Fisher information*; note that it is evaluated at the MLE, so it is a number rather than a function.

Example 2.9: Let $x_1, \ldots, x_n$ be an iid sample from $N(\theta, \sigma^2)$. For the moment assume that σ^2 is known. Ignoring irrelevant constant terms

$$\log L(\theta) = -\frac{1}{2\sigma^2}\sum_{i=1}^{n}(x_i - \theta)^2,$$

so we immediately get

$$S(\theta) = \frac{\partial}{\partial\theta}\log L(\theta) = \frac{1}{\sigma^2}\sum_{i=1}^{n}(x_i - \theta).$$

Solving $S(\theta) = 0$ produces $\widehat{\theta} = \overline{x}$ as the MLE of θ. The second derivative of the log-likelihood gives the observed Fisher information

$$I(\widehat{\theta}) = \frac{n}{\sigma^2}.$$

Here $\text{var}(\widehat{\theta}) = \sigma^2/n = I^{-1}(\widehat{\theta})$, so larger information implies a smaller variance. Furthermore, the standard error of $\widehat{\theta}$ is $\text{se}(\widehat{\theta}) = \sigma/\sqrt{n} = I^{-1/2}(\widehat{\theta})$.

This is an important example, for it is a common theme in statistics that many properties which are exactly true in the normal case are approximately true in regular problems. □

Example 2.10: Based on x from the binomial(n, θ) the log-likelihood function is

$$\log L(\theta) = x\log\theta + (n - x)\log(1 - \theta).$$

We can first find the score function

$$\begin{aligned} S(\theta) &\equiv \frac{\partial}{\partial\theta}\log L(\theta) \\ &= \frac{x}{\theta} - \frac{n - x}{1 - \theta}, \end{aligned}$$

giving the MLE $\widehat{\theta} = x/n$ and

$$\begin{aligned} I(\theta) &\equiv -\frac{\partial^2}{\partial\theta^2}\log L(\theta) \\ &= \frac{x}{\theta^2} + \frac{n - x}{(1 - \theta)^2}, \end{aligned}$$

so at the MLE we have the Fisher information

$$I(\widehat{\theta}) = \frac{n}{\widehat{\theta}(1 - \widehat{\theta})}. \quad \square$$

Example 2.11: In realistic problems we do not have a closed form solution to the score equation. Suppose an iid sample of size $n = 5$ is taken from $N(\theta, 1)$, and only the maximum $x_{(5)} = 3.5$ is reported. From Example 2.4 we have

$$L(\theta) = 5\{\Phi(x_{(5)} - \theta)\}^4 \phi(x_{(5)} - \theta).$$

It is best to use a numerical optimization procedure to find $\widehat{\theta}$ directly from $L(\theta)$; in practice we do not even need to find $S(\theta)$ analytically, and the procedure can also provide $I(\widehat{\theta})$ numerically. In this example

$$\widehat{\theta} = 2.44, \quad I(\widehat{\theta}) = 2.4.$$

Informally, we might say the maximum carries the same information as 2.4 observations from $N(\theta, 1)$. □

Using a second-order Taylor's expansion around $\widehat{\theta}$

$$\log L(\theta) \approx \log L(\widehat{\theta}) + S(\widehat{\theta})(\theta - \widehat{\theta}) - \frac{1}{2} I(\widehat{\theta})(\theta - \widehat{\theta})^2$$

we get

$$\log \frac{L(\theta)}{L(\widehat{\theta})} \approx -\frac{1}{2} I(\widehat{\theta})(\theta - \widehat{\theta})^2, \qquad (2.4)$$

providing a quadratic approximation of the normalized log-likelihood around $\widehat{\theta}$.

We can judge the quadratic approximation by plotting the true log-likelihood and the approximation together. In a log-likelihood plot, we set the maximum of the log-likelihood to zero and check a range of θ such that the log-likelihood is approximately between -4 and 0. In the normal example above (Example 2.9) the quadratic approximation is exact:

$$\log \frac{L(\theta)}{L(\widehat{\theta})} = -\frac{1}{2} I(\widehat{\theta})(\theta - \widehat{\theta})^2,$$

so a quadratic approximation of the log-likelihood corresponds to a normal approximation of $\widehat{\theta}$. We have here a practical rule in all likelihood applications: a reasonably regular likelihood means $\widehat{\theta}$ is approximately normal, so statements which are exactly true for the normal model will be approximately true for $\widehat{\theta}$.

Alternatively, in terms of the score function, we can take the derivative of the quadratic approximation (2.4) to get

$$S(\theta) \approx -I(\widehat{\theta})(\theta - \widehat{\theta})$$

or

$$-I^{-1/2}(\widehat{\theta}) S(\theta) \approx I^{1/2}(\widehat{\theta})(\theta - \widehat{\theta}).$$

The latter has the advantage of being dimensionless, in the sense that it is not affected by the scale of the parameter θ. So, a quadratic approximation can be checked graphically by plotting $-I^{-1/2}(\widehat{\theta})S(\theta)$ against $I^{1/2}(\widehat{\theta})(\theta - \widehat{\theta})$, which should be linear with unit slope. We can check that it is exactly true in the normal case. Any smooth function is locally linear, so the question is how wide we should expect to see the linearity. In the ideal (normal) case $I^{1/2}(\widehat{\theta})(\theta - \widehat{\theta})$ is a $N(0,1)$-variate, so intuitively we should check at least between -2 and 2.

Example 2.12: Figure 2.5(a) shows the log-likelihood function of the binomial parameter θ based on $n = 10$ trials and $x = 8$ successes. Here $\widehat{\theta} = 0.8$ and $I(\widehat{\theta}) = 62.5$. Both Figure 2.5(a) and (b) show a poor quadratic approximation. In Figure 2.5(c) we have the log-likelihood of a much larger sample size $n = 100$, but the same estimate $\widehat{\theta} = 0.8$; the Fisher information is $I(\widehat{\theta}) = 625$. Now the quadratic approximation is more successful. □

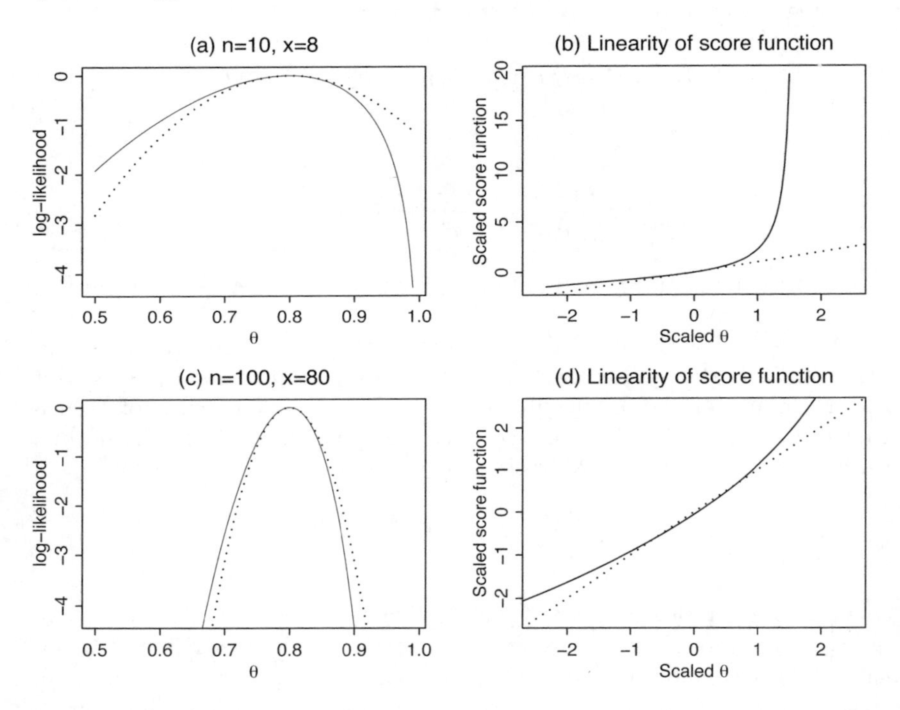

Figure 2.5: *Quadratic approximation of the log-likelihood function. (a) The true log-likelihood (solid) and the approximation (dotted) for the binomial parameter θ. (b) Linearity check of the score function, showing a poor approximation; (c)–(d) The same as (a)–(b) for a bigger experiment.*

In cases where the MLE $\widehat{\theta}$ and the curvature $I(\widehat{\theta})$ can represent the likelihood function, one can simply report the pair $\widehat{\theta}$ and $I(\widehat{\theta})$ instead of showing the graph, though this still leaves the question of interpreting $I(\widehat{\theta})$. In the normal case $\text{var}(\widehat{\theta}) = I^{-1}(\widehat{\theta})$, or the standard error $\text{se}(\widehat{\theta}) = I^{-1/2}(\widehat{\theta})$. This is approximately true in nonnormal cases, so $\text{se}(\widehat{\theta}) = I^{-1/2}(\widehat{\theta})$ is the most commonly used quantity to supplement the MLE. We will come back to this in Section 2.7.

If the likelihood function is not regular, then the curvature of the log-likelihood at the MLE or the standard error is not meaningful. In this case, a set of likelihood or confidence intervals described in the coming section is a better supplement to the MLE.

2.6 Likelihood-based intervals

How do we communicate the statistical evidence using the likelihood? We can simply show the likelihood function and, based on it, state our conclusion regarding the question of interest, or let others draw their own conclusion. We adopt this approach in many of our examples, but such an approach can be very impractical, especially when we are dealing with many parameters.

In Section 2.5 we show that in regular cases we can simply present the MLE and its standard error. In less regular cases we can construct intervals that still acknowledge the existing uncertainty, while simplifying the communication of the likelihood function.

Pure likelihood inference

In his last book Fisher (1973, pages 75–78) proposed that in some problems we interpret the observed likelihood function directly to communicate our uncertainty about θ. These problems include those where exact probability-based inference is not available, while the sample size is too small to allow large-sample results to hold. A likelihood interval is defined as a set of parameter values with high enough likelihood:

$$\left\{\theta, \frac{L(\theta)}{L(\widehat{\theta})} > c\right\},$$

for some cutoff point c, where $L(\theta)/L(\widehat{\theta})$ is the normalized likelihood.Among modern authors, Barnard *et al.* (1962), Sprott (1975, 2000), Edwards (1992), Royall (1997) and Lindsey (1996, 1999a,b) are proponents of direct likelihood inference.

Fisher gave a specific example in the case of a binomial parameter. The question of how to choose the cutoff point c is left open, but he suggested that parameter values with less than $1/15$ or 6.7% likelihood 'are obviously open to grave suspicion'. This prescription only works for scalar parameters; in general there is a calibration issue we have to deal with.

Example 2.13: In the binomial example where we observe $x = 8$ out of $n = 10$, the likelihood intervals for θ at $c = 15\%$ and 4% are (0.50,0.96) and (0.41,0.98), shown in Figure 2.6. Typically there will not be any closed form formula for the interval, but in practice it can be found quite easily using numerical methods. For scalar parameters, we can use a simple grid search. □

Probability-based inference

While convenient, the pure likelihood inference suffers a serious weakness: there is no externally validated way to justify the cutoff point c, since a

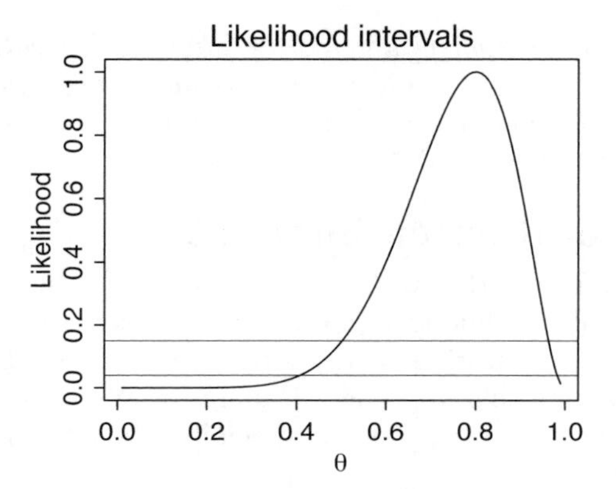

Figure 2.6: *Likelihood intervals at 15% and 4% cutoff for the binomial parameter θ are (0.50,0.96) and (0.41,0.98).*

chosen value c does not refer to anything observable. This is a general calibration problem associated with the likelihood: a 5% likelihood does not have a strict meaning (it depends on the size of the parameter space). In contrast, a 5% probability is always meaningful as a long-term frequency, so one way to 'calibrate' the likelihood is via probability. This is in fact the traditional likelihood-based inference in statistics. Fisher maintained that whenever possible we should use probability-based inference; here he included cases where an exact confidence level is available and the large-sample cases.

Traditional (frequentist) inference on an unknown parameter θ relies on the distribution theory of its estimate $\widehat{\theta}$. A large-sample theory is needed in the general case, but it is simple in the normal mean model. From Example 2.9 we have

$$\log \frac{L(\theta)}{L(\widehat{\theta})} = -\frac{n}{2\sigma^2}(\overline{x} - \theta)^2.$$

Now, we know $\overline{x}$ is $N(\theta, \sigma^2/n)$, so

$$\frac{n}{\sigma^2}(\overline{x} - \theta)^2 \sim \chi^2_1,$$

or

$$W \equiv 2 \log \frac{L(\widehat{\theta})}{L(\theta)} \sim \chi^2_1. \tag{2.5}$$

W is called Wilk's likelihood ratio statistic. Its χ^2 distribution is exact in the normal mean model, and as will be shown in Chapter 9, it is approximately true in general cases. A practical guide to use the approximation is that the likelihood is reasonably regular.

This is the key distribution theory needed to calibrate the likelihood. In view of (2.5), for an unknown but fixed θ, the probability that the likelihood interval covers θ is

$$\begin{aligned} P\left(\frac{L(\theta)}{L(\widehat{\theta})} > c\right) &= P\left(2\log\frac{L(\widehat{\theta})}{L(\theta)} < -2\log c\right) \\ &= P(\chi_1^2 < -2\log c). \end{aligned}$$

So, if for some $0 < \alpha < 1$ we choose a cutoff

$$c = e^{-\frac{1}{2}\chi^2_{1,(1-\alpha)}}, \tag{2.6}$$

where $\chi^2_{1,(1-\alpha)}$ is the $100(1-\alpha)$ percentile of χ^2_1, we have

$$P\left(\frac{L(\theta)}{L(\widehat{\theta})} > c\right) = P(\chi_1^2 < \chi^2_{1,(1-\alpha)}) = 1 - \alpha.$$

This means that by choosing c in (2.6) the likelihood interval

$$\left\{\theta, \frac{L(\theta)}{L(\widehat{\theta})} > c\right\}$$

is a $100(1-\alpha)\%$ confidence interval for θ.

In particular, for $\alpha = 0.05$ and 0.01 formula (2.6) gives $c = 0.15$ and 0.04. So, we arrive at the important conclusion that, in the normal mean case, we get an exact 95% or 99% confidence interval for the mean by choosing a cutoff of 15% or 4%, respectively. This same confidence interval interpretation is approximately true for reasonably regular problems.

When can we use a pure likelihood interval?

A likelihood interval represents a set of parameter values which are well supported by, or consistent with, the data. Given a model, a likelihood interval is an objective interval in the sense that it does not involve any subjective choice of prior. Fisher was clear, however, that the likelihood on its own provides only a *weaker form of inference than probability-based inference.* Unlike the confidence interval, a pure likelihood interval does not have a repeated sampling interpretation; i.e. it is silent regarding its long-term properties if the experiment is repeated a large number of times. These long-term properties provide a (potential) external validity to probability-based confidence intervals.

Names such as 'likelihood-based interval' can be confusing, since both pure likelihood intervals and traditional likelihood-based confidence intervals are derived from the same likelihood function. In fact, numerically they can be the same. What we are discussing here is *the sense of uncertainty* associated with the interval. Traditionally it is only available in terms of probability (or confidence level), but in the Fisherian view it can also be reported in terms of likelihood. From here on, if a likelihood-based interval has a theoretically justified confidence level it is called a 'confidence interval', otherwise it is called a 'likelihood interval'.

It is well known that generally a confidence level does not actually apply to an observed interval, as it only makes sense in the long run. If we think of an interval as a guess of where θ is, a 95% probability of being correct does not apply to a particular guess. (In contrast, the sense of uncertainty provided by the likelihood does apply to a particular guess.) The following example is adapted from Berger and Wolpert (1988).

Example 2.14: Someone picks a fixed integer θ and asks you to guess it based on some data as follows. He is going to toss a coin twice (you do not see the outcomes), and from each toss he will report $\theta + 1$ if it turns out heads, or $\theta - 1$ otherwise. Hence the data x_1 and x_2 are an iid sample from a distribution that has probability 0.5 on $\theta - 1$ or $\theta + 1$. For example, he may report $x_1 = 5$ and $x_2 = 5$.

The following guess will have 75% probability of being correct:

$$C(x_1, x_2) = \begin{cases} \frac{1}{2}(x_1 + x_2) & \text{if } x_1 \neq x_2 \\ x_1 - 1 & \text{if } x_1 = x_2. \end{cases}$$

According to the standard logic of confidence procedure, the above guess has 75% 'confidence level'. But if $x_1 \neq x_2$ we should be '100% confident' that the guess is correct, otherwise we are only '50% confident'. It will be absurd to insist that on observing $x_1 \neq x_2$ you only have 75% confidence in $\{(x_1 + x_2)/2\}$. A pure likelihood approach here would match our common sense: it would report at each observed $\{x_1, x_2\}$ what our uncertainty is about θ. It would not say anything, however, about the long-term probability of being correct. □

In Section 5.10 we will discuss additional statistical examples that have similar problems. Fisher himself added an extra requirement for the use of probability *for inference*: it should *not* be possible to recognize a subset of the sample space, for which we can make an equally valid but different (conditional) probability statement. The literature on these 'recognizable subsets' is rather esoteric, and so far there has not been any impact on statistical practice.

In general we will interpret a likelihood interval this way:

- as the usual confidence interval if an exact or a large-sample approximate justification is available. This covers most of routine data analysis where parameters are chosen so that the likelihood is reasonably regular.

- as a pure likelihood interval if there is no exact probability-based justification and the large-sample theory is suspect. This usually involves small-sample problems with nonnormal or complicated distributions, where the likelihood is decidedly not regular. It also includes cases where a probability-based statement is obviously absurd as in the previous example.

In a real data analysis there is always an inevitable judgement call; acknowledging the current dominance of confidence procedure, we will err on the side of regularity and tend to report approximate confidence intervals. This is mitigated by the fact that the regularity requirement for likelihood-based confidence intervals is quite forgiving; see Section 2.9.

Example 2.15: Let $x_1, \ldots, x_n$ be a sample from Uniform$(0, \theta)$ for some $\theta > 0$. Let $x_{(n)}$ be the maximum of $x_1, \ldots, x_n$. The likelihood function is

$$\begin{aligned} L(\theta) &= \theta^{-n}, \qquad \text{for } x_i < \theta \text{ for all } i \\ &= \theta^{-n}, \qquad \text{for } \theta > x_{(n)}, \end{aligned}$$

and equal to zero otherwise. For example, given data 2.85, 1.51, 0.69, 0.57 and 2.29, we get $x_{(n)} = 2.85$ and the likelihood is shown in Figure 2.7. Asymmetric likelihood typically occurs if θ is a boundary parameter. The likelihood interval at 5% cutoff is (2.85,5.19).

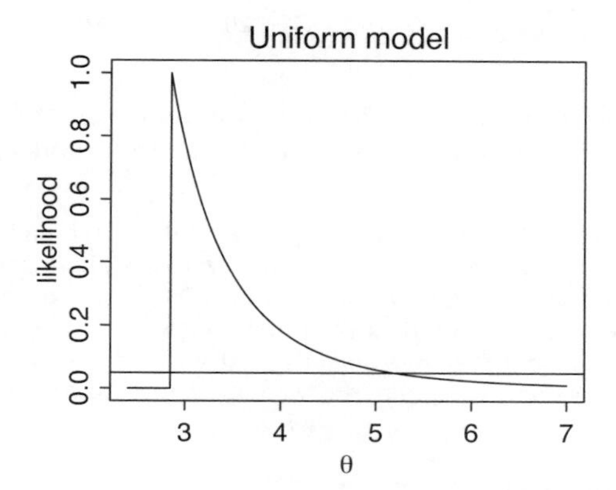

Figure 2.7: *Likelihood function of θ in Uniform$(0, \theta)$ based on $x_{(5)} = 2.85$.*

While the likelihood is not regular, it is still possible to provide an exact theoretical justification for a confidence interval interpretation. Now

$$\begin{aligned} P\left(\frac{L(\theta)}{L(\widehat{\theta})} > c\right) &= P\left(\frac{X_{(n)}}{\theta} > c^{1/n}\right) \\ &= 1 - P\left(\frac{X_{(n)}}{\theta} < c^{1/n}\right) \\ &= 1 - (c^{1/n})^n \\ &= 1 - c. \end{aligned}$$

So the likelihood interval with cutoff c is a $100(1-c)\%$ confidence interval. □

Example 2.16: This is a continuation of Example 2.1. Suppose 100 seeds were planted and it is recorded that $x < 11$ seeds germinated. Assuming a binomial model we obtain the likelihood in Figure 2.8 from

$$L(\theta) = \sum_{x=0}^{10} \binom{n}{x} \theta^x (1-\theta)^{n-x}. \qquad (2.7)$$

The MLE is $\widehat{\theta} = 0$ and its standard error is not well defined, while the likelihood

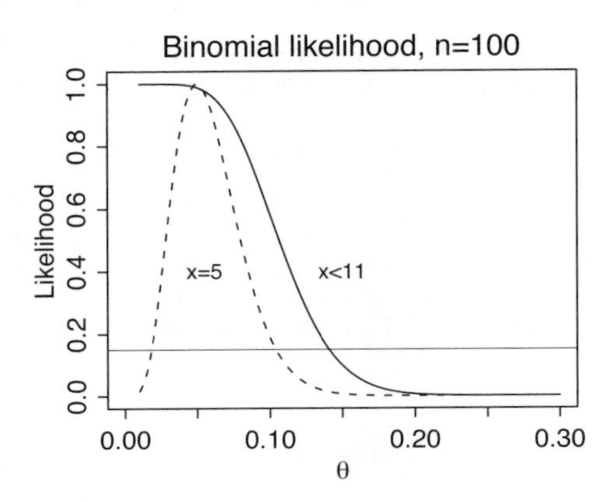

Figure 2.8: *Likelihood functions from two binomial experiments:* $n = 100$ *and* $x < 11$*, and* $n = 100$ *and* $x = 5$*. The latter is reasonably regular.*

interval at 15% cutoff is (0,0.14). Since the likelihood is irregular, and we do not have any theoretical justification, it is not clear how to assign a confidence level for this interval. To get a confidence level we also need to make an extra assumption about how data are to be collected in future (hypothetical) experiments.

Had we observed $x = 5$ the information on θ would have been more precise, and the likelihood reasonably regular. So we can report an approximate 95% CI $0.02 < \theta < 0.10$. □

Likelihood ratio test

To use the likelihood directly for hypothesis testing, for example to test a null hypothesis H_0: $\theta = \theta_0$, we can report the likelihood of H_0 as the normalized likelihood of θ_0

$$\frac{L(\theta_0)}{L(\widehat{\theta})}.$$

We can 'reject H_0', declaring it 'unsupported by data', if its likelihood is 'too small', indicating there are other hypotheses which are much better supported by the data.

How small is too small can be left arbitrary, depending on the application or other considerations that may include informal prior knowledge. In court cases a low level of likelihood may be set for the hypothesis 'the defendant is innocent' before we can reject it, while the hypothesis 'genetically engineered food has no side effects' can be rejected more readily. However, the issue of calibration of the likelihood is also relevant here as a likelihood of 5%, say, does not have a fixed meaning.

In regular one-parameter problems, a probability-based calibration is available to produce a P-value via the distribution of Wilk's statistic. From (2.5), on observing $L(\theta_0)/L(\widehat{\theta}) = c$, the approximate P-value is

$$P(\chi_1^2 > -2\log c).$$

This shows that there is typically some relationship between likelihood ratio and P-value, so when P-value is used it is used as a measure of support. However, since P-value depends on the sample space of the experiment, the relationship depends on the experiment; see Chapter 5 for more discussion. Even though it is under constant criticism from statisticians, the P-value is still widely used in practice, and it seems unlikely to disappear.

In non-regular problems, where we do not know how to calibrate the likelihood, we may have to use the pure likelihood ratio as a measure of support. This is in line with the use of pure likelihood intervals when there is no justification for probability-based inference. For example, in Example 2.16 it is not clear how to define a P-value, say to test $\theta = 0.5$. Another example, discussed in more detail in Section 5.4, is a general situation where the distribution of the test statistic is asymmetric: the 'exact' two-sided P-value is ambiguous and there are several competing definitions.

2.7 Standard error and Wald statistic

The likelihood or confidence intervals are a useful supplement to the MLE, acknowledging the uncertainty in a parameter θ; they are simpler to communicate than the likelihood function. In Section 2.5 we also mention the observed Fisher information $I(\widehat{\theta})$ as a supplement to the MLE. What is its relationship with the likelihood-based interval?

In regular cases where a quadratic approximation of the log-likelihood works well and $I(\widehat{\theta})$ is meaningful, we have

$$\log\frac{L(\theta)}{L(\widehat{\theta})} \approx -\frac{1}{2}I(\widehat{\theta})(\theta-\widehat{\theta})^2$$

so the likelihood interval $\{\theta,\ L(\theta)/L(\widehat{\theta}) > c\}$ is approximately

$$\widehat{\theta} \pm \sqrt{-2\log c} \times I(\widehat{\theta})^{-1/2}.$$

In the normal mean model in Example 2.9 this is an exact CI with confidence level

$$P(\chi_1^2 < -2\log c).$$

For example,

$$\widehat{\theta} \pm 1.96\ I(\widehat{\theta})^{-1/2}$$

is an exact 95% CI. In nonnormal cases this is an approximate 95% CI. In these cases, note the two levels of approximation to set up this interval:

the log-likelihood is approximated by a quadratic and the confidence level is approximate.

Also in analogy with the normal mean model, in general $I(\widehat{\theta})^{-1/2}$ provides the *standard error* of $\widehat{\theta}$. It is common practice to report the pair in the form 'MLE(standard error)'. Its main use is to test H_0: $\theta = \theta_0$ using the *Wald statistic*

$$z = \frac{\widehat{\theta} - \theta_0}{\text{se}(\widehat{\theta})},$$

or to compute *Wald confidence intervals*. For example, the Wald 95% CI for θ is

$$\widehat{\theta} \pm 1.96\text{se}(\widehat{\theta}). \tag{2.8}$$

Under H_0 in the normal mean model the statistic z has an exact standard normal distribution, and approximately so in the nonnormal case. A large value of $|z|$ is associated with a low likelihood of H_0: $\theta = \theta_0$. For example, $|z| > 2$ is associated with a likelihood less than 15%, or P-value less than 5%.

What if the log-likelihood function is far from quadratic? See Figure 2.9. From a likelihood point of view the Wald interval is deficient since it includes values with lower likelihood compared to values outside the interval.

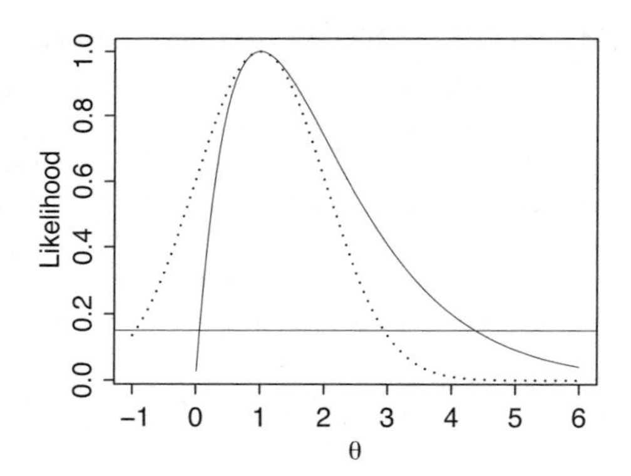

Figure 2.9: *Poor quadratic approximation (dotted) of a likelihood function (solid).*

Wald intervals might be called MLE-based intervals. To be clear, confidence intervals based on $\{\theta,\ L(\theta)/L(\widehat{\theta}) > c\}$ will be called *likelihood-based* confidence intervals. Wald intervals are always symmetric, but likelihood-based intervals can be asymmetric. Computationally the Wald interval is much easier to compute than the likelihood-based interval. If the likelihood is regular the two intervals will be similar. However, if they are not similar a likelihood-based CI is preferable; see the binomial example below and the discussion in Section 2.9. If the likelihood is available we will usually report the likelihood-based intervals.

Example 2.17: In the binomial example with $n = 10$ and $x = 8$ the quadratic approximation was poor. The standard error of $\widehat{\theta}$ is $I(\widehat{\theta})^{-1/2} = 1/\sqrt{62.5} = 0.13$, so the Wald 95% CI is

$$0.8 \pm 1.96/\sqrt{62.5},$$

giving $0.55 < \theta < 1.05$, clearly not appropriate. For $n = 100$ and $x = 80$, the standard error for $\widehat{\theta}$ is $I(\widehat{\theta})^{-1/2} = 1/\sqrt{625} = 0.04$. Here we have a much better quadratic approximation, with the Wald 95% CI

$$0.8 \pm 1.96/\sqrt{625}$$

or $0.72 < \theta < 0.88$, compared with $0.72 < \theta < 0.87$ from the exact likelihood. □

To use the 'MLE(standard error)' pair to represent the likelihood function, should we always check whether the likelihood is regular? In principle yes, but in practice we learn that certain problems or parameters tend to have better behaviour than others. Plotting the likelihood function in unfamiliar problems is generally advised. This is why in many of our examples we regularly show the likelihood function. When the data are sparse or if the parameter estimate is near a boundary, such as 0 or 1 for a probability parameter, then the quadratic approximation is not appropriate. It is almost never meaningful to report the standard error of odds-ratios or correlation coefficients.

2.8 Invariance principle

The likelihood function represents the uncertainty for a fixed parameter, but it is not a probability density function. How do we deal with parameter transformation? We will assume a one-to-one transformation, but the idea applies generally. In the first binomial example with $n = 10$ and $x = 8$, the likelihood ratio of $\theta_1 = 0.8$ versus $\theta_2 = 0.3$ is

$$\frac{L(\theta_1 = 0.8)}{L(\theta_2 = 0.3)} = \frac{\theta_1^8(1-\theta_1)^2}{\theta_2^8(1-\theta_2)^2} = 208.7,$$

i.e. given the data $\theta = 0.8$ is about 200 times more likely than $\theta = 0.3$.

Suppose we are interested in expressing θ on the logit scale as

$$\psi \equiv \log\{\theta/(1-\theta)\},$$

then 'intuitively' our relative information about $\psi_1 = \log(0.8/0.2) = 1.39$ versus $\psi_2 = \log(0.3/0.7) = -0.85$ should be

$$\frac{L^*(\psi_1)}{L^*(\psi_2)} = \frac{L(\theta_1)}{L(\theta_2)} = 208.7.$$

That is, our information should be *invariant* to the choice of parameterization.

This is not the case in the Bayesian formulation. Suppose θ has a 'non-informative' prior $f(\theta) = 1$; the posterior is

$$f(\psi|x) \quad = \quad f(\theta(\psi)|x) \times \left|\frac{\partial\theta}{\partial\psi}\right| \qquad (2.9)$$

$$= \quad f(\theta(\psi)|x)\frac{e^{\psi}}{(1+e^{\psi})^2}. \qquad (2.10)$$

In the new ψ-scale the relative information about ψ_1 versus ψ_2 is now equal to

$$\begin{aligned}\frac{f(\psi_1|x)}{f(\psi_2|x)} &= \frac{L(\theta_1)}{L(\theta_2)} \times \frac{e^{\psi_1}(1+e^{\psi_2})^2}{e^{\psi_2}(1+e^{\psi_1})^2} \\ &= 208.7 \times 0.81 = 169.0.\end{aligned}$$

Thus the invariance property of the likelihood ratio is incompatible with the Bayesian habit of assigning a probability distribution to a parameter.

It seems sensible that, if we do not know where θ is, then we should not know where $\log\{\theta/(1-\theta)\}$ or θ^2 or $1/\theta$ are; in other words, we should be equally ignorant regardless of how we model our problem. This is not true in the Bayesian world: if we assume θ is uniform between zero and one, then θ^2 is more likely to be closer to zero than to one.

Note that the invariance property of the likelihood ratio is not an implication of the repeated sampling principle, so it is not a frequentist requirement. However, frequentists generally adopt the invariance without question. This is in line with the frequentist refusal to accept the distributional reality for a fixed parameter, so a Jacobian term to account for parameter transformation is not meaningful.

There is actually some element of truth in the Bayesian position. It seems pedantic to think that we should be equally ignorant about the unknown probability θ, which is known to be between 0 and 1, as we are about θ^{100}. I would bet the latter would be closer to zero than to one (if 100 is not big enough for you, make it 1000 or 10,000), thus violating the invariance principle. The only way to accommodate the changing degree of ignorance after transformation is by adopting a probability density for the parameter, i.e. being a fully-fledged Bayesian. However, the loss of the invariance property of likelihood ratio would be a substantial loss in practice, since

- we lose the invariance property of the MLE (see Section 2.9)
- the likelihood of every parameterization would then require a Jacobian term, which must be computed analytically starting from a prior density, and consequently
- we are cornered into having to specify the prior density axiomatically.

These reasons alone may be enough to justify the utilitarian value of the invariance principle. However, the invariant property of the likelihood ratio should be seen only as a convenient axiom, rather than a self-evident truth.

(These discussions do not apply in random effects models where a parameter can have an objective distribution. Here the invariance property of the likelihood is not needed, as the likelihood can function like a density; see Section 16.1 and Section 16.2.)

2.9 Practical implications of invariance principle

Computing the likelihood of new parameters

The invariance principle implies that plotting the likelihood of a new parameter $\psi \equiv g(\theta)$ is automatic, a useful fact when performing likelihood-based inference for ψ. Consider the pair $\{\theta, L(\theta)\}$ as the graph of the likelihood of θ; then the graph of the likelihood of ψ is simply

$$\begin{aligned}\{\psi, L^*(\psi)\} &= \{g(\theta), L(g(\theta))\} \\ &= \{g(\theta), L(\theta)\}.\end{aligned}$$

In effect, with fixed parameters, the likelihood function is treated like a probability mass function of a discrete distribution, i.e. no Jacobian is used to account for the transformation. This does not create any inconsistency since the likelihood is not a density function, so it does not have to integrate to one.

If $g(\theta)$ is not one-to-one we need to modify the technique slightly. For example, consider

$$\psi = g(\theta) = \theta^2,$$

so $\theta = \pm 1$ implies $\psi = 1$. If $L(\theta = 1) = 0.5$ and $L(\theta = -1) = 0.3$ what is $L^*(\psi = 1)$? In this case, we define

$$\begin{aligned}L^*(\psi = 1) &\equiv \max_{\{\theta, g(\theta)=1\}} L(\theta) \\ &= \max\{0.5, 0.3\} \\ &= 0.5.\end{aligned}$$

In general

$$L^*(\psi) = \max_{\{\theta, g(\theta)=\psi\}} L(\theta). \qquad (2.11)$$

Invariance property of the MLE

An important implication of the invariance of likelihood ratio is the so-called invariance property of the MLE.

Theorem 2.1 *If $\widehat{\theta}$ is the MLE of θ and $g(\theta)$ is a function of θ, then $g(\widehat{\theta})$ is the MLE of $g(\theta)$.*

The function $g(\theta)$ does not have to be one-to-one, but the definition of the likelihood of $g(\theta)$ must follow (2.11). The proof is left as an exercise (Exercise 2.19).

It seems intuitive that if $\widehat{\theta}$ is most likely for θ and our knowledge remains the same then $g(\widehat{\theta})$ is most likely for $g(\theta)$. In fact, we would find it strange if $\widehat{\theta}$ is an estimate of θ, but $\widehat{\theta}^2$ is not an estimate of θ^2. In the binomial example with $n = 10$ and $x = 8$ we get $\widehat{\theta} = 0.8$, so the MLE of $g(\theta) = \theta/(1-\theta)$ is

$$g(\widehat{\theta}) = \widehat{\theta}/(1-\widehat{\theta}) = 0.8/0.2 = 4.$$

This convenient property is not necessarily true with other estimates. For example, if $\widehat{\theta}$ is the minimum variance unbiased estimate (MVUE) of θ, then $g(\widehat{\theta})$ is generally not MVUE for $g(\theta)$.

Improving the quadratic approximation

In practice we often consider a transformation $\psi = g(\theta)$ to 'improve' the likelihood function so that it is more regular. We know that in such a case we can rely on the MLE and the standard error. Given such a transform, confidence intervals of θ can be first set for $g(\theta)$ using the Wald interval:

$$g(\widehat{\theta}) \pm 1.96\ \text{se}\{g(\widehat{\theta})\},$$

then retransformed back to the original θ-scale. For scalar θ we can show (Exercise 2.20) that

$$\text{se}\{g(\widehat{\theta})\} = \text{se}(\widehat{\theta}) \left| \frac{\partial g}{\partial \widehat{\theta}} \right|.$$

Example 2.18: Suppose $x = 8$ is a sample from the binomial$(n = 10, \theta)$. The MLE of θ is $\widehat{\theta} = 0.8$. Consider the log-odds

$$\psi = g(\theta) = \log \frac{\theta}{1-\theta}.$$

Figure 2.10(b) shows the likelihood of ψ is more regular than in the original θ-scale.

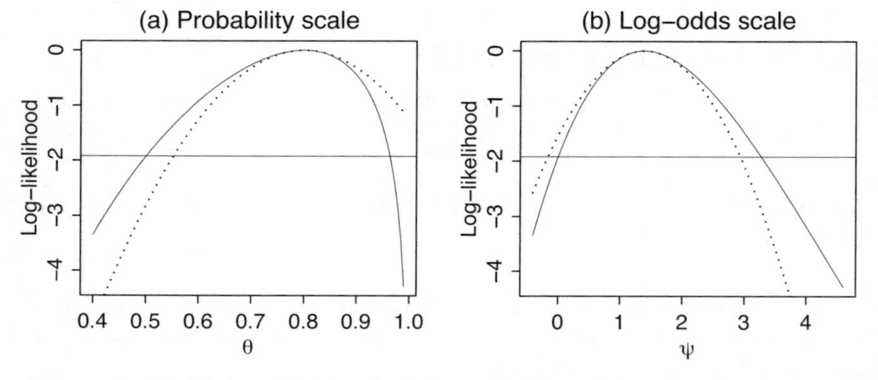

Figure 2.10: *(a) Log-likelihood of the probability θ from binomial data $n = 10$ and $x = 8$. (b) Log-likelihood of the log-odds ψ is more regular.*

The MLE of ψ is

$$\widehat{\psi} = \log \frac{0.8}{0.2} = 1.39.$$

To get the standard error of ψ, we first compute

$$\frac{\partial g}{\partial \theta} = \frac{1}{\theta} + \frac{1}{1-\theta},$$

so

$$\begin{aligned} \text{se}(\widehat{\psi}) &= \text{se}(\widehat{\theta}) \left(\frac{1}{\widehat{\theta}} + \frac{1}{1-\widehat{\theta}} \right) \\ &= \left(\frac{1}{x} + \frac{1}{n-x} \right)^{0.5} \\ &= 0.79. \end{aligned}$$

The Wald 95% CI for ψ is

$$1.39 \pm 1.96 \times 0.79,$$

giving $-0.16 < \psi < 2.94$. Transforming this back to the original scale yields

$$0.46 < \theta < 0.95.$$

This compares with $0.5 < \theta < 0.96$ using the likelihood-based 95% CI, and $0.55 < \theta < 1.05$ using the Wald 95% CI on the original scale θ (Example 2.17). □

Likelihood-based CI is better than Wald CI

We do not always know how to transform a parameter to get a more regular likelihood. This difficulty is automatically overcome by likelihood-based intervals. Let us consider a scalar parameter case. The likelihood-based approximate 95% CI is

$$\left\{ \theta, 2 \log \frac{L(\widehat{\theta})}{L(\theta)} \leq 3.84 \right\}$$

and the Wald interval is

$$\widehat{\theta} \pm 1.96 \text{ se}(\widehat{\theta}).$$

While both are based on a similar normal-approximation theory, the likelihood-based interval is better than the Wald interval.

The Wald interval is correct only if

$$\frac{\widehat{\theta} - \theta}{\text{se}(\widehat{\theta})} \sim N(0,1).$$

In contrast, because of invariance of the likelihood ratio, the likelihood-based interval is correct as long as *there exists* a one-to-one transformation $g(\cdot)$, *which we do not need to know*, so that

$$\frac{g(\widehat{\theta}) - g(\theta)}{\text{se}(g(\widehat{\theta}))} \sim N(0,1). \tag{2.12}$$

To show this, suppose $L < g(\theta) < U$ is the 95% CI for $g(\theta)$ based on $g(\widehat{\theta})$. From normality of $g(\widehat{\theta})$, the interval (L, U) is a likelihood interval at 15%

cutoff. The invariance of the likelihood ratio implies that the likelihood interval of θ at 15% cutoff is

$$g^{-1}(L) < \theta < g^{-1}(U),$$

which then has exactly the same 95% confidence level. The proof is clear if you draw the likelihood functions of θ and $g(\theta)$ next to each other. It should be emphasized that $g(\theta)$ need not be known, only that it exists.

With the same argument, if an exact normalizing transform does not exist, the likelihood-based CI still has the same confidence level as the transform that makes the likelihood most regular. In other words, for the purpose of probability calibration, the likelihood-based CI is automatically employing the best possible normalizing transform.

The main source of problems with the Wald interval is that $\widehat{\theta}$ may be far from normal, and if we want to transform it to improve the normality we need to know what transform to use. The likelihood interval does the required transform automatically. Hence the applicability of the likelihood-based CI is much wider and, consequently, it is much safer to use than the Wald interval.

2.10 Exercises

Exercise 2.1: To estimate the proportion of luxury cars θ in a city, a group of students stands on a street corner and counts 213 cars by the time they see the 20th luxury car, at which point they stop counting. Draw the likelihood of θ. What assumption do you need to make in your analysis? Should it make a difference (in terms of the information about θ) whether the number 20 was decided in advance or decided in the 'spur of the moment'?

Exercise 2.2: N runners participate in a marathon and they are each assigned a number from 1 to N. From one location we record the following participants:

218 88 254 33 368 235 294 115 9

Make a reasonable assumption about the data and draw the likelihood of N.

Exercise 2.3: The train service between cities A and B announces scheduled departures at 07:00 and arrivals at 10:05. Assume that the train always leaves on time. The arrival time is based on the earliest arrival recorded on 10 sampled journeys. Assume journey time is normally distributed with standard deviation 5 minutes. Draw the likelihood of the average journey time. What is a reasonable guess for a late arrival?

Exercise 2.4: Suppose $X_{(1)}, \ldots, X_{(n)}$ are the order statistics of an iid sample from a continuous distribution $F(x)$.

(a) Show that

$$P(X_{(k)} \leq x) = P\{N(x) \geq k\}$$

where $N(x)$, the number of sample values less than x, is binomial with parameters n and probability $p = F(x)$.

(b) Use (a) to show that the density of $X_{(k)}$ is

$$p(x) = \frac{n!}{(k-1)!(n-k)!}\{F(x)\}^{k-1}\{1 - F(x)\}^{n-k} f(x)$$

where $f(x)$ is the density from $F(x)$. Verify the density using the multinomial argument in Example 2.4.

(c) Use the multinomial argument to find the joint density of $\{X_{(j)}, X_{(k)}\}$ for any j and k.

Exercise 2.5: The following shows the heart rate (in beats/minute) of a person, measured throughout the day:

```
73 75 84 76 93 79 85 80 76 78 80
```

Assume the data are an iid sample from $N(\theta, \sigma^2)$, where σ^2 is known at the observed sample variance. Denote the ordered values by $x_{(1)}, \ldots, x_{(11)}$. Draw and compare the likelihood of θ if

(a) the whole data $x_1, \ldots, x_{11}$ are reported.

(b) only the sample mean $\overline{x}$ is reported.

(c) only the sample median $x_{(6)}$ is reported.

(d) only the minimum $x_{(1)}$ and maximum $x_{(11)}$ are reported.

(e) only the lowest two values $x_{(1)}$ and $x_{(2)}$ are reported.

Exercise 2.6: Given the following data

```
0.5   -0.32   -0.55   -0.76    -0.07  0.44   -0.48
```

draw the likelihood of θ based on each of the following models:

(a) The data are an iid sample from a uniform distribution on $(\theta - 1, \theta + 1)$.

(b) The data are an iid sample from a uniform distribution on $(-\theta, \theta)$.

(c) The data are an iid sample from $N(0, \theta)$.

Exercise 2.7: Given the following data

```
2.32  3.98  3.41  3.08  2.51  3.01  2.31 3 .07  2.97  3.86
```

draw the likelihood of θ assuming the data are an iid sample from a uniform distribution on $(\theta, 2\theta)$. Find the MLE of θ, and report a likelihood-based interval for θ. Comment on the use of standard error here. Comment on the use of probability-based inference here.

Exercise 2.8: For the following paired data (x_i, y_i)

x_i	-0.18	-0.16	-0.73	0.80	-0.41	0.00	-0.08
y_i	0.18	-0.51	-0.62	-0.32	0.55	0.57	-0.32

assume that they are an iid sample from a uniform distribution *in* a circle with mean $(0, 0)$ and radius θ. Draw the likelihood of θ, find the MLE of θ, and report a likelihood-based interval for θ.

Exercise 2.9: For the data in the previous exercise assume that they are an iid sample from a uniform distribution *in* a square $(\theta - 1, \theta + 1) \times (\theta - 1, \theta + 1)$. Draw the likelihood of θ, find the MLE of θ, and report a likelihood-based interval for θ.

Exercise 2.10: Suppose $x_1, \ldots, x_n$ are an iid sample from the exponential distribution with density

$$p(x) = \lambda^{-1} e^{-x/\lambda}.$$

Derive the MLE and its standard error.

Exercise 2.11: Ten light bulbs are left on for 30 days. One fails after 6 days, another one after 28 days, but the others are still working after 30 days. Assume the lifetime is exponential with density

$$p(x) = \lambda^{-1} e^{-x/\lambda}.$$

(a) Given λ, what is the probability of a light bulb working more than 30 days?

(b) Derive and draw the likelihood of λ. (Hint: only the first two order statistics are reported.)

(c) Derive the MLE of λ.

(d) Estimate how long it takes before 90% of the light bulbs will fail.

(e) Suppose the only information available is that two have failed by 30 days, but not their exact failure times. Draw the likelihood of λ and compare with (b).

Exercise 2.12: A communication device transmits a sequence of 0 and 1 digits. To ensure correctness a sequence of length n is transmitted twice. Let θ be the probability of error during transmission of a single digit and let X be the number of digits that differ in the two transmissions. For example, for $n = 8$ and if the two received messages are 00011000 and 00010001, then $x = 2$. Write down the likelihood in general, and draw it for this simple example. Interpret why the likelihood is bimodal.

Exercise 2.13: Suppose the following observed data

```
0.5   -0.32   -0.55   -0.76   -0.07   0.44   -0.48
```

are an iid sample from the double-exponential (Laplacian) distribution with density

$$p_\theta(x) = \frac{1}{2} e^{|x-\theta|}, \quad -\infty < x < \infty.$$

(a) Draw the log-likelihood of θ.

(b) Find the MLE of θ, and report a likelihood-based interval for θ.

(c) Change the largest value in the data (0.5) to 2.5, and redraw the log-likelihood function. Comment about the inference for θ.

Exercise 2.14: To estimate the proportion θ of business people who cheat on their taxes, a randomized response survey is tried. Each person is asked to secretly flip a fair coin and answer question A if it comes out heads, or answer question B otherwise:

A. Did you cheat on your last year's tax form?

B. Was your mother born between January and June? (Assume that the probability of this event is 0.5.)

The data show 15 said Yes and 30 No.

(a) Find the probability of a 'Yes' answer.

(b) Draw the likelihood of θ. Report the MLE, its standard error, and likelihood-based interval for θ.

(c) Compare the method-of-moment estimate of θ, and its 95% CI based on the Wald CI from the sample proportion.

(d) Repeat (a) to (c) based on 10 Yes's and 40 No's. Describe the advantage of the likelihood approach over the method of moments here.

Exercise 2.15: Modify Example 2.14 by changing the coin toss with a card draw (with replacement), where each card c_i is valued from 1 to 13 (Ace=1 ... King=13) and after each draw the sum $x_i = \theta + c_i$ is reported.

(a) Based on the following report from 10 draws

```
16 22 15 17 12 20 15 23 16 23
```

present your guess in terms of likelihood. Split the data into two sets of five observations and repeat the exercise for each set. (Hint: construct the likelihood based on each observation and find the overlap.)

(b) Discuss pure likelihood and probability-based inference as used in this case.

Exercise 2.16: According to the Hardy–Weinberg law, if a population is in equilibrium, the frequencies of genotypes AA, Aa and aa are θ^2, $2\theta(1-\theta)$ and $(1-\theta)^2$. Suppose we type N subjects and find a breakdown of (n_1, n_2, n_3) for the three genotypes.

(a) Find the MLE of θ and the Fisher information $I(\theta)$ as a function of (n_1, n_2, n_3).

(b) Given data $n_1 = 125$, $n_2 = 34$ and $n_3 = 10$, draw the likelihood of θ, and report the MLE and the standard error.

(c) Compare the 95% likelihood-based interval for θ with the Wald interval.

(d) If A is a dominant trait, then the first two groups will show the same phenotype, and there are only two observable groups. Repeat the previous exercises based on phenotype data only. (From part (b): add n_1 and n_2.)

Exercise 2.17: For the previous Exercise 2.5:

(i) Compare the MLE $\widehat{\theta}$ based on the different available data (a) to (e). For (a) and (b) derive the estimate theoretically. For (c), (d) and (e), use a numerical optimization program to get the estimate.

(ii) Compare the Fisher information $I(\widehat{\theta})$ based on the different available data (a) to (e). For (a) and (b) derive the Fisher information theoretically. For (c), (d) and (e), use a numerical optimization program to get the Fisher information.

(iii) Compute and compare the 99% likelihood-based CIs for θ.

(iv) Compute and compare the 99% Wald CIs for θ.

Exercise 2.18: Find the approximate likelihood of H_0: $\theta = \theta_0$ from the 'MLE(standard error)' pair. Apply it in Example 2.17 to test $H_0 : \theta = 0.5$, and compare it with the exact result. When is the approximate likelihood larger or smaller than the exact value?

Exercise 2.19: Prove Theorem 2.1.

Exercise 2.20: For scalar θ show that the Fisher information on $g(\theta)$ is

$$I^*\{g(\widehat{\theta})\} = I(\widehat{\theta}) \left| \frac{\partial g(\widehat{\theta})}{\partial \widehat{\theta}} \right|^{-2}.$$

So the standard error of $g(\widehat{\theta})$ is

$$\text{se}\{g(\widehat{\theta})\} = \text{se}(\widehat{\theta}) \left| \frac{\partial g}{\partial \widehat{\theta}} \right|.$$

Exercise 2.21: Suppose we observe $x = 3$ misprints on a newspaper page and consider a Poisson model with mean θ.

(a) Find a transformation of θ so that the log-likelihood is well approximated by a quadratic. (Hint: consider transformations of the form θ^a for small a and $\log\theta$.)

(b) Compute the Wald-based CI based on a transform with reasonably regular likelihood. Report the interval for the original θ.

(c) Compare the interval with the likelihood-based interval.

Exercise 2.22: We measure some scores from $n = 10$ subjects

```
0.88 1.07 1.27 1.54 1.91 2.27 3.84 4.50 4.64 9.41
```

and assume they are an iid sample from $N(\mu, \sigma^2)$; for simplicity assume σ^2 is known at the observed value. Draw the likelihood, and find the MLE and the standard errors of the following parameters. Comment on the quadratic approximation of the likelihood.

(a) $e^{-\mu}$.

(b) The threshold probability $P(X > 3)$.

(c) The coefficient of variation σ/μ.

Exercise 2.23: For binomial data with $n = 10$ and $x = 8$ repeat the previous exercise for

(a) $g(\theta) = \theta/(1-\theta)$

(b) $g(\theta) = \log\{\theta/(1-\theta)\}$

(c) $g(\theta) = \sin^{-1}\sqrt{\theta}$.

Check for which parameter the standard error quantity is most meaningful. Compute the Wald CI for θ based on each transform, and compare with the likelihood-based CI.

3

More properties of likelihood

3.1 Sufficiency

The idea of Fisher information (Section 2.5) captures the notion of 'information' only roughly. Owing to its association with a quadratic approximation of the likelihood, it is typically meaningful only in large samples. The qualitative concept of *sufficiency* (Fisher 1922) captures precisely the idea of information in the data. An estimate $T(x)$ is sufficient for θ if it summarizes all relevant information contained in the data about θ. This is true if for any other estimate $U(x)$, the distribution of $U(x)$ given $T(x)$ is free of θ. So, once $T(x)$ is known $U(x)$ does not carry any additional information.

It is a fundamental result that the likelihood function is *minimal sufficient.* This means the likelihood function captures all of the information in the data, and *anything less involves some loss of information.*

First we define an experiment E to be a collection $\{x, \theta, p_\theta(x)\}$. Such a definition is completely general, since the data x can be of any complexity. The probability model $p_\theta(x)$ describes how the data are generated, such as an iid sampling, or sequential experiment, etc.

Definition 3.1 *A statistic $T(X)$ is sufficient for θ in an experiment E if the conditional distribution of $X|T = t$ is free of θ.*

Note first the intuitive content of the definition: if $X|T = t$ is free of θ, then once we know $T = t$ we could have simulated X from the conditional distribution so that, unconditionally, X still follows the model p_θ with the true (but unknown) θ. Since $X|T = t$ itself does not involve θ, then it cannot carry any information about θ and all information about it must be contained in T.

Technically the definition does include the full description of the experiment, in particular the model $p_\theta(x)$, so sufficiency is only meaningful in this context. It is wrong and meaningless to state that the sample mean is sufficient for the population mean without any reference to the model $p_\theta(x)$.

Example 3.1: Suppose $X_1, \ldots, X_n$ are an iid sample from a Poisson distribution with mean θ. Then $T(X) = \sum X_i$ is sufficient for θ.

Proof: We want to show that the conditional distribution of $X_1, \ldots, X_n$ given $T = t$ is free of θ. First we can show, for example using the moment generating function technique, that T is Poisson with parameter $n\theta$. Then, using the Bayes theorem, we have

$$\begin{aligned} P(X_1 = x_1, \ldots, X_n = x_n | T = t) &= \frac{P(X_1 = x_1, \ldots, X_n = x_n, T = t)}{P(T = t)} \\ &= \frac{P(X_1 = x_1, \ldots, X_n = x_n)}{P(T = t)} \\ &= \frac{e^{-n\theta}\theta^{\sum x_i}/\prod x_i!}{e^{-n\theta}(n\theta)^t/t!} \\ &= \frac{t!}{\prod x_i!}\prod\left(\frac{1}{n}\right)^{x_i}, \end{aligned}$$

for $x_1, \ldots, x_n$ such that $\sum x_i = t$, and the probability is zero otherwise. Note that any one-to-one function of $\sum x_i$ is also sufficient, so in particular $\overline{x}$ is a sufficient estimate of θ.

For example, given $n = 2$ and $x_1 + x_2 = 10$, a further determination whether we have observed $(0, 10)$ or $(1, 9)$ ... or $(10, 0)$ involves an event whose probability is free of θ. So this process can add no information about θ, and inference about θ should be the same regardless of which event is obtained. This means that all information about θ is contained in $x_1 + x_2$. □

From the example, it is clear that sufficiency is a useful concept of data reduction: $x_1, \ldots, x_n$ are summarized to $\overline{x}$ only. As stated earlier the notion of sufficiency is *tied* with or *conditional* on an assumed model. Knowing $\overline{x}$ alone is not enough to capture the uncertainty in the population mean; we also need to know the probability model such as $x_1, \ldots, x_n$ are an iid sample from Poisson(θ).

Proving the sufficiency of a statistic from first principles is usually not very illuminating. A much more direct way is given by the so-called factorization theorem.

Theorem 3.1 *$T(X)$ is sufficient for θ in an experiment E if and only if the model $p_\theta(x)$ can be factorized in the form*

$$p_\theta(x) = g(t(x), \theta)\ h(x),$$

where $h(x)$ is free of θ.

Example 3.2: Suppose $x_1, \ldots, x_n$ are an iid sample from $N(\mu, \sigma^2)$ and let $\theta = (\mu, \sigma^2)$. The density is

$$\begin{aligned} p_\theta(x) &= (2\pi\sigma^2)^{-n/2} \exp\left\{-\frac{1}{2\sigma^2}\sum_i (x_i - \mu)^2\right\} \\ &= (2\pi\sigma^2)^{-n/2} \exp\left\{-\frac{\sum_i x_i^2}{2\sigma^2} + \frac{\mu\sum_i x_i}{\sigma^2} - \frac{n\mu^2}{2\sigma^2}\right\}. \end{aligned}$$

From the factorization theorem, it is clear that

(a) if σ^2 is known $\sum_i x_i$ is sufficient for μ;

(b) if μ is known $\sum_i (x_i - \mu)^2$ is sufficient for σ^2;

(c) if (μ, σ^2) is unknown $(\sum_i x_i, \sum_i x_i^2)$ is sufficient.

Knowing $(\sum_i x_i, \sum_i x_i^2)$ is sufficient means that the rest of the data does not add any more information about (μ, σ^2). If normality has been established, this means we only need to keep $(\sum_i x_i, \sum_i x_i^2)$ for further analysis. (In practice we do not routinely throw most of the data away, since instinctively we do not believe in any specific model, in which case $(\sum_i x_i, \sum_i x_i^2)$ is not sufficient.) □

3.2 Minimal sufficiency

Fisher originally defined sufficiency as a property of an estimate, but currently, as defined above, the concept applies to any set of values computed from the data. For a particular dataset there are infinitely many sufficient statistics; for example, the whole dataset is always sufficient for any model. If $x_1, \ldots, x_n$ are an iid sample from $N(\theta, 1)$, then the following statistics are sufficient for μ: $\sum x_i$, $\overline{x}$, $(\sum_{i=1}^m x_i, \sum_{i=m+1}^n x_i)$ for any m, or the set of order statistics $(x_{(1)}, \ldots, x_{(n)})$. Notice that $\overline{x}$ is a function of the other sufficient statistics, while the set of order statistics, for example, is not a function of $\overline{x}$.

From the definition we can immediately see that if T is sufficient then any one-to-one function of T is also sufficient. There is no such guarantee if the function is many-to-one, as in this case there is a reduction in the sample space of the statistic. It is then useful to consider the following concept:

Definition 3.2 *A sufficient statistic $T(X)$ is* minimal sufficient *if it is a function of any other sufficient statistic.*

So a statistic is minimal sufficient if no further data reduction is allowed. Generally, if the dimension of the sufficient statistic is the same as that of the parameter space, then the statistic is minimal sufficient; or, *if an estimate is sufficient then it is minimal sufficient.* Any further reduction of the data from the minimal sufficient statistic would involve some loss of information. Establishing minimal sufficiency relies on the connection between sufficiency and likelihood.

Example 3.3: Suppose $x_1, \ldots, x_n$ are an iid sample from $N(\theta, 1)$, so by the factorization theorem, $\overline{x}$ is sufficient. Based on $x_1, \ldots, x_n$ we have

$$L(\theta; x_1, \ldots, x_n) = \left(\frac{1}{\sqrt{2\pi}}\right)^n \exp\left\{-\frac{1}{2}\sum_i (x_i - \theta)^2\right\}.$$

The dependence on the data is made explicit for later comparison. With some algebra

$$\log L(\theta; x_1, \ldots, x_n) = -\frac{1}{2}\sum_i (x_i - \overline{x})^2 - \frac{n}{2}(\overline{x} - \theta)^2$$

$$= \text{constant} - \frac{n}{2}(\overline{x} - \theta)^2,$$

so the likelihood depends on the data only through the sample mean $\overline{x}$. Suppose, from this same experiment, only the sample mean $\overline{x}$ is recorded. Then, since $\overline{x}$ is $N(\theta, 1/n)$, the likelihood based on $\overline{x}$ is

$$L(\theta; \overline{x}) = \frac{1}{\sqrt{2\pi/n}} \exp\left\{-\frac{n}{2}(\overline{x} - \theta)^2\right\},$$

or

$$\log L(\theta; \overline{x}) = \text{constant} - \frac{n}{2}(\overline{x} - \theta)^2.$$

Therefore the likelihood based on the whole dataset $x_1, \ldots, x_n$ is the same as that based on $\overline{x}$ alone. □

The preceding result is true for any sufficient statistic. If t is any function of the data x, the likelihood based on x is the same as the likelihood based on both x and t. So, if t is sufficient

$$\begin{aligned} L(\theta; x) &= L(\theta; x, t) = p_\theta(x, t) = p_\theta(t) p(x|t) \\ &= \text{constant} \times p_\theta(t) \\ &= \text{constant} \times L(\theta; t), \end{aligned}$$

meaning that $L(\theta; x)$ can be computed based on t alone.

The likelihood function itself is a sufficient statistic. If this sounds surprising, note that there is an ambiguity with the function notation: $L(\theta)$ can mean the *entire function* over all possible θ, or the function at a particular value of θ. To make an explicit distinction, let $L(\cdot; x)$ be the entire likelihood function based on x; it is a statistic since it can be computed from the data alone, no unknown value is involved. Then, for any choice of θ_0,

$$t(x) = \frac{L(\cdot; x)}{L(\theta_0; x)}$$

is sufficient. To prove this, we use the factorization theorem by defining

$$g(t, \theta) = \frac{L(\theta; x)}{L(\theta_0; x)}$$

and $h(x) = L(\theta_0; x)$. Similarly, the normalized likelihood function is also a sufficient statistic. Hence we arrive at a fundamental property of the likelihood function:

Theorem 3.2 *If T is sufficient for θ in an experiment E then the likelihood of θ based on the whole data x is the same as that based on T alone. Therefore, the likelihood function is minimal sufficient.*

Recall that we say two likelihoods are the same (or equivalent) if they are proportional. The proof of the second part is immediate: that any sufficient statistic would yield the same likelihood function implies that the likelihood function is a function of any other sufficient statistic. The theorem implies that *any statistic that is a one-to-one map with the likelihood function is minimal sufficient.* This occurs if different values of the statistic lead to different (normalized) likelihood functions.

Example 3.4: Based on $x_1, \ldots, x_n$ from $N(\mu, 1)$ the likelihood function

$$L(\mu; x_1, \ldots, x_n) = \text{constant} \times \exp\{-\frac{n}{2}(\overline{x} - \mu)^2\}$$

is a one-to-one map with $\overline{x}$: if we have two different samples $x_1, \ldots, x_n$ and $y_1, \ldots, y_n$ then

$$L(\mu; x_1, \ldots, x_n) = \text{constant} \times L(\mu; y_1, \ldots, y_n) \;\; \text{iff} \;\; \overline{x} = \overline{y}.$$

This establishes $\overline{x}$ as minimal sufficient. With a similar argument we can show that $(\overline{x}, S^2)$ is minimal sufficient for an iid sample from $N(\mu, \sigma^2)$, where S^2 is the sample variance. □

Example 3.5: Suppose $x_1, \ldots, x_n$ are an iid sample from the uniform distribution on $(\theta - 1, \theta + 1)$. The likelihood function is

$$L(\theta) = 2^{-n} I(x_{\max} - 1 < \theta < x_{\min} + 1),$$

so $(x_{\min}, x_{\max})$ is minimal sufficient for θ. Since the minimal sufficient statistic is two-dimensional no estimate of θ is sufficient; the MLE of θ is not even well defined.

The nonexistence of a sufficient estimate does not mean there is no sensible inference for θ. We can simply report that θ must be between $x_{\max} - 1$ and $x_{\min} + 1$. For example, given data

```
1.07 1.11 1.31 1.51 1.69 1.72 1.92 2.24 2.62 2.98
```

we are certain that $(2.98 - 1) < \theta < (1.07 + 1)$. If a point estimate is required we can take the midpoint of the interval, which in this case is a better estimate than the sample mean or median. Here $\overline{x} = 1.817$ and the sample median 1.705 have zero likelihood, so they are not even sensible estimates. □

Monotone likelihood ratio property

If the parameter θ is not a boundary parameter, minimal sufficiency can be expressed in terms of likelihood ratio. Being a one-to-one map with the likelihood function means that a statistic $t(x)$ is minimal sufficient iff, for any choice of θ_0 and θ_1, the likelihood ratio $L(\theta_1)/L(\theta_0)$ is a one-to-one function of $t(x)$. If $t(x)$ is scalar, this is called a monotone likelihood ratio property. Furthermore, if $L(\theta)$ is smooth enough, as θ_1 approaches θ_0,

$$\begin{aligned}\frac{L(\theta_1)}{L(\theta_0)} &\approx \frac{L(\theta_0) + L'(\theta_0)(\theta_1 - \theta_0)}{L(\theta_0)} \\ &= 1 + \frac{\partial \log L(\theta_0)}{\partial \theta_0}(\theta_1 - \theta_0),\end{aligned}$$

for any θ_0. Hence we get a simple characterization that $t(x)$ is minimal sufficient iff the score statistic is a one-to-one function of $t(x)$.

3.3 Multiparameter models

Most real problems require multiparameter models; for example, the normal model $N(\mu, \sigma^2)$ needs two parameters for the mean and the variance. Model complexity is determined by the number of parameters, not by the number of observations. No new definition of likelihood is required. Given data x, the likelihood is

$$L(\theta) = p_\theta(x),$$

where $p_\theta(x)$ is the probability of the observed data. With the same argument as the one in Section 2.1, it is sufficient to use the density function for purely continuous models.

Example 3.6: Let $x_1, \ldots, x_n$ be an iid sample from $N(\mu, \sigma^2)$. Ignoring irrelevant constant terms, we can write

$$\log L(\mu, \sigma^2) = -\frac{n}{2} \log \sigma^2 - \frac{1}{2\sigma^2} \sum_i (x_i - \mu)^2. \tag{3.1}$$

Suppose $n_1, \ldots, n_k$ are a sample from the multinomial distribution with known $N = \sum_i n_i$ and unknown probabilities $p_1, \ldots, p_k$. There are $(k-1)$ free parameters since $\sum_i p_i = 1$. Ignoring irrelevant terms, the log-likelihood of the parameters $(p_1, \ldots, p_k)$ is

$$\log L(p_1, \ldots, p_k) = \sum_i n_i \log p_i.$$

For $k = 3$, the parameter space is a 2D simplex satisfying $p_1 + p_2 + p_3 = 1$, which can be represented in a triangle. □

We have limited ability to view or communicate $L(\theta)$ in high dimensions. If θ is two dimensional, we can represent $L(\theta)$ in a contour plot or a perspective plot. In general a mathematical analysis of the likelihood surface is essential for its description. Let $\theta = (\theta_1, \ldots, \theta_p)$. Assuming $\log L(\theta)$ is differentiable, the score function is the first derivative vector

$$S(\theta) = \frac{\partial}{\partial \theta} \log L(\theta),$$

and the MLE $\widehat{\theta}$ is the solution of the score equation $S(\theta) = 0$. The Fisher information $I(\widehat{\theta})$ is a matrix of second derivatives with elements

$$I_{ij}(\widehat{\theta}) = -\frac{\partial^2}{\partial \theta_i \partial \theta_j} \log L(\theta)\bigg|_{\theta=\widehat{\theta}}.$$

The original idea of the Fisher information introduced in Section 2.5 is related to a quadratic approximation of the log-likelihood function. Such an idea has a natural extension in the multiparameter case. In regular cases the likelihood function can be represented by the MLE $\widehat{\theta}$ and information matrix $I(\widehat{\theta})$ via a second-order approximation

$$\log L(\theta) \approx \log L(\widehat{\theta}) + S(\widehat{\theta})(\theta - \widehat{\theta}) - \frac{1}{2}(\theta - \widehat{\theta})' I(\widehat{\theta})(\theta - \widehat{\theta})$$

$$= \log L(\widehat{\theta}) - \frac{1}{2}(\theta - \widehat{\theta})' I(\widehat{\theta})(\theta - \widehat{\theta}).$$

Example 3.7: Let $x_1, \ldots, x_n$ be an iid sample from $N(\mu, \sigma^2)$. Taking derivatives of the log-likelihood (3.1), we obtain the score functions

$$\begin{aligned} S_1(\mu, \sigma^2) &= \frac{\partial}{\partial \mu} \log L(\mu, \sigma^2) = \frac{1}{\sigma^2} \sum_i (x_i - \mu) \\ S_2(\mu, \sigma^2) &= \frac{\partial}{\partial \sigma^2} \log L(\mu, \sigma^2) = -\frac{n}{2\sigma^2} + \frac{1}{2\sigma^4} \sum_i (x_i - \mu)^2. \end{aligned}$$

Equating these to zero yields the MLEs

$$\begin{aligned} \widehat{\mu} &= \overline{x} \\ \widehat{\sigma}^2 &= \frac{1}{n} \sum_i (x_i - \overline{x})^2. \end{aligned}$$

Note the $1/n$ divisor for the variance estimate, which is different from the usual $1/(n-1)$ divisor for the sample variance S^2.

Taking further derivatives of the log-likelihood gives the Fisher information matrix

$$I(\widehat{\mu}, \widehat{\sigma}^2) = \begin{pmatrix} n/\widehat{\sigma}^2 & 0 \\ 0 & n/(2\widehat{\sigma}^4) \end{pmatrix}. \quad \Box$$

Example 3.8: Suppose $x_1, \ldots, x_n$ are an iid sample from the gamma distribution with density

$$p(x) = \frac{1}{\Gamma(\alpha)} \lambda^\alpha x^{\alpha - 1} e^{-\lambda x}, \quad x > 0.$$

The log-likelihood of $\theta \equiv (\lambda, \alpha)$ is

$$\begin{aligned} \log L(\theta) &= \sum_i \{-\log \Gamma(\alpha) + \alpha \log \lambda + (\alpha - 1) \log x_i - \lambda x_i\} \\ &= -n \log \Gamma(\alpha) + n\alpha \log \lambda + (\alpha - 1) \sum_i \log x_i - \lambda \sum_i x_i. \end{aligned}$$

The score function is a vector with elements

$$\begin{aligned} S_1(\theta) &= \frac{\partial \log L(\lambda, \alpha)}{\partial \lambda} = \frac{n\alpha}{\lambda} - \sum_i x_i, \\ S_2(\theta) &= \frac{\partial \log L(\lambda, \alpha)}{\partial \alpha} = -n\psi(\alpha) + n \log \lambda + \sum_i \log x_i, \end{aligned}$$

where $\psi(\alpha) \equiv \partial \log \Gamma(\alpha) / \partial \alpha$. The MLE $(\widehat{\lambda}, \widehat{\alpha})$ satisfies

$$\widehat{\lambda} = \frac{n\widehat{\alpha}}{\sum_i x_i} = \frac{\widehat{\alpha}}{\overline{x}}$$

and

$$-n\psi(\widehat{\alpha}) + n\log(\widehat{\alpha}/\overline{x}) + \sum_i \log x_i = 0. \tag{3.2}$$

A numerical procedure is needed to solve (3.2).

The elements of the Fisher information matrix $I(\widehat{\theta})$ are

$$\begin{aligned} I_{11} &= \frac{n\widehat{\alpha}}{\widehat{\lambda}^2} \\ I_{12} &= -\frac{n}{\widehat{\lambda}} \\ I_{21} &= I_{12} \\ I_{22} &= n\psi'(\widehat{\alpha}). \ \square \end{aligned}$$

Example 3.9: Suppose $n_1, \ldots, n_k$ are a sample from the multinomial distribution with known $N = \sum_i n_i$ and unknown probabilities $p_1, \ldots, p_k$. The MLE is $\widehat{p}_i = n_i/N$. We can show this, using the Lagrange multiplier technique, by maximizing

$$Q(p_1, \ldots, p_k, \lambda) = \sum_i n_i \log p_i + \lambda\left(\sum_i p_i - 1\right)$$

with respect to $p_1, \ldots, p_k$ and λ (Exercise 3.12). The estimate is intuitive, since we can collapse the categories into two: the i'th category and the rest, hence creating binomial data.

Often there are theories on how the probabilities vary with a certain parameter θ, so $p_i \equiv p_i(\theta)$. The log-likelihood of θ is

$$\log L(\theta) = \sum_i n_i \log p_i(\theta).$$

The following table shows the grouped data of 100 measurements of the speed of light (as deviations from 299 in 1000's km/s) from Michelson's experiment. (The original data are given in Example 4.8.)

Intervals	Counts	Probability
$x \le 0.75$	$n_1 = 9$	p_1
$0.75 < x \le 0.85$	$n_2 = 46$	p_2
$0.85 < x \le 0.95$	$n_3 = 33$	p_3
$0.95 < x \le 1.05$	$n_4 = 11$	p_4
$x > 1.05$	$n_5 = 1$	p_5

Suppose the original data $x_1, \ldots, x_{100}$ are an iid sample from $N(\mu, \sigma^2)$. A particular probability p_i is given by a normal probability; for example,

$$p_1(\mu, \sigma) = P(X \le 0.75) = \Phi\left(\frac{0.75 - \mu}{\sigma}\right)$$

where $\Phi(\cdot)$ is the standard normal distribution function.

There is no closed formula for the MLEs. Numerical optimization of the likelihood yields the MLEs

$$\widehat{\mu} = 0.8485, \quad \widehat{\sigma} = 0.0807.$$

This compares to $\widehat{\mu} = 0.8524$ and $\widehat{\sigma} = 0.0790$ from the original data; see Example 4.8 for further discussion. The Fisher information based on the grouped data

$$I(\widehat{\mu}, \widehat{\sigma}) = \begin{pmatrix} 13507.70 & 513.42 \\ 513.42 & 21945.03 \end{pmatrix}$$

is also found numerically. □

3.4 Profile likelihood

While the definition of likelihood covers the multiparameter models, the resulting multidimensional likelihood function can be difficult to describe or to communicate. Even when we are interested in several parameters, it is always easier to describe one parameter at a time. The problem also arises in cases where we may be interested in only a subset of the parameters; in the normal model, we might only be interested in the mean μ, while σ^2 is a 'nuisance', being there only to make the model able to adapt to data variability. A method is needed to 'concentrate' the likelihood on a single parameter by eliminating the nuisance parameter.

Accounting for the extra uncertainty due to unknown nuisance parameters is an essential consideration, especially in small-sample cases. Almost all of the analytical complexities in the theory and application of the likelihood are associated with this problem. Unfortunately, *there is no single technique that is acceptable in all situations* (see, for example, Bayarri *et al.* 1987). Speaking of the uncertainty about one parameter independently from the other is not always meaningful.

The likelihood approach to eliminate a nuisance parameter is to replace it by its MLE at each fixed value of the parameter of interest. The resulting likelihood is called the *profile likelihood.* It is a pragmatic approach that leads to a reasonable answer. We will discuss the problem of nuisance parameters more responsibly in Chapter 10.

Bayesians eliminate all unwanted parameters by integrating them out; that is consistent with their view that parameters have regular density functions. However, the likelihood function is not a probability density function, and it does not obey probability laws (see Section 2.8), so integrating out a parameter in a likelihood function is not a meaningful operation. It turns out, however, there is a close connection between the Bayesian integrated likelihood and a modified profile likelihood (Section 10.6).

For the moment we will only introduce the bare minimum to be able to deal with the basic models in the next chapter. Specifically, let (θ, η) be the full parameter and θ is the parameter of interest.

Definition 3.3 *Given the joint likelihood* $L(\theta, \eta)$ *the* profile likelihood *of* θ *is*

$$L(\theta) = \max_{\eta} L(\theta, \eta),$$

where the maximization is performed at fixed value of θ.

It should be emphasized that at fixed θ the MLE of η is generally a function of θ, so we can also write

$$L(\theta) = L(\theta, \widehat{\eta}_\theta).$$

The profile likelihood is then treated like a regular likelihood; for example, we can normalize it, display it, calibrate it and compute likelihood intervals from it. Note that we have adopted a convenient generic notation for the likelihood function $L(\cdot)$, where the *named argument* and context determine the meaning of the function. This is similar to using the symbol $P(\cdot)$ to indicate the 'probability of' the event inside the bracket. If there is any danger of confusion, we use $L_p(\theta)$ to indicate a profile likelihood.

Example 3.10: Suppose $x_1, \ldots, x_n$ are an iid sample from $N(\mu, \sigma^2)$ with both parameters unknown. The likelihood function of (μ, σ^2) is given by

$$L(\mu, \sigma^2) = \left(\frac{1}{\sqrt{2\pi\sigma^2}}\right)^n \exp\left\{-\frac{1}{2\sigma^2}\sum_i (x_i - \mu)^2\right\}.$$

A likelihood of μ without reference to σ^2 is not an immediately meaningful quantity, since it is very different at different values of σ^2. As an example, suppose we observe

```
0.88 1.07 1.27 1.54 1.91 2.27 3.84 4.50 4.64 9.41.
```

The MLEs are $\widehat{\mu} = 3.13$ and $\widehat{\sigma}^2 = 6.16$. Figure 3.1(a) plots the contours of the likelihood function at 90%, 70%, 50%, 30% and 10% cutoffs. There is a need to plot the likelihood of each parameter individually: it is more difficult to describe or report a multiparameter likelihood function, and usually we are not interested in a simultaneous inference of μ and σ^2.

The profile likelihood function of μ is computed as follows. For fixed μ the maximum likelihood estimate of σ^2 is

$$\widehat{\sigma}^2_\mu = \frac{1}{n}\sum_i (x_i - \mu)^2,$$

so the profile likelihood of μ is

$$L(\mu) \quad = \quad \text{constant} \times (\widehat{\sigma}^2_\mu)^{-n/2}.$$

This is not the same as

$$L(\mu, \sigma^2 = \widehat{\sigma}^2) = \text{constant} \times \exp\left\{-\frac{1}{2\widehat{\sigma}^2}\sum_i (x_i - \mu)^2\right\},$$

the slice of $L(\mu, \sigma^2)$ at $\sigma^2 = \widehat{\sigma}^2$; this is known as an estimated likelihood. Both likelihoods will be close if σ^2 is well estimated, otherwise the profile likelihood is preferred.

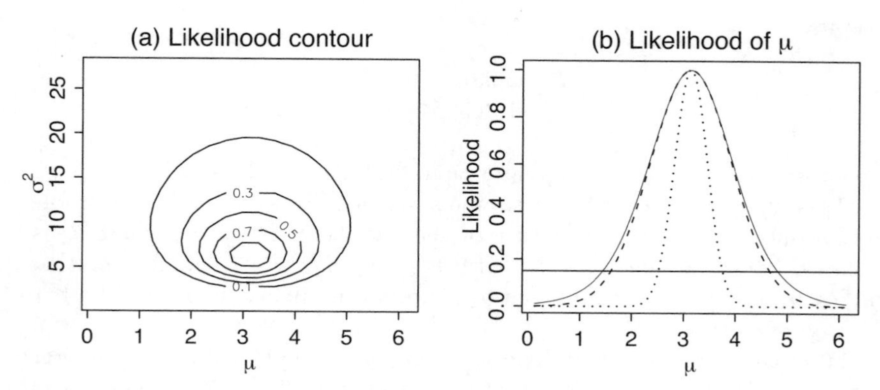

Figure 3.1: *(a) Likelihood function of* (μ, σ^2)*. The contour lines are plotted at 90%, 70%, 50%, 30% and 10% cutoffs; (b) Profile likelihood of the mean* μ *(solid),* $L(\mu, \sigma^2 = \widehat{\sigma}^2)$ *(dashed), and* $L(\mu, \sigma^2 = 1)$ *(dotted).*

For the observed data $L(\mu)$ and $L(\mu, \sigma^2 = \widehat{\sigma}^2)$ are plotted in Figure 3.1(b). It is obvious that ignoring the unknown variability, e.g. by assuming $\sigma^2 = 1$, would give a wrong inference. So, in general a nuisance parameter is needed to allow for a better model, but it has to be eliminated properly in order to concentrate the inference on the parameter of interest.

The profile likelihood of σ^2 is given by

$$\begin{aligned} L(\sigma^2) &= \text{constant} \times (\sigma^2)^{-n/2} \exp\left\{-\frac{1}{2\sigma^2}\sum_i (x_i - \overline{x})^2\right\} \\ &= \text{constant} \times (\sigma^2)^{-n/2} \exp\{-n\widehat{\sigma}^2/(2\sigma^2)\}. \ \square \end{aligned}$$

Curvature of profile likelihood

The curvature of a profile log-likelihood is related to the elements of the Fisher information matrix; the proof is given in Section 9.11. Suppose we are interested in θ_1 from the total parameter $\theta = (\theta_1, \theta_2)$. We partition the information matrix $I(\widehat{\theta})$ as

$$I(\widehat{\theta}) \equiv \begin{pmatrix} I_{11} & I_{12} \\ I_{21} & I_{22} \end{pmatrix}$$

and its inverse

$$I^{-1}(\widehat{\theta}) \equiv \begin{pmatrix} I^{11} & I^{12} \\ I^{21} & I^{22} \end{pmatrix}.$$

Then the curvature of the profile log-likelihood of θ_1 is *not* I_{11}, but $(I^{11})^{-1}$. The latter is, in general, smaller than the former; the proof is given in Section 8.7. As an example, consider the following matrix:

$$I(\widehat{\theta}) = \begin{pmatrix} 2.3 & 2.8 \\ 2.8 & 4.3 \end{pmatrix}.$$

The inverse is

$$I^{-1}(\widehat{\theta}) = \begin{pmatrix} 2.097561 & -1.365854 \\ -1.365854 & 1.121951 \end{pmatrix}$$

so, for example, $I_{11} = 2.3$ is greater than $(I^{11})^{-1} = 1/2.097561 = 0.48$.

This numerical result has a simple statistical interpretation: the information index I_{11} is the curvature of the log-likelihood of θ_1, where θ_2 is assumed known at the observed MLE $\widehat{\theta}_2$, while $(I^{11})^{-1}$ is the information on θ_1 that takes into account that θ_2 is unknown. Hence, it is sensible that I_{11} is greater than $(I^{11})^{-1}$.

From the result above, individual profile likelihoods for each parameter can be quadratically approximated using the pair $\{\widehat{\theta}_i, (I^{ii})^{-1}\}$ via

$$\log L(\theta_i) \approx -\frac{1}{2}(I^{ii})^{-1}(\theta_i - \widehat{\theta}_i)^2,$$

and the standard error of $\widehat{\theta}_i$ is simply

$$\text{se}(\widehat{\theta}_i) = \sqrt{I^{ii}},$$

namely the square root of the diagonal elements of $I^{-1}(\widehat{\theta})$. For Michelson's data in Example 3.9 we obtain

$$I^{-1}(\widehat{\mu}, \widehat{\sigma}) = \begin{pmatrix} 7.409771 \times 10^{-05} & -1.733578 \times 10^{-06} \\ -1.733578 \times 10^{-06} & 4.560895 \times 10^{-05} \end{pmatrix}$$

and

$$\text{se}(\widehat{\mu}) = \sqrt{7.409771 \times 10^{-05}} = 0.0086.$$

3.5 Calibration in multiparameter case

A fundamental question in likelihood inference is how one should react to an observed likelihood. In the one parameter case we can declare a hypothesis is doubtful if its likelihood is less than 15%, say. We will now show that, in repeated sampling sense, this prescription does not work in higher dimensions.

Consider the case of several normal means; the results are approximately true in the general case if the likelihood is regular. Let $x_1, \ldots, x_p$ be an independent sample from the normal distribution with mean $\mu_1, \ldots, \mu_p$, respectively, and *known* common variance σ^2. Suppose we are interested to test H_0: $\mu_1 = \cdots = \mu_p = 0$ against the alternative that at least one μ_i is not zero. The standard test statistic in this case is Wilk's likelihood ratio statistic

$$W \equiv 2 \log \frac{L(\widehat{\theta})}{L(\theta_0)}$$

$$= \sum_{i=1}^{p} \frac{x_i^2}{\sigma^2},$$

where $\theta = (\mu_1, \ldots, \mu_p)$, and $\theta_0 = (0, \ldots, 0)$. Under H_0 the random variate W has a χ^2 distribution with p degrees of freedom. In classical (frequentist) computations we use this distribution theory to calibrate the observed likelihood ratio.

A classical test of level α is to reject H_0: $\theta = \theta_0$ if

$$2 \log \frac{L(\widehat{\theta})}{L(\theta_0)} > \chi^2_{p,(1-\alpha)} \tag{3.3}$$

or, in terms of likelihood, if

$$\frac{L(\theta_0)}{L(\widehat{\theta})} < e^{-\frac{1}{2}\chi^2_{p,(1-\alpha)}}. \tag{3.4}$$

An observed $L(\theta_0)/L(\widehat{\theta}) = r$ corresponds to $w = -2 \log r$, and

$$\text{P-value} = P(W \geq -2 \log r),$$

which depends on the degrees of freedom or the number of parameters being tested.

An observed $w = 3.84$ is associated with a fixed likelihood of 15% and the following P-values depending on p:

p	1	2	3	4	5	6	7	8	9	10
P-value	0.05	0.15	0.28	0.43	0.57	0.70	0.80	0.87	0.92	0.95

(It can be verified generally for $p = 2$ that the likelihood of H_0 is the same as the P-value.) The table indicates that in high dimensions it is not at all unusual to get a likelihood of 15% or less. So declaring significant evidence with 15% critical value will lead to many spurious results, and it is essential to calibrate an observed likelihood differently depending on the dimensionality of the parameter.

When there are nuisance parameters, we have stated in Section 3.4 that a profile likelihood of the parameters of interest is to be treated like a regular likelihood. This means that the calibration of a multiparameter profile likelihood also follows the result above.

Likelihood-based confidence region

From (3.4) it is immediate that the set

$$\left\{ \theta, \frac{L(\theta)}{L(\widehat{\theta})} > e^{-\frac{1}{2}\chi^2_{p,(1-\alpha)}} \right\}$$

is a $100(1 - \alpha)\%$ confidence region for θ. In the normal case the confidence level is exact, otherwise it is only approximate.

Because of display problems confidence regions are only used (if at all) for two-parameter cases. At $p = 2$ we can show this interesting relationship:

$$\left\{\theta, \frac{L(\theta)}{L(\widehat{\theta})} > \alpha\right\}$$

is a $100(1-\alpha)\%$ confidence region for θ. For example, Figure 3.1(a) shows approximate 90%, ... ,10% confidence regions for (μ, σ^2).

AIC-based calibration

In Section 2.6 we discuss pure likelihood and probability-based inference; for multiparameter problems the latter has just been described. How can we account for parameter dimension in pure likelihood inference, without appealing to probability considerations? One solution, proposed by Lindsey (1999b), leads to the Akaike information criterion (AIC), a general criterion for model selection. To get the idea in the simplest case, suppose we have two parameters θ_1 and θ_2 such that

$$L(\theta_1, \theta_2) = L(\theta_1)L(\theta_2).$$

Suppose, individually, the likelihood of H_0: $\theta_1 = 0$ and H_0: $\theta_2 = 0$ are both 0.2, so the joint likelihood of H_0: $\theta_1 = \theta_2 = 0$ is $0.2^2 = 0.04$. A simplistic prescription that a likelihood less than 15% indicates evidence against a hypothesis leads to a logical problem: H_0: $\theta_1 = \theta_2 = 0$ is rejected, but individually we do not have evidence to reject either H_0: $\theta_1 = 0$ or H_0: $\theta_1 = 0$.

An immediate solution is to compare the likelihood with c^p, where c is the critical value for a single parameter and p is the dimension of the parameter space. So, in the above example, if we set $c = 0.15$, then the joint likelihood of H_0: $\theta_1 = \theta_2 = 0$ must be compared against $0.15^2 = 0.0225$. The evaluation of hypotheses in different dimensions is then logically compatible, in the sense that if H_0: $\theta_1 = \theta_2 = 0$ is rejected, then one of H_0: $\theta_1 = 0$ or H_0: $\theta_1 = 0$ must be rejected.

The method to account for the dimension of the parameter space above is closely related to the AIC, discussed in detail in Section 13.5. The AIC of a model with free parameter θ is defined as

$$\text{AIC} = -2\log L(\widehat{\theta}) + 2p, \tag{3.5}$$

where $\widehat{\theta}$ is the MLE and p is the dimension of θ. Using the AIC, the log-likelihood of a model is penalized by the number of parameters in the model, which makes for a fairer comparison between models or hypotheses. The prescription is simple: the model with a smaller AIC wins.

A particular hypothesis H_0: $\theta = \theta_0$ describes a model of zero dimension (it has no free parameter), so its AIC is

$$\text{AIC}(H_0) = -2\log L(\theta_0).$$

Based on the AIC, the full model with free parameter θ is preferable to the simple model H_0: $\theta = \theta_0$ (i.e. evidence against H_0) if

$$2\log\frac{L(\widehat{\theta})}{L(\theta_0)} > 2p \tag{3.6}$$

or

$$\frac{L(\theta_0)}{L(\widehat{\theta})} < e^{-p}; \tag{3.7}$$

that is, if the likelihood of θ_0 is less than c^p, using the critical value $c = e^{-1} = 0.37$, which fits the method we just describe above. It is instructive to compare (3.3) against (3.6), and (3.4) against (3.7).

One controversial aspect of the use of AIC is in how to attach significance (in the classical P-value sense) to an observed difference. There is no easy answer to this problem. While it is possible to derive the AIC using large-sample theory, Lindsey's idea of compatible inference implies that (3.7) can be justified from the likelihood alone. This means the AIC has the same logical level as a pure likelihood inference: it is weaker than probability-based inference, and it is especially useful if probability-based inference is not available.

At a fixed level of α, the critical value $\chi^2_{p,(1-\alpha)}$ increases with p, but not as fast as $2p$ in the AIC-based calibration. For $\alpha = 0.05$ we have:

p	1	2	3	4	5	7	8	10
$2p$	2	4	6	8	10	14	16	20
$\chi^2_{p,1-\alpha}$	3.84	5.99	7.81	9.49	11.07	14.07	15.51	18.31

Compared with the probability-based χ^2 criterion, the AIC would allow a model to grow ($p > 1$), but it does not permit too many parameters ($p > 6$). Using the AIC is equivalent to changing the α-level depending on the number of parameters.

3.6 Exercises

Exercise 3.1: Prove Theorem 3.1.

Exercise 3.2: Use the factorization theorem to find the (nontrivial) sufficient statistics based on an iid sample from the following distributions:

(a) Uniform on $(0, \theta)$.

(b) Uniform on $(\theta - 1, \theta + 1)$.

(c) Uniform on (θ_1, θ_2) with both parameters unknown.

(d) Uniform on $(-\theta, \theta)$.

(e) Uniform on $(\theta, 2\theta)$.

(f) Uniform *in* the circle with mean $(0, 0)$ and radius θ.

(g) Uniform *in* the square $(\theta - 1, \theta + 1) \times (\theta - 1, \theta + 1)$.

(h) Exponential with mean θ.

(i) Gamma with density

$$p(x) = \frac{1}{\Gamma(\alpha)} \lambda^\alpha x^{\alpha-1} e^{-\lambda x}, \quad x > 0,$$

first assuming one parameter is known, then assuming both parameters are unknown.

(j) Weibull with distribution function

$$F(x) = 1 - e^{-(\lambda x)^\alpha}, \quad x > 0,$$

first assuming one parameter is known, then assuming both parameters are unknown.

(k) Beta with density

$$p(x) = \frac{\Gamma(\alpha+\beta)}{\Gamma(\alpha)\Gamma(\beta)} x^{\alpha-1}(1-x)^{\beta-1}, \quad 0 < x < 1,$$

first assuming one parameter is known, then assuming both parameters are unknown.

Exercise 3.3: We have shown that if $x_1, \ldots, x_n$ are an iid sample from $N(\theta, 1)$ then $\overline{x}$ is sufficient for θ. What is the conditional distribution of the data given $\overline{x}$? (It must be free of θ.) How would you simulate a dataset from the conditional distribution?

Exercise 3.4: Suppose the model P_θ is the class of all continuous distributions; this is called a 'nonparametric family', where the unknown parameter θ is the whole distribution function. Let $x_1, \ldots, x_n$ be an iid sample from P_θ. Show that the order statistics are sufficient for P_θ.

Exercise 3.5: Suppose $x_1, \ldots, x_n$ are an iid sample from the double-exponential distribution with density

$$p_\theta(x) = (2\theta)^{-1} e^{-|x|/\theta}.$$

It is obvious from the factorization theorem that $t(x) = \sum_i |x_i|$ is sufficient. What is the conditional distribution of the data given $T = t$?

Exercise 3.6: By definition, if T is sufficient for θ then $p(x|T = t)$ is free of θ. We can use this theoretical fact to remove nuisance parameters in hypothesis testing. For example, suppose n_{ij}, for $i = 1, \ldots, I$ and $j = 1, \ldots, J$, form a two-way contingency table. We are interested to test the independence between the row and column characteristics. Show that, under independence, the set of marginal totals is sufficient. Explain how you might test the hypothesis.

Exercise 3.7: Establish if the statistics you found in Exercise 3.2 are minimal sufficient.

Exercise 3.8: Find the minimal sufficient statistics in the following cases:

(a) y_i, for $i = 1, \ldots, n$, are independent Poisson with mean μ_i, with

$$\log \mu_i = \beta_0 + \beta_1 x_i,$$

where x_i's are known predictors, and (β_0, β_1) are unknown parameters.

(b) y_i, for $i = 1, \ldots, n$, are independent Poisson with mean μ_i, with

$$\mu_i = \beta_0 + \beta_1 x_i,$$

where x_i's are known predictors, and (β_0, β_1) are unknown parameters.

(c) $y_1, \ldots, y_n$ are an iid sample from $N(\theta, \theta)$.

(d) $y_1, \ldots, y_n$ are an iid sample from $N(\theta, \theta^2)$.

(e) y_i, for $i = 1, \ldots, n$, are independent normal with mean μ_i and variance σ^2, with

$$\mu_i = \beta_0 + \beta_1 x_i,$$

where x_i's are known predictors, and $(\beta_0, \beta_1, \sigma^2)$ are unknown parameters.

(f) y_i, for $i = 1, \ldots, n$, are independent normal with mean μ_i, with

$$\mu_i = \beta_0 + \beta_1 e^{\beta_2 x_i},$$

where x_i's are known predictors, and $(\beta_0, \beta_1, \beta_2, \sigma^2)$ are unknown parameters.

Exercise 3.9: Suppose $(x_1, y_1), \ldots, (x_n, y_n)$ are an iid sample from the bivariate normal distribution with mean (μ_x, μ_y) and covariance matrix

$$\Sigma = \begin{pmatrix} \sigma_x^2 & \rho\sigma_x\sigma_y \\ \rho\sigma_x\sigma_y & \sigma_y^2 \end{pmatrix}.$$

Find the minimal sufficient statistics under each of the following conditions:

(a) All parameters are unknown.

(b) All parameters are unknown, but $\sigma_x^2 = \sigma_y^2$.

(c) Assuming $\sigma_x^2 = \sigma_y^2 = 1$, but all the other parameters are unknown.

Exercise 3.10: Let $x_1, \ldots, x_n$ be an iid sample. Show that for the Cauchy and double-exponential model with location θ the entire order statistics are minimal sufficient. The Cauchy density is

$$p_\theta(x) = \frac{1}{\pi\{1 + (x - \theta)^2\}}, \quad -\infty < x < \infty$$

and the double-exponential density is

$$p_\theta(x) = \frac{1}{2} e^{-|x-\theta|}, \quad -\infty < x < \infty.$$

Exercise 3.11: Suppose $x_1, \ldots, x_n$ are an iid sample from the inverse Gaussian distribution $\mathrm{IG}(\mu, \lambda)$ with density

$$p(x) = \left(\frac{\lambda}{2\pi x^3}\right)^{1/2} \exp\left\{-\frac{\lambda}{2\mu^2}\frac{(x-\mu)^2}{x}\right\}, \qquad x > 0.$$

It has mean μ and variance μ^3/λ.

(a) Derive the score equations for the MLE.

(b) Find the Fisher information matrix.

Exercise 3.12: For the multinomial model in Example 3.9 show that $\widehat{p_i} = n_i/N$. Find the Fisher information for $(p_1, \ldots, p_{k-1})$.

Exercise 3.13: Suppose we have multinomial data $n_1, \ldots, n_5$, but the last probability is known to be $p_5 = 0.25$. Find the MLE of the unknown probabilities $p_1, \ldots, p_4$, and interpret the result.

Exercise 3.14: In the ABO blood grouping, each person can be classified into group A, B, O, or AB. Because of recessive and dominant characteristics, the phenotypes consist of various genotypes according to A = {AA, AO}, B= {BB, BO}, AB={AB} and O={OO}. Suppose that in a population the frequencies of the ABO genes are p, q and r, respectively, with $p + q + r = 1$.

(a) Assuming random mixing, show that the proportions of groups A, B, O and AB are, respectively,

$$(p^2 + 2pr), \ (q^2 + 2qr), \ r^2, \ 2pq.$$

(b) From a sample of 435 people we observe the following frequencies (Rao 1973, page 372):

$$\text{A} = 182, \ \text{B} = 60, \ \text{O} = 176, \ \text{AB} = 17.$$

Find the MLEs of the parameters. (Use an optimization program.) Compare the observed and estimated blood group frequencies.

(c) Report the standard errors of the estimates and the 95% Wald CIs.

(d) Find the profile likelihood of q, the proportion of gene B.

Exercise 3.15: Use the following data

```
2.85 1.51 0.69 0.57 2.29
```

to find the profile likelihood of the mean μ, where it is assumed that the data are uniform on $(\mu - \sigma, \mu + \sigma)$ with unknown σ. Compare with the profile likelihood assuming $N(\mu, \sigma^2)$ model. Modify a single value of the data so that the profiles become dramatically different.

Exercise 3.16: The following are the first 20 measurements of the speed of light (as deviations from 299 in 1000's km/s) from Michelson's speed-of-light experiment (see Example 4.8):

```
0.85 0.74 0.90 1.07 0.93 0.85 0.95 0.98 0.98 0.88
1.00 0.98 0.93 0.65 0.76 0.81 1.00 1.00 0.96 0.96
```

Denote the data by $x_1, \ldots, x_n$, and assume these are an iid sample from $N(\theta, \sigma^2)$ with both parameters unknown. Let $x_{(1)}, \ldots, x_{(n)}$ be the order statistics.

(a) Draw the likelihood of (θ, σ^2) based only on $x_{(1)}$ and $x_{(n)}$. (See Example 2.4.)

(b) Compare the profile likelihood of μ based on the whole data, and the one based only on $x_{(1)}$ and $x_{(n)}$.

(c) Repeat (a) and (b), where only the minimum, median and maximum values are available.

(d) Repeat (a) and (b), where only the first, second and third quartiles are available.

(e) Repeat (a) and (b), where the data are only available in grouped form: $x \leq 0.75$, $0.75 < x \leq 0.85$, $0.85 < x \leq 0.95$, $0.95 < x \leq 1.05$ and $x > 1.05$. The numbers in each category are (2,4,5,8,1).

Exercise 3.17: Plot the profile likelihood of σ^2 for the data in Example 3.10. The sample variance is commonly defined as

$$s^2 = \frac{1}{n-1}\sum_{i=1}^{n}(x_i - \overline{x})^2.$$

It is known that $(n-1)s^2/\sigma^2$ is χ^2_{n-1}, which would generate a likelihood of σ^2 free of μ; Fisher called this a second-stage likelihood. Using the data example, compare this likelihood with the profile likelihood.

Exercise 3.18: Generalize the capture-recapture model in Example 2.2 to the case of more than one unknown N. For example, suppose in the whole population we have marked 55 deer of species A. In a sample of size 60 we identified 5 marked deer of species A, 25 unmarked deer of species A, and 30 unmarked deer of species B.

(a) Draw the joint likelihood of N_A and N_B, the population sizes of species A and B. (Hint: use the hypergeometric probability involving more than two types of objects.)

(b) Find the MLE of N_A and N_B, the Fisher information matrix, and the standard errors.

(c) Compute the profile likelihood of N_A and N_B.

Exercise 3.19: Because of a possible connection with bovine tuberculosis it is of interest to estimate the number N of badgers in a certain area. The traps set in the area over five consecutive week periods caught 31, 15, 22, 19 and 6 badgers each week, and each time they were removed from the area. Give a sensible model to describe the number of catches, and present a likelihood inference on N. (Hint: assume a Poisson model, where the catch rate is a function of the current number of badgers. For example, y_k is Poisson with mean λN_k, where N_k is the existing number of badgers, and λ is a nuisance parameter.)

Exercise 3.20: From Section 3.3, under what condition is $I_{11} = (I^{11})^{-1}$? How would you interpret this statistically?

Exercise 3.21: Let $x_1, \ldots, x_n$ be an iid sample from $N(\mu, \sigma^2)$. Show that the Fisher information matrix for the parameter $\theta = (\mu, \sigma^2)$ is

$$I(\widehat{\theta}) = \begin{pmatrix} n/\widehat{\sigma}^2 & 0 \\ 0 & n/(2\widehat{\sigma}^4) \end{pmatrix}$$

so the standard error of $\overline{x}$ is $\widehat{\sigma}/\sqrt{n}$. For the observed data in Example 3.10, check the quadratic approximation for the profile likelihood of μ and σ^2. Verify that the curvatures at the maximum correspond to the appropriate entries of $I^{-1}(\widehat{\theta})$.

Exercise 3.22: Generalize the result in Exercise 2.20 for vector parameter θ, and find the Fisher information of $g(\widehat{\theta})$, where $\widehat{\theta} \in R^p$. In particular, show that for a fixed vector a the standard error of $a'\widehat{\theta}$ is

$$\text{se}(a'\widehat{\theta}) = \{a' I^{-1}(\widehat{\theta}) a\}^{1/2}.$$

As an important special case, if the Fisher information is diagonal, show that the standard error of a contrast $\widehat{\theta}_1 - \widehat{\theta}_2$ is

$$\text{se}(\widehat{\theta}_1 - \widehat{\theta}_2)^2 = \text{se}(\widehat{\theta}_1)^2 + \text{se}(\widehat{\theta}_2)^2.$$

Exercise 3.23: Let $(x_1, y_1), \ldots, (x_n, y_n)$ be an iid sample from the bivariate normal distribution with mean zero and covariance matrix

$$\Sigma = \sigma^2 \begin{pmatrix} 1 & \rho \\ \rho & 1 \end{pmatrix}.$$

(a) Verify the Fisher information

$$I(\sigma^2, \rho) = \begin{pmatrix} \frac{n}{\sigma^4} & -\frac{n\rho}{\sigma^2(1-\rho^2)} \\ -\frac{n\rho}{\sigma^2(1-\rho^2)} & \frac{n(1+\rho^2)}{(1-\rho^2)^2} \end{pmatrix},$$

and find a standard error formula for $\widehat{\rho}$.

(b) For the following IQ scores, where x = verbal thinking and y = mathematical thinking, jointly and individually, draw the likelihood of σ^2 and ρ. Compare I_{22} with $(I^{22})^{-1}$ and comment. (For simplicity, assume the mean is known at the observed value, so the data can be centred before analysis.)

x	y	x	y
109	116	85	91
88	77	100	88
96	95	113	115
96	79	117	119
109	113	107	100
116	122	104	115
114	109	101	95
96	94	81	90

(c) Investigate how well the quadratic approximation works. Experiment with transformations of ρ and σ^2 to get a more regular likelihood.

(d) The so-called Fisher's z transform of the sample correlation

$$z = \frac{1}{2} \log \frac{1+\widehat{\rho}}{1-\widehat{\rho}}$$

is approximately normal with mean

$$\psi = \frac{1}{2} \log \frac{1+\rho}{1-\rho}$$

and variance $1/(n-3)$. Compare the likelihood of ρ based on Fisher's z transform with the profile likelihood.

4

Basic models and simple applications

We use models to represent reality in simpler ways, so we can understand, describe, predict or control it better. In statistics, the reality is variability, without which there would be no statistics (and life would be rather dull). In modelling we typically separate a systematic pattern from the stochastic variation. The systematic pattern may describe a relationship between variables of interest, while the stochastic variation usually represents the unaccounted part of the total variation. Probability models are used to represent this stochastic variation.

From a statistical point of view, any probability model has the potential for data analysis, although we tend to concentrate on models that are convenient from analytical and computational aspects, and have parameters that are easy to interpret. The models covered in this chapter are the basic building blocks for real data analytic models. Only the simplest systematic structure is presented, mostly in terms of one-sample or one-group structure. The exceptions are the comparison of two binomial proportions and two Poisson means.

4.1 Binomial or Bernoulli models

The Bernoulli model is useful for experiments with dichotomous outcomes. Each experimental unit is thought of as a trial; in their simplest form Bernoulli trials are assumed to be independent, each with probability θ for a successful outcome. Some simple examples are a coin tossing experiment, the sex of children in a family, the success or failure of business enterprises, the success or failure of a medical procedure, etc. In studies of twins it is common to collect data where at least one of the twins has a particular condition; in these studies a pair of twins is the 'trial' and the 'success' occurs if both twins have the condition. Such a study usually tries to establish if there is a genetic factor in the condition under examination.

Suppose we observe $x \equiv (x_1, \ldots, x_n)$, which are a realization of a Bernoulli experiment with $P(X_i = 1) = \theta$ and $P(X_i = 0) = 1 - \theta$. The likelihood function is

$$L(\theta) = \prod_{i=1}^{n} \theta^{x_i}(1-\theta)^{1-x_i}$$

$$= \theta^{\sum x_i}(1-\theta)^{n-\sum x_i}.$$

Example 4.1: For example, we observe a sequence 0111110111; then $n = 10$, $\sum x_i = 8$, and the likelihood function for θ is given in Figure 4.1. It is easy to show that the MLE is $\widehat{\theta} = \sum x_i/n = 0.8$. The likelihood interval at 15% cutoff (computed numerically) is (0.50,0.96); since the likelihood is reasonably regular this is an approximate 95% CI for θ. (From here on, we will report only one interval or the other depending on the regularity of the likelihood, and we will err on the side of regularity.) □

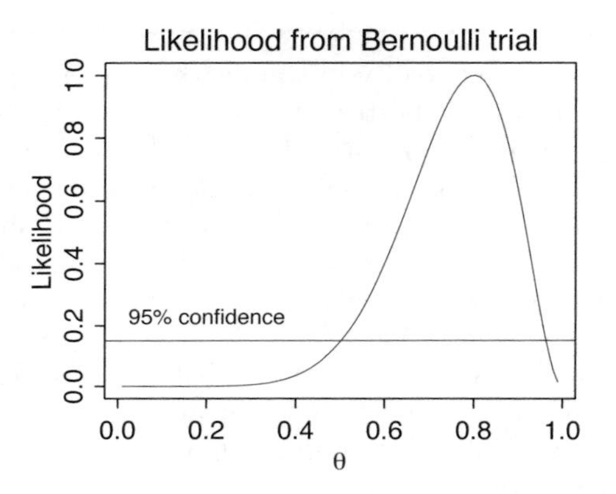

Figure 4.1: *Likelihood function of θ from a Bernoulli experiment with outcome 0111110111, so $n = 10$ and $\sum x_i = 8$. This is the same likelihood from a binomial experiment with $n = 10$ trials and $x = 8$ successes. The approximate 95% CI is (0.50,0.96).*

We have described the likelihood construction for a simple binomial observation in Section 2.1. For completeness, suppose $X \sim \text{binomial}(n, \theta)$ and we observe $X = x$; then

$$L(\theta) = \binom{n}{x} \theta^x (1-\theta)^{n-x}.$$

If we observe $x = 8$ successes in $n = 10$ trials, the likelihood function and other quantities are exactly as obtained above. This is sensible since knowing the order of the 0–1 outcomes in an independent Bernoulli experiment should not add any information about θ; this would not be the case if, for example, the sequence is dependent and the dependence is also a function of θ.

Discrete data are usually presented in grouped form. For example, suppose $x_1, \ldots, x_N$ are an iid sample from binomial(n, θ) with n known. We first summarize the data as

k	0	1	$\cdots$	n
n_k	n_0	n_1	$\cdots$	n_n .

where n_k is the number of x_i's equal to k, so $\sum_k n_k = N$. We can now think of the data $(n_0, \ldots, n_n)$ as having a multinomial distribution with probabilities $(p_0, \ldots, p_n)$ given by the binomial probabilities

$$p_k = \binom{n}{k} \theta^k (1-\theta)^{n-k}.$$

The log-likelihood is given by

$$\log L(\theta) = \sum_{k=0}^{n} n_k \log p_k.$$

We can show that the MLE of θ is

$$\widehat{\theta} = \frac{\sum_k k n_k}{Nn}$$

with standard error

$$\text{se}(\widehat{\theta}) = \sqrt{\frac{\widehat{\theta}(1-\widehat{\theta})}{Nn}}.$$

Example 4.2: In a classic study of human sex ratio in 1889, based on hospital records in Saxony, Geissler reported the distribution of the number of boys per family. Among 6115 families with 12 children he observed:

No. boys k	0	1	2	3	4	5	6
No. families n_k	3	24	104	286	670	1033	1343

No. boys k	7	8	9	10	11	12
No. families n_k	1112	829	478	181	45	7

The estimated proportion of boys is

$$\widehat{\theta} = \frac{\sum_k k n_k}{6115 \times 12} = \frac{38,100}{6115 \times 12} = 0.5192,$$

with standard error $\text{se}(\widehat{\theta}) = 0.0018$. (For comparison, in Ireland in 1994 the proportion of boys among 48,255 births was 0.5172.) □

Negative binomial model

In the so-called negative or inverse binomial experiment we continue a Bernoulli trial with parameter θ until we obtain x successes, where x is fixed in advance. Let n be the number of trials needed; the likelihood function is

$$\begin{aligned} L(\theta) &= P_\theta(N = n) \\ &= \binom{n-1}{x-1} \theta^x (1-\theta)^{n-x}. \end{aligned}$$

Again here we find the same likelihood function as the one from the binomial experiment, even though the sampling property is quite different. This is an example where the likelihood function ignores the sampling scheme. Further uses of the negative binomial model are given in Section 4.5, which connects it with the Poisson model.

4.2 Binomial model with under- or overdispersion

For modelling purposes the binomial model has a weakness in that it specifies a rigid relationship between the mean and the variance.

Example 4.3: In Example 4.2, using the estimate $\widehat{\theta} = 0.5192$ we can compute the expected frequencies

$$e_k = N\widehat{p}_k,$$

given in the table below. We then obtain the goodness-of-fit statistic

$$\chi^2 = \sum_k \frac{(n_k - e_k)^2}{e_k} = 110.5,$$

which is highly significant at 11 degrees of freedom. To see the nature of the model violation, define a residual

$$r_k = \frac{n_k - e_k}{\sqrt{e_k}}$$

such that $\chi^2 = \sum_k r_k^2$.

No. boys k	0	1	2	3	4	5	6
Observed n_k	3	24	104	286	670	1033	1343
Expected e_k	1	12	72	258	628	1085	1367
Residual r_k	2.1	3.4	3.8	1.7	1.7	−1.6	−0.7

No. boys k	7	8	9	10	11	12
Observed n_k	1112	829	478	181	45	7
Expected e_k	1266	854	410	133	26	2
Residual r_k	−4.3	−0.9	3.4	4.2	3.7	3.0

Hence the observed frequencies tend to be larger at the edges, and lower in the middle of the distribution, indicating greater variability than expected under the binomial model. □

We now describe theoretically how the standard binomial model might fail. Suppose $X_1, \ldots, X_n$ are independent Bernoulli trials with probabilities $p_1, \ldots, p_n$. Let $X = \sum_i X_i$; then

$$E(X) = \sum p_i \equiv n\theta$$

where $\theta = \sum p_i/n$, but

$$\begin{aligned} \text{var}(X) &= \sum_i \text{var}(X_i) \\ &= \sum_i p_i(1 - p_i) \end{aligned}$$

$$\begin{aligned}
&= \sum_i p_i - \sum_i p_i^2 \\
&= \sum_i p_i - (\sum_i p_i)^2/n - \{\sum_i p_i^2 - (\sum_i p_i)^2/n\} \\
&= n\theta - n\theta^2 - n\sigma_p^2 \\
&= n\theta(1-\theta) - n\sigma_p^2,
\end{aligned}$$

where we have defined the variance among the p_i's as

$$\sigma_p^2 = \frac{1}{n}\left\{\sum_i p_i^2 - \frac{(\sum_i p_i)^2}{n}\right\}.$$

So, allowing individual Bernoulli probabilities to vary produces less-than-standard binomial variance.

Now, suppose X_i's, for $i = 1, \ldots, m$, are independent binomial(n, p_i), and let $X = X_I$ be a random choice from one of these X_i's, i.e. the random index $I = i$ has probability $1/m$. This process produces a mixture of binomials with marginal probability

$$\begin{aligned}
P(X = x) &= E\{P(X_I = x|I)\} \\
&= \frac{1}{m}\sum_{i=1}^{m} P(X_i = x) \\
&= \frac{1}{m}\sum_i \binom{n}{x} p_i^x (1-p_i)^{n-x},
\end{aligned}$$

which does not simplify further, but

$$\begin{aligned}
E(X) &= E\{E(X_I|I)\} \\
&= \frac{1}{m}\sum_i E(X_i) \\
&= \frac{n}{m}\sum_i p_i \equiv n\theta
\end{aligned}$$

where we set $\theta = \sum p_i/m$. The variance is

$$\begin{aligned}
\text{var}(X) &= E\{\text{var}(X_I|I)\} + \text{var}\{E(X_I|I)\} \\
&= \frac{1}{m}\sum_i \text{var}(X_i) + \frac{1}{m}\sum_i E^2(X_i) - (\sum_i E(X_i)/m)^2 \\
&= \frac{1}{m}\sum_i np_i(1-p_i) + \frac{1}{m}\sum_i (np_i)^2 - (n\theta)^2 \\
&= n\theta(1-\theta) + n(n-1)\sigma_p^2
\end{aligned}$$

where σ_p^2 is the variance among the p_i's as defined previously. So, here we have greater-than-standard binomial variation.

The equal sampling probability simplifies the previous evaluation of mean and variance, but the extra-binomial variance is always observed when we sample from a mixed population. For example, if families in the population have different probabilities p_i's for an outcome of interest x_i, then a random sample of families will exhibit values with extra-binomial variance.

A wrong binomial assumption can adversely impact inference by producing wrong standard errors (i.e. the assumed likelihood is too narrow or too wide). Section 4.9 describes a general model that allows under- or overdispersion within the general exponential family models. In random effects modelling (Section 17.1) the individual estimates of p_i's may be of interest, in which case we typically put more structure on how they vary. Lindsey and Altham (1998) analyse Geissler's (complete) data taking the overdispersion into account; see also Exercise 4.4.

4.3 Comparing two proportions

Comparing two binomial proportions is probably the most important statistical problem in epidemiology and biostatistics.

Example 4.4: A geneticist believes she has located a gene that controls the spread or metastasis of breast cancer. She analyzed the expression pattern of the gene in the cells of 15 patients whose cancer had spread (metastasized) and 10 patients whose cancer did not spread. The first group had 5 patients with the gene overexpressed, while one patient in the second group had the gene overexpressed. Such data are usually presented in a 2×2 table:

Overexpression	Spread	Localized	Total
Present	5(33%)	1(10%)	6
Absent	10	9	19
Total	15	10	25

Is the evidence strong enough to justify her belief? □

It is instructive to start with the (large-sample) frequentist solution. First put names on the elements of the 2×2 table

	Spread	Localized	Total
Present	x	y	$t = x + y$
Absent	$m - x$	$n - y$	$N - t$
Total	m	n	$N = m + n$

The standard test of equality of proportions is the famous χ^2 test given by

$$\chi^2 = \frac{N\{x(n-y) - y(m-x)\}^2}{mnt(N-t)},$$

which, under the null hypothesis, has χ^2_1 distribution.

For the observed data

$$\chi^2 = \frac{25(5 \times 9 - 1 \times 10)^2}{15 \times 10 \times 6 \times 19} = 1.79,$$

producing a P-value of 0.18; this is a two-sided P-value for the hypothesis of equal proportion. This method has an appealing simplicity, but note that in small samples its validity is doubtful, and it does not give full information about the parameter of interest (e.g. it is not clear how to get a confidence interval).

In small-sample situations we commonly use the so-called Fisher's exact test. Under the null hypothesis and conditional on the observed margins, the probability of an observed table is a hypergeometric probability

$$p(x) = \frac{\binom{m}{x}\binom{n}{y}}{\binom{m+n}{x+y}}.$$

Fisher's exact P-value is then computed as the probability of the observed or more-extreme tables. For the above example, there is only one more-extreme table, namely when we get 0 'present' out of 10 localized cases. The one-sided P-value is

$$\text{P-value} = \frac{\binom{15}{5}\binom{10}{1}}{\binom{25}{6}} + \frac{\binom{15}{6}\binom{10}{0}}{\binom{25}{6}} = 0.17 + 0.03 = 0.20.$$

To proceed with a likelihood analysis, suppose the number of successes X in the first group is binomial $B(m, \pi_x)$, and, independently, Y in the second group is $B(n, \pi_y)$. On observing x and y the joint likelihood of (π_x, π_y) is

$$L(\pi_x, \pi_y) = \pi_x^x (1-\pi_x)^{m-x} \pi_y^y (1-\pi_y)^{n-y}.$$

The comparison of two proportions can be expressed in various ways, for example using the difference $\pi_x - \pi_y$, the relative risk π_x/π_y or the log odds-ratio θ defined by

$$\theta = \log \frac{\pi_x/(1-\pi_x)}{\pi_y/(1-\pi_y)}.$$

In terms of θ the null hypothesis of interest H_0: $\pi_x = \pi_y$ is equivalent to H_0: $\theta = 0$. Each parameterization has its own advantage/disadvantage in terms of interpretation and statistical properties.

The reader can check that, in small samples, the likelihood of the log odds-ratio is more regular than the likelihood of the other parameters. So, we will consider the log odds-ratio parameter θ as the parameter of interest. Any other parameter can be used as a nuisance parameter, but for convenience we will use the log odds η defined by

$$\eta = \log \frac{\pi_y}{1-\pi_y}.$$

Some simple algebra would show that

$$\pi_y = \frac{e^\eta}{1+e^\eta}$$
$$\pi_x = \frac{e^{\theta+\eta}}{1+e^{\theta+\eta}}.$$

Therefore, we get the joint likelihood

$$\begin{aligned} L(\theta,\eta) &= \left(\frac{\pi_x}{1-\pi_x}\right)^x (1-\pi_x)^m \left(\frac{\pi_y}{1-\pi_y}\right)^y (1-\pi_y)^n \\ &= \left(\frac{\pi_x/(1-\pi_x)}{\pi_y/(1-\pi_y)}\right)^x \left(\frac{\pi_y}{1-\pi_y}\right)^{x+y} (1-\pi_x)^m(1-\pi_y)^n \\ &= e^{\theta x} e^{\eta(x+y)}(1+e^{\theta+\eta})^{-m}(1+e^\eta)^{-n}. \end{aligned}$$

The MLE of θ is available directly from the invariance property:

$$\widehat{\theta} = \log \frac{x/(m-x)}{y/(n-y)}.$$

The standard error has an interesting formula

$$\text{se}(\widehat{\theta}) = \left(\frac{1}{x} + \frac{1}{y} + \frac{1}{m-x} + \frac{1}{n-y}\right)^{1/2}.$$

Its derivation is left as an exercise.

To get the profile likelihood of θ we can compute the MLE of η at each fixed value of θ, but there is no closed form formula for the MLE. The profile likelihood has to be computed numerically according to the definition:

$$L(\theta) = \max_\eta L(\theta,\eta).$$

Example 4.4: continued. Figure 4.2(a) shows the contours of the joint likelihood at 10%, 30%, 50%, 70% and 90% cutoffs. The profile likelihood in Figure 4.2(b) shows little evidence of the gene associated with the spread of cancer. The likelihood-based 95% CI for θ is (−0.54,4.52), with the corresponding 95% CI (0.58,92.22) for the odds ratio. The MLE of θ is

$$\widehat{\theta} = \log\frac{5/10}{1/9} = 1.50$$

with $\text{se}(\widehat{\theta}) = 1.19$.

Now consider the analysis of the more extreme table: $x = 6$ out of $m = 15$, versus $y = 0$ out of $n = 10$. The MLE of θ is $\widehat{\theta} = \infty$, and the likelihood is irregular. Figure 4.2(c) shows the joint likelihood, and Figure 4.2(d) the profile likelihood. The 15% likelihood interval for θ is now a naturally one-sided interval $\theta > 0.93$. The likelihood of H_0: $\theta = 0$ is around 2.5%, indicating stronger evidence of association. □

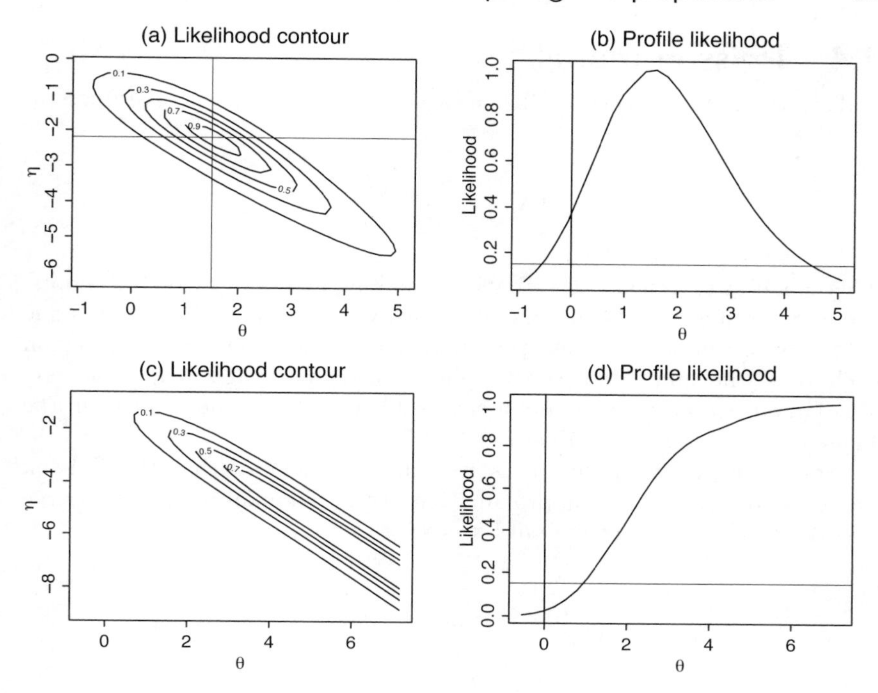

Figure 4.2: *(a) Joint likelihood of θ and η for the original data. (b) Profile likelihood of θ. (c) For extreme data: 6 out of 15 versus 0 out of 10. (d) Profile likelihood from (c).*

A series of 2×2 tables

In practice, we often stratify the study groups according to some risk factor if the risk factor is not balanced. For example, the evidence of cancer depends on age, and the two groups being compared have different age distributions. In epidemiology, age is called a confounding variable. Stratifying by age reduces the comparison bias due to confounding. Thus we divide the study subjects into young and old strata, and for each stratum we construct a 2×2 table.

Assuming the tables have a *common* odds-ratio parameter θ, we can compute the profile likelihood $L_i(\theta)$ from each table, and combine them using

$$\log L(\theta) = \sum_i \log L_i(\theta).$$

This method works for a small number of strata, otherwise there can be a serious bias problem; the proper method for combining information from many sparse tables is given in Section 10.5.

4.4 Poisson model

A discrete random variable X has a Poisson distribution with parameter θ if

$$P(X = x) = e^{-\theta}\frac{\theta^x}{x!}.$$

This model is extremely versatile for modelling any data involving counts: the number of accidents on a highway each year, the number of deaths due to a certain illness per week, the number of insurance claims in a region each year, the number of typographical errors per page in a book, etc. The famous example of the number of soldiers killed by horse kicks in the Prussian army is given in Exercise 4.12.

For modelling purposes, it is fruitful to remember that the Poisson model with mean θ is an approximation of the binomial model with large n and a small success probability $\pi = \theta/n$:

$$\begin{aligned} P(X = x) &= \binom{n}{x}\pi^x(1-\pi)^{n-x} \\ &= \frac{n!}{(n-x)!x!}\left(\frac{\theta}{n}\right)^x\left(1-\frac{\theta}{n}\right)^{n-x} \\ &= \frac{n\cdot(n-1)\cdots(n-x+1)}{n^x}\times\frac{\theta^x}{x!}\left(1-\frac{\theta}{n}\right)^{n-x} \\ &\to \frac{\theta^x}{x!}e^{-\theta}. \end{aligned}$$

On observing an iid sample $x_1, \ldots, x_n$ from Poisson(θ), the likelihood of θ is given by

$$L(\theta) = e^{-n\theta}\theta^{\sum x_i}.$$

The maximum likelihood estimate is $\widehat{\theta} = \overline{x}$ with standard error $\text{se}(\widehat{\theta}) = \sqrt{\overline{x}/n}$. We can check that the likelihood is quite regular if $\sum x_i$ is large enough; this is true even for $n = 1$.

Example 4.5: For each year in the past 5 years, a town recorded 3, 2, 5, 0 and 4 earthquakes (of at least a certain magnitude). Assuming a Poisson model, the likelihood function of the earthquake frequency θ is given in Figure 4.3. There was an average of $\widehat{\theta} = 2.8$ (se $= 0.75$) earthquakes per year, with approximate 95% CI $1.6 < \theta < 4.5$. □

Example 4.6: The Poisson assumption can be checked if we have grouped data. Jenkins and Johnson (1975) reported 64 incidents of international terrorism in the USA between January 1968 and April 1974. The data are categorized according to the monthly number of incidents as follows:

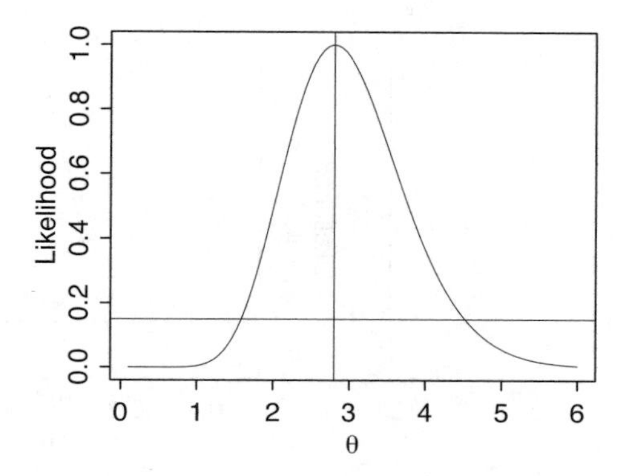

Figure 4.3: *Likelihood function of earthquake frequency.*

Number of incidents k	Number of months n_k
0	38
1	26
2	8
3	2
4	1
12	1

The idea of a Poisson plot (Hoaglin and Tukey 1985) is to plot k against a function of n_k such that it is expected to be a straight line for Poisson data. Since $p_k = P(X = k) = e^{-\theta}\theta^k/k!$, then

$$\log p_k + \log k! = -\theta + k \log \theta,$$

indicating that we should plot $\log(p_k k!)$ versus k, where p_k is estimated by $n_k/\sum n_k$. Figure 4.4 shows that the months with $k = 0, \ldots, 3$ incidents follow the Poisson prescription, but the month with 4 incidents is rather unusual and the one with 12 incidents is extremely unusual. (It turns out that 11 of those 12 incidents were carried out by an anti-Castro group. This raises inductive questions as discussed in Section 1.3: how should we define an event? Should we instead count the groups involved?)

In general, for a Poisson model with mean θ the likelihood based on observing n_k, for $k = 0, 1, \ldots$, is

$$L(\theta) = \prod_k p_k^{n_k},$$

where $p_k = P(X = k) = e^{-\theta}\theta^k/k!$, and the log-likelihood is

$$\log L(\theta) = -\theta \sum_k n_k + \sum_k k n_k \log \theta.$$

We can show that the MLE $\widehat{\theta} = \sum_k k n_k / \sum_k n_k$. Figure 4.4 shows the likelihood functions including and excluding the month with $k = 12$. It is clear that the likelihood is sensitive to the outlier. The mean number of incidents is 0.84 and 0.69, with approximate 95% CIs given by $0.65 < \theta < 1.06$ and $0.53 < \theta < 0.90$. □

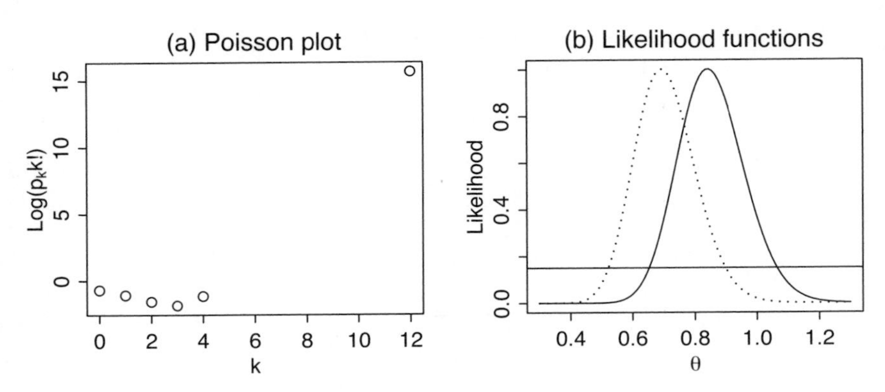

Figure 4.4: *(a) The Poisson plot of the incidents of international terrorism in the USA. The months with 4 and 12 incidents are unusual. (b) Likelihood function of the mean number of incidents including (solid) and excluding (dotted) the month with* $k = 12$ *incidents.*

4.5 Poisson with overdispersion

Like the binomial model, the Poisson model imposes a strict relationship between the mean and variance that may not be appropriate for the data.

Example 4.7: The first two columns of the following table summarizes the number of accidents among factory workers (Greenwood and Yule 1920). For example, 447 workers did not have any accident.

Number of accidents k	Number of workers n_k	Fitted number Poisson	Negative binomial
0	447	406.3	446.2
1	132	189.0	134.1
2	42	44.0	44.0
3	21	6.8	14.9
4	3	0.8	5.1
> 4	2	0.1	2.7

Assuming a Poisson model, the average accident rate is

$$\sum_k kn_k / \sum_k n_k = 0.47.$$

To simplify the computation, assume that the last category is represented by five accidents. Comparing the observed and fitted frequencies, the Poisson model is clearly inadequate. The χ^2 statistic is 107.3 with four degrees of freedom. The observed data show an overdispersion: there are more accident-free and accident-prone workers than predicted by the Poisson model. □

Since the Poisson model is a limit of the binomial, models for under- or overdispersion described in Section 4.2 also apply. General modelling of Poisson-type data with extra variation is described under the general exponential family models in Section 4.9.

Overdispersion can occur if, conditionally on μ, an outcome X_μ is Poisson with mean μ, and μ is random with mean $E\mu$ and variance σ^2. For example, individuals vary in their propensity to have accidents, so even if the number of accidents per individual is Poisson, the marginal distribution will show some overdispersion. In this setup, the marginal distribution of X_μ has

$$\begin{aligned} E(X_\mu) &= E\{E(X_\mu|\mu)\} = E\mu \\ \text{var}(X_\mu) &= E\{\text{var}(X_\mu|\mu)\} + \text{var}\{E(X_\mu|\mu)\} \\ &= E\mu + \text{var}(\mu) \\ &= E\mu + \sigma^2, \end{aligned}$$

showing an extra variability compared with the standard Poisson model.

If μ has a gamma distribution we will get a closed form formula for the marginal probabilities. Specifically, let X_μ be Poisson with mean μ, where μ has density

$$f(\mu) = \frac{1}{\Gamma(\alpha)}\mu^{\alpha-1}\lambda^\alpha e^{-\lambda\mu}.$$

A random sample of X from the mixture of X_μ for all μ has mean

$$E(X) = E\mu = \frac{\alpha}{\lambda}$$

and variance

$$\begin{aligned} \text{var}(X) &= E\mu + \text{var}(\mu) \\ &= \frac{\alpha}{\lambda} + \frac{\alpha}{\lambda^2}. \end{aligned}$$

For likelihood modelling we need to compute the marginal probability, for $x = 0, 1, \ldots,$

$$\begin{aligned} P(X = x) &= E\{P(X_\mu = x|\mu)\} \\ &= E\left(e^{-\mu}\frac{\mu^x}{x!}\right) \\ &= \frac{\lambda^\alpha}{\Gamma(\alpha)x!}\int e^{-\mu}\mu^x\mu^{\alpha-1}e^{-\lambda\mu}\,d\mu \\ &= \frac{\lambda^\alpha\Gamma(x+\alpha)}{(\lambda+1)^{x+\alpha}\Gamma(\alpha)x!} \\ &\equiv \binom{x+\alpha-1}{\alpha-1}\left(\frac{\lambda}{\lambda+1}\right)^\alpha\left(1-\frac{\lambda}{\lambda+1}\right)^x. \qquad (4.1) \end{aligned}$$

using as a definition $(x + \alpha - 1)! \equiv \Gamma(x + \alpha)$. For integer α we have a negative binomial distribution: the outcome x is the number of failures recorded when we get exactly α successes, and the probability of success is

$$\pi = \frac{\lambda}{\lambda+1}.$$

Note, however, that as a parameter of the gamma distribution α does not have to be an integer. Given the probability formula, and on observing data $x_1, \ldots, x_n$, we can construct the likelihood of (α, λ) or (α, π).

Example 4.7: continued. To fit the negative binomial model to the data, the log-likelihood of (α, π) is

$$\log L(\alpha, \pi) = \sum_{k=0}^{4} n_k \log P(X = k) + n_{[>4]} \log\{1 - P(X > 4)\}$$

where $P(X = k)$ is given by (4.1). Numerical optimization of the log-likelihood yields

$$\widehat{\alpha} = 0.84, \quad \widehat{\pi} = 0.64.$$

The fitted frequencies shown in the previous table are now much closer to the observed data; the χ^2 statistic is 3.7, with 3 degrees of freedom, indicating a very good fit. The estimated accident rate is $0.84(1 - 0.64)/0.64 = 0.47$; this is the same as in Poisson model, but has different profile likelihood and standard error (Exercise 4.18). □

4.6 Traffic deaths example

In our prototypical problem suppose the number of traffic deaths increases from $x = 170$ last year to $y = 190$ this year. Is this a significant increase? Let us assume that the number of deaths X and Y are independent Poisson with parameters λ_x and λ_y. Then the likelihood function is

$$L(\lambda_x, \lambda_y) = e^{-(\lambda_x + \lambda_y)} \lambda_x^x \lambda_y^y.$$

Here we are only interested in comparing the two rates, not in the absolute level.

We define the parameter of interest $\theta = \lambda_y/\lambda_x$, and consider λ_x as the nuisance parameter. (Alternatively one might consider λ_y as nuisance.) We have reparameterized (λ_x, λ_y) as $(\lambda_x, \theta\lambda_x)$, so using the invariance principle we obtain

$$L(\theta, \lambda_x) = e^{-\lambda_x(1+\theta)} \lambda_x^{x+y} \theta^y.$$

For each fixed θ the MLE for λ_x is $\widehat{\lambda}_x(\theta) = (x + y)/(1 + \theta)$, so the profile likelihood of θ is

$$\begin{aligned} L(\theta) &= e^{-\widehat{\lambda}_x(\theta)(1+\theta)} \widehat{\lambda}_x(\theta)^{x+y} \theta^y \\ &= \text{constant} \times \left(\frac{\theta}{1+\theta}\right)^y \left(\frac{1}{1+\theta}\right)^x. \end{aligned}$$

The MLE of θ is $\widehat{\theta} = y/x = 1.12$. Figure 4.5 shows the profile likelihood and the approximate 95% CI of θ is (0.91,1.37). So, there is not enough evidence to claim that there is an increase.

Suppose the death count is now redefined as the number of drivers involved in the accidents (several deaths may be associated with one driver), and this is split according to the age of drivers. Table 4.1 summarizes the

Age group	Last year	Current year	$\widehat{\theta}$	Approx. 95% CI
Under 20	20	35	1.75	(1.05,3.05)
20–30	40	54	1.35	(0.93,2.03)
40–50	75	65	0.87	(0.65,1.20)
Over 50	10	2	0.20	(0.05,0.73)

Table 4.1: *Traffic deaths data according to the number of drivers involved, categorized by the drivers' age, and the summaries from the likelihood functions.*

data. The profile likelihood functions for the four age groups are shown in Figure 4.5. We see here the need to summarize the plot: it is too busy and contains too much information. In the table, each likelihood is represented by the MLE and the 95% CI. The data show that the accident rate ratio θ is a function of age and it appears that the accident rate has increased among the 'under 20' group, but has not increased for other age groups and in fact has dropped significantly for the 'over 50' group.

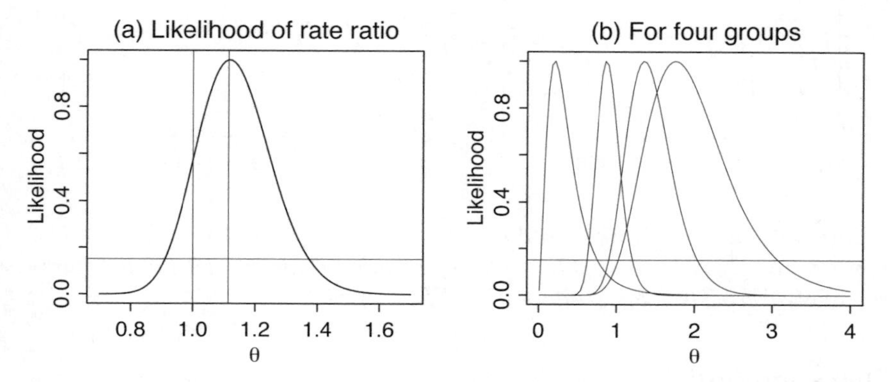

Figure 4.5: *(a) Profile likelihood of rate ratio θ. (b) Profile likelihood of θ for the four age groups shown in Table 4.1.*

4.7 Aspirin data example

The analysis of the aspirin data example (Section 1.1) is similar to that of the traffic death example. Suppose that the number of heart attacks in the active group X_a is binomial(n_a, θ_a) and that in the placebo group X_p is binomial(n_p, θ_p). We observed $x_a = 139$ from a total $n_a = 11,037$ subjects, and $x_p = 239$ from a total of $n_p = 11,034$. The parameter of interest is $\theta = \theta_a/\theta_p$. Since n_a and n_p are large, while the event rates are small, X_a and X_p are approximately Poisson with parameter $n_a\theta_a$ and $n_p\theta_p$. The analysis can be further simplified by using $n_a \approx n_p$, though the simplification is minor. The likelihood of θ_p and $\theta = \theta_a/\theta_p$ is

$$\begin{aligned} L(\theta, \theta_p) &= e^{-(n_a\theta_a + n_p\theta_p)}(n_a\theta_a)^{x_a}(n_p\theta_p)^{x_p} \\ &= \text{constant} \times e^{-\theta_p(n_a\theta + n_p)}\theta^{x_a}\theta_p^{x_a+x_p} \end{aligned}$$

and one proceeds in the usual way to produce the profile likelihood of θ (Exercise 4.21)

$$L(\theta) = \text{constant} \times \left(\frac{n_a\theta}{n_a\theta + n_b}\right)^{x_a}\left(1 - \frac{n_a\theta}{n_a\theta + n_b}\right)^{x_p}.$$

Exactly the same theory also applies for the number of strokes. The profile likelihoods of θ for heart attacks and stroke are shown in Figure 4.6. The approximate 95% CIs for θ are (0.47,0.71) and (0.93,1.59), respectively. There is a significant benefit of aspirin in reducing heart attacks, but the evidence for increased rate of stroke is not significant.

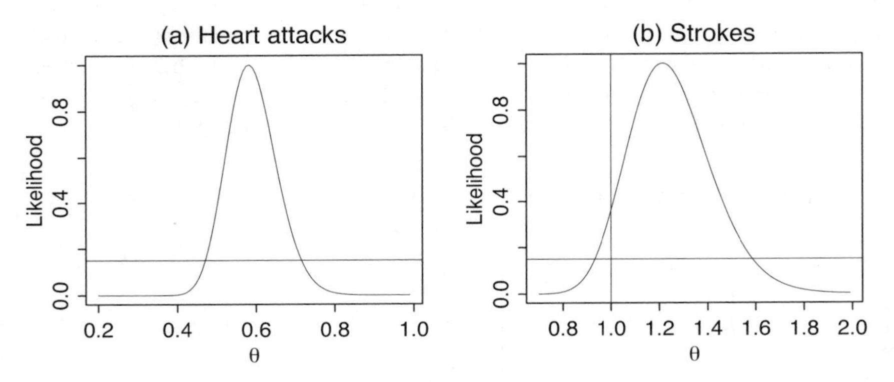

Figure 4.6: *(a) Profile likelihood of θ for the number of heart attacks. The approximate 95% CI is (0.47,0.71). (b) Profile likelihood of θ for the number of strokes. The approximate 95% CI is (0.93,1.59).*

Delta method

As a comparison we will analyse the aspirin data in a more ad hoc way, without using any likelihood. We will use the Delta method, one of the most useful classical tools to derive distributions of statistical estimates.

Assuming that X_a and X_p are Poisson with rate $n_a\theta_a$ and $n_p\theta_p$, we can show (Exercise 4.22) that the conditional distribution of X_a given $X_a + X_p$ is binomial with parameters $X_a + X_p$ and $\pi = n_a\theta_a/(n_a\theta_a + n_b\theta_b)$. Since $n_a = 11,307 \approx n_p = 11,304$, we approximately have $\pi = \theta/(\theta+1)$ or $\theta = \pi/(1-\pi)$. On observing $x_a = 139$ and $x_a + x_p = 378$, we get

$$\begin{aligned} \widehat{\pi} &= \frac{139}{378} = 0.368 \\ \widehat{\theta} &= \frac{\widehat{\pi}}{1-\widehat{\pi}} = 0.58. \end{aligned}$$

We then use the following theorem to get a confidence interval for θ.

Theorem 4.1 (Delta method) *Let $\widehat{\psi}$ be an estimate of ψ based on a sample of size n such that*

$$\sqrt{n}(\widehat{\psi} - \psi) \to N(0, \sigma^2).$$

Then, for any function $h(\cdot)$ that is differentiable around ψ and $h'(\psi) \neq 0$, we have

$$\sqrt{n}(h(\widehat{\psi}) - h(\psi)) \to N(0, \sigma^2 |h'(\psi)|^2).$$

In view of the central limit theorem, the Delta method applies to functions of the sample mean. Informally we say that $h(\widehat{\psi})$ is approximately normal with mean $h(\psi)$ and variance $|h'(\psi)|^2 \text{var}(\widehat{\psi})$.

To apply the method here, recall that $\widehat{\pi}$ is approximately $N(\pi, \sigma^2/378)$, where

$$\sigma^2 = \pi(1 - \pi).$$

So, from $h(\pi) = \pi/(1 - \pi)$, we have

$$\begin{aligned} h'(\pi) &= \frac{1}{(1-\pi)^2} \\ \text{var}(\widehat{\theta}) &= \frac{1}{(1-\pi)^4} \frac{\pi(1-\pi)}{378} \\ &= \frac{\pi}{378(1-\pi)^3} \\ &\approx \frac{139/378}{378(1 - 139/378)^3} = 0.3849 \times 10^{-2}. \end{aligned}$$

So, the approximate 95% CI for θ, given by $\widehat{\theta} \pm 1.96\sqrt{\text{var}(\widehat{\theta})}$, is

$$0.46 < \theta < 0.70,$$

shifted slightly to the left of the likelihood-based interval. (Of course in practice the difference is not important; what matters for discussion here is how we arrive at the intervals.)

We may interpret the Delta method this way: we obtain a single observation $\widehat{\theta} = 0.58$ from $N(\theta, 0.3849 \times 10^{-2})$. This produces an approximate likelihood of θ, differing slightly from the likelihood we derived before; see Figure 4.7. Hence the Delta method can be seen to produce a quadratic approximation of the likelihood function.

4.8 Continuous data

Normal models

We have discussed the normal model $N(\mu, \sigma^2)$ in Section 2.5 as the ideal or exactly regular case for likelihood inference. It is one of the most commonly used models for analysing continuous outcomes. Many results in classical statistics are derived for normal data. Let $x_1, \ldots, x_n$ be an iid sample

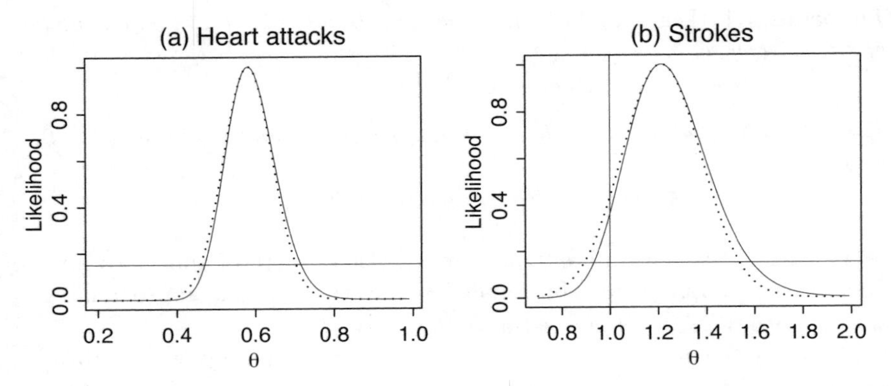

Figure 4.7: *(a) Profile likelihood of θ for the number of heart attacks in the aspirin data example and its normal approximation (dotted line). (b) The same as (a) for strokes.*

from $N(\mu, \sigma^2)$ with both parameters unknown. Ignoring irrelevant constant terms, we can write

$$\log L(\mu, \sigma^2) = -\frac{n}{2}\log\sigma^2 - \frac{1}{2\sigma^2}\sum_i (x_i - \mu)^2,$$

so maximizing the likelihood function for μ is equivalent to finding the least-squares estimate. The MLEs are given by

$$\begin{aligned} \widehat{\mu} &= \overline{x} \\ \widehat{\sigma}^2 &= \frac{1}{n}\sum_i (x_i - \overline{x})^2. \end{aligned}$$

Note the $1/n$ divisor for the variance estimate, which is different from the usual $1/(n-1)$ divisor for the sample variance s^2.

Example 4.8: The following are 100 measurements of the speed of light (in km/s, minus 299,000) made by A. Michelson between 5 June and 2 July 1879 (Stigler 1977). The data were based on five series of experiments, each with 20 runs. The first two lines of data (read by row) are from the first experiment, etc.

850	740	900	1070	930	850	950	980	980	880
1000	980	930	650	760	810	1000	1000	960	960
960	940	960	940	880	800	850	880	900	840
830	790	810	880	880	830	800	790	760	800
880	880	880	860	720	720	620	860	970	950
880	910	850	870	840	840	850	840	840	840
890	810	810	820	800	770	760	740	750	760
910	920	890	860	880	720	840	850	850	780
890	840	780	810	760	810	790	810	820	850
870	870	810	740	810	940	950	800	810	870

The MLEs of the normal parameters are $\widehat{\mu} = 852.4$ and $\widehat{\sigma}^2 = 6180.24$. The sample variance is $s^2 = 6242.67$; it is common practice to use $s/\sqrt{n}$ as the standard error for $\widehat{\mu}$. The 95% CI for μ is

$$\widehat{\mu} \pm 1.96 s/\sqrt{n},$$

yielding $803.4 < \mu < 901.4$; this does not cover the 'true' value $\mu = 734.5$ (based on the currently known speed of light in vacuum, 299,792.5 km/s, corrected for the air condition at the time of the experiment). There are several problems: the means of the five experiments are 909, 856, 845, 820.5, 831.5, so there seems to be a problem with the first experiment (Exercise 4.24). The average from the last four experiments is 838.3 (se = 7.2). Also, there is a correlation of 0.54 between consecutive measurements, so the data are not independent, and the standard error is wrong (too small). □

A normal assumption can be checked using a QQ-plot, which is a general method to see the shape and texture of a distribution. If X has a continuous distribution $F(x)$, then

$$P\{F(X) \le u\} = P\{X \le F^{-1}(u)\} = F(F^{-1}(u)) = u.$$

This means $F(X)$ is a standard uniform variate. Conversely, if U is standard uniform, then $F^{-1}(U)$ is a random variable with distribution function $F(x)$. If $X_1, \ldots, X_n$ are an iid sample from $F(x)$, then $F(X_1), \ldots, F(X_n)$ are an iid sample from $U(0,1)$.

Suppose $x_1, \ldots, x_n$ have a hypothesized distribution F. Intuitively the ordered values $x_{(1)}, \ldots, x_{(n)}$ should behave like order statistics $s_1, \ldots, s_n$ simulated from F. We can plot s_i against $x_{(i)}$: if the data come from F, we can expect to see a straight line. To remove the randomness of $s_1, \ldots, s_n$ we might use the median of the order statistics. Denoting by $U_{(i)}$ the i'th order statistic from $U_1, \ldots, U_n$, we have

$$\begin{aligned} \text{median}(s_i) &= \text{median}\{F^{-1}(U_{(i)})\} \\ &= F^{-1}(\text{median}\{U_{(i)}\}) \\ &\approx F^{-1}\left(\frac{i - 1/3}{n + 1/3}\right). \end{aligned}$$

The approximate median of $U_{(i)}$ is accurate even for very small n (Hoaglin 1985); for example, for $n = 5$ the median of $U(1)$ is 0.129, and the approximation is $(1 - 1/3)/(5 + 1/3) = 0.125$. The QQ-plot is a plot of

$$F^{-1}\left(\frac{i - 1/3}{n + 1/3}\right) \quad \text{versus } x_{(i)}.$$

As a visual aid, we can draw a line going through the first and third quartiles of the distribution.

If $X = \sigma X_0 + \mu$ for some standard variate X_0 and some unknown location μ and scale σ, we do not need to estimate μ and σ. We can use the distribution of X_0 as the basis of the QQ-plot. Here we have

$$\text{median}(s_i) \approx \sigma F_0^{-1}\left(\frac{i - 1/3}{n + 1/3}\right) + \mu$$

where F_0 is the distribution of X_0. For example, to check normality we use the standard normal distribution function $\Phi(x)$ in place of $F_0(x)$. The

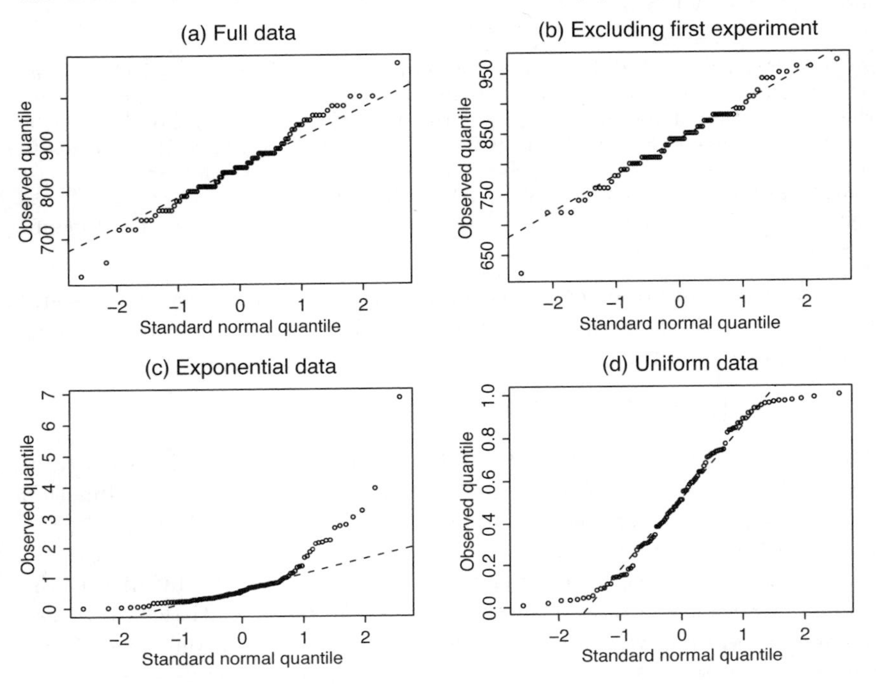

Figure 4.8: *(a) The normal QQ-plot indicates that Michelson's data are not normal. (b) Excluding data from the first experiment, the rest of the data appears normal. (c) Simulated exponential data, showing skewed distribution. (d) Simulated uniform data, showing short-tailed distribution.*

QQ-plot of Michelson's data in Figure 4.8(a) indicates some nonnormality: notice the deviation from the guideline, especially of the upper half. This occurs because the first experiment is different from the rest. As shown in Figure 4.8(b), excluding data from the first experiment, the remaining data look normal, except for rounding and possibly a single outlier.

The normal QQ-plot is a useful exploratory tool even for nonnormal data. The plot shows skewness, heavy-tailed or short-tailed behaviour, digit preference, or outliers and other unusual values. Figure 4.8(c) shows the QQ-plot of simulated exponential data (skewed to the right) and uniform data (short tail).

The case $n = 1$

In the extreme case $n = 1$ the likelihood function becomes unbounded at $\mu = x_1$, so $\widehat{\mu} = x_1$, $\widehat{\sigma}^2 = 0$, and the estimated distribution is degenerate at x_1. The maximized likelihood is $L(\widehat{\mu}, \widehat{\sigma}^2) = \infty$. This has been cited as a weakness of the likelihood method: the likelihood suggests there is very strong evidence that $\widehat{\sigma}^2 = 0$, which is 'surprising' as we feel a single observation cannot tell us anything about variability.

The infinite likelihood can be avoided by using a finite precision model and the original definition of likelihood as probability (although we will end up with the same result that the best model is 'degenerate' at x_1, where 'x_1' represents an interval $(x - \epsilon, x + \epsilon)$). This example is a valid caveat about over-interpretation of the (observed maximized) likelihood: even an enormous likelihood ratio does not establish the objective truth (of a parameter value). Unlike the likelihood, our sense of surprise is affected not just by the data, but also by our prior expectation (e.g. the data are continuous). The likelihood evidence will not feel as surprising if we know beforehand that a discrete point model $\mu = x_1$ and $\sigma^2 = 0$ is a real possibility.

Two-sample case

Suppose $x_1, \ldots, x_m$ are an iid sample from $N(\mu, \sigma_x^2)$, and $y_1, \ldots, y_n$ from $N(\mu + \delta, \sigma_y^2)$. If we are only interested in the mean difference δ, there are three nuisance parameters. Computation of the profile likelihood of δ is left as an exercise.

In practice it is common to assume that the two variances are equal. Under this assumption, the classical analysis is very convenient; the t-statistic

$$t = \frac{\overline{y} - \overline{x} - \delta}{s_p \sqrt{\frac{1}{m} + \frac{1}{n}}},$$

where s_p^2 is the pooled variance

$$s_p^2 = \frac{(m-1)s_x^2 + (n-1)s_y^2}{m+n-2},$$

has a t-distribution with $m+n-2$ degrees of freedom. We can use this for testing hypotheses or setting CIs on δ.

For Michelson's data, suppose we want to compare the first experiment with the remaining data. We obtain

$$\begin{aligned} m &= 20, \quad \overline{x} = 909, \quad s_x^2 = 11009.47 \\ n &= 80, \quad \overline{y} = 838.2, \quad s_y^2 = 4161.5 \end{aligned}$$

and $t = -3.58$ to test $\delta = 0$, so there is strong evidence that the first experiment differs from the subsequent ones.

Nonnormal models

As an example of a nonnormal continuous model, the gamma model is useful for positive outcome data such as measures of concentration, weight, lifetime, etc. The density of the gamma(α, λ) model is

$$f(x) = \frac{1}{\Gamma(\alpha)} \lambda^\alpha x^{\alpha-1} e^{-\lambda x}, \quad x > 0.$$

The parameter α is called the *shape* parameter and λ^{-1} is the *scale* parameter. One important property of the gamma model is that the mean $E(X) = \alpha/\lambda$ is quadratically related to the variance $\text{var}(X) = \alpha/\lambda^2 = E^2(X)/\alpha$. It is sometimes useful to reparameterize using $\mu = \alpha/\lambda$ and $\phi = 1/\alpha$, so that we have gamma(μ, ϕ) with mean $E(X) = \mu$, variance $\phi\mu^2$ and coefficient of variation $\sqrt{\phi}$. The density can be written as

$$f(x) = \frac{1}{x\Gamma(1/\phi)} \left(\frac{x}{\phi\mu}\right)^{1/\phi} \exp\left(-\frac{x}{\phi\mu}\right), \qquad x > 0.$$

If the shape parameter $\alpha = 1$ we obtain the exponential model with density

$$f(x) = \mu^{-1} e^{-x/\mu}.$$

Example 4.9: The following data are the duration of service (in minutes) for 15 bank customers. Of interest is the average length of service μ.

```
23.91 27.33 0.15 3.65 5.99 0.88 0.93 0.53
  0.17 14.17 6.18 0.05 3.89 0.24 0.08.
```

Assuming the gamma model, Figure 4.9 shows the joint likelihood of μ and ϕ; the contour lines are drawn at 90%, 70%, 50%, 30% and 10% cutoffs. The profile likelihood for μ can be computed numerically from the joint likelihood. Also shown are the gamma likelihood assuming ϕ is known at $\widehat{\phi} = 2.35$ and the exponential likelihood by assuming $\phi = 1$. From Figure 4.9(a) it is clear that the exponential model ($\phi = 1$) is not well supported by the data. □

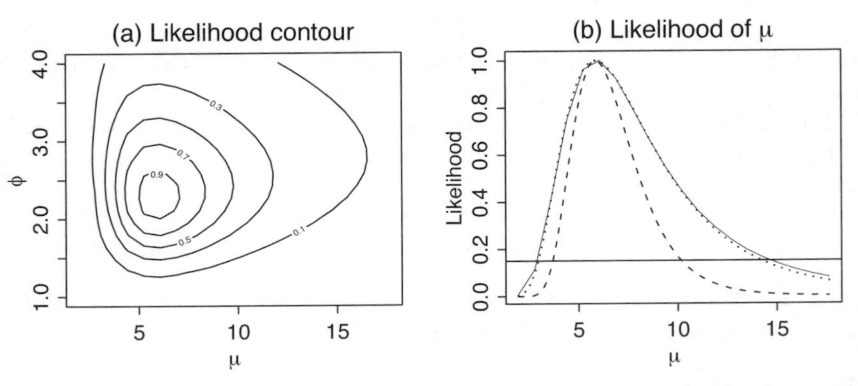

Figure 4.9: *(a) Joint likelihood of μ and ϕ using the gamma model. (b) The profile likelihood of μ from the gamma model (solid line), the estimated likelihood using the gamma model assuming $\phi = \widehat{\phi} = 2.35$ (dotted line) and the likelihood using an exponential model (dashed line), which is equivalent to the gamma model with $\phi = 1$.*

4.9 Exponential family

The exponential family is a very wide class of models that includes most of the commonly used models in practice. It is an unfortunate family name, since it is mixed up with the standard exponential model, which is a member of the exponential family. However, we will stick with the terminology as no one has come up with a better name, and usually what we mean is clear from the context. A general p-parameter exponential family has $\theta \equiv (\theta_1, \ldots, \theta_p)$ and its log-density is of the form

$$\log p_\theta(x) = \sum_{i=1}^{p} \eta_i(\theta) T_i(x) - A(\theta) + c(x) \qquad (4.2)$$

for known functions $A(\theta)$ and $c(x)$, and $\eta_i(\theta)$ and $T_i(x)$ for each i. Furthermore the support of the distribution must not involve the unknown parameter θ. The parameters η_i's are called the natural parameters of the family, and T_i's the natural statistics. To avoid trivialities we assume that there is no linear dependence between η_i's nor that between T_i's. Under this assumption, the T_i's can be shown to be minimal sufficient (Exercise 4.28).

The family is called 'full rank' or simply 'full' if the natural parameter space contains an open set; for example, a 2D square in 2D space contains an open set, but a curve in 2D-space does not. Typically the family is full if the number of unknown parameters is the same as the number of natural sufficient statistics.

If there is a nonlinear relationship among the natural parameters, the number of natural sufficient statistics is greater than the number of free parameters, and the family is called a *curved* exponential family. The distinction is important, since many theoretical results are true only for full exponential families. Compared with those of a full family, problems involving the curved exponential family are somewhat 'more nonlinear'.

The exponential family includes both discrete and continuous random variables. The normal, binomial, Poisson or gamma models are in the exponential family, but the Cauchy and t-distributions are not.

Example 4.10: For the normal model with $\theta = (\mu, \sigma^2)$, with both parameters unknown,

$$\log p_\theta(x) = \frac{\mu x}{\sigma^2} - \frac{x^2}{2\sigma^2} - \frac{\mu^2}{2\sigma^2} - \frac{1}{2}\log(2\pi\sigma^2)$$

is a two-parameter full exponential family model with natural parameters $\eta_1 = \mu/\sigma^2$ and $\eta_2 = -1/(2\sigma^2)$, and natural statistics $T_1(x) = x$ and $T_2(x) = x^2$.

If x has a known coefficient of variation c, such that it is $N(\mu, c^2\mu^2)$, with unknown $\mu > 0$, then x is a curved exponential family. Even though there is only one unknown parameter, the natural statistics are still $T_1(x) = x$ and $T_2(x) = x^2$, and they are minimal sufficient. □

Example 4.11: For the Poisson model with mean μ we have

$$\log p_\mu(x) = x \log \mu - \mu - \log x!;$$

therefore it is a one-parameter exponential family.

If x is a truncated Poisson distribution in the sense that $x = 0$ is not observed, we have, for $x = 1, 2, \ldots$,

$$P(X = x) = \frac{e^{-\mu}\mu^x/x!}{1 - e^{-\mu}},$$

so

$$\log P(X = x) = x \log \mu - \mu - \log(1 - e^{-\mu}) - \log x!.$$

This also defines a one-parameter exponential family with the same canonical parameter and statistic, but a slightly different $A(\mu)$ function. □

The joint distribution of an iid sample from an exponential family is also in the exponential family. For example, let $x_1, \ldots, x_n$ be an iid sample from $N(\mu, \sigma^2)$. The joint density is

$$\log p_\theta(x_1, \ldots, x_n) = \frac{\mu \sum_i x_i}{\sigma^2} - \frac{\sum_i x_i^2}{2\sigma^2} - \frac{n\mu^2}{2\sigma^2} - \frac{n}{2}\log(2\pi\sigma^2).$$

This has the same natural parameters $\eta_1 = \mu/\sigma^2$ and $\eta_2 = -1/(2\sigma^2)$, and natural statistics $T_1 = \sum_i x_i$ and $T_2 = \sum_i x_i^2$.

To illustrate the richness of the exponential family and appreciate the special role of the function $A(\theta)$, suppose X is any random variable with density $\exp\{c(x)\}$ and moment generating function

$$m(\theta) = Ee^{\theta X}.$$

Let $A(\theta) \equiv \log m(\theta)$, usually called the cumulant generating function. Then

$$\int e^{\theta x + c(x)} dx = e^{A(\theta)}$$

or

$$\int e^{\theta x - A(\theta) + c(x)} dx = 1$$

for all θ. So

$$p_\theta(x) \equiv e^{\theta x - A(\theta) + c(x)} \tag{4.3}$$

defines an exponential family model with parameter θ. Such a construction is called an 'exponential tilting' of the random variable X; the original X corresponds to $\theta = 0$.

For any exponential model of the form (4.3) we can show (Exercise 4.30) that

$$\mu = E_\theta(X) = A'(\theta)$$

and

$$\text{var}_\theta(X) = A''(\theta) = \frac{\partial}{\partial\theta} E_\theta(X) = v(\mu).$$

Therefore $A(\theta)$ implies a certain relationship between the mean and variance.

Exponential dispersion model

For statistical modelling it is often adequate to consider a two-parameter model known as the exponential dispersion model (Jørgensen 1987). Based on an observation x the log-likelihood of the scalar parameters θ and ϕ is of the form

$$\log L(\theta, \phi) = \frac{x\theta - A(\theta)}{\phi} + c(x, \phi), \tag{4.4}$$

where $A(\theta)$ and $c(x, \phi)$ are assumed known functions. In this form the parameter θ is called the canonical parameter and ϕ the *dispersion parameter.* Since $c(x, \phi)$ or $A(\theta)$ can be anything there are infinitely many submodels in this exponential family, though of course the density must satisfy

$$\sum_x \exp\left\{\frac{x\theta - A(\theta)}{\phi} + c(x, \phi)\right\} = 1,$$

which forces a certain relationship between $A(\theta)$ and $c(x, \phi)$.

The dispersion parameter allows the variance to vary freely from the mean:

$$\mu = E_\theta(X) = A'(\theta)$$

and

$$\begin{aligned} \text{var}_\theta(X) &= \phi A''(\theta) \\ &= \phi \frac{\partial}{\partial\theta} E_\theta(X) = \phi v(\mu). \end{aligned}$$

This is the biggest advantage of the exponential dispersion model over the more rigid model (4.3).

In practice the form of $A(\theta)$ in (4.4) is usually explicitly given from the standard models, while $c(x, \phi)$ is left implicit. This is not a problem as far as estimation of θ is concerned, since the score equation does not involve $c(x, \phi)$. However, without explicit $c(x, \phi)$, a likelihood-based estimation of ϕ and a full likelihood inference on both parameters are not available. One might compromise by using the method of moments to estimate ϕ, or the approximation of the likelihood given later.

Example 4.12: For the normal model $N(\mu, \sigma^2)$ we have

$$\log L(\mu, \sigma^2) = \frac{x\mu - \mu^2/2}{\sigma^2} - \frac{1}{2}\log\sigma^2 - \frac{x^2}{2\sigma^2},$$

so the normal model is in the exponential family with a canonical parameter $\theta = \mu$, dispersion parameter $\phi = \sigma^2$, and $A(\theta) = \theta^2/2$ and $c(x, \phi) = -\frac{1}{2}\log\phi - \frac{1}{2}x^2/\phi$. This is a rare case where $c(x, \phi)$ is available explicitly. □

Example 4.13: For the Poisson model with mean μ

$$\log L(\mu) = x\log\mu - \mu - \log x!$$

so we have a canonical parameter $\theta = \log\mu$, dispersion parameter $\phi = 1$ and $A(\theta) = \mu = e^\theta$.

By keeping $A(\theta) = e^\theta$, but varying the dispersion parameter, we would generate a Poisson model with under- or overdispersion: with the same $EX = \mu$ we have

$$\text{var}(X) = \phi A''(\theta) = \phi\mu.$$

The function $c(x, \phi)$ has to satisfy

$$\sum_{x=0}^{\infty} \exp\left[\left\{\frac{x\theta - e^\theta}{\phi}\right\} + c(x,\phi)\right] = 1,$$

for all θ and ϕ. To compute the likelihood we need $c(x, \phi)$ explicitly, but finding the solution $c(x, \phi)$ that satisfies such an equation is not trivial. □

Example 4.14: The following data shows the number of paint defects recorded in a sample of 20 cars using an experimental process:

```
0 10  1  1  1  2  1  4 11  0  5  2  5  2  0  2  0  1  3  0
```

The sample mean is $\overline{x} = 2.55$ and the sample variance is $s^2 = 9.84$, indicating overdispersion. Using the Poisson-type model above $A(\theta) = e^\theta$, so

$$\text{var}_\theta(X) = \phi E_\theta(X),$$

and the method-of-moments estimate of ϕ is $\widehat{\phi} = 9.84/2.55 = 3.86$. Obviously there is some variability in $\widehat{\phi}$; is it significantly away from one? An exact likelihood analysis is difficult, but instead we can do a Monte Carlo experiment to test $\phi = 1$ (Exercise 4.33):

1. Generate $x_1^*, \ldots, x_{20}^*$ as an iid sample from the Poisson distribution with mean $\mu = \overline{x} = 2.55$.
2. Compute $\widehat{\phi}^* = (s^2)^*/\overline{x}^*$ from the data in part 1.
3. Repeat 1 and 2 a large number of times and consider the $\widehat{\phi}^*$'s as a sample from the distribution of $\widehat{\phi}$.
4. Compute P-value = the proportion of $\widehat{\phi}^* >$ the observed $\widehat{\phi}$.

Figure 4.10 shows the QQ-plot of 500 $\widehat{\phi}^*$'s and we can see that P-value ≈ 0, confirming that $\phi > 1$.

An approximate likelihood inference for μ can proceed based on assuming that ϕ is known at $\widehat{\phi}$; this is made particularly simple since the term $c(x_i, \phi)$ is constant relative to μ or θ. Figure 4.10 compares the standard Poisson likelihood with the estimated likelihood using $\widehat{\phi} = 3.86$. For small samples this approach is not satisfying as the uncertainty in $\widehat{\phi}$ is not accounted for in the inference for μ. □

Approximate likelihood*

While an exact likelihood of the exponential dispersion model (4.4) might not be available, there is a general approximation that can be computed easily. The main advantage of the approximate likelihood is that it puts the whole inference of the mean and dispersion parameters within the likelihood context. It also facilitates further modelling of the dispersion parameter.

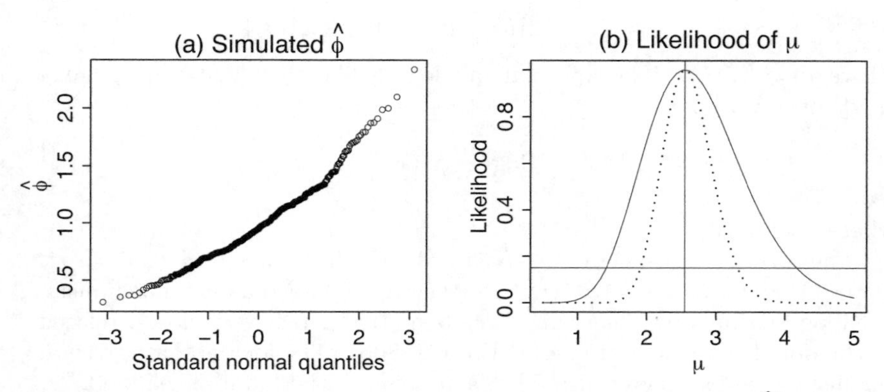

Figure 4.10: *(a) Monte Carlo estimate of the distribution of* $\widehat{\phi}$ *under the standard Poisson model. (b) Likelihood of* μ *based on the standard Poisson model (*$\phi = 1$*, dotted line) and the model with extra Poisson variability (*$\phi = \widehat{\phi} = 3.86$*, solid line).*

At fixed ϕ the MLE of θ is the solution of

$$A'(\widehat{\theta}) = x.$$

Alternatively $\widehat{\mu} = x$ is the MLE of $\mu = A'(\theta)$. The Fisher information on θ based on x is

$$I(\widehat{\theta}) = A''(\widehat{\theta})/\phi.$$

From Section 9.8, at fixed ϕ, the approximate density of $\widehat{\theta}$ is

$$p(\widehat{\theta}) \approx (2\pi)^{-1/2} I(\widehat{\theta})^{1/2} \frac{L(\theta, \phi)}{L(\widehat{\theta}, \phi)}.$$

Since $\widehat{\mu} = A'(\widehat{\theta})$, the density of $\widehat{\mu}$, and hence of x, is

$$\begin{aligned} p(x) = p(\widehat{\mu}) &= p(\widehat{\theta}) \left| \frac{\partial \widehat{\theta}}{\partial \widehat{\mu}} \right| \\ &= p(\widehat{\theta}) \{A''(\widehat{\theta})\}^{-1} \\ &\approx \{2\pi\phi A''(\widehat{\theta})\}^{-1/2} \frac{L(\theta, \phi)}{L(\widehat{\theta}, \phi)}. \end{aligned}$$

The potentially difficult function $c(x, \phi)$ in (4.4) cancels out in the likelihood ratio term, so we end up with something simpler. In commonly used distributions, the approximation is highly accurate for a wide range of parameter values.

Let us define the deviance function

$$D(x, \mu) = 2 \log \frac{L(\widehat{\theta}, \phi = 1)}{L(\theta, \phi = 1)}$$

$$= 2\{\widehat{\theta}x - \theta x - A(\widehat{\theta}) + A(\theta)\},$$

where $\mu = A'(\theta)$. The approximate log-likelihood contribution from a single observation x is

$$\log L(\theta, \phi) \approx -\frac{1}{2}\log\{2\pi\phi v(x)\} - \frac{1}{2\phi}D(x, \mu), \tag{4.5}$$

where $v(x) = A''(\widehat{\theta})$.

The formula is exact if x is $N(\mu, \sigma^2)$, in which case $\theta = \mu$, $\phi = \sigma^2$, $v(x) = 1$, and $D(x, \mu) = (x - \mu)^2$. In general it is called an extended quasi-likelihood formula by Nelder and Pregibon (1987). It is very closely related to the double-exponential family (Efron 1986b). The idea is that, given θ, the dispersion parameter also follows an exponential family likelihood.

Given an iid sample $x_1, \ldots, x_n$, at fixed ϕ, the estimate of μ is the minimizer of the total deviance

$$\sum_i D(x_i, \mu).$$

The approximate profile log-likelihood of ϕ is

$$\log L(\phi) \approx -\frac{1}{2}\sum_i \log\{2\pi\phi v(x_i)\} - \frac{1}{2\phi}\sum_i D(x_i, \widehat{\mu})$$

and the approximate MLE of ϕ is the average deviance

$$\widehat{\phi} = \frac{1}{n}\sum_i D(x_i, \widehat{\mu}).$$

Example 4.15: In Example 4.13 we cannot provide an explicit two-parameter Poisson log-likelihood

$$\log L(\mu, \phi) = \frac{x\log\mu - \mu}{\phi} + c(x, \phi),$$

since $c(x, \phi)$ is not available. Using

$$\theta = \log\mu, \quad A(\theta) = e^\theta = \mu$$
$$\widehat{\theta} = \log x, \quad A(\widehat{\theta}) = x$$
$$D(x, \mu) = 2(x\log x - x\log\mu - x + \mu)$$
$$v(x) = A''(\widehat{\theta}) = e^{\widehat{\theta}} = x$$

we get an approximate likelihood

$$\log L(\mu, \phi) \approx -\frac{1}{2}\log\{2\pi\phi x\} - \frac{1}{\phi}(x\log x - x\log\mu - x + \mu).$$

Nelder and Pregibon (1987) suggest replacing $\log\{2\pi\phi x\}$ by

$$\log\{2\pi\phi(x + 1/6)\}$$

to make the formula work for $x \geq 0$.

For the data in Example 4.14 the estimate of the mean μ is $\overline{x} = 2.55$, and

$$\widehat{\phi} = \frac{1}{20}\sum_i D(x_i, \overline{x}) = 3.21,$$

where we have used $0 \log 0 = 0$; recall $\widehat{\phi} = 3.86$ using the method of moments. The approximate profile likelihood of ϕ can also be computed from (4.5). □

Minimal sufficiency and the exponential family*

There is a surprising connection between minimal sufficiency and the exponential family: under mild conditions, a (minimal) sufficient estimate exists if and only if the model $p_\theta(x)$ is in the full exponential family. The exponential family structure in turn implies that the likelihood function has a unique maximum, and the MLE is sufficient. *Therefore, if a sufficient estimate exists, it is provided by the MLE.*

It is easy to show that the MLE of θ in a full exponential family model is sufficient (Exercise 4.28). To show that the existence of a sufficient estimate implies the exponential family, we follow the development in Kendall *et al.* (1977) for the scalar case. Assume that we have an iid sample $x_1, \ldots, x_n$ from $p_\theta(x)$, $t(x)$ is a sufficient estimate (i.e. it is minimal sufficient for θ), and that the support of $p_\theta(x)$ does not depend on θ. By the factorization theorem,

$$\frac{\partial \log L(\theta)}{\partial \theta} = \sum_i \frac{\partial \log p_\theta(x_i)}{\partial \theta} = K(t, \theta)$$

for some function $K(t, \theta)$. From Section 3.2, minimal sufficiency implies $K(t, \theta)$ is a one-to-one function of t. Since this is true for any θ, choose one value of θ, so t must be of the form

$$t = M\{\sum_i k(x_i)\}$$

for some function $M(\cdot)$ and $k(\cdot)$. Defining $w(x) = \sum_i k(x_i)$, then $K(t, \theta)$ is a function of w and θ only, say $N(w, \theta)$. Now

$$\frac{\partial^2 \log L(\theta)}{\partial x_i \partial \theta} = \frac{\partial N}{\partial w}\frac{\partial w}{\partial x_i}.$$

Since $\partial^2 \log L(\theta)/\partial x_i \partial \theta$ and $\partial w/\partial x_i$ only depend on x_i and θ, it is clear that $\partial N/\partial w$ is also a function of θ and x_i. To be true for all x_i then $\partial N/\partial w$ must be a function of θ alone, which means

$$N(w, \theta) = w(x)u(\theta) + v(\theta)$$

for some function $u(\theta)$ and $v(\theta)$, or, in terms of the log-likelihood,

$$\frac{\partial}{\partial \theta} \log L(\theta) = u(\theta)\sum_i k(x_i) + v(\theta).$$

This implies the model must be in the exponential family with log-density of the form

$$\log p_\theta(x) = \eta(\theta)T(x) - A(\theta) + c(x).$$

Now it is straightforward to show that the likelihood is unimodal, so it has a unique maximum.

4.10 Box–Cox transformation family

Box–Cox (1964) transformation family is another useful class of nonnormal models for positive-valued outcomes. Typical characteristics of such outcomes are skewness, and some relationship between the mean and variance. In the Box–Cox transformation model it is assumed that there is $\lambda \neq 0$ such that a transformation of the observed data y according to

$$y_\lambda = \frac{y^\lambda - 1}{\lambda}$$

has a normal model $N(\mu, \sigma^2)$. As λ approaches zero $y_\lambda \to \log y$, so this family includes the important log-transformation. For convenience we will simply write $\lambda = 0$ to represent the log-transformation. The log-likelihood contribution of a single observation y is

$$\begin{aligned}\log L(\lambda, \mu, \sigma^2) &= \log p(y_\lambda) + (\lambda - 1)\log y \\ &= -\frac{1}{2}\log\sigma^2 - \frac{(y_\lambda - \mu)^2}{2\sigma^2} + (\lambda - 1)\log y,\end{aligned}$$

where $p(y_\lambda)$ is the density of y_λ and $(\lambda - 1)\log y$ is the log of the Jacobian.

Note that if y is positive then y_λ cannot be strictly normal for any $\lambda \neq 0$. The possibility of a truncated normal may be considered, and we should check the normal plot of the transformed data. If $y_\lambda > 0$ has a truncated normal distribution, it has a density

$$p(y_\lambda) = \frac{\sigma^{-1}\phi\{(y_\lambda - \mu)/\sigma\}}{1 - \Phi(-\mu/\sigma)}$$

where $\phi(\cdot)$ and $\Phi(\cdot)$ are the standard normal density and distribution functions. So we only need to modify the log-likelihood above by adding $\log\{1 - \Phi(-\mu/\sigma)\}$. If μ/σ is quite large, $\Phi(-\mu/\sigma) \approx 0$, so we can simply ignore the effect of truncation.

For data with a skewed distribution sometimes a transformation is more successful when applied to shifted values, i.e.

$$y_\lambda = \frac{(y + c)^\lambda - 1}{\lambda}.$$

The shift parameter c can be estimated from the data by maximizing the likelihood above.

Interpretation of the parameters is an important practical issue when using a transformation. After transformation μ and σ^2 may not have a simple interpretation in terms of the original data. However, since the Box–Cox transformation is monotone, *the median of* y_λ *is the transformed median of the original data* y. For example, if $\lambda = 0$ and the median of y_λ is $\mu = 0$, then the median of the original data y is $e^\mu = e^0 = 1$. It is sensible to limit the choice of λ to a few meaningful values.

The main reason to consider the Box–Cox transformation family is to analyse the data in a more normal scale. Such a scale typically leads to models that are simpler and easier to interpret, have well-behaved residuals, and clearer inference.

To decide what transformation parameter λ to use in a dataset $y_1, \ldots, y_n$, it is convenient to compare a range of values or simply several sensible values such as $\lambda = -1,\ 0, 0.5, 1$ and 2. In practice, it will not be very sensible to transform data using, say, $\lambda = 0.5774$. The natural criterion to compare is the profile likelihood of λ, which is very simple to compute. At fixed λ we simply transform the data to get y_λ, and the MLEs of μ and σ^2 are simply the sample mean and variance of the transformed data:

$$\widehat{\mu}(\lambda) = \frac{1}{n} \sum_i y_{\lambda i}$$

and

$$\widehat{\sigma}^2(\lambda) = \frac{1}{n} \sum_i \{y_{\lambda i} - \widehat{\mu}(\lambda)\}^2.$$

The profile log-likelihood for λ is

$$\log L(\lambda) = -\frac{n}{2} \log \widehat{\sigma}^2(\lambda) - \frac{n}{2} + (\lambda - 1) \sum_i \log y_i.$$

One needs to be careful in the inference on μ based on the transformed data $y_{\widehat{\lambda}}$. Mathematically we expect to 'pay' for having to estimate λ (Bickel and Doksum 1981), for example by taking a profile likelihood over λ. However, such an approach can be meaningless. A serious problem in the use of transformation is the logical meaning of y_λ and the parameter μ. Here is a variation of a simple example from Box and Cox (1982): suppose we have a well-behaved sample from a normal population with mean around 1000 and a small variance. Then there is an extremely wide range of λ for which the transformation is essentially linear and the transformed data are close to normal. This means λ is poorly determined from the data, and its profile likelihood is flat as it has a large uncertainty. Propagating this large uncertainty to the estimation of μ produces a very wide CI for μ. That is misleading, since in fact the untransformed data ($\lambda = 1$) provide the best information.

So, a rather flat likelihood of λ indicates that the Box–Cox transformation family is not appropriate; it warrants a closer look at the data. If we limit the use of the Box–Cox transformation model to cases where λ is well determined from the data, then it is reasonable to perform inference on μ assuming that λ is known at $\widehat{\lambda}$, i.e. we do not have to pay for having to estimate λ.

Example 4.16: The following figures are the population of 27 countries already in or wishing to join the European Union:

```
82 59 59 57 39 38 22 16 10 10 10 10
10 9 9 8 5 5 5 4 4 2 2 1 0.7 0.4 0.4
```

Figure 4.11(a) shows the population data are skewed to the right. The profile log-likelihood of λ is maximized at $\widehat{\lambda} = 0.12$, pointing to the log-transform $\lambda = 0$ as a sensible transformation. The QQ-plots of the square-root and the log-transform indicate we should prefer the latter. (The reader can verify that $\log(y+1)$ is a better normalizing transform). The use of log-transform is, for example, for better plotting of the data to resolve the variability among low as well as high population counts. □

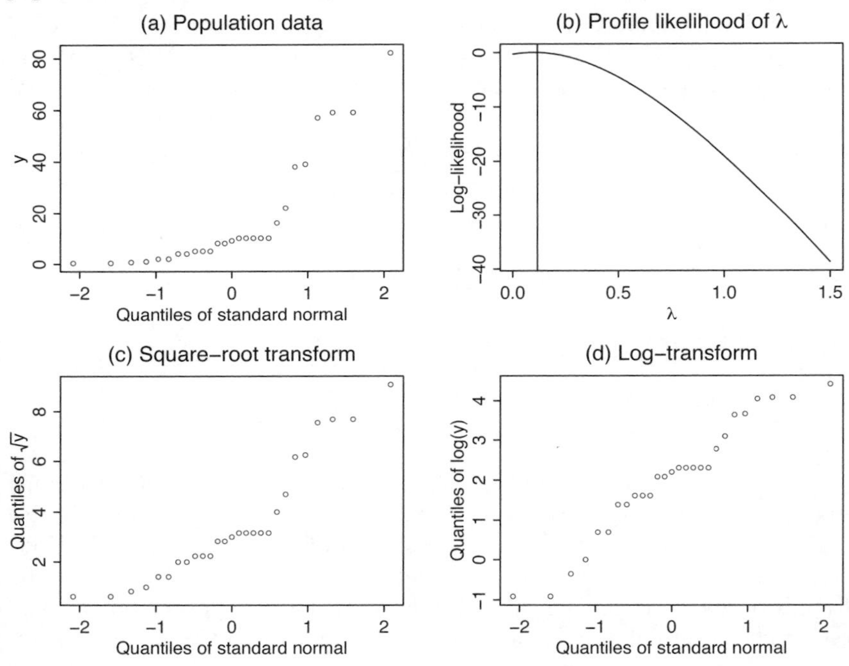

Figure 4.11: *(a) Normal plot of the population data. (b) the profile likelihood function of the transformation parameter λ. (c) and (d) QQ-plots of the log- and square-root transforms of the original data.*

4.11 Location-scale family

The location-scale family is a family of distributions parameterized by μ and σ, and a known density $f_0(\cdot)$, such that any member of the family has a density of the form

$$f(x) = \frac{1}{\sigma} f_0\left(\frac{x-\mu}{\sigma}\right).$$

The parameter μ is called the location parameter and σ the scale parameter, and $f_0(\cdot)$ is called the *standard density* of the family. For example, $N(\mu, \sigma^2)$ is a location-scale family with standard density

$$f_0(x) = \frac{1}{\sqrt{2\pi}} e^{-x^2/2}.$$

Another famous example is the Cauchy(μ, σ) with standard density

$$f_0(x) = \frac{1}{\pi(1+x^2)}.$$

The Cauchy model is useful as a model for data with heavy tails, characterized by the presence of outliers. Furthermore, in theoretical studies the Cauchy model is a good representative of complex models.

Example 4.17: Recording of the difference in maximal solar radiation between two geographical regions over a time period produced the following (sorted) data:

```
-26.8 -3.5 -3.4 -1.2 0.4 1.3 2.3 2.7 3.0 3.2 3.2 3.5
3.6 3.9 4.2 4.4 5.0 6.5 6.7 7.1 8.1 10.5 10.7 24.0 32.8
```

The normal plot in Figure 4.12(a) shows clear outliers or a heavy-tailed behaviour. The mean and median of the data are 3.6 and 4.5 respectively; they are not dramatically different since there are outliers on both ends of the distribution. The Cauchy model with location μ and scale σ has a likelihood function

$$L(\mu, \sigma) = \prod_i \frac{1}{\sigma} \left\{ 1 + \frac{(x_i - \mu)^2}{\sigma^2} \right\}^{-1},$$

from which we can compute a profile likelihood for μ, shown in Figure 4.12(b); there is no closed form solution to get the MLE of σ at a fixed value of μ, so it needs to be found numerically. The MLE from the Cauchy model is $\widehat{\mu} = 3.7$, closer to the median, and $\widehat{\sigma} = 2.2$.

To illustrate the potential of robust modelling, the profile likelihood is compared with that assuming a normal model. The normal likelihood is centered at the sample mean. There is clearly a better precision for the location parameter μ using the Cauchy model than using the normal model. (Note that in these two models the parameter μ is comparable as the median of the distributions, but the scale parameter σ in the Cauchy model does not have meaning as a standard deviation.)

What if the data are closer to normal? The following data are from the last two series of Michelson's speed of light measurements in Example 4.8.

```
890 810 810 820 800 770 760 740 750 760
910 920 890 860 880 720 840 850 850 780
890 840 780 810 760 810 790 810 820 850
870 870 810 740 810 940 950 800 810 870
```

Figure 4.12(c) indicates that, except for some repeat values (at 810), the data are quite normal. As shown in Figure 4.12(d) the normal and Cauchy models produce comparable results. Generally, there will be some loss of efficiency if we use the Cauchy model for normal data and vice versa. However, as we see in this example, the amount of loss is not symmetrical; it is usually larger if we wrongly

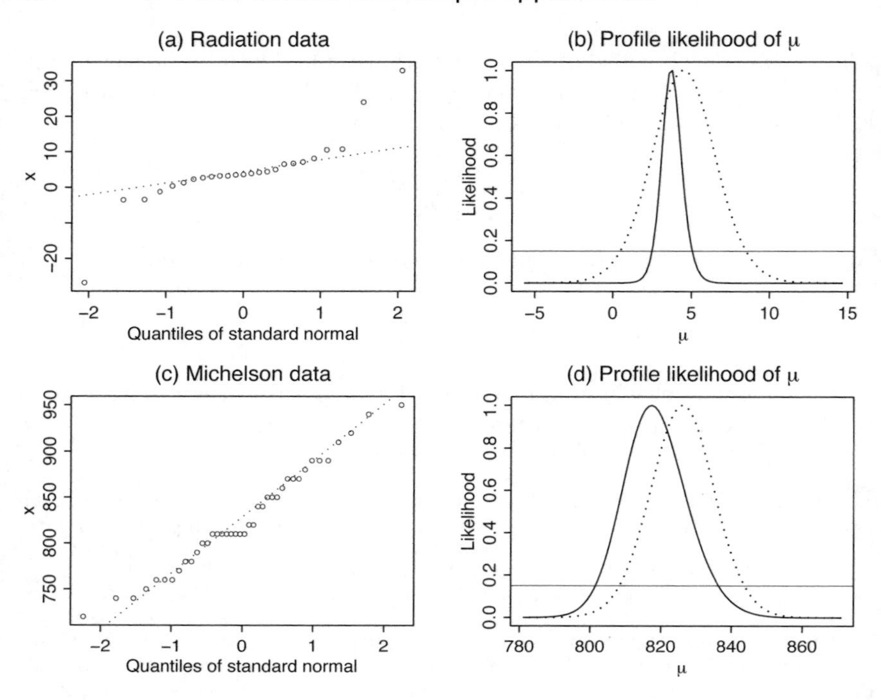

Figure 4.12: *(a) Normal plot of radiation data. (b) The profile likelihood function of the location parameter assuming the normal model (dashed line) and assuming the Cauchy model (solid line). (c)–(d) the same as (a)–(b) for Michelson's last two experiments.*

assume a normal model, so it is not a bad strategy to err on a heavier-tailed model than the normal. □

We can use the AIC (Section 3.5) to choose between the normal and Cauchy models. For the radiation data, the AIC of the normal model is

$$\begin{aligned} \text{AIC} &= -2\log L(\widehat{\mu},\widehat{\sigma}^2)+4 \\ &= n\log(2\pi\widehat{\sigma}^2)+n+4 = 189.7, \end{aligned}$$

where $n = 25$, and $\widehat{\sigma}^2 = 98.64$. For the Cauchy model, using the MLEs $\widehat{\mu} = 3.7$ and $\widehat{\sigma} = 2.2$, the AIC is 169.3. Therefore, as expected from the QQ-plot, the Cauchy model is preferred. For Michelson's data, the AICs are 439.6 and 456.3, now preferring the normal model.

Another useful general technique is to consider a larger class of location models that includes both the Cauchy and the normal families. Such a class is provided by the t-family with ν degrees of freedom. The standard density is given by

$$f_0(x) = \frac{\Gamma\{(\nu+1)/2\}}{\Gamma(1/2)\Gamma(\nu/2)\nu^{1/2}}(1+x^2/\nu)^{-(\nu+1)/2}.$$

If $\nu = 1$ we obtain the Cauchy model, while as $\nu \to \infty$ we get the normal model. So as the parameter ν is allowed to vary we can adapt it to the tail behaviour of the data.

To find which degree of freedom ν is appropriate for particular data, one can maximize the profile likelihood of ν at several values such as $\nu = 1$, 2, 4, 8 and ∞. Inference on μ can be performed at the observed value of $\widehat{\nu}$. As is commonly the case with complex models, all computations must be done numerically. The application of the t-family models on the above datasets is left as an exercise.

Other useful location-scale families of distributions include the logistic family, with the standard density

$$f_0(x) = \frac{e^x}{(1+e^x)^2}, \quad -\infty < x < \infty,$$

and the double-exponential or Laplacian family, with density

$$f_0(x) = \frac{1}{2}e^{-|x|}, \quad -\infty < x < \infty.$$

Both of these distributions have heavier tails than the normal, so they offer protection against outliers if we are interested in the location parameter. The MLE of the location parameter μ under the Laplacian assumption is the sample median, an estimate known to be resistant to outliers.

The existence of a wide variety of models tends to overwhelm the data analyst. It is important to reflect again that model selection is generally harder than model fitting once a model is chosen.

4.12 Exercises

Exercise 4.1: The following table shows Geissler's data for families of size 2 and 6.

No. boys k	0	1	2	3	4	5	6
No. families n_k	42,860	89,213	47,819				
No. families n_k	1096	6233	15,700	22,221	17,332	7908	1579

(a) For each family size, assuming a simple binomial model(n, θ) for the number of boys, write down the likelihood of θ.

(b) Combining the data from families of size 2 and 6, draw the likelihood of θ; find the MLE of θ and its standard error.

(c) Based on (b), examine if the binomial model is a good fit to the data. Describe the nature of the model violation and discuss what factors might cause it.

Exercise 4.2: Now suppose that (in another experiment on sex ratio) the family information is based on questioning 100 boys from families of size k that attend a boys-only school. Explain what is unusual in this sampling scheme and what needs to be modified in the model.

Exercise 4.3: Simulate binomial-type data with under- and overdispersion. Compare the realizations with simulated binomial data. Identify real examples where you might expect to see each case.

Exercise 4.4: Another example of a binomial mixture is the so-called beta–binomial distribution. Conditional on p let X_p be binomial(n, p), where p itself is random with beta distribution with density

$$f(p) = \frac{1}{B(\alpha, \beta)} p^{\alpha-1}(1-p)^{\beta-1}.$$

If $\alpha = \beta = 1$ then p has uniform distribution. Let X be a sample from the mixture distribution.

(a) Simulate data from the mixture for $n = 10$, $\alpha = 1$ and $\beta = 3$.

(b) Find the mean and variance formula for X, and show that there is an extra-binomial variation.

(c) Verify the marginal probability

$$P(X = x) = \frac{B(\alpha + x, \beta + n - x)}{(n+1)\ B(x+1, n-x+1)B(\alpha, \beta)},$$

for $x = 0, 1, \ldots, n$.

(d) Using Geissler's family data of size 6 in Exercise 4.1, find the joint likelihood of $\theta = (\alpha, \beta)$ as well as the profile likelihood of α, β and the mean of X.

(e) Compare the expected frequencies under the beta–binomial model with the observed frequency, and perform a goodness-of-fit test.

(f) Compare the profile likelihood of the mean of X in part (d) with the likelihood assuming the data are simple binomial. Comment on the result.

Exercise 4.5: Generalize the negative binomial distribution along the same lines as the beta–binomial distribution, where the success probability p is a random parameter.

Exercise 4.6: A total of 678 women, who got pregnant under planned pregnancies, were asked how many cycles it took them to get pregnant. The women were classified as smokers and nonsmokers; it is of interest to compare the association between smoking and probability of pregnancy. The following table (Weinberg and Gladen 1986) summarizes the data.

Cycles	Smokers	Nonsmokers
1	29	198
2	16	107
3	17	55
4	4	38
5	3	18
6	9	22
7	4	7
8	5	9
9	1	5
10	1	3
11	1	6
12	3	6
> 12	7	12

(a) Fit a geometric model to each group and compare the estimated probability of pregnancy per cycle.

(b) Check the adequacy of the geometric model. (Hint: devise a plot, or run a goodness-of-fit test.)

(c) Allow the probability to vary between individuals according to a beta distribution (see Exercise 4.4), and find the beta–geometric probabilities.

(d) Fit the beta–geometric model to the data and check its adequacy.

Exercise 4.7: The Steering Committee of the Physicians' Health Study (1989) also reported the following events during follow-up. Recall that 11,037 subjects were assigned to the aspirin group, and 11,034 to the placebo. For each end-point,

Events	Aspirin	Placebo
Total cardiac deaths	81	83
Fatal heart attacks	10	28
Sudden deaths	22	12

report the relative risk of aspirin versus placebo, the profile likelihood, and the approximate 95% CI.

Exercise 4.8: To investigate the effect of race in the determination of death penalty a sociologist M. Wolfgang in 1973 examined the convictions on 55 rapes in 19 Arkansas counties between 1945 and 1965.

(a) Out of 34 black defendants 10 received the death penalty, compared with 4 out of 21 white defendants.

(b) Out of 39 cases where the victims were white, 13 led to the death penalty, compared with 1 out 15 cases with black victims.

Find in each case the significance of the race factor, and state your overall conclusion.

Exercise 4.9: Breslow and Day (1980) reported case-control data on the occurrence of esophageal cancer among French men. The main risk factor of interest is alcohol consumption, where 'high' is defined as over one litre of wine per day. The data are stratified into six age groups.

Age	Consumption	Cancer	No cancer
25–34	High	1	9
	Low	0	106
35–44	High	4	26
	Low	5	164
45–54	High	25	29
	Low	21	138
55–64	High	42	27
	Low	34	139
65–74	High	19	18
	Low	36	88
≥ 75	High	5	0
	Low	8	31

(a) Assess the association between alcohol consumption and cancer using the unstratified data (by accumulating over the age categories).

(b) Compare the result using the stratified data, assuming that the odds ratio is common across strata.

(c) Compare the profile likelihoods of odds-ratio across strata and assess the common odds ratio assumption.

Exercise 4.10: We showed that the one-sided P-value from Fisher's exact test of the data in Example 4.4 is 0.20. What do you think is the exact two-sided P-value?

Exercise 4.11: Let θ be the odds-ratio parameter for a 2×2 table as described in Section 4.3. Use the Delta method to show that the standard error of $\widehat{\theta}$ is

$$\text{se}(\widehat{\theta}) = \left(\frac{1}{x} + \frac{1}{y} + \frac{1}{m-x} + \frac{1}{n-y}\right)^{1/2}$$

The Wald statistic $z = \widehat{\theta}/\text{se}(\widehat{\theta})$ satisfies $z^2 \approx \chi^2$. Verify this for several 2×2 tables. This also means that, if the likelihood is regular, the approximate likelihood of H_0 is

$$e^{-\chi^2/2}.$$

Verify this for several 2×2 tables.

Exercise 4.12: A famous example of Poisson modelling was given by L.J. Bortkiewicz (1868–1931). The data were the number of soldiers killed by horse kicks per year per Prussian army corp. Fourteen corps were examined with varying number of years, resulting in a total 200 corp-year combinations.

Number of deaths k	0	1	2	3	4	≥ 5
Number of corp-years n_k	109	65	22	3	1	0

Fit a simple Poisson model and evaluate the goodness of fit of the model.

Exercise 4.13: To 'solve' the authorship question of 12 of the so-called 'Federalist' papers between Madison or Hamilton, the statisticians Mosteller and Wallace (1964) collated papers of already known authorship, and computed the appearance of some keywords in blocks of approximately 200 words. From 19 of Madison's papers, the appearance of the word 'may' is given in the following table.

Number of occurrences k	0	1	2	3	4	5	6
Number of blocks n_k	156	63	29	8	4	1	1

Check the Poissonness of the data and test whether the observed deviations are significant. Fit a Poisson model and compute a goodness-of-fit test.

Exercise 4.14: Modify the idea of the Poisson plot to derive a binomial plot, i.e. to show graphically if some observed data are binomial. Try the technique on Geissler's data in Exercise 4.1.

Exercise 4.15: Another way to check the Poisson assumption is to compare consecutive frequencies (Hoaglin and Tukey 1985). Under the Poisson assumption with mean θ show that

$$kp_k/p_{k-1} = \theta$$

and in practice we can estimate p_k/p_{k-1} by n_k/n_{k-1}.

(a) Apply the technique to the datasets in Example 4.6 and Exercise 4.13.

(b) Develop a method to test whether an observed deviation is significant.

(c) Modify the technique for the binomial and negative binomial distributions, and apply them to the datasets in Exercises 4.13 and 4.1, respectively.

Exercise 4.16: This is an example of a binomial–Poisson mixture. Suppose X_n is binomial with parameters n and p, but n itself is a Poisson variate with mean μ. Find the marginal distribution of X_n. Discuss some practical examples where this mixture model is sensible.

Exercise 4.17: The annual report of a pension fund reported a table of the number of children of 4075 widows, who were supported by the fund (Cramér 1955).

Number of children	0	1	2	3	4	5	6
Number of widows	3062	587	284	103	33	4	2

(a) Fit a simple Poisson model and indicate why it fails.

(b) Fit a negative binomial model to the data, and check its adequacy using the χ^2 goodness-of-fit test.

(c) Consider a mixture model where there is a proportion θ of widows without children, and another proportion $(1-\theta)$ of widows with x children, where x is Poisson with mean λ. Show that the marginal distribution of the number of children X follows

$$P(X=0) = \theta + (1-\theta)e^{-\lambda}$$

and, for $k > 0$,

$$P(X=k) = (1-\theta)e^{-\lambda}\frac{\lambda^k}{k!}.$$

This model is called the 'zero-inflated' Poisson model.

(d) Fit the model to the data, and compare its adequacy with the negative binomial model.

Exercise 4.18: Refer to the data in Example 4.7.

(a) Check graphically the two candidate models, the Poisson and negative binomial models, using the technique described in Exercise 4.15.

(b) Verify the fit of the Poisson and negative binomial models to the data.

(c) Compare the likelihood of the mean parameter under the two models. (It is a profile likelihood under the negative binomial model.) Comment on the results.

Exercise 4.19: Repeat the previous exercise for the dataset in Exercise 4.13.

Exercise 4.20: The following table (from Evans 1953) shows the distribution of a plant species *Glaux maritima* (from the primrose family) in 500 contiguous areas of 20cm squares.

count k	0	1	2	3	4	5	6	7
Number of squares n_k	1	15	27	42	77	77	89	57

count k	8	9	10	11	12	13	14
Number of squares n_k	48	24	14	16	9	3	1

Compare Poisson and negative binomial fits of the data. Interpret the result in terms of clustering of the species: do the plants tend to cluster?

Exercise 4.21: Verify the profile likelihood in Section 4.7 for the parameter of interest $\theta = \theta_a/\theta_p$.

Exercise 4.22: Assuming that X and Y are Poisson with rate λ_x and λ_y, show that the conditional distribution of X given $X+Y$ is binomial with parameters $n = X+Y$ and $\pi = \lambda_x/(\lambda_x + \lambda_y)$.

Exercise 4.23: Prove Theorem 4.1.

Exercise 4.24: Verify that Michelson's first experiment is out of line with the rest. If you know the analysis of variance technique, use it to analyse all experiments together, and then repeat the analysis after excluding the first experiment. Otherwise, use a series of two-sample t-tests.

Exercise 4.25: Suppose $U_1, \ldots, U_n$ are an iid sample from the standard uniform distribution, and let $U_{(1)}, \ldots, U_{(n)}$ be the order statistics. Investigate the approximation

$$\text{median}\{U_{(i)}\} \approx \frac{i - 1/3}{n + 1/3}$$

for $n = 5$ and $n = 10$. (Hint: use the distribution of the order statistics in Exercise 2.4.)

Exercise 4.26: Suppose $x_1, \ldots, x_m$ are an iid sample from $N(\mu, \sigma_x^2)$, and $y_1, \ldots, y_n$ from $N(\mu+\delta, \sigma_y^2)$. Define Michelson's first experiment as the x-sample, and the subsequent experiment as the y-sample.

(a) Compute the profile likelihood for δ assuming that $\sigma_x^2 = \sigma_y^2$.

(b) Compute the profile likelihood for δ without further assumption.

(c) Compute the profile likelihood for σ_x^2/σ_y^2.

Exercise 4.27: The following is the average amount of rainfall (in mm/hour) per storm in a series of storms in Valencia, southwest Ireland. Data from two months are reported below.

```
January 1940
      0.15 0.25 0.10 0.20 1.85 1.97 0.80 0.20 0.10 0.50 0.82 0.40
      1.80 0.20 1.12 1.83 0.45 3.17 0.89 0.31 0.59 0.10 0.10 0.90
      0.10 0.25 0.10 0.90

July 1940
      0.30 0.22 0.10 0.12 0.20 0.10 0.10 0.10 0.10 0.10 0.10 0.17
      0.20 2.80 0.85 0.10 0.10 1.23 0.45 0.30 0.20 1.20 0.10 0.15
      0.10 0.20 0.10 0.20 0.35 0.62 0.20 1.22 0.30 0.80 0.15 1.53
      0.10 0.20 0.30 0.40 0.23 0.20 0.10 0.10 0.60 0.20 0.50 0.15
      0.60 0.30 0.80 1.10 0.20 0.10 0.10 0.10 0.42 0.85 1.60 0.10
      0.25 0.10 0.20 0.10
```

(a) Compare the summary statistics for the two months.

(b) Look at the QQ-plot of the data and, based on the shape, suggest what model is reasonable.

(c) Fit a gamma model to the data from each month. Report the MLEs and standard errors, and draw the profile likelihoods for the mean parameters. Compare the parameters from the two months.

(d) Check the adequacy of the gamma model using a gamma QQ-plot.

Exercise 4.28: In a p-parameter exponential family (4.2) show that the natural statistics $T_1(x), \ldots, T_p(x)$ are minimal sufficient. If the family is full rank, explain why these statistics are the MLE of their own mean vector, which is a one-to-one function of θ, so the MLE of θ is sufficient.

Exercise 4.29: Let $x_1, \ldots, x_n$ be an iid sample from uniform$(0, \theta)$. The maximum order statistic $x_{(n)}$ is a sufficient estimate of θ, yet the uniform is not in the exponential family. This does not conform to the basic theorem about minimal sufficiency. What condition is not satisfied?

Exercise 4.30: In Section 4.9 verify the general mean and variance formulas as derivatives of $A(\theta)$.

Exercise 4.31: Identify the exponential and gamma models as exponential family models; in particular, find the canonical and dispersion parameters, and the function $c(x, \phi)$.

Exercise 4.32: Identify the inverse Gaussian model as an exponential family model. The density is

$$f(y) = \left(\frac{\lambda}{2\pi y^3}\right)^{1/2} \exp\left\{-\frac{\lambda}{2\mu^2}\frac{(y-\mu)^2}{y}\right\}, \qquad y > 0.$$

Verify that it has mean μ and variance μ^3/λ.

Exercise 4.33: Perform the Monte Carlo test described in Example 4.14 and verify the results.

Exercise 4.34: Compare the likelihood analysis of the mean parameter for the dataset in Exercise 4.18 using the negative binomial model and the general exponential family model. Discuss the advantages and disadvantages of each model.

Exercise 4.35: Compare the likelihood analysis of the mean parameter for the dataset in Exercise 4.4 using the beta–binomial model and the general exponential family model.

Exercise 4.36: Suppose x is in the exponential family with log-density

$$\log p_\mu(x) = x\theta - A(\theta) + c(x),$$

where $\mu = A'(\theta) = EX$ is used as the index; the variance is $v(\mu) = A''(\theta)$. The double-exponential family (Efron 1986b) is defined as

$$g(x) = b(\theta, \alpha)\alpha^{1/2}\{p_\mu(x)\}^\alpha\{p_x(x)\}^{1-\alpha}.$$

Efron shows that the normalizing constant $b(\theta, \alpha) \approx 1$, and the mean and variance are approximately μ and $v(\mu)/\alpha$.

(a) Starting with a standard Poisson model with mean μ, find the density of a double Poisson model.

(b) Show that the likelihood from part (a) is approximately the same as the extended quasi-likelihood formula in Example 4.15. Hint: use Stirling's approximation for the factorial:

$$n! = \sqrt{2\pi}e^{-n}n^{n+1/2}. \tag{4.6}$$

Exercise 4.37: Starting with a standard binomial model, find the density function of a double binomial model.

Exercise 4.38: The following dataset is the sulphur dioxide (SO_2) content of air (in 10^{-6} g/m^3) in 41 US cities averaged over 3 years:

```
10  13  12  17  56  36  29  14  10  24 110  28  17
 8  30   9  47  35  29  14  56  14  11  46  11  23
65  26  69  61  94  10  18   9  10  28  31  26  29
31  16
```

Use the Box–Cox transformation family to find which transform would be sensible to analyse the data. Plot the profile likelihood of the mean SO_2 content. Check the normal plot of the data before and after the transformation.

Exercise 4.39: The following data are the average adult weights (in kg) of 28 species of animals.

```
  0.4 1.0 1.9 3.0 5.5  8.1 12.1 25.6     50.0  56.0 70.0  115.0
115.0    119.5    154.5  157.0    175.0    179.0     180.0   406.0
419.0    423.0    440.0  655.0    680.0   1320.0    4603.0  5712.0
```

Use the Box–Cox transformation family to find which transform would be sensible to analyse or present the data.

Exercise 4.40: In a study of the physiological effect of stress, Miralles *et al.* (1983) measured the level of beta endorphine in patients undergoing surgery. (Beta endorphine is a morphine-like chemical with narcotic effects found in the brain.) The measurements were taken at 12–14 hours and at 10 minutes before surgery.

Patient	12–14 h	10 min	Patient	12–14 h	10 min
1	10.0	6.5	11	4.7	25.0
2	6.5	14.0	12	8.0	12.0
3	8.0	13.5	13	7.0	52.0
4	12.0	18.0	14	17.0	20.0
5	5.0	14.5	15	8.8	16.0
6	11.5	9.0	16	17.0	15.0
7	5.0	18.0	17	15.0	11.5
8	3.5	42.0	18	4.4	2.5
9	7.5	7.5	19	2.0	2.0
10	5.8	6.0			

(a) Draw the QQ-plot of the data at 12–14 hours and 10 minutes prior to surgery, and comment on the normality.

(b) Draw the QQ-plot of the change in the beta endorphine level. To get a measure of change note that the beta endorphine level is a positive variate, so it may be more meaningful to use a ratio or to consider a log-transformation. Examine the need to add a shift parameter.

(c) Compare the likelihood of the location parameter assuming normal and Cauchy models for the three variables above.

Exercise 4.41: Consider the Box–Cox tranformation family for the data from Michelson's fifth experiment:

```
890  840  780  810  760  810  790  810  820  850
870  870  810  740  810  940  950  800  810  870
```

(a) Plot the profile likelihood of λ.

(b) Plot the profile likelihood of μ assuming $\lambda = 1$.

(c) Compare the plot in (b) with the profile likelihood of μ by profiling over λ and σ^2. Comment on the result.

Exercise 4.42: Suppose $x_1, \ldots, x_n$ are an iid sample from the Laplacian distribution with parameters μ and σ. Find the MLE of the parameters and find the general formula for the profile likelihood of μ.

Exercise 4.43: Use the t-family model for a range of degrees of freedom ν for the datasets in Example 4.17; in particular plot the profile likelihood over ν.

Exercise 4.44: Repeat the likelihood analysis given for the datasets in Example 4.17 using logistic and Laplacian models, and compare the results.

5

Frequentist properties

Frequentist or repeated sampling properties of sample statistics form a basis of probability-based inference. These properties also indicate a potential objective verification of our statistical procedures. In this chapter we will expand our discussion on important frequentist properties such as bias and variance of point estimates, calibration of likelihood using P-values, and coverage probability of CIs. We will introduce a powerful computational method called the bootstrap, and cover the so-called 'exact' confidence procedures for specific models.

5.1 Bias of point estimates

Definition 5.1 *Suppose $T(X)$ is an estimate of θ.* Bias *and* mean square error *of T are defined as*

$$\begin{aligned} b(\theta) &= E_\theta T - \theta \\ MSE(\theta) &= E_\theta (T-\theta)^2 = var_\theta(T) + b^2(\theta). \end{aligned}$$

We say T is unbiased *for θ if $E_\theta T = \theta$.*

The subscript θ in E_θ means that the expected value is taken with respect to the probability model $p_\theta(x)$. Why do we want a small (or no) bias in our estimates?

- It gives a sense of objectivity, especially if $\text{var}(T)$ is small. This is closely related to the idea of *consistency* of point estimates: in large samples our estimate should be close to the true parameter. Generally, it is sensible to require that bias does not dominate variability. Or, the bias and variability components in the MSE should be balanced.
- If we are pooling information from many relatively small samples, unbiasedness of the estimate from each sample is vital to avoid an accumulation of bias (the pooled dataset is said to be highly stratified, discussed in detail in Section 10.1). This is also true if point estimates are used for further modelling. For example, in repeated measures experiments we sometimes need to simplify the data by taking summary statistics from each subject.

The above reasons do not point to the need for an *exact unbiasedness*, since it is not always practical. For example, suppose $x_1, \ldots, x_n$ are an iid sample from $N(\mu, \sigma^2)$ with both parameters unknown. Do we really use the sample variance

$$s^2 = \frac{1}{n-1}\sum_i (x_i - \overline{x})^2$$

because it is unbiased? If we do, why do we not worry that the sample standard deviation is *biased*? From the fact that $(n-1)s^2/\sigma^2$ is χ^2_{n-1}, we can show that

$$E(s) = \frac{\Gamma(n/2)\sqrt{2}}{\Gamma((n-1)/2)\sqrt{n-1}}\sigma,$$

where $\Gamma(\cdot)$ is the gamma function. We can correct the bias, but in practice we almost never do so. (The likelihood-based reason to use the $(n-1)$-divisor for the variance is given in Chapter 10.)

As discussed by Hald (1999), Fisher had two aversions: (arguments based on) unbiasedness, and lack of invariance. Nonetheless, a lot of classical estimation theory is built around unbiasedness. Here are further examples showing the problem with the exact unbiasedness requirement.

Producing an unbiased estimate is never automatic

Suppose $x_1, \ldots, x_n$ are an iid sample from $N(\mu, \sigma^2)$, and we want to estimate the threshold probability $\theta = P(X_1 > 2)$. It is not obvious how to obtain an unbiased estimate of θ (see Lehmann 1983, pages 86–87, for an answer). In contrast, the MLE is immediate:

$$\widehat{\theta} = P\left(Z > \frac{2-\overline{x}}{\widehat{\sigma}}\right),$$

where Z is the standard normal variate, and $\overline{x}$ and $\widehat{\sigma}$ are the sample mean and standard deviation of the data.

In general, except for linear $g(T)$, if T is unbiased for θ then $g(T)$ is a biased estimate of $g(\theta)$; this is a lack of invariance with regards to the choice of parameterization.

Not all parameters have an unbiased estimator

In complex problems it is not always clear whether there even exists an unbiased estimator for a parameter of interest. For example, let $X \sim$ binomial(n, π); there is no unbiased estimate for the odds $\theta = \pi/(1-\pi)$. To show this, consider any statistic $T(X)$. We have

$$\begin{aligned} ET(X) &= \sum_{x=0}^{n} t(x)\binom{n}{x}\pi^x(1-\pi)^{n-x} \\ &= \sum_{x=0}^{n} c(x)\pi^x \end{aligned}$$

for some function $c(x)$. But, while ET is a polynomial of maximum degree n,

$$\theta = \frac{\pi}{1-\pi} = \sum_{k=1}^{\infty} \pi^k$$

is an infinite-degree polynomial. So, there is no $T(X)$ such that $ET(X) = \theta$. This does not mean that there is no reasonable inference for θ. If $\widehat{\pi}$ is a reasonable estimate for π, then the MLE $\widehat{\theta} = \widehat{\pi}/(1-\widehat{\pi})$, while biased, is also reasonable.

The unbiasedness requirement can produce terrible estimates

Let X be a sample from a Poisson distribution with mean μ. Suppose the parameter of interest is $\theta = e^{-a\mu}$. If $T(X)$ is unbiased for θ then

$$ET = \sum_{x=0}^{\infty} t(x) e^{-\mu} \frac{\mu^x}{x!} = e^{-a\mu},$$

or

$$\begin{aligned} \sum_{x=0}^{\infty} t(x) \frac{\mu^x}{x!} &= e^{(1-a)\mu} \\ &= \sum_{x=0}^{\infty} (1-a)^x \frac{\mu^x}{x!} \end{aligned}$$

so $t(x) = (1-a)^x$; see also Lehmann (1983, page 114). If $a = 1$ then $\theta = P(X = 0)$ is estimated by $t(x) = I(x = 0)$, which is equal to one if $x = 0$, and zero otherwise. Even worse: if $a = 2$ then $\theta = e^{-2\mu}$ is estimated by $t(x) = (-1)^x$, which is equal to 1 if x is even, and -1 if x is odd.

5.2 Estimating and reducing bias

The bias of the MLE of σ^2 based on a random sample $x_1, \ldots, x_n$ from $N(\mu, \sigma^2)$ is

$$-\frac{\sigma^2}{n}.$$

Generally, in regular estimation problems the bias of an estimate of θ is usually of the form

$$\frac{b_1}{n} + \frac{b_1}{n^2} + \frac{b_3}{n^3} + \cdots,$$

where b_k's are functions of θ only, but not of sample size n. The standard deviation of an estimate is typically of the form

$$\frac{s_1}{n^{1/2}} + \frac{s_2}{n} + \frac{s_3}{n^{3/2}} + \cdots.$$

Therefore bias is usually smaller than the stochastic uncertainty in the estimate. It may not be worth all that much to try to correct for bias in this general case. Correction is important in less regular cases, where bias is known to be large relative to the variance such as when we pool information over many small samples or when the data are highly stratified.

When an exact theoretical result is not available or too complicated, there are three general, but approximate, methods to remove or reduce the bias of a point estimate:

1. Taylor series method.
2. Jackknife or leave-one-out cross-validation method.
3. Bootstrap method.

Taylor series method for functions of the mean

This is closely related to the Delta method in Section 4.7. The proof of the following result is left as an exercise.

Theorem 5.1 *Suppose we estimate $h(\mu)$ by $h(\overline{x})$. If the second derivative $h''(\cdot)$ is continuous around μ, then*

$$Eh(\overline{X}) \approx h(\mu) + \frac{1}{2}h''(\mu)var(\overline{X}),$$

with the last term providing an approximate bias.

Example 5.1: The following is the number of accidents recorded on a stretch of highway for the past 10 months:

1 1 0 0 2 4 1 2 0 3

Assume that the accidents are an iid sample from a Poisson distribution with mean μ. Let the parameter of interest be

$$\begin{aligned} \theta &= P(\text{no accident}) \\ &= P(X=0) = e^{-\mu}. \end{aligned}$$

From the data we get $\overline{x} = 1.4$, so the MLE of θ is $\widehat{\theta} = e^{-1.4} = 0.25$. To obtain a bias-corrected estimate, let $h(\mu) = e^{-\mu}$, so $h''(\mu) = e^{-\mu}$. Hence the bias is

$$b(\mu) \approx \frac{1}{2}e^{-\mu}\frac{\mu}{10}$$

and the bias-corrected estimate is

$$\widehat{\theta} = e^{-\overline{x}} - \frac{1}{2}e^{-\overline{x}}\frac{\overline{x}}{10} = 0.23.$$

Using the Delta method (Section 4.7) the standard error of $\widehat{\theta}$ is

$$e^{-\overline{x}}\sqrt{\overline{x}/10} = 0.09,$$

which is much larger than the bias. □

Jackknife or cross-validation method

Suppose we want to predict an outcome y, where a large number of predictors $x_1, \ldots, x_K$ are available, and we want to choose an optimal subset of the predictors. We cannot use as our criterion the usual residual variance from each possible regression model, since it is a biased estimate. In particular, a larger model will have a smaller residual variance, even though it is not a better model. Consider instead the following commonly used validation technique: split the data into two, and

- use half of the dataset (called the training set) to develop the competing regression models
- use the other half of the data (the validation set) to estimate the prediction error variance of each model. This variance can then be used for model selection or comparison.

The jackknife or leave-one-out cross-validation method is an extreme version of the above technique; a lot of computation is typically required, so if there are enough subjects the above method is much more practical. For clarity we will continue to use the error variance estimation to describe the general methodology:

- use $(n-1)$ units/subjects to develop the regression models
- use the one left-out for validation; i.e. compute the prediction error using each of the models
- (this is the laborious part) cycle around every single unit of the data as the one left-out.

In the end, we can compute for each competing model an unbiased estimate of its prediction error variance, and the best model can be chosen as the minimizer of the crossvalidated error variance. Simpler methods, requiring much less computations, are available for regression model selection.

For bias estimation the general prescription is as follows. Suppose we have data $x_1, \ldots, x_n$, and an estimate T for a parameter θ. The leave-x_i-out data is

$$x_{-i} \equiv (x_1, \ldots, x_{i-1}, x_{i+1}, \ldots, x_n).$$

Let T_i be the 'jackknife replicate' of T based on x_{-i}; then its bias is estimated by

$$\widehat{b} = (n-1)(\overline{T} - T),$$

where $\overline{T} = \frac{1}{n}\sum T_i$, and the bias-corrected estimate is $T - \widehat{b}$. Exercise 5.5 shows that this correction procedure removes bias of order $1/n$.

Example 5.2: Let $T = \frac{1}{n}\sum(x_i - \overline{x})^2$ be an estimate of the variance σ^2. The jackknife replicate is

$$T_i = \frac{1}{n-1}\sum_{j \neq i}(x_j - \overline{x}_i)^2$$

where

$$\overline{x}_i = \frac{1}{n-1}\sum_{j\neq i} x_j.$$

Some algebra will show that

$$\widehat{b} = -\frac{1}{n}\left\{\frac{1}{n-1}\sum (x_i - \overline{x})^2\right\},$$

so that the bias-corrected estimate is

$$T - \widehat{b} = \frac{1}{n-1}\sum (x_i - \overline{x})^2,$$

which is the usual sample variance formula. □

Bootstrap method

The most general way of estimating bias is provided by the bootstrap method. It was first presented by Efron in 1977 Rietz lecture as an alternative to the jackknife; a general exposition is given by Efron and Tibshirani (1993). In its evolution the bootstrap becomes the main frequentist tool to deal with complicated sampling distributions. In practice there is almost no analytical work required by the method, but, in exchange, we must perform a lot of computations.

Underlying the bootstrap is the plug-in principle: replace the unknown distribution F in a quantity of interest by an estimate $\widehat{F}$. Thus, for example, the bias of T as an estimate of θ is

$$b(\theta) = E_F T - \theta,$$

where $E_F T$ means that the expectation is taken with respect to the distribution F. Intuitively, using the plug-in principle, an estimate of the bias is

$$\widehat{b} = E_{\widehat{F}} T - \widehat{\theta}.$$

The estimate $\widehat{F}$ may be parametric or nonparametric. In almost all bootstrap applications the evaluation of expected values with respect to $\widehat{F}$ is done using Monte Carlo simulations. In the following example the method is parametric, as $\widehat{F}$ is a member of the parametric family. In nonparametric bootstrap applications we use the empirical distribution function (EDF). Sampling from the EDF is equivalent to sampling with replacement from the sample itself, hence the name 'bootstrap'.

Example 5.3: In the Poisson example above $\overline{x} = 1.4$, so $\widehat{F}$ is Poisson with mean 1.4. The estimated bias of $T = e^{-\overline{x}}$ is

$$\widehat{b} = E_{\widehat{F}} e^{-\overline{X}} - e^{-\overline{x}}.$$

In this case the term $E_{\widehat{F}} e^{-\overline{X}}$ can be computed analytically. Since X_i is Poisson with mean μ we have

$$Ee^{tX_i} = \exp(-\mu + \mu e^t),$$

and

$$E_F e^{-\overline{X}} = \exp(-n\mu + n\mu e^{-1/n}).$$

Using $\widehat{\mu} = \overline{x} = 1.4$ we obtain

$$E_{\widehat{F}} e^{-\overline{X}} = \exp(-14 + 14e^{-1/10}) = 0.264.$$

To illustrate a general scheme of bootstrap simulation, without any analytical derivation this quantity can be computed as follows:

- Generate $x_1^*, \ldots, x_{10}^*$ iid from Poisson(mean=1.4). These are called the bootstrap sample.
- Compute $\overline{x}^*$ and $e^{-\overline{x}^*}$ from the bootstrap sample.
- Repeat a large number of times B, and take the average of $e^{-\overline{x}^*}$. The collection of $e^{-\overline{x}^*}$ is an approximate random sample from the distribution of $e^{-\overline{X}}$, and forms the bootstrap distribution of $e^{-\overline{X}}$.

Figure 5.1 shows the histogram of $e^{-\overline{x}^*}$ using $B = 1000$. The average $e^{-\overline{x}^*}$ is 0.263, close to the theoretically derived value above, and the estimated bias is

$$\begin{aligned} \widehat{b} &= \text{average}(e^{-\overline{x}^*}) - e^{-1.4} \\ &= 0.263 - 0.246 = 0.017, \end{aligned}$$

so the bias-corrected estimate is $e^{-1.4} - 0.017 = 0.23$, the same as the estimate found using the Taylor series method. □

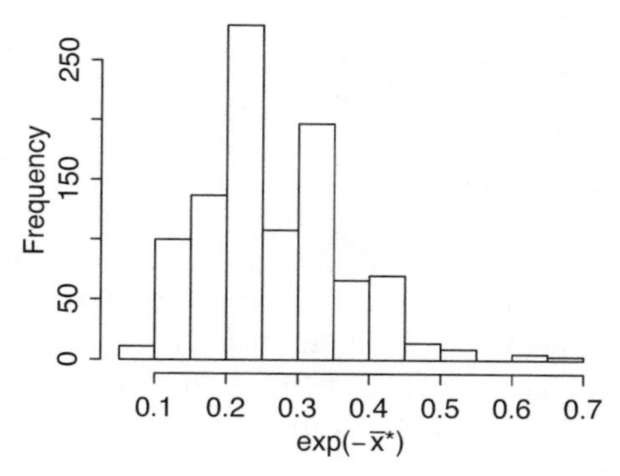

Figure 5.1: *Bootstrap distribution of* $e^{-\overline{X}}$.

5.3 Variability of point estimates

Compare the following statements:

1. 'Aspirin reduces heart attacks by 42%.'
2. 'We are 95% confident that the risk of heart attacks in the aspirin group is between 47% to 71% of the risk in the placebo group.'

When a parameter has some physical meaning it is usually desirable to report its estimate. While appealing, the first statement does not convey the sense of uncertainty found in the second statement. The uncertainty is embodied by statistical variability, so the deficiency is usually remedied by also reporting the standard error (or 'margin of error') of the point estimate: the relative risk is 58% (standard error= 6%).

More generally, in classical (frequentist) statistics, probability statements made regarding the uncertainty of a point estimate are based on the sampling distribution. This is the distribution of the estimate as it varies from sample to sample. Except in some specific cases deriving exact sampling distributions is notoriously difficult; in practice we often rely on the normal approximation. We will refer to *regular cases* as those where the normal approximation is reasonable. (This is a parallel sampling-based argument to representing the likelihood with the MLE and its curvature.) In such cases we only need to compute the estimate and its variance; inference follows from these two quantities alone. For example, an approximate 95% CI is

$$\widehat{\theta} \pm 1.96\text{se}(\widehat{\theta}),$$

where the standard error $\text{se}(\widehat{\theta})$ is the estimated standard deviation of $\widehat{\theta}$. While $\widehat{\theta}$ does not have to be an MLE, we will still refer to this CI as a Wald(-type) CI.

Estimating variance

The general methods for estimating the variance of an estimate are similar to those for estimating bias. The most important classical method is the Delta method, used in the analysis of the aspirin data in Section 4.7. The method is based on the Taylor series expansion and generally works for estimates which are functions of the sample mean.

The jackknife procedure to estimate the variance of a statistic T is given, for example, in Miller (1964). Practical statistics seems to have ignored this method. The problem is that we need the procedure in complicated cases where we cannot use the Delta method, but in such cases there is no guarantee that the procedure works. See Efron (1979) for further discussion.

The most general method is again provided by the bootstrap method. We are interested in

$$v = \text{var}_F(T).$$

Using the plug-in principle we replace the unknown distribution F by its estimate $\widehat{F}$, so the variance estimate is

$$\widehat{v} = \text{var}_{\widehat{F}}(T).$$

The variance evaluation is usually done using Monte Carlo sampling from $\widehat{F}$, though in some cases it might be possible to do it analytically.

Example 5.4: Continuing the analytical result in Example 5.3,

$$\begin{aligned}\text{var}_F(e^{-\overline{X}}) &= E_F e^{-2\overline{X}} - \left(E_F e^{-\overline{X}}\right)^2 \\ &= \exp(-n\mu + n\mu e^{-2/n}) - \exp(-2n\mu + 2n\mu e^{-1/n}).\end{aligned}$$

Using $n = 10$ and $\overline{x} = 1.4$, the variance is estimated by

$$\begin{aligned}\text{var}_{\widehat{F}}(e^{-\overline{X}}) &= \exp(-n\overline{x} + n\overline{x}e^{-2/n}) - \exp(-2n\overline{x} + 2n\overline{x}e^{-1/n}) \\ &= 0.079 - 0.264^2 = 0.0093\end{aligned}$$

It is intuitive that the bootstrap replications of $e^{-\overline{x}^*}$ contain much more information than bias alone. Such replications are an approximate random sample from the distribution of $e^{-\overline{X}}$. We can simply compute the sample variance as an estimate of the population variance.

From the same 1000 bootstrap replications in Example 5.3, the estimated variance of $e^{-\overline{X}}$ is

$$\text{var}(e^{-\overline{X}^*}) = 0.0091,$$

close to the analytically derived estimate. For comparison, using the Delta method, the estimated variance is

$$\widehat{\text{var}}(e^{-\overline{X}}) = (e^{-\overline{x}})^2 \frac{\overline{x}}{n} = 0.0085. \ \square$$

Example 5.5: The bootstrap method is indispensible for more complicated statistics. Consider the IQ data in Exercise 3.23, giving paired data (x_i, y_i) for $i = 1, \ldots, 16$. The sample correlation between the verbal (x) and mathematical thinking (y) is $\widehat{\rho} = 0.83$. Without making further assumptions on the bivariate distribution of (x, y), we can perform a nonparametric bootstrap by:

- taking a sample of size $n = 16$ *with replacement* from the paired data
- computing $\widehat{\rho}^*$ from the new dataset
- repeating this B times.

Figure 5.2 shows the plot of the data and the histogram of the bootstrap replicates $\widehat{\rho}^*$ using $B = 500$. The estimated variance of $\widehat{\rho}$ is 0.00397, so the standard error of $\widehat{\rho}$ is 0.063. $\square$

To end with a note of caution, in practice the bootstrap method is used in complex situations where it may not be common knowledge whether asymptotic normality holds. Some care is required if the observed bootstrap distribution is decidedly nonnormal. In Sections 5.6 and 15.3 we will discuss some bootstrap-based inference that uses the whole bootstrap distribution rather than just the variance information.

5.4 Likelihood and P-value

P-value is the most common measure of evidence, although it is more correct to say that P-value measures the 'extremeness' or 'unusualness' (*not* the probability or likelihood) of the observed data given a null hypothesis. The null hypothesis is doubtful if it is associated with a small P-value.

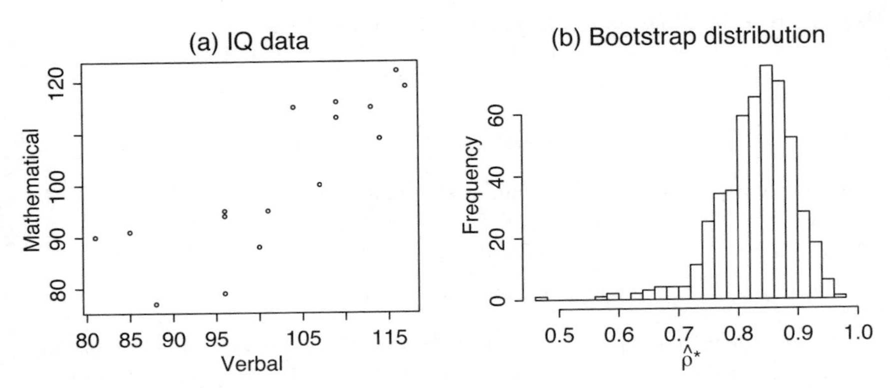

Figure 5.2: *IQ data and the bootstrap distribution of the sample correlation.*

While we can construct examples where P-value is meaningless (e.g. Royall 1997, Chapter 3), in many cases of practical interest it does measure evidence, since there is a close relationship between likelihood and P-value. Throughout, when we say 'likelihood' we mean 'normalized likelihood'. However, since P-value depends on the sample and the parameter spaces, the relationship is not the same over different experiments.

Even though P-value is the most rudimentary form of inference, it has some attractive properties:

- Only the 'null model' is needed to compute a P-value, but producing a likelihood requires a model specification over the whole parameter space. There are practical situations (e.g. Example 4.14) where a statistical analysis can readily produce a P-value, but not a likelihood.
- There are simple adjustments of P-value to account for multiple testing.
- P-value provides a way of calibrating the likelihood, especially for high-dimensional problems (Section 3.5).

The weakness of P-value is apparent, even in one-parameter problems, when the sampling distribution of the test statistic is not symmetric. While the one-sided P-value is obvious and essentially unique (at least for continuous distributions), there are many ways of defining a two-sided P-value, none of which is completely satisfactory (Example 5.7).

For testing simple-versus-simple hypotheses, namely H_0: $\theta = \theta_0$ versus H_1: $\theta = \theta_1$, an observed likelihood ratio r (think of a small $r < 1$) satisfies

$$P_{\theta_0}\left\{\frac{L(\theta_0)}{L(\theta_1)} \le r\right\} \le r. \tag{5.1}$$

If we treat the probability as a P-value, we have a simple result that the *P-value is always smaller than or equal to the likelihood ratio.* This is not true for general alternatives, although it is true for many one-parameter models. To prove the relationship

$$\begin{aligned} P_{\theta_0}\left\{\frac{L(\theta_0)}{L(\theta_1)} \le r\right\} &= \int_{[L(\theta_0)\le rL(\theta_1)]} p_{\theta_0}(x)\,dx \\ &\le \int_{[L(\theta_0)\le rL(\theta_1)]} rL(\theta_1)\,dx \\ &= \int_{[L(\theta_0)\le rL(\theta_1)]} rp_{\theta_1}(x)\,dx \\ &\le r\int p_{\theta_1}(x)\,dx \\ &= r. \end{aligned}$$

The next example, first discussed by Edwards *et al.* (1963), is more typical.

Example 5.6: Suppose $x_1, \ldots, x_n$ are an iid sample from $N(\mu, \sigma^2)$, with σ^2 known and where we are interested to test H_0: $\mu = \mu_0$. The likelihood of μ is

$$L(\mu) = \text{constant} \times \exp\left\{-\frac{n}{2\sigma^2}(\overline{x} - \mu)^2\right\}.$$

Since the MLE of μ is $\overline{x}$, the normalized likelihood of H_0 is

$$\frac{L(\mu_0)}{L(\overline{x})} = e^{-z^2/2},$$

where

$$z = \frac{\overline{x} - \mu_0}{\sigma/\sqrt{n}}$$

is the usual Wald or z-statistic. The two-sided P-value associated with an observed z is

$$P(|Z| > |z|),$$

where Z has the standard normal distribution.

Figure 5.3(a) shows the plot of P-value and likelihood as a function of z. Figure 5.3(b) shows P-value as a function of likelihood. For example, a P-value of 5% is equivalent to a 15% likelihood. □

Since in the normal model there is no complication in the definition of P-value, we may view the relationship between likelihood and P-value here as an ideal relationship. This will help us decide which two-sided P-value is best in the following Poisson example.

Example 5.7: Suppose x ($x > 1$) is a single observation from the Poisson distribution with mean θ. We want to test H_0: $\theta = 1$. The one-sided P-value is

$$P_{\theta=1}(X \ge x) = \sum_{k \ge x} \frac{e^{-1}}{k!}.$$

The asymmetric sampling distribution, however, presents a problem in defining a two-sided P-value. Fisher once recommended simply doubling the one-sided

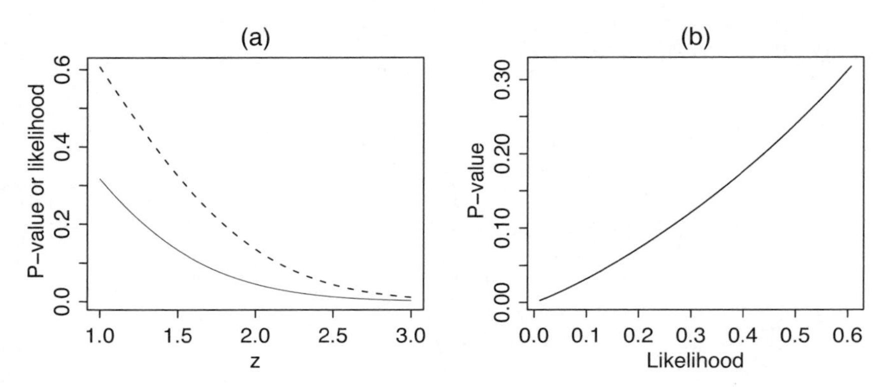

Figure 5.3: *(a) Plot of P-value (solid) and likelihood (dashed) as a function of z. (b) Relationship between likelihood and P-value.*

P-value, a proposal commonly used in practice, but it is not without its defects. The (normalized) likelihood of H_0 is

$$\frac{L(1)}{L(x)} = e^{x-1}x^{-x}.$$

Figure 5.4(a) shows that the P-value is still smaller than the likelihood. The solid line in Figure 5.4(b) shows that, as a function of likelihood, the P-value is higher than in the ideal normal case (dashed line). Hence doubling the P-value seems to produce a bigger P-value than necessary.

As an alternative, some authors, for example Lancaster (1961) or Agresti (1996), suggested a one-sided *mid-P-value* defined as

$$\frac{1}{2}P(X = x) + P(X > x).$$

The two-sided version is simply double this amount. Figure 5.4(b) shows that this definition matches very closely the likelihood–P-value relationship obtained in the normal case. This result holds across a very wide range of null parameter values.

Note, however, that these definitions of two-sided P-value are not satisfactory for $x = 0$. To test the same H_0: $\theta = 1$, the one-sided P-value is

$$P(X = 0) = e^{-1} = 0.37,$$

exactly the same as the likelihood of H_0. □

5.5 CI and coverage probability

We have used the idea of confidence procedure by appealing to the simple normal model: if $x_1, \ldots, x_n$ are an iid sample from $N(\theta, \sigma^2)$ with known σ^2, then the 95% CI for θ is

$$\overline{x} - 1.96\ \sigma/\sqrt{n} < \theta < \overline{x} + 1.96\ \sigma/\sqrt{n}.$$

In general, the $100(1-\alpha)\%$ CI for θ is a random interval $L < \theta < U$ satisfying, for each fixed θ,

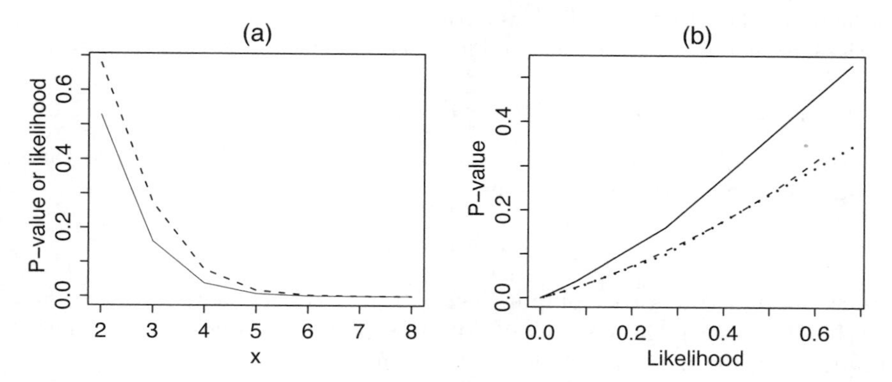

Figure 5.4: *(a) Plot of two-sided P-value (solid) and likelihood (dashed) as a function of x to test H_0: $\theta = 1$ for the Poisson model. (b) Relationship between likelihood and two-sided P-value using doubling method (solid), for the normal model (dashed) and using mid-P-value method (dotted).*

$$P_\theta(L < \theta < U) = 1 - \alpha.$$

The probability $P_\theta(L < \theta < U)$ as a function of θ is called the *coverage probability* of the interval. It is the probability of a correct guess of where the unknown parameter is.

For a simple normal mean model, the coverage probability of a 95% CI is exactly 95%. For nonnormal models the coverage probability rarely matches the advertised (claimed) confidence level. For some specific models, such as Poisson or binomial models, it is sometimes possible to guarantee a minimum coverage; for example, a $100(1-\alpha)\%$ CI for θ satisfies

$$P_\theta(L < \theta < U) \geq 1 - \alpha.$$

In more complex models we can only get an approximate coverage with no guaranteed minimum.

A general procedure to construct CIs can be motivated based on a close connection between CIs and hypothesis testing. Given a test procedure, a $100(1-\alpha)\%$ CI is the set of null hypotheses that are not rejected at level α. Specifically, if we compute a two-sided P-value to test H_0: $\theta = \theta_0$, the $100(1-\alpha)\%$ CI is the set

$$\{\theta_0, \text{ P-value} \geq \alpha\}.$$

In particular, the lower and upper limits L and U are the parameter values with P-value equals α. We say we 'invert' the test into a CI; it is intuitive that a 'good' test procedure will generate a 'good' CI.

Depending on the shape of the likelihood function, likelihood-based intervals are naturally either one sided or two sided; CIs do not have such restriction. We can obtain one-sided CIs using the connection with one-sided tests. Suppose T is a sensible estimate of θ, and t is the observed value of T. The $100(1-\alpha)\%$ lower confidence bound L for θ is the value of the parameter associated with a one-sided (right-side) P-value equals α, i.e.

$$P_L(T \geq t) = \alpha. \qquad (5.2)$$

The $100(1-\alpha)\%$ upper confidence bound U can be defined similarly as the solution of the left-side P-value equation

$$P_U(T \leq t) = \alpha.$$

Figure 5.5 illustrates the computation of L and U. The curve $t_{1-\alpha}(\theta)$ is the $100(1-\alpha)$ percentile of the distribution of T as a function of θ. (The interval $L < \theta < U$ is the $100(1-2\alpha)\%$ CI for θ).

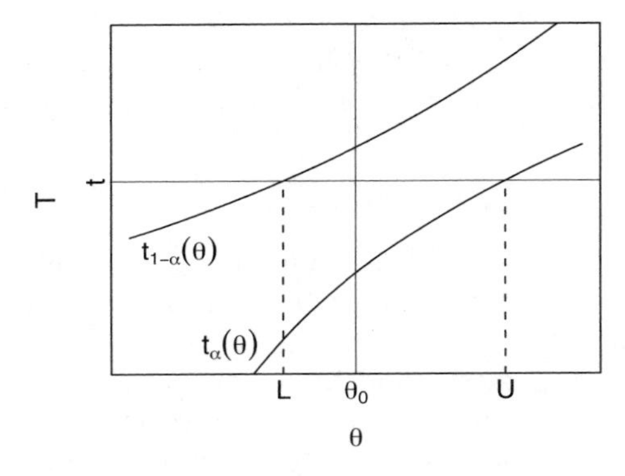

Figure 5.5: *Finding the lower and upper confidence bounds from $T = t$.*

Example 5.8: Suppose $x_1, \ldots, x_n$ are an iid sample from $N(\mu, \sigma^2)$, with σ^2 known. The $100(1-\alpha)\%$ lower confidence bound for μ is

$$\overline{x} - z_{1-\alpha}\, \sigma/\sqrt{n}.$$

For Michelson's data in Example 4.8 we observe $\overline{x} = 852.4(\text{se} = 7.9)$, so we are 95% confident that μ is larger than

$$852.4 - 1.65 \times 7.9 = 839.4.$$

The upper confidence bound can be found similarly. □

If T is continuous, the $100(1-\alpha)\%$ lower confidence bound L is a random quantity that satisfies

$$P_{\theta_0}(L < \theta_0) = 1 - \alpha, \qquad (5.3)$$

where θ_0 is the true parameter value, which means L has a correct coverage probability. This is true, since the event $[L < \theta_0]$ is equivalent to $[T < t_{1-\alpha}(\theta_0)]$; see Figure 5.5. By definition of $t_{1-\alpha}(\theta)$, we have

$$P_{\theta_0}(T < t_{1-\alpha}(\theta_0)) = 1 - \alpha.$$

Note that finding L in (5.2) is a sample-based computation, no unknown parameter is involved. However, the probability in (5.3) is a theoretical quantity.

5.6 Confidence density, CI and the bootstrap

Let T be a sensible estimate of a scalar parameter θ, and t represent the observed value of T. For a fixed t, as a function of θ, the one-sided P-value

$$\alpha(\theta) \equiv P_\theta(T \geq t),$$

looks like a distribution function: it is monotone increasing and ranging from zero to one. It is a statistic in the sense that no unknown parameter is involved in its computation. Fisher (1930) called it the fiducial distribution of θ (though he required T to be sufficient and continuous). To follow the current confidence-based procedures, we will call it the *confidence distribution* of θ (Efron 1998), and we do not require T to be sufficient nor continuous. Its derivative

$$c(\theta) = \frac{\partial \alpha(\theta)}{\partial \theta}$$

will be called the confidence density of θ. To put it more suggestively, we might write

$$C(\theta < a) = \int_{-\infty}^{a} c(u)du,$$

where $C(\cdot)$ is read as 'the confidence of', to represent the confidence level of the statement $\theta < a$. Remember, however, that θ is not a random variable. The confidence distribution is a distribution of our (subjective) confidence about where θ is. It is an alternative representation of uncertainty other than the likelihood function.

Example 5.9: In the normal case, suppose we observe $X = x$ from $N(\theta, \sigma^2)$ with known σ^2. The confidence distribution is

$$\begin{aligned} P_\theta(X \geq x) &= P(Z \geq (x - \theta)/\sigma) \\ &= 1 - \Phi\left(\frac{x-\theta}{\sigma}\right), \end{aligned}$$

and the confidence density is the derivative with respect to θ:

$$c(\theta) = \frac{1}{\sigma}\phi\left(\frac{x-\theta}{\sigma}\right).$$

This is exactly the normal density centred at the observed $\widehat{\theta} = x$, which is also the likelihood function (if normalized to integrate to one). The equality between the confidence density and such normalized likelihood holds generally for symmetric location problems (Section 9.8). □

The confidence distribution can be viewed simply as a collection of P-values across the range of null hypotheses. Alternatively, P-value can be represented as the tail area of the confidence density. The main use of the confidence density is to derive and show graphically CIs with certain confidence level. In the continuous case, let $\theta_\alpha \equiv L$ be a $100(1-\alpha)\%$ lower confidence bound θ such that

$$C(\theta < \theta_\alpha) = P_{\theta_\alpha}(T \geq t) = \alpha.$$

This means θ_α behaves like the 100α percentile of the confidence distribution. Furthermore, from (5.3), θ_α has coverage probability

$$P_\theta(\theta < \theta_\alpha) = \alpha. \tag{5.4}$$

Therefore, the confidence density works intuitively like a Bayesian density on θ, but it can be justified by standard probability arguments without invoking any prior probability. This was Fisher's original motivation for the fiducial distribution in 1930.

For a two-sided $100(1-\alpha)\%$ CI we can simply set

$$\theta_{\alpha/2} < \theta < \theta_{1-\alpha/2},$$

which satisfies the confidence requirement

$$C(\theta_{\alpha/2} < \theta < \theta_{1-\alpha/2}) = 1 - \alpha$$

as well as the probability requirement, from (5.4),

$$P_\theta(\theta_{\alpha/2} < \theta < \theta_{1-\alpha/2}) = 1 - \alpha. \tag{5.5}$$

In the normal example above it is easy to see how to construct CIs from the confidence density.

The use of confidence density is attractive when the confidence and probability statements match, but that does not hold generally in discrete cases or when there are nuisance parameters. We will discuss the Poisson and binomial models in the coming sections.

Example 5.10: Suppose we observe $X = x$ from a Poisson distribution with mean θ. The commonly defined one-sided P-value is

$$\text{P-value} = P(X \geq x),$$

so the confidence distribution is

$$P_\theta(X \geq x) = \sum_{k=x}^{\infty} e^{-\theta}\theta^k/k!$$

and the confidence density is

$$c(\theta) = \sum_{k=x}^{\infty}(ke^{-\theta}\theta^{k-1} - e^{-\theta}\theta^k)/k! = e^{-\theta}\theta^{x-1}/(x-1)!,$$

coincidentally the same as the likelihood based on observing $X = x - 1$. See Figures 5.6(a)–(b). For nonlocation parameters the confidence density is generally different from the likelihood function (normalized to integrate to one).

If $x = 0$ the above definition of P-value produces a degenerate confidence density at $\theta = 0$, which is not sensible.

A two-sided CI for θ can be derived by appropriately allocating the tail probabilities of the confidence density; see Figure 5.6(b). This interval, however, does not match the standard definition to be discussed in the next section. □

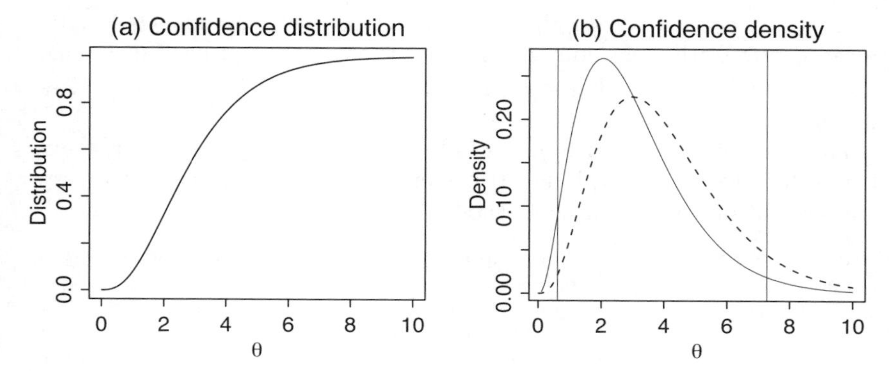

Figure 5.6: *(a) The confidence distribution of the Poisson mean θ based on observing $x = 3$. (b) The corresponding confidence density (solid) and likelihood (dashed) functions; the vertical lines mark the 2.5% tail probabilities of the confidence distribution.*

Bootstrap density as confidence density

Let $\psi \equiv g(\theta)$ be a one-to-one increasing function of θ, and θ_α be the $100(1-\alpha)$% lower confidence bound for θ. For $\psi_\alpha \equiv g(\theta_\alpha)$ we have

$$C(\psi < \psi_\alpha) = C\{g(\theta) < g(\theta_\alpha)\} = \alpha,$$

which means ψ_α is the 100α percentile of the confidence distribution of ψ. In other words the rule of transformation for confidence distributions follows the usual rule for probability distributions.

We can exploit this result to get an inference directly from the bootstrap distribution of an estimate $\widehat{\theta}$. Suppose there exists a transformation $g(\widehat{\theta})$ such that, for all θ, the random variable $g(\widehat{\theta})$ is normally distributed with mean $g(\theta)$ and constant variance. From Example 5.9 the confidence density of $g(\theta)$ matches the probability density of $g(\widehat{\theta})$. By back-transformation, this means the confidence density of θ matches the probability density of $\widehat{\theta}$. The latter is exactly what we get from the bootstrap! To emphasize, we can interpret the bootstrap distribution as a confidence distribution of θ.

Therefore, *without* knowing $g(\cdot)$, but only assuming that it exists, the $100(1-\alpha)$% CI of θ is the appropriate percentiles of the bootstrap distribution. For example, the 95% CI for θ is

$$\widehat{\theta}_{0.025} < \theta < \widehat{\theta}_{0.975},$$

where $\widehat{\theta}_{0.025}$ and $\widehat{\theta}_{0.975}$ are the 2.5 and 97.5 percentiles of the bootstrap distribution. This is called the bootstrap percentile method (Efron 1982). Using a similar argument as that used in Section 2.9 for the likelihood-based intervals, *the bootstrap CI is safer to use than the Wald CI,* especially if $\widehat{\theta}$ is nonnormal. In effect the bootstrap CI automatically employs the best normalizing transform.

Alternatively, by viewing the bootstrap density as a confidence density, we can compute the one-sided (right-side) P-value for testing H$_0$: $\theta = \theta_0$, simply by finding the proportion of bootstrap replicates $\widehat{\theta}^*$'s that are less than θ_0

The percentile method is not valid if the required $g(\theta)$ does not exist. Efron (1987) proposed the BC$_a$ method (Section 15.3) as a general improvement of the percentile method.

Example 5.11: From the boostrap distribution of the sample correlation in Example 5.5, using $B = 500$ replications, we obtain the 95% CI

$$0.68 < \rho < 0.93$$

directly from the bootstrap distribution. Note that it is an asymmetric interval around the estimate $\widehat{\rho} = 0.83$. For comparison, using the previously computed standard error, the Wald 95% CI is $0.83 \pm 1.95 \times 0.063$, or $0.71 < \rho < 0.95$. The one-sided P-value to test H$_0$: $\rho = 0.5$ is 1/500, since there is just one bootstrap replicate $\widehat{\rho}^*$ less than 0.5. □

5.7 Exact inference for Poisson model

Suppose we observe x from a Poisson distribution with mean θ. If we want a $100(1-\alpha)\%$ CI for θ, then the standard construction is based on performing *two one-sided hypothesis tests.* The upper limit U is chosen so that the left-side P-value is

$$P_U(X \le x) = \alpha/2$$

where X is Poisson with mean U. The lower limit L is chosen so that

$$P_L(X \ge x) = \alpha/2$$

where X is Poisson with mean L. This technique is clearly seen graphically in the normal case, where the limits match the standard formula.

In the Poisson case a problem occurs again at $x = 0$: naturally we only get a one-sided interval, which is associated with one-sided tests. Should we allow a full α for the one-sided P-value to find U, namely by solving

$$P(X = 0) = e^{-U} = \alpha,$$

or still only allow $\alpha/2$ and solve

$$P(X = 0) = e^{-U} = \alpha/2?$$

In practice the latter is used.

A CI found using the method above is called an 'exact' interval, but that is a misnomer. The discreteness and asymmetry in the sampling distribution creates inexact coverage: 95% CIs do not have 95% coverage probability. In this case, we can show (Exercise 5.13) that the true coverage probability is larger than the stated or nominal level.

Example 5.12: For $x = 1$, the upper limit of 95% CI for θ is the solution of

$$P(X \leq 1) = e^{-\theta} + \theta e^{-\theta} = 0.025,$$

which yields $\theta = U = 5.57$. The lower limit is the solution of

$$P(X \geq 1) = 1 - e^{-\theta} = 0.025,$$

which yields $\theta = L = 0.0253$. The following table shows the 95% confidence limits for a range of x.

x	0	1	2	3	4	5	6	7	8
$L(x)$	0	0.03	0.24	0.62	1.09	1.60	2.20	2.82	3.45
$U(x)$	3.69	5.57	7.22	8.77	10.24	11.67	13.06	14.42	15.76

The CIs are shown in Figure 5.7(a). The coverage probability is based on numerically computing

$$P_\theta\{L(X) \leq \theta \leq U(X)\} = \sum_{x=0}^{\infty} I\{L(x) \leq \theta \leq U(x)\}\, e^{-\theta}\theta^x/x!,$$

where $I\{\cdot\}$ is one if the condition in the bracket is true, and zero otherwise. For example, for $\theta = 1$,

$$P_\theta\{L(X) \leq \theta \leq U(X)\} = P_\theta(X = 0, 1, 2, 3) = 0.98.$$

Figure 5.7(b) shows that the true coverage probability is higher than 95% for all values of θ. □

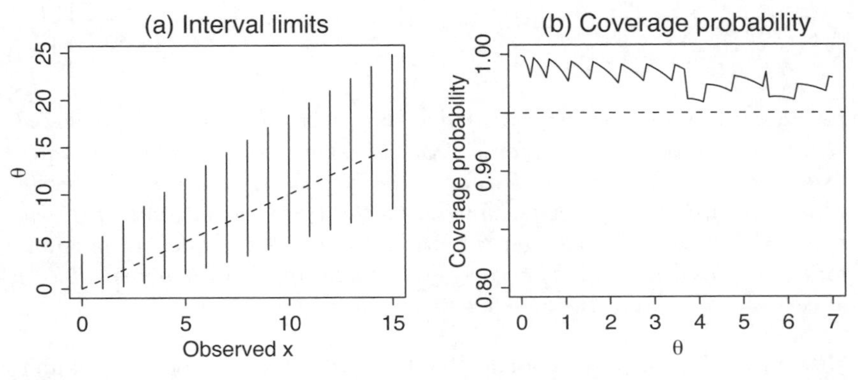

Figure 5.7: *(a) CIs for the Poisson mean θ based on observing x. (b) The true coverage probability of the 'exact' 95% CI.*

In terms of confidence density

The discreteness of Poisson data creates a slight complication. Define a confidence density $c_L(\theta)$ based on the P-value formula

$$\text{P-value} = P_\theta(X \geq x).$$

The $100\alpha/2$ percentile of this density gives the lower limit L. To get the upper limit U, we compute another confidence density $c_U(\theta)$ based on the P-value formula

$$\text{P-value} = P_\theta(X > x)$$

and find U as the $100(1-\alpha/2)$ percentile of this density. Figures 5.8(a)–(b) show the confidence distributions and densities based on these P-values.

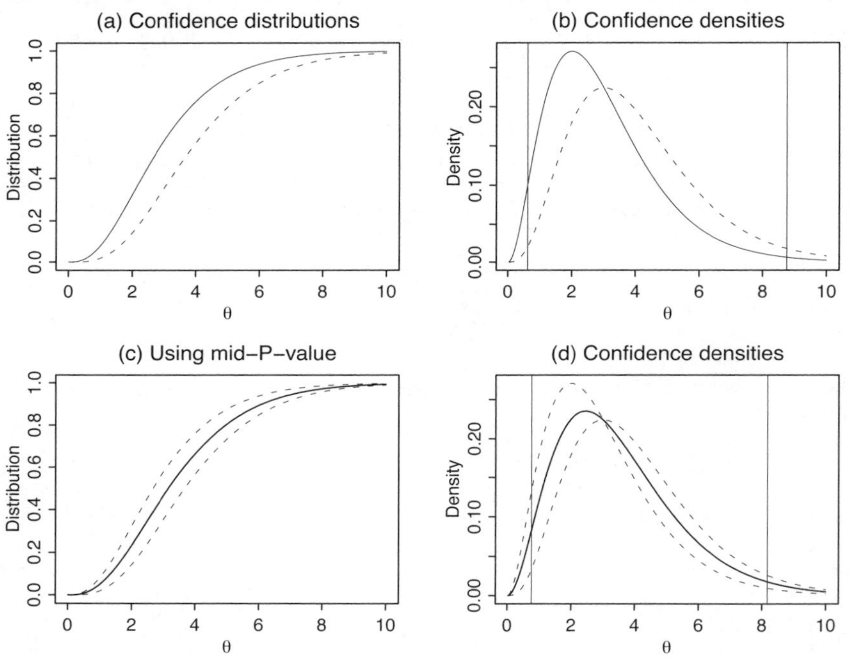

Figure 5.8: *(a) The confidence distributions of the Poisson mean θ, based on two definitions of P-value for $x = 3$. (b) The corresponding confidence densities $c_L(\theta)$ and $c_U(\theta)$ in solid and dashed curves; both are required to get the standard two-sided CIs, marked by the two vertical lines. (c) The confidence distribution based on the mid-P-value (solid) and the previous definitions (dashed). (d) The corresponding confidence densities of (c). The vertical lines mark the 95% CI based on the mid-P-value.*

It is desirable to have a single P-value that works for both the lower and upper confidence bounds. Such a P-value is given by the mid-P-value

$$\text{P-value} = P_\theta(X > x) + \frac{1}{2}P_\theta(X = x).$$

Figures 5.8(c)–(d) show the confidence distribution and density based on the mid-P-value for $x = 3$; for example, a 95% CI derived from the confidence density is $0.76 < \theta < 8.17$. The CIs derived from this definition, however, no longer have guaranteed minimum coverage.

Exact coverage of likelihood-based intervals

In the normal case the likelihood interval at 15% cutoff has an exact 95% coverage probability. We do not expect such a simple relationship in the Poisson case, but we expect it to be approximately true. In general when x is observed from a Poisson distribution with mean θ, the likelihood interval at cutoff α is the set of θ such that

$$\frac{L(\theta)}{L(x)} = e^{x-\theta}\left(\frac{\theta}{x}\right)^x > \alpha.$$

Example 5.13: To be specific let us compare the case when $x = 3$. The exact 95% CI is $0.62 < \theta < 8.76$. The likelihood interval at 15% cutoff (hence an approximate 95% CI) is $0.75 < \theta < 7.77$. Figure 5.9 shows that the exact CI includes values of θ that have lower likelihood than some values outside the interval. □

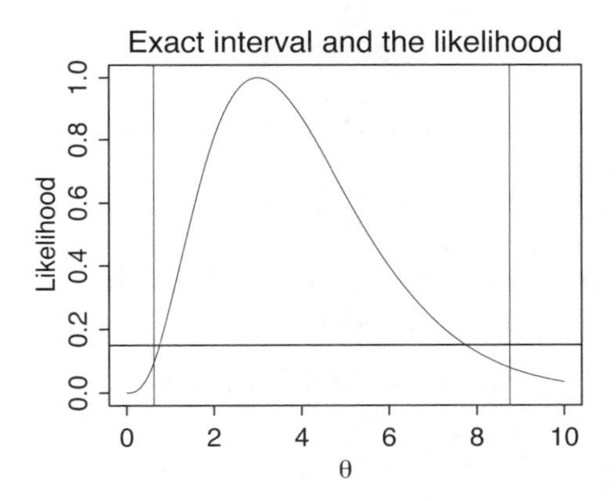

Figure 5.9: *The exact 95% CI for Poisson mean θ (marked by the vertical lines) and the likelihood based on $x = 3$.*

Example 5.14: For comparison with the exact CIs the following table shows the likelihood intervals at 15% cutoff for the same values of x shown in Example 5.12. For convenience we also name the interval limits as $L(x)$ and $U(x)$:

x	0	1	2	3	4	5	6	7	8
$L(x)$	0	0.06	0.34	0.75	1.25	1.80	2.39	3.01	3.66
$U(x)$	1.92	4.40	6.17	7.77	9.28	10.74	12.15	13.54	14.89

Figure 5.10(a) shows that there is very little practical difference between the two types of intervals, except at $x = 0$. The coverage probability plot in Figure 5.10(b), however, shows that the likelihood interval may have less than 95%

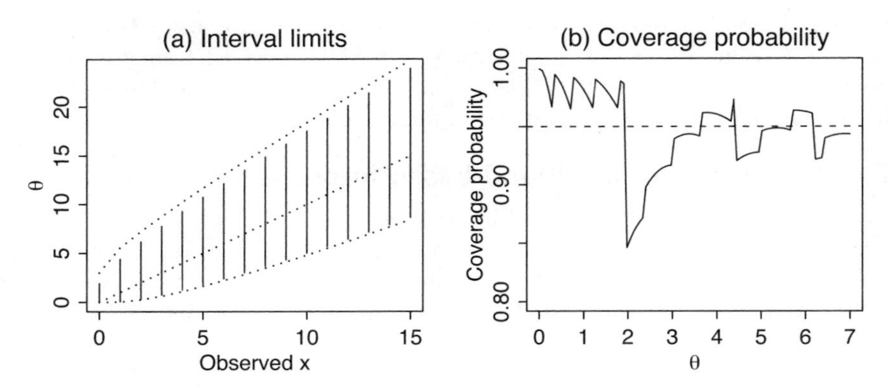

Figure 5.10: *(a) Likelihood intervals for the Poisson mean θ based on observing x; the dotted lines are the limits of 95% exact confidence intervals. (b) The coverage probability of the likelihood-based intervals at 15% cutoff.*

coverage. The minimum coverage of 85% is observed at θ around 1.95, just outside the upper limit for $x = 0$. □

Now note that the likelihood function based on $x = 0$ is

$$L(\theta) = e^{-\theta},$$

which is not regular, so we cannot relate 15% cutoff with the usual (approximate) 95% confidence level. The problem of how to construct a sensible interval when $x = 0$ arises also with the exact method. It is interesting to see the dramatic change in the coverage probability plot if we simply 'fix' the likelihood interval at $x = 0$ to match the CI; that is, change the upper limit from 1.92 to 3.69 (corresponding to a cutoff of 2.5%). Figure 5.11

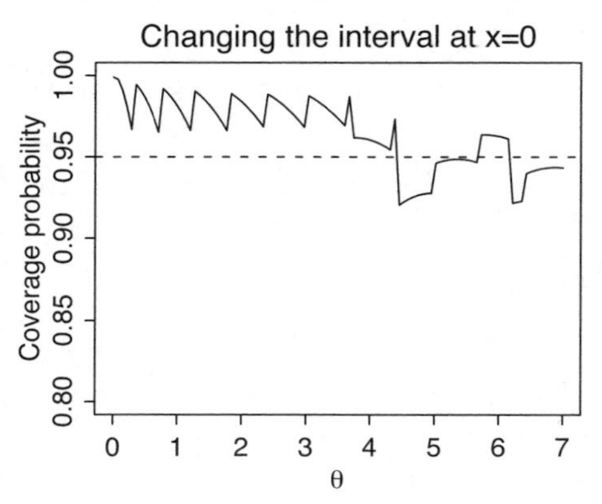

Figure 5.11: *The coverage probability of the likelihood-based intervals where the upper limit at $x = 0$ is changed from 1.92 to 3.68.*

shows that the coverage probability is now mostly above 95%, except for some small intervals of θ. The minimum coverage is now 92%; the low coverage is mainly caused by the interval for $x = 1$ (Exercise 5.12).

5.8 Exact inference for binomial model

Suppose we observe x from a binomial distribution with a known n and an unknown probability θ. The normal approximation of the sample proportion $\widehat{\theta}$ gives the Wald CI formula

$$\widehat{\theta} \pm z_{\alpha/2}\,\mathrm{se}(\widehat{\theta})$$

where the standard error is

$$\mathrm{se}(\widehat{\theta}) = \sqrt{\widehat{\theta}(1-\widehat{\theta})/n}.$$

This approximate interval works well if n is large enough and θ is far from zero or one. For small n, Agresti and Coull (1998) suggest adding '2 successes and 2 failures' to the observed data before using the Wald interval formula.

There is a large literature on the 'exact' CI for the binomial proportion. Many texts recommend the Clopper–Pearson (1934) interval, similar to the one described previously for the Poisson mean. A $100(1-\alpha)\%$ CI for θ based on observing x is $L(x)$ to $U(x)$, where they satisfy two one-sided P-value conditions

$$P_{\theta=L}(X \geq x) = \alpha/2$$

and

$$P_{\theta=U}(X \leq x) = \alpha/2.$$

At $x = 0$ and $x = n$ the CI is naturally one sided, which is associated with one-sided tests, so, as in the Poisson case, there is the question whether we allow the full α or still $\alpha/2$ in the computation above. However, to guarantee a coverage probability of at least the nominal (claimed) value, we must use $\alpha/2$.

Example 5.15: For $n = 10$ the confidence limits based on observing $x = 0, \ldots, 10$ are given in the table and plotted in Figure 5.12(a).

x	0	1	2	3	4	5	6	7	8	9	10
$L(x)$	0	.01	.03	.07	.13	.19	.26	.35	.44	.56	.70
$U(x)$	.30	.44	.56	.65	.74	.81	.87	.93	.97	.99	1

As before the coverage probability is computed according to

$$\sum_{x=0}^{10} I\{L(x) \leq \theta \leq U(x)\} \binom{10}{x} \theta^x (1-\theta)^{10-x}.$$

The coverage probability is plotted in Figure 5.12(b), showing that the procedure is quite conservative. A simple average of the coverage probability is 98%, much higher than the intended 95% level. □

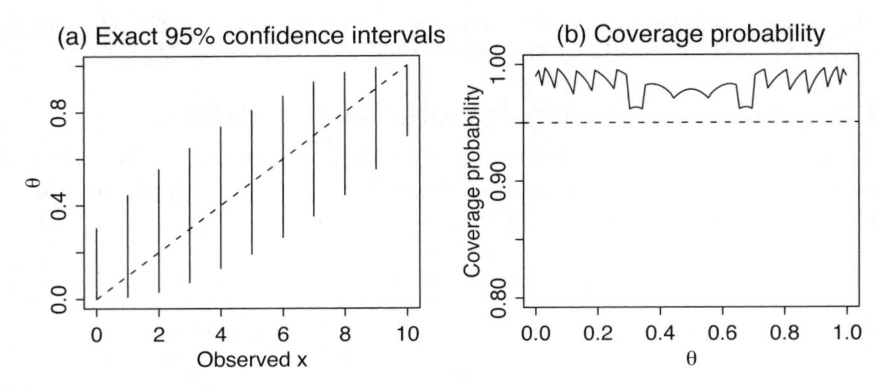

Figure 5.12: *(a) Clopper–Pearson 95% CIs for binomial proportion. (b) The coverage probability of the 'exact' 95% CI.*

5.9 Nuisance parameters

We will discuss very briefly the classical confidence procedures when there are nuisance parameters. We want a CI with the correct coverage at all possible values of the nuisance parameters. There are two classical ways to deal with the nuisance parameters: using pivotal statistics and conditioning.

Pivotal quantity

A *pivot* is a random variable whose distribution is free of any unknown parameter; typically it is a function of both the data and the unknown parameters. Freedom from the nuisance parameter implies that a CI constructed from a pivot has a simple coverage probability statement, true for all values of the unknown parameter. This is an ideal route to arrive at a CI, but there is no guarantee that we can find such a statistic for the problem at hand. Pivots are usually available in normal theory linear models, which form a large part of classical statistics.

One-sample problems

If $x_1, \ldots, x_n$ are an iid sample from $N(\mu, \sigma^2)$, then

$$\frac{\overline{X} - \mu}{s/\sqrt{n}} \sim t_{n-1}$$

for all μ and σ^2. So the $100(1-\alpha)\%$ CI for μ given by

$$\overline{x} - t_{n-1,\alpha/2}\frac{s}{\sqrt{n}} < \mu < \overline{x} + t_{n-1,\alpha/2}\frac{s}{\sqrt{n}}$$

has an exact coverage of $100(1-\alpha)\%$ for all μ and σ^2.

Two-sample problems

If $x_1, \ldots, x_m$ are an iid sample from $N(\mu_x, \sigma^2)$ and $y_1, \ldots, y_n$ from $N(\mu_y, \sigma^2)$, then

$$\frac{\overline{x} - \overline{y} - (\mu_x - \mu_y)}{s_p\sqrt{1/m + 1/n}} \sim t_{m+n-2}$$

for all μ_x, μ_y and σ^2, where

$$s_p^2 = \frac{(m-1)s_x^2 + (n-1)s_y^2}{m+n-2}$$

and s_x^2 and s_y^2 are the sample variances. We can then derive the CI for $\mu_x - \mu_y$.

One-sample variance

If $x_1, \ldots, x_n$ are an iid sample from $N(\mu, \sigma^2)$, then

$$\frac{(n-1)s^2}{\sigma^2} \sim \chi^2_{n-1}.$$

So the $100(1-\alpha)\%$ CI for σ^2 is

$$\frac{(n-1)s^2}{\chi^2_{n-1,1-\alpha/2}} < \sigma^2 < \frac{(n-1)s^2}{\chi^2_{n-1,\alpha/2}}.$$

For Michelson's first twenty measurements of the speed of light in Example 4.8 we get $s^2 = 11,009.5$, so the 95% CI for σ^2 is

$$0.58s^2 < \sigma^2 < 2.13s^2,$$

giving

$$6,367 < \sigma^2 < 23,486.$$

Conditioning

Conditioning on the minimal sufficient statistics for the nuisance parameters is used to reduce the unknown parameter space to the parameter of interest only. This method usually works for models in the exponential family.

Comparing two Poisson means

We have already seen this problem in the aspirin data example (Section 4.7); our current discussion provides a repeated sampling interpretation to the usual likelihood intervals found by conditioning.

Suppose x and y are independent Poisson samples with means λ_x and λ_y. We are not interested in the magnitude of λ_x or λ_y, but in the relative size $\theta = \lambda_x/\lambda_y$. The conditional distribution of x given $x + y = n$ is binomial with parameters n and

$$\pi = \frac{\lambda_x}{\lambda_x + \lambda_y} = \frac{\theta}{\theta + 1}.$$

Using the method given in the previous section we can construct a CI for π, which can then be transformed to an interval for θ. Suppose we set the $100(1-\alpha)\%$ CI for θ such that

$$P(\theta \in \mathrm{CI}|X + Y) \geq 1 - \alpha,$$

for all λ_x and λ_y. That is, the coverage probability is greater than $1 - \alpha$ for every value of $X + Y$. So, unconditionally,

$$P(\theta \in \mathrm{CI}) \geq 1 - \alpha.$$

Preference test

Suppose we asked 50 people for their preferences for a cola drink of brand A or B, and we obtained the following result:

Prefer A	17
Prefer B	10
No preference	23

Is A really preferred over B? The 'no preference' group is a nuisance in this comparison. Suppose we model the number of responses (n_A, n_B, n_C) in the three groups as multinomial with probability p_A, p_B and $p_C \equiv 1 - p_A - p_B$. Then the conditional distribution of n_A given $n_C = 23$ is binomial with parameters $50 - 23 = 27$ and $p_A/(p_A + p_B) = \theta/(\theta + 1)$, from which we can derive a conditional inference for θ (Exercise 5.18).

5.10 Criticism of CIs

A fundamental issue in statistical inference is how to attach a relevant measure of uncertainty to a sample-based statement such as a confidence interval. In practice the confidence level is usually interpreted as such a measure: an observed 95% CI has '95% level of certainty' of covering the true parameter. However, using a long-run frequency property for such a purpose sometimes leads to logical contradictions. A weakness of the traditional confidence theory is that, even when there is an obvious contradiction, no alternative measure of uncertainty is available.

First, it is sometimes possible to know that a particular interval does not contain the true parameter. This creates a rather absurd situation, where we are '95% confident' in an interval that we know is wrong, just because in the long run we are 95% right.

Example 5.16: Suppose $x_1, \ldots, x_{10}$ are an iid sample from $N(\mu, 1)$, where μ is known to be nonnegative. On observing $\overline{x} = -1.0$, the 95% CI for μ is

$$\overline{x} \pm 1.96/\sqrt{10},$$

yielding $-1.62 < \mu < -0.38$, which cannot cover a nonnegative μ. (In a single experiment, there is a 2.5% chance for this to happen if $\mu = 0$.) We can

- either take the interval at face value and start questioning the assumption that $\mu \geq 0$,
- or still believe $\mu \geq 0$ to be true, but in this case it only makes sense to report a (likelihood) interval over $\mu \geq 0$.

How we react depends on our relative strength of belief about the data and the assumption. □

The logical problem is captured by a quality control anecdote. A US car company ordered parts from a Japanese supplier and stated that they would accept 1% defectives. Sure enough, the parts came in two separate boxes, a large one marked 'perfect' and a small one marked 'defective'.

The moral of the story is that a global confidence statement such as 'we are 95% confident' can be meaningless when there are *recognizable or relevant subsets*, for which we can make an equally valid but different statement. This is a Fisherian insight that has not become common knowledge in statistics.

The previous problem tends to happen if we restrict the parameter space, but a similar phenomenon can also happen without parameter restriction. The following example from Buehler and Fedderson (1963) is of fundamental importance, since it strikes at the t-based CIs. See also Lehmann (1986, page 557).

Example 5.17: Suppose x_1, x_2 are an iid sample from $N(\mu, \sigma^2)$, where σ^2 is unknown. The exact 50% CI for μ is

$$x_{(1)} < \mu < x_{(2)},$$

where $x_{(1)}$ and $x_{(2)}$ are the order statistics; such an interval is a likelihood interval at 16% cutoff. We can show, however, that the conditional coverage

$$P(X_{(1)} < \mu < X_{(2)} | C) > 2/3$$

for all μ and σ^2, where C is a set of the form

$$C = \{(x_1, x_2), \text{coefficient of variation} > \sqrt{2}/(\sqrt{2}+1) = 0.59\}, \qquad (5.6)$$

and the coefficient of variation is

$$\frac{s}{|\overline{x}|} = \frac{\sqrt{2}|x_1 - x_2|}{|x_1 + x_2|}.$$

So, if the observations are rather far apart in the sense of C, which can be readily checked, then we know that the coverage is greater than 67%, and vice versa if the

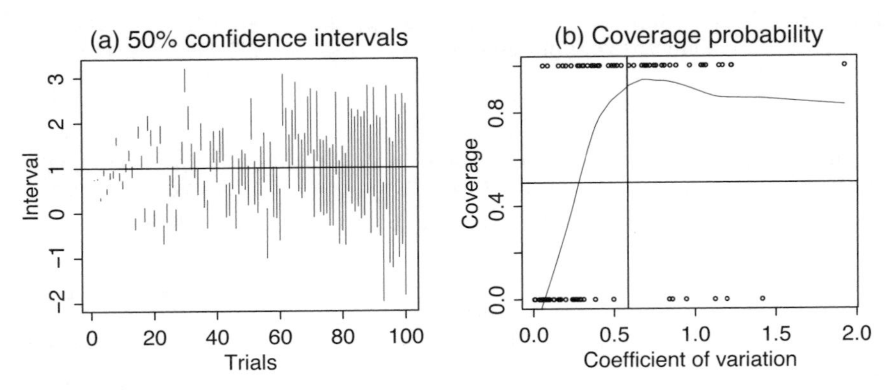

Figure 5.13: *(a) One hundred simulated intervals $x_{(1)} < \mu < x_{(2)}$ arranged by the coefficient of variation. (b) The estimated coverage probability as a function of the coefficient of variation. The vertical line at the coefficient of variation of 0.59 corresponds to value in (5.6).*

observations are closer together. What we have shown here is that it is possible to group the intervals according to the recognizable subset C, so that different groups have a different coverage probability, but overall they are 50% correct.

Figure 5.13(a) shows 100 simulated intervals $x_{(1)} < \mu < x_{(2)}$ arranged by the corresponding coefficient of variation; the true mean is $\mu_0 = 1$. The coverage probability in Figure 5.13(b) is based on smoothing the 100 pairs of values {coefficient of variation, y}, where $y = 1$ if the interval covers μ_0, and zero otherwise. The coverage is lower than the nominal 50% if the coefficient of variation is small, and can be greater otherwise.

The question is, if we observe (x_1, x_2) in C, say $(-2, 10)$, what 'confidence' do we have in the CI $-2 < \mu < 10$? How should we report the CI? Is it a 50% or 67% CI? Suppose (x_1, x_2) is a member of other relevant subsets with different coverage probabilities; which one should we attach to a particular interval? Note that in this example $x_{(1)} < \mu < x_{(2)}$ is a unique likelihood interval with 16% cutoff, unaffected by the existence of C. □

The phenomenon in the example is true for the general one-sample problem: 'wide' CIs in some sense have larger coverage probability than the stated confidence level, and vice versa for 'short' intervals. Specifically, let $x_1, \ldots, x_n$ be an iid sample from $N(\mu, \sigma^2)$ with both parameters unknown, and

$$C = \{(x_1, \ldots, x_n), s/|\overline{x}| > k\},$$

for some k, be a set where $\overline{x}$ is small relative to s, or where the hypothesis $\mu = 0$ is *not rejected* at a certain level. The standard CI for μ is

$$\overline{x} - t_{n-1,\alpha/2}\frac{s}{\sqrt{n}} < \mu < \overline{x} + t_{n-1,\alpha/2}\frac{s}{\sqrt{n}}.$$

It has been shown (Lehmann 1986, Chapter 10) that, for some $\epsilon > 0$,

$$P(\mu \in \text{CI}|C) > (1 - \alpha) + \epsilon$$

for all μ and σ^2. This means C is a relevant subset.

There has been no satisfactory answer to this problem from the frequentist quarters. In fact, Lehmann (1986, page 558) declared that the existence of certain relevant subsets is 'an embarrassment to confidence theory'. The closest thing to an answer is the area of conditional inference (Reid 1995), which has produced many theoretical results, but as yet no ready methodology or rules for routine data analysis. We should bear in mind, however, that these criticisms are not directed at the interval itself, but at the relevance of long-run frequentist properties as a measure of uncertainty for an observed interval.

5.11 Exercises

Exercise 5.1: Let $T = \overline{x}$ be an estimate of the mean μ. Show that the jackknife estimate of the bias is $\widehat{b} = 0$.

Exercise 5.2: Let $T = \frac{1}{n}\sum(x_i - \overline{x})^2$ be an estimate of the variance σ^2. Show that the jackknife estimate of the bias is

$$\widehat{b} = -\frac{1}{n}\left\{\frac{1}{n-1}\sum(x_i - \overline{x})^2\right\},$$

so that the corrected estimate is the unbiased estimate

$$T - \widehat{b} = \frac{1}{n-1}\sum(x_i - \overline{x})^2.$$

Exercise 5.3: Use the jackknife method to estimate the bias of $e^{-\overline{x}}$ in Example 5.1.

Exercise 5.4: Let X be a sample from the binomial distribution with parameters n and θ, and the estimate $\widehat{\theta} = X/n$. Find the general bias formula for $\widehat{\theta}^2$ as an estimate of θ^2. Find the jackknife estimate of θ^2 and show that it is unbiased.

Exercise 5.5: Suppose the bias of T as an estimate of θ is of the form

$$b(\theta) = \sum_{k=1}^{\infty} a_k/n^k,$$

where a_k may depend on θ, but not on n. Show that the corrected estimate using the jackknife has a bias of order n^{-2}. That is, the jackknife procedure removes the bias of order n^{-1}.

Exercise 5.6: Investigate the relationship between likelihood and two-sided P-value in the binomial model with $n = 10$ and success probability θ. Test $H_0: \theta = \theta_0$ for some values of θ_0.

Exercise 5.7: Efron (1993) defines an implied likelihood from a confidence density as

$$L(\theta) = \frac{c_{xx}(\theta)}{c_x(\theta)}$$

where $c_{xx}(\theta)$ is the confidence density based on doubling the observed data x, and $c_x(\theta)$ is based on the original data x. For example, if we observe $x = 5$ from Poisson(θ), doubling the data means observing $(x, x) = (5, 5)$ as an iid sample. Using simple observations, compare the exact and implied likelihoods in the normal, Poisson and binomial cases.

Exercise 5.8: Verify the bootstrap variance and CI for the correlation coefficient given in Example 5.5 and its continuation in Section 5.6. Compare the results of the nonparametric bootstrap with a parametric bootstrap that assumes bivariate normal model.

Exercise 5.9: Suppose x is $N(\mu_x, 1)$ and y is $N(\mu_y, 1)$, and they are independent. We are interested in the ratio $\theta = \mu_y/\mu_x$. Define $z \equiv y - \theta x$, so z is $N(0, 1+\theta^2)$, which depends only on θ and can be a basis for inference for θ. The so-called Fieller's CI is based on

$$P\left(\frac{(y-\theta x)^2}{1+\theta^2} < \chi^2_{1-\alpha}\right) = 1 - \alpha.$$

Find the general conditions so that the 95% CI for θ is (i) an interval, (ii) two disjoint intervals, or (iii) the whole real line. Discuss how we should interpret part (iii). As a separate exercise, given $x = -1$ and $y = 1.5$,

(a) find Fieller's 95% CI for θ.

(b) plot the likelihood function of θ.

(c) find the $100(1-\alpha)$% CI for θ at various values of α, so you obtain the conditions that satisfy (i), (ii) or (iii) above. Explain the result in terms of the likelihood function.

(d) Discuss the application of confidence density concept to this problem.

Exercise 5.10: For the simple Poisson mean model, compute and plot the coverage probability of the two-sided intervals based on the mid-P-value.

Exercise 5.11: It is known that the Poisson distribution with a large mean θ is approximately normal.

(a) Derive an approximate confidence interval based on this result.

(b) Show that it is equivalent to a likelihood-based interval using quadratic approximation on the log-likelihood function. Discuss the problem at $x = 0$.

(c) Compare the intervals we get using the approximation with the intervals in the text.

(d) Find the coverage probability for θ between 0 and 7.

Exercise 5.12: The likelihood of the Poisson mean θ based on $x = 1$ is also quite asymmetric. Revise the likelihood-based interval to match the exact confidence interval, and recompute the coverage probability plot.

Exercise 5.13: For the exact $100(1-\alpha)$% CI defined in Section 5.7 show that the coverage probability is at least $(1-\alpha)$. (Hint: draw a plot similar to Figure 5.5, and express the probability of $L(x) < \theta < U(x)$ in terms of the random variable x. Note that x is a discrete random variable.)

Exercise 5.14: Investigate and compare the coverage probability of the Wald and likelihood-based CIs for θ based on x from binomial$(n = 10, \theta)$.

Exercise 5.15: For inference of binomial θ for n small, consider transforming the parameter to the log-odds

$$\psi = \log\frac{\theta}{1-\theta}.$$

Show that the standard error of $\widehat{\psi}$ is

$$\text{se}(\widehat{\psi}) = \left(\frac{1}{x} + \frac{1}{n-x}\right)^{1/2}.$$

We can construct a 95% confidence interval for ψ, and then transform back to get a 95% CI for θ. (For $x = 0$ and $x = n$, use the exact intervals.) For $n = 10$, compare and investigate the coverage probability of the resulting intervals.

Exercise 5.16: For inference of binomial θ, Agresti and Coull (1998) suggest adding '2 successes and 2 failures' to the observed x and then using the Wald interval formula. For $n = 10$, investigate this interval and compare it with the Clopper–Pearson and the likelihood-based intervals. Discuss the advantages.

Exercise 5.17: Using Michelson's first twenty measurements of the speed of light in Example 4.8, compute the confidence density of variance parameter. Hint: use the marginal distibution of the sample variance

$$\frac{(n-1)s^2}{\sigma^2} \sim \chi^2_{n-1}.$$

Compare the confidence density with the (normalized) likelihood function based on the same distribution. Compute the 95% CI for σ^2 from the confidence density and show that it matches the 95% CI using the pivotal statistic method.

Exercise 5.18: Compute the likelihood-based and exact CIs for θ in the preference data in Section 5.9.

6

Modelling relationships: regression models

Modelling relationships and comparing groups is the essence of statistical modelling. Separate descriptions of individual groups are usually less interesting than group comparisons. Most scientific knowledge is about relationships between various measurements; for example, $E = mc^2$, or 'population grows exponentially but resources only grow linearly', etc.

In this chapter we will learn that any of the basic models in Chapter 4 can be extended to a regression model. The outcome variable is no longer an iid observation, but a function of some predictor variable(s). What model to use and when is generally determined by the nature of the data; it is knowledge that we acquire by working through many examples.

Example 6.1: To allow an estimate of altitude without carrying a barometer, in the mid-19th century the physicist James Forbes conducted experiments relating the water boiling point T and barometric pressure p. The latter had a known relationship to altitude. The measurements are shown in first two columns of Table 6.1; the last column is computed according to a known formula (6.1) below. The barometric pressure p has a physical relationship with altitude A according to differential equation

$$\frac{dp}{dA} = -cp,$$

where c is a known constant. The solution is

$$\log(p/p_o) = -cA$$

where p_o is the pressure at sea level. If p is in mmHg and A is in metres, then $p_o = 760$ and $c^{-1} = 8580.71$, so

$$\begin{aligned} A &= -8580.71 \log(p/760) \\ &\equiv \beta_0 + \beta_1 \log p \end{aligned} \qquad (6.1)$$

with $\beta_0 = 8580.71 \log 760$ and $\beta_1 = -8580.71$. Except for one possible outlier, Figure 6.1 shows a clear linear relationship between boiling point and altitude. The dashed line is a regression line, also of the form

$$A = b_0 + b_1 T,$$

for appropriate regression coefficients b_0 and b_1. It is rare in statistical applications to see such a perfect relationship. (In fact physicists now have a determin-

Boiling point (^{o}F)	Pressure (in Hg)	Altitude (m)
194.5	20.79	3124.21
194.3	20.79	3124.21
197.9	22.40	2484.18
198.4	22.67	2381.37
199.4	23.15	2201.59
199.9	23.35	2127.77
200.9	23.89	1931.59
201.1	23.99	1895.75
201.4	24.02	1885.03
201.3	24.01	1888.60
203.6	25.14	1493.98
204.6	26.57	1019.27
209.5	28.49	420.59
208.6	27.76	643.32
210.7	29.04	256.52
211.9	29.88	11.84
212.2	30.06	-39.69

Table 6.1: *Forbes' data on boiling point, barometric pressure and altitude (Weisberg 1985).*

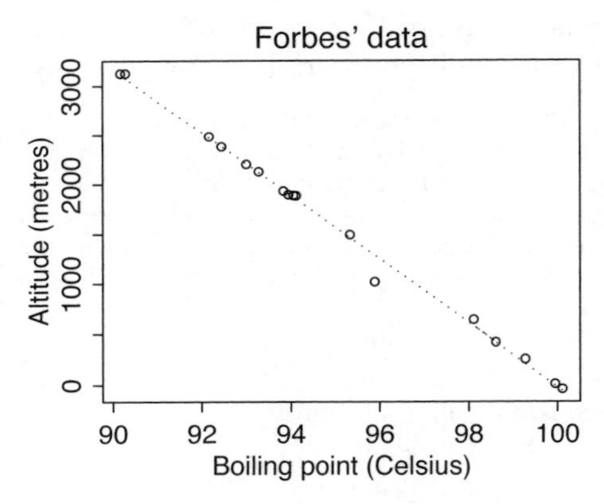

Figure 6.1: *Relationship between boiling point and altitude. The dashed line is the regression line.*

istic phase diagram of water, which has a curve for boiling point as a function of pressure.) □

6.1 Normal linear models

The normal regression model is the basis of classical statistical modelling. It is natural for models in the natural sciences, where the outcome variables are usually continuous and the error variable usually represents measurement noise.

In the basic model the outcomes $y_1, \ldots, y_n$ are iid observations from

$N(\mu, \sigma^2)$. The nontrivial extension that allows some modelling of the relationship with predictors is to drop the identical requirement from the iid: y_i's are independent $N(\mu_i, \sigma^2)$. The mean μ_i is then modeled as a function of the predictors. We say the mean vector follows a particular structure; for example, given a predictor x_i we might specify

$$\mu_i = h(x_i, \beta)$$

for some function h depending on unknown an vector parameter β. The simplest structure is the linear model

$$\mu_i = x_i'\beta,$$

where $\beta \in R^p$. Traditionally the model is written in matrix form as

$$y = X\beta + e,$$

where X is an $n \times p$ design matrix. Now e_i's are iid $N(0, \sigma^2)$, but note that this explicit error specification is not necessary. It is sufficient to specify that y_i is $N(\mu_i, \sigma^2)$ and $\mu_i = x_i'\beta$. This is especially relevant for nonnormal regression models, since in this case there is no explicit error term.

Let the parameter $\theta = (\beta, \sigma^2)$. The likelihood is

$$L(\theta) = \left(\frac{1}{2\pi\sigma^2}\right)^{n/2} \exp\left\{-\frac{1}{2\sigma^2}\sum_{i=1}^{n}(y_i - x_i'\beta)^2\right\}.$$

It is straightforward to show (Exercise 6.1) that the MLEs of β and σ^2 are

$$\begin{aligned} \widehat{\beta} &= (X'X)^{-1}X'Y \\ \widehat{\sigma}^2 &= \frac{1}{n}\sum(y_i - x_i'\widehat{\beta})^2, \end{aligned}$$

and the observed Fisher information for β is

$$I(\widehat{\beta}) = \widehat{\sigma}^{-2}(X'X).$$

The standard errors of the regression estimates are the square root of the diagonal of

$$I^{-1}(\widehat{\beta}) = \widehat{\sigma}^2(X'X)^{-1}.$$

To make these correspond exactly with standard practice in regression analysis, we can use the $(n - p)$ divisor for $\widehat{\sigma}^2$. A likelihood justification for this divisor is given in Example 10.11 in Section 10.6.

The standard regression models are usually solved using the least-squares (LS) principle, i.e. we estimate β by minimizing

$$\sum_i (y_i - x_i'\beta)^2.$$

This is equivalent to the likelihood approach as far as the computation is concerned, but there is a great difference in the statistical content of the two methods. To use the likelihood method we start by making a distributional assumption on the outcome y, while the LS method does not make such an assumption. The implication is that with the likelihood approach inference on β is already implicit; it is a matter of further computation. In contrast, inference on LS estimates requires further analytical work in terms of distribution or sampling theory, at which point we need some distributional assumptions.

If σ^2 is unknown, as is commonly the case, it is straightforward to compute the profile likelihood for β. For fixed β the MLE of σ^2 is

$$\widehat{\sigma}^2(\beta) = \frac{1}{n}\sum(y_i - x_i'\beta)^2,$$

so we obtain the profile

$$L(\beta) = \text{constant} \times \{\widehat{\sigma}^2(\beta)\}^{-n/2}.$$

Profile likelihood of individual regression parameters can be also computed analytically (Exercise 6.2).

Example 6.2: Plutonium has been produced in Hanford, Washington State, since World War II. It is believed that radioactive waste has leaked into the water table and the Columbia River, which flows through parts of Oregon on its way to the Pacific Ocean. Fadeley (1965) reported the following data:

County	Index of exposure	Cancer mortality
Clatsop	8.34	210.3
Columbia	6.41	177.9
Gilliam	3.41	129.9
Hood River	3.83	162.2
Morrow	2.57	130.1
Portland	11.64	207.5
Sherman	1.25	113.5
Umatilla	2.49	147.1
Wasco	1.62	137.5

Key to this study are the choice of counties, the definition of 'index of exposure' and the cancer classification; these carry an inductive uncertainty and are open to controversy. The chosen counties in this study have a waterfront on the Columbia River or the Pacific Ocean; the index of exposure was computed from several factors, for example the average distance of the population from the waterfront, and the cancer mortality is the number of cancer deaths per 100,000 person-years between 1959–1964.

Figure 6.2(a) shows that a linear model

$$y_i = \beta_0 + \beta_1\ (x_i - \overline{x}) + e_i$$

is quite sensible to describe the relationship, where β_1 is the parameter of interest. Assuming a normal model Figure 6.2(b) shows the likelihood contour of (β_0, β_1) at 90% to 10% cutoffs; these define the approximate 10% to 90% confidence regions. It is convenient to summarize the MLEs in the following table:

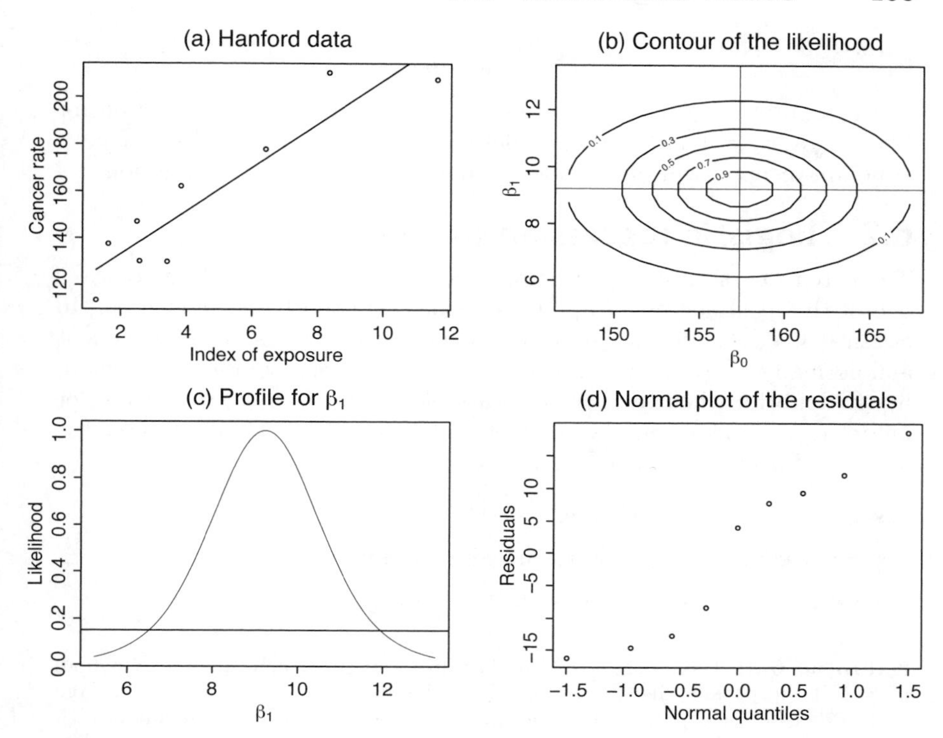

Figure 6.2: *(a) The scatter plot of Hanford data. (b) The likelihood contour of the regression parameters* (β_0, β_1). *(c) The profile likelihood of* β_1*; the approximate 95% CI for* β_1 *is (6.5,11.9). (d) The residual plot indicates some nonnormal behaviour.*

Effect	Parameter	Estimate	se
Intercept	β_0	157.33	4.67
Exposure	β_1	9.23	1.42

Figure 6.2(c) shows the profile likelihood for β_1; the Wald statistic for testing H_0: $\beta_1 = 0$ is $z = 9.23/1.42 = 6.5$, so there is a strong evidence that exposure to radioactive waste is associated with increased cancer rate. The QQ-plot of the residuals in Figure 6.2(d) shows evidence of non-normality. Some extra data and modelling may be warranted; for example, the counties have different populations, so the observed cancer rates have different precision. There could be other factors that need to enter the model, such as age distribution and gender balance, etc. □

Nonlinear regression models

The normal models can be extended to cover nonlinear relationships between the outcome and the predictors. Suppose we believe that

$$y_i = f(x_i, \beta) + e_i$$

where $f(\cdot)$ is a known function up to the regression parameter β; for example,

$$f(x_i, \beta) = \beta_0 + \beta_1 e^{-\beta_2 x_i},$$

and e_i's are iid $N(0, \sigma^2)$. Then we can derive the likelihood function for $\theta = (\beta_0, \beta_1, \beta_2, \sigma^2)$ and compute inferential quantities for β (Exercise 6.4). A nonlinear optimization routine is required for parameter estimation.

6.2 Logistic regression models

The extension of classical linear models to cover non-normal outcomes is one of the most successful applications of likelihood-based modelling. In classical linear models we usually assume the outcomes are independent and normally distributed with equal variance. These assumptions are manifestly doubtful when the outcome variable is, for example, the success or failure of an operation, the number of hits per hour for a website, the number of insurance claims per month, etc. In these cases

- normality is not plausible,
- a linear model $Ey_i = x_i'\beta$ is usually not natural,
- variance generally depends on the mean.

Example 6.3: Table 6.2 shows the data from an experimental surgery, where $y_i = 1$ if the patient died within 30 days of surgery and zero otherwise. Age is recorded for each patient and the question is whether age is associated with survival rate. There is a total of $n = 40$ patients and $\sum y_i = 14$ deaths. The pattern is shown in Figure 6.3(a). □

Patient	Age	y_i	Patient	Age	y_i
1	50	0	21	61	0
2	50	0	22	61	1
3	51	0	23	61	1
4	51	0	24	62	1
5	53	0	25	62	1
6	54	0	26	62	0
7	54	0	27	62	1
8	54	0	28	63	0
9	55	0	29	63	0
10	55	0	30	63	1
11	56	0	31	64	0
12	56	0	32	64	1
13	56	0	33	65	0
14	57	1	34	67	1
15	57	1	35	67	1
16	57	0	36	68	0
17	57	0	37	68	1
18	58	0	38	69	0
19	59	1	39	70	1
20	60	0	40	71	0

Table 6.2: *Surgical mortality and age information on 40 patients*

It is natural to model y_i as a Bernoulli event with probability θ_i depending on age. We might, for example, consider

(i) $\theta_i = \beta_0 + \beta_1 \text{ Age}_i$. This simple choice is not very natural, since θ_i is not constrained to be between 0 and 1.

(ii) $\theta_i = F(\beta_0 + \beta_1 \text{ Age}_i)$, where $0 \le F(\cdot) \le 1$. In principle, any distribution function $F(\cdot)$ will work. For example, the choice of the normal distribution function gives the so-called *probit regression.*

(iii) the *logistic regression* model:

$$\theta_i = \frac{\exp(\beta_0 + \beta_1 \text{ Age}_i)}{1 + \exp(\beta_0 + \beta_1 \text{ Age}_i)},$$

or

$$\log \frac{\theta_i}{1 - \theta_i} = \beta_0 + \beta_1 \text{ Age}_i,$$

i.e. the log odds is linear in age, or the effect of age on the odds (loosely means risk) of death is multiplicative. For example, $\beta_1 = 0.1$ means that for every year increase in age the odds of death increases by a factor of $e^{0.1} = 1.11$. For ease of interpretation and computation logistic regression is more commonly used than probit regression.

For the logistic regression model, given the observed data, the likelihood function of the parameters (β_0, β_1) is

$$\begin{aligned} L(\beta_0, \beta_1) &= \prod_{i=1}^{n} \theta_i^{y_i} (1 - \theta_i)^{1 - y_i} \\ &= \prod_i \left(\frac{\theta_i}{1 - \theta_i} \right)^{y_i} (1 - \theta_i). \end{aligned}$$

To reduce the correlation between the estimates of β_0 and β_1, we centre the age by recomputing Age $\leftarrow$ Age $-$ mean(Age). This redefinition only affects β_0, but does not change the magnitude or meaning of β_1. The log-likelihood is

$$\log L(\beta_0, \beta_1) = \sum_i [(\beta_0 + \beta_1 \text{Age}_i) y_i - \log\{1 + \exp(\beta_0 + \beta_1 \text{Age}_i)\}].$$

In principle the statistical problem is over: the rest is a matter of computing or finding summaries from the likelihood. The numerical method to obtain the MLE is discussed in Section 6.7.

The contours of this likelihood function are given in Figure 6.3(b). These contour lines represent the approximate 10% to 90% confidence region for the parameters. A summary of the estimates is given in the following table:

Effect	Parameter	Estimate	se
Intercept	β_0	−0.723	0.367
Age	β_1	0.160	0.072

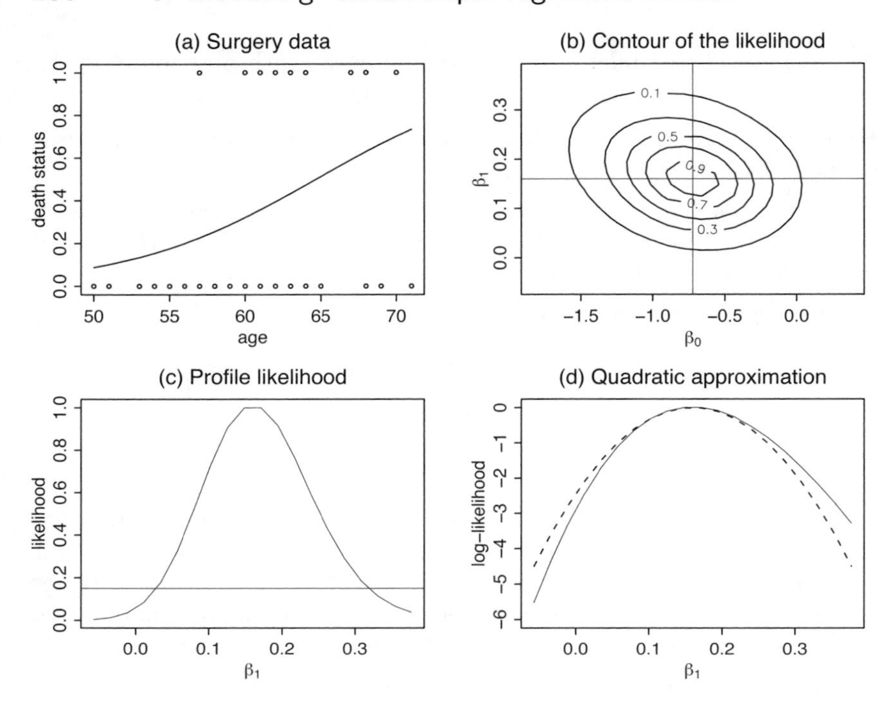

Figure 6.3: *(a) The surgical mortality data with the fitted logistic regression line. (b) The contours of the likelihood function. (c) The profile likelihood of β_1. (d) Quadratic approximation of the log-likelihood of β_1.*

The profile likelihood of β_1 is shown in Figure 6.3(c). The approximate 95% CI for β_1, computed from the profile likelihood, is (0.03,0.32), indicating some evidence of association between age and surgical mortality. Figure 6.3(d) shows a good quadratic approximation of the profile likelihood. So, alternatively we can report the Wald statistic $z = 0.160/0.072 = 2.22$.

In contrast with the normal model, note that there is no explicit variance parameter in the logistic regression model. The Bernoulli model automatically specifies a relationship between the mean and the variance; in engineering terms we use a 'Bernoulli noise'. Of course this implied specification might be wrong, for example the observed variance is inconsistent with the Bernoulli variance. An extension of the Bernoulli model that allows a more flexible variance term is the exponential family model in Section 6.5.

Grouped data

Suppose the i'th outcome consists of the number of successes y_i in n_i trials. A sensible model for such an outcome is that y_i is binomial(n_i, θ_i), where the success probability θ_i is a function of some predictors x_i. Consideration of the model and derivation of the likelihood (Exercise 6.8) are similar to the preceding development, which is only a special case for $n_i = 1$.

Example 6.4: Table 6.3, from Crowder (1978), shows the results of a 2×2 factorial experiment on seed variety and type of root extract. The outcome y_i is the number of seeds that germinated out of n_i planted seeds.

Seed A				Seed B			
Extract 1		Extract 2		Extract 1		Extract 2	
y_i	n_i	y_i	n_i	y_i	n_i	y_i	n_i
10	39	5	6	8	16	3	12
23	62	53	74	10	30	22	41
23	81	55	72	8	28	15	30
26	51	32	51	23	45	32	51
17	39	46	79	0	4	3	7
		10	13				

Table 6.3: *Seed germination data from Crowder (1978). The outcome y_i is the number seeds that germinated out of n_i planted seeds.*

The average germination rates for the four treatments are 0.36, 0.68, 0.39 and 0.53. The effect of root extract appears to be larger for seed A, so there is an indication of interaction. Assuming that y_i is binomial(n_i, p_i), consider the logistic regression

$$\text{logit } p_i = x_i'\beta$$

where β contains the constant term, the main effects for seed and root extract, and their interaction. The appropriate design matrix X is given by

$$\begin{pmatrix} 1 & 0 & 0 & 0 \\ 1 & 0 & 0 & 0 \\ 1 & 0 & 0 & 0 \\ 1 & 0 & 0 & 0 \\ 1 & 0 & 0 & 0 \\ 1 & 0 & 1 & 0 \\ 1 & 0 & 1 & 0 \\ 1 & 0 & 1 & 0 \\ 1 & 0 & 1 & 0 \\ 1 & 0 & 1 & 0 \\ 1 & 0 & 1 & 0 \\ 1 & 1 & 0 & 0 \\ 1 & 1 & 0 & 0 \\ 1 & 1 & 0 & 0 \\ 1 & 1 & 0 & 0 \\ 1 & 1 & 0 & 0 \\ 1 & 1 & 1 & 1 \\ 1 & 1 & 1 & 1 \\ 1 & 1 & 1 & 1 \\ 1 & 1 & 1 & 1 \\ 1 & 1 & 1 & 1 \end{pmatrix}.$$

Table 6.4 confirms the significant interaction term. Separate analyses within each seed group show that root extract has a significant effect. □

6.3 Poisson regression models

Example 6.5: A health insurance company is interested in studying how age is associated with the number of claims y filed for the previous year. The data from a sample of 35 customers are given in Table 6.5. Figure 6.4(a) shows the scatter plot of the number of claims versus age. □

Effects	Estimate	se
Constant	−0.56	0.13
Seed variety	0.15	0.22
Root extract	1.32	0.18
Interaction	−0.78	0.31

Table 6.4: *Summary analysis of germination data.*

Customer	Age	Number of claims	Customer	Age	Number of claims
1	18	0	19	31	0
2	20	1	20	31	3
3	22	1	21	32	4
4	23	0	22	33	2
5	23	0	23	33	0
6	24	0	24	33	1
7	24	1	25	34	2
8	25	0	26	34	3
9	25	5	27	34	0
10	27	0	28	35	1
11	28	1	29	35	2
12	28	2	30	35	1
13	28	2	31	37	2
14	29	4	32	37	5
15	30	2	33	37	1
16	30	1	34	39	2
17	30	3	35	40	4
18	30	1			

Table 6.5: *Health insurance claim data.*

It is sensible in this case to start with the assumption that y_i is Poisson with mean θ_i, where θ_i is a function of age. For example,

(i) $\theta_i = \beta_0 + \beta_1 \text{Age}_i$. Again, this simple linear model has a weakness in that it is not constrained to the range of $\theta_i > 0$. This is especially important if θ_i is near zero.

(ii) $\theta_i = \exp(\beta_0 + \beta_1 \text{Age}_i)$ or $\log \theta_i = \beta_0 + \beta_1 \text{Age}_i$. This log-linear structure is the most commonly used model for Poisson regression; it overcomes the weakness of the simple linear model.

Assuming a Poisson log-linear model, the log-likelihood of the parameters is given by

$$\begin{aligned}\log L(\beta_0, \beta_1) &= \sum_{i=1}^{n} \{-\theta_i + y_i \log \theta_i\} \\ &= \sum_i \{-\exp(\beta_0 + \beta_1 \text{Age}_i) + y_i(\beta_0 + \beta_1 \text{Age}_i)\}\end{aligned}$$

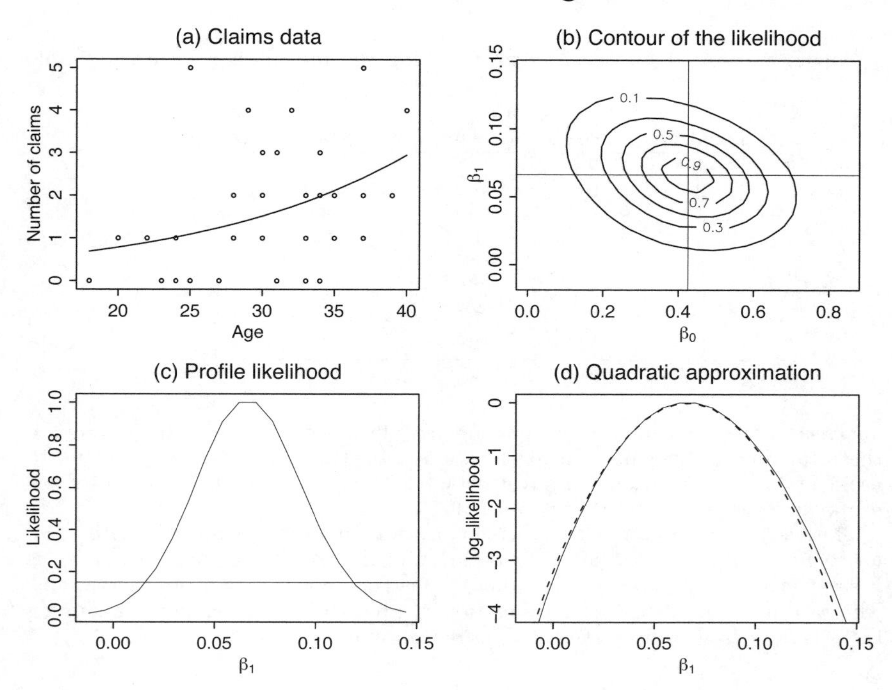

Figure 6.4: *(a) The customer claims data with the Poisson regression estimate. (b) The contours of the likelihood function. (c) The profile likelihood of β_1. (d) The quadratic approximation of the log-likelihood of β_1.*

where θ_i is a function of (β_0, β_1). To reduce the correlation between the estimates of β_0 and β_1 we centre the age data by setting Age $\leftarrow$ Age $-$ mean(Age). Figure 6.4(b) shows the contours at the approximate 10% to 90% confidence regions for the parameters. The MLEs of the parameters are summarized in the following table:

Effect	Parameter	Estimate	se
Intercept	β_0	0.43	0.14
Age	β_1	0.066	0.026

The profile likelihood of β_1 and its quadratic approximation are shown in Figures 6.4(c) and (d). Here the quadratic approximation is excellent. The Wald statistic to test H_0: $\beta_1 = 0$ is $z = 0.066/0.026 = 2.54$, so there is evidence that the number of claims is associated with age. The approximate 95% CI for the claim rate is 0.02 to 0.12 claims per customer per year.

Example 6.6: The following table shows the number of accidents at eight different locations, over a number of years, before and after installation of some traffic control measures. The question is whether there has been a significant change in the rate of accidents. For example, in location 1, before the traffic control measure was installed, there were 13 accidents occurring in 9 years; no accidents transpired for the 2 years following the installation. With a simple

	Before		After	
Location	Years	Accidents	Years	Accidents
1	9	13	2	0
2	9	6	2	2
3	8	30	3	4
4	8	20	2	0
5	9	10	2	0
6	8	15	2	6
7	9	7	2	1
8	8	13	3	2

analysis, if the accident rate is constant over locations, we can simply compare the total of 114 accidents over 68 location-years (rate of 1.676/year) versus 15 accidents over 18 location-years (rate of 0.833/year). This indicates the rate has dropped (rate ratio = 0.833/1.676 = 0.497).

Let y_{ij} be the number of accidents in location i under 'treatment' j, with $j = 0$ for 'before' and $j = 1$ for 'after' installation of the traffic control. Assume y_{ij} is Poisson with mean $\mu_{ij} = p_{ij}\lambda_{ij}$, where p_{ij} is the known period of observations. The rate λ_{ij} is modelled as the function of predictors. For example, assuming there is no location effect, we can consider a log-linear model

$$\begin{aligned}\log \mu_{ij} &= \log p_{ij} + \log \lambda_{ij}\\ &= \log p_{ij} + \lambda_0 + \tau_j,\end{aligned}$$

where τ_j is the effect of treatment j; assume that $\tau_0 = 0$, so τ_1 is the treatment contrast. The special predictor $\log p_{ij}$ is called an *offset term*; we can think of it as a predictor with known coefficient (equal to one in this case).

Computing the Poisson regression as in the previous example, we obtain a summary table:

Effect	Parameter	Estimate	se	z
Constant	λ_0	0.517	0.094	
Treatment	τ_1	−0.699	0.274	−2.55

So, the observed drop in accident rate appears to be significant. Note that the relative drop is $e^{-0.699} = 0.497$, matching the previous simple computation.

The main advantage of the Poisson regression model is that it can be easily extended if we believe there are other factors associated with accident rates. For example, we might consider

$$\log \lambda_{ij} = \lambda_0 + \ell_i + \tau_j, \tag{6.2}$$

where ℓ_i is the effect of location i; for identifiability assume, for example, that $\ell_0 = 0$. Estimation of this model is left as an exercise. □

6.4 Nonnormal continuous regression

Example 6.7: In a study of plant competition, a certain species was planted in 10 plots using various densities d; the density is measured by the number of plants per unit area. The outcome of interest y is the average yield per plant. The data are given in Table 6.6 and plotted in Figure 6.5(a). □

Plot	Density	Yield
1	5	122.7
2	10	63.0
3	15	32.5
4	20	34.5
5	30	31.4
6	40	17.7
7	60	21.9
8	80	21.3
9	100	18.4

Table 6.6: *Plant competition data.*

In view of Figure 6.5(a) and (b) it is sensible to model yield as inversely related to plant density. As a first approach and for future comparison we will analyse the data according to a normal linear model, where

$$1/y_i = \beta_0 + \beta_1 \log d_i + e_i,$$

but note that the errors appear to have larger variance for larger values of $1/y_i$. As before, to avoid the correlation between the estimates of β_0 and β_1 the log-density is centred: $\log d_i \leftarrow \log d_i - \text{mean}(\log d_i)$. The results of the regression analysis are summarized in the following table.

Effect	Parameter	Estimate	se
Intercept	β_0	0.0355	0.0023
Log-density	β_1	0.0155	0.0024

The unbiased estimate of error variance is $\widehat{\sigma}^2 = 0.0000468$. The Wald statistic here is $z = 0.0155/0.0024 = 6.46$, which, as expected, confirms a strong competition effect.

To account for unequal variability in the outcome values we now assume that y_i is exponential with mean μ_i, where

$$1/\mu_i = \beta_0 + \beta_1 \log d_i.$$

The variance of y_i is μ_i^2; using the Delta method the variance of $1/y_i$ is approximately $1/\mu_i^2$, consistent with the pattern in Figure 6.5(b). The exponential density is given by

$$p_{\mu_i}(y_i) = \mu_i^{-1} e^{-y_i/\mu_i}.$$

Given the observed data, the log-likelihood for the parameter $\beta = (\beta_0, \beta_1)$ is

$$\log L(\beta) = \sum_i \{-\log \mu_i - y_i/\mu_i\}$$

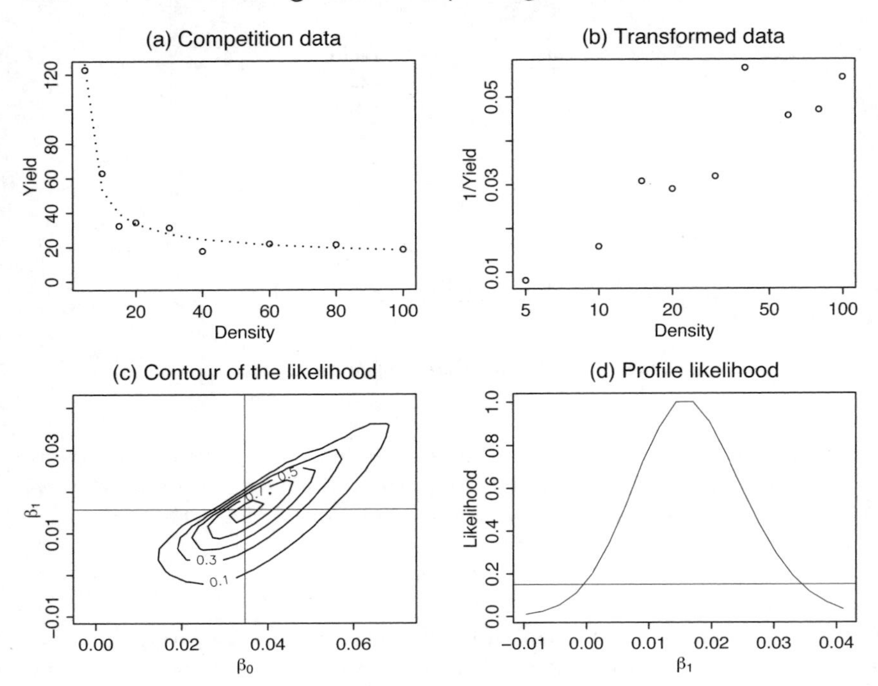

Figure 6.5: *(a) Competition data showing yield as a function of plant density. (b) Inverse yield is approximately linear in log density, but note the increasing variance. (c) Contours of the likelihood function. (d) The profile likelihood of* β_1.

$$= \sum_i \{\log(\beta_0 + \beta_1 \ \log d_i) - y_i(\beta_0 + \beta_1 \ \log d_i)\}.$$

The Fisher information of regression parameter $\beta = (\beta_0, \beta_1)$ is

$$\begin{aligned} I(\widehat{\beta}) &= -\frac{\partial^2}{\partial\beta\partial\beta'} \log L(\beta)\Big|_{\beta=\widehat{\beta}} \\ &= \sum_i \widehat{\mu}_i^2 x_i x_i' \end{aligned}$$

where the vector $x_i' \equiv (1, \log d_i)$. Defining the design matrix X appropriately, and setting a diagonal weight matrix $W = \text{diag}\{\widehat{\mu}_i^2\}$, we can write

$$I(\widehat{\beta}) = XWX,$$

so the estimated variance of $\widehat{\beta}$ is $(XWX)^{-1}$.

Figure 6.5(c) shows the contours of the likelihood function with the usual approximate 10% to 90% confidence regions. Figure 6.5(d) shows the profile likelihood of β_1. A summary of the parameter estimates is

Effect	Parameter	Estimate	se
Intercept	β_0	0.0347	0.0123
Log-density	β_1	0.0157	0.0085

The parameter estimates are similar to the normal-based estimates, but there is a dramatic change in the standard errors. The Wald statistic is $z = 0.0157/0.0085 = 1.85$, so now there is only moderate evidence for association. There is obviously something wrong with this approach.

The exponential model implies that the variance is the square of the mean. How do we check this assumption? If $\text{var}(y_i) = \mu_i^2$, then we should expect

$$\text{var}\left(\frac{y_i - \mu_i}{\mu_i}\right) = 1.$$

From the estimated model we can compute

$$\frac{1}{n-2}\sum_i \frac{(y_i - \widehat{\mu}_i)^2}{\widehat{\mu}_i^2} = 0.025,$$

where we use the $(n-p)$ divisor to get a less biased estimate. This suggests that the exponential model is not appropriate. The problem with the fixed mean–variance relationship is the same as that in the Poisson and logistic regression models. A larger family of models that overcomes this general weakness is given by the exponential family model discussed in the coming section.

6.5 Exponential family regression models

As discussed in Section 4.9, in these models the log-likelihood contribution of an outcome y_i is of the form

$$\log L(\theta_i, \phi) = \frac{y_i\theta_i - A(\theta_i)}{\phi} + c(y_i, \phi),$$

where $A(\theta_i)$ and $c(y_i, \phi)$ are known functions; the latter does not need to be explicit. The parameter ϕ is the dispersion parameter that allows a more flexible relationship between the mean and variance. For an exponential family model we have

$$Ey_i = A'(\theta_i) \equiv \mu_i$$

and

$$\text{var}(y_i) = \phi A''(\theta_i) \equiv \phi v(\mu_i).$$

Suppose we are interested to analyse the association between an outcome y and a predictor vector x. Using the general exponential family for a regression analysis requires two specifications. Firstly we need to specify $A(\theta_i)$, which is usually chosen from among the standard models: $A(\theta_i) = \theta_i^2/2$ from the normal model, $A(\theta_i) = e^{\theta_i}$ from the Poisson model, etc.; see Section 4.9. Hence we usually refer to this specification as the choice of distribution or family. The choice of $A(\theta_i)$ implies a certain mean-variance relationship.

Secondly we need to specify a *link function* $h(\mu_i)$ so that

$$h(\mu_i) = x_i'\beta.$$

In the previous sections we have used:

1. the identity link $h(\mu_i) = \mu_i$ for normal data
2. the logistic link $h(\mu_i) = \log\{\mu_i/(1-\mu_i)\}$ for Bernoulli data
3. the log link $h(\mu_i) = \log \mu_i$ for Poisson data
4. the inverse link $h(\mu_i) = 1/\mu_i$ for exponential data.

These cover most of the link functions used in practice; other possible links are, for example, the probit, complementary log-log and square-root. The choice of link function is usually determined by some subject matter or other theoretical considerations. With these two specifications, we might fit 'a normal model with an identity link', or 'a gamma model with a log link', etc.

Since $\mu_i = A'(\theta_i)$, there is an implied relationship

$$g(\theta_i) = x_i'\beta$$

between θ_i and β. The choice of $h(\mu_i)$ such that $\theta_i = h(\mu_i)$ or

$$\theta_i = x_i'\beta$$

is called the *canonical-link function.* We can check that the link functions listed above are the canonical link for the corresponding distributions. By choosing a canonical link we need only specify $A(\theta_i)$ or a distribution of y_i. While convenient, there is no reason why the canonical link is necessarily an appropriate link. For example, in some applications we may need to model a Poisson outcome using identity link.

The class of linear models under the general exponential family is called *generalized linear models* (GLM). Most of the nonnormal regression models performed in practice, such as logistic or Poisson regressions, are instances of GLM, so the class constitutes one of the most important frameworks for data analysis. One might argue that it will be easier to specify the mean–variance relationship directly, and, furthermore, one can check the relationship graphically. Such an approach is provided by the estimating equation approach, which is discussed in Chapter 14.

To apply the model for the competition data from Example 6.7, first look at the basic exponential model

$$\log L(\mu_i) = -y_i/\mu_i - \log \mu_i.$$

To extend this model we simply state that we use the exponential family model with $\theta_i = 1/\mu_i$, and

$$A(\theta_i) = \log \mu_i = -\log \theta_i.$$

The standard exponential model corresponds to $\phi = 1$; it can be shown that extending the standard model with free parameter ϕ is equivalent to using the gamma(μ_i, ϕ) model.

Let us continue with the inverse relationship

$$\mu_i = \frac{1}{\beta_0 + \beta_1 \log d_i},$$

or

$$1/\mu_i = \beta_0 + \beta_1 \log d_i,$$

i.e. using the inverse-link function $h(\mu_i) = 1/\mu_i$. Since $1/\mu_i = \theta_i$ the inverse link is the canonical-link function in this case; the possibility of the log-link function is given in Exercise 6.21. The total log-likelihood of $\beta = (\beta_0, \beta_1)$ is

$$\log L(\beta, \phi) = \sum_i \left\{ \frac{-y_i\theta_i + \log\theta_i}{\phi} + c(y_i, \phi) \right\},$$

where the explicit form of $c(y_i, \phi)$ is given by the gamma model (Section 4.8). For fixed value of ϕ, the estimation of β is exactly the same as that for the basic exponential model, so we will get the same MLE.

The dispersion parameter ϕ only modifies the standard errors. The Fisher information of the regression parameter is

$$\begin{aligned} I(\widehat{\beta}) &= -\frac{\partial^2}{\partial\beta\partial\beta'} \log L(\beta, \phi)\bigg|_{\beta=\widehat{\beta}} \\ &= -\phi^{-1}\frac{\partial^2}{\partial\beta\partial\beta'}\sum_i \log\theta_i\bigg|_{\beta=\widehat{\beta}} \\ &= \phi^{-1}\sum_i \widehat{\mu}_i^2 x_i x_i' \end{aligned}$$

where the vector $x_i' \equiv (1, \log d_i)$. Defining the design matrix X appropriately like before, and setting a diagonal weight matrix $W = \text{diag}\{\widehat{\mu}_i^2\}$, we can write

$$I(\widehat{\beta}) = \phi^{-1}(XWX).$$

Thus the Fisher information is modified by a factor ϕ^{-1} compared with the basic exponential model, or the standard errors of the MLEs are modified by a factor of $\sqrt{\phi}$.

In this example it is possible to estimate ϕ using an exact gamma likelihood; the approximate likelihood method is given in the next section. Alternatively, we might use the method of moments estimate: from

$$\text{var}(y_i) = \phi\mu_i^2,$$

or $\text{var}\{(y_i - \mu_i)/\mu_i\} = \phi$, we have

$$\widehat{\phi} = \frac{1}{n-2}\sum_i \frac{(y_i - \widehat{\mu}_i)^2}{\widehat{\mu}_i^2}.$$

Using a standard statistical package that allows GLM with the gamma family, we obtain the following summary table for the model estimates:

Effect	Parameter	Estimate	se
Intercept	β_0	0.0347	0.0019
Log-density	β_1	0.0157	0.0013

The estimated value of ϕ is $\widehat{\phi} = 0.025$. The standard error of $\widehat{\beta}_1$ is now more in line with the value from the normal regression; in fact, here we have a better standard error as a reward for better modelling of the unequal variances.

6.6 Deviance in GLM

One of the most important applications of the likelihood ratio statistic is in the concept of *deviance* in GLM. Its main use is for comparison nested models: analysis of deviance is a generalization of the classical analysis of variance. In some special cases, deviance also works as a measure of lack-of-fit.

For the moment assume a dispersion model with $\phi = 1$, so by definition the contribution of an outcome y_i to the log-likelihood is

$$\log L(\mu_i; y_i) = y_i\theta_i - A(\theta_i) + c(y_i, \phi = 1),$$

where $\mu_i = A'(\theta_i)$. Given outcome data $y = (y_1, \ldots, y_n)$ and a model for the mean $\mu = Ey$, let $L(\mu; y)$ be the likelihood of μ based on data y. For independent outcomes $\log L(\mu; y) = \sum_i \log L(\mu_i; y_i)$. The model μ might depend on further parameters; for example:

1. $\mu = \beta_0$, the constant model, also known as the 'null model'. It has one free parameter.
2. $h(\mu) = X\beta$, a general model with p unknown regression parameters. (The link function $h(\mu)$ applies element-wise to μ.) Given an estimate $\widehat{\beta}$, we compute $\widehat{\mu} = h^{-1}(X\widehat{\beta})$.

If μ does not follow any regression model, so $\widehat{\mu} = y$, it is called the 'saturated model', having n free parameters.

The deviance of a model for μ is defined as the likelihood ratio of the saturated model versus the particular model:

$$D(y, \mu) = 2 \log \frac{L(y; y)}{L(\mu; y)}.$$

It is a measure of distance between a particular model μ and the observed data y or the saturated model. The deviance of the null model is called the null deviance.

The term 'deviance' covers both the theoretical $D(y, \mu)$ and the observed $D(y, \widehat{\mu})$. We can also define an individual deviance

$$D(y_i, \mu_i) = 2 \log \frac{L(y_i; y_i)}{L(\mu_i; y_i)}.$$

where $L(\mu_i; y_i)$ is the likelihood contribution of y_i given a mean model μ_i. For independent data

$$D(y, \mu) = \sum_i D(y_i, \mu_i).$$

Example 6.8: Suppose y_i is independent $N(\mu_i, \sigma^2 = 1)$ for $i = 1, \ldots, n$. We have

$$\log L(\mu; y) = -\frac{1}{2} \sum_i (y_i - \mu_i)^2,$$

so the deviance of a model μ is

$$D(y, \mu) = \sum_i (y_i - \mu_i)^2,$$

which is equal to the error sum of squares, and motivates calling the observed deviance $D(y, \widehat{\mu})$ a 'residual deviance'. The individual deviance

$$D(y_i, \mu_i) = (y_i - \mu_i)^2$$

suggests a concept of 'deviance residual'

$$r_{Di} = \text{sign}(y_i - \mu_i) \sqrt{D(y_i, \mu_i)}$$

that might be useful for residual analysis.

Suppose we model $\mu = X\beta$, where X is of rank p. Then the observed deviance $D(y, \widehat{\mu})$ is χ^2 with $n - p$ degrees of freedom. Note that this assumes $\sigma^2 = 1$, and the χ^2 distribution is not generally true for nonnormal models. However, the degrees of freedom $n - p$ is deemed applicable in all cases. □

Example 6.9: Suppose y_i is binomial(n_i, p_i), where $\mu_i = n_i p_i$. Then

$$\begin{aligned} \log L(\mu) &= \sum_i \{y_i \log p_i + (n_i - y_i) \log(1 - p_i)\} \\ &= \sum_i \left\{ y_i \log \frac{\mu_i}{n_i} + (n_i - y_i) \log \frac{n_i - \mu_i}{n_i} \right\}, \end{aligned}$$

so

$$D(y, \widehat{\mu}) = 2 \sum_i \left\{ y_i \log \frac{y_i}{\widehat{\mu}_i} + (n_i - y_i) \log \frac{n_i - y_i}{n_i - \widehat{\mu}_i} \right\}.$$

In the extreme case $n_i \equiv 1$, which usually happens if we perform logistic regression on a continuous predictor, y_i is a zero-one outcome. Defining $0 \log 0 \equiv 0$, we get

$$D(y, \widehat{\mu}) = -2 \log L(\widehat{\mu}; y). \ \Box$$

Example 6.10: Suppose y_i is independent Poisson with mean μ_i for $i = 1, \ldots, n$. The individual deviance is

$$D(y_i, \mu_i) = 2 \left\{ y_i \log \frac{y_i}{\mu_i} - (y_i - \mu_i) \right\}.$$

From independence, the total deviance is $D(y, \mu) = \sum_i D(y_i, \mu_i)$.

Model comparison

Deviance is used mainly to compare two nested models. Suppose we have:

$$\begin{aligned} A: \quad \mu_A &= X_1\beta_1 \\ B: \quad \mu_B &= X_1\beta_1 + X_2\beta_2, \end{aligned}$$

where X_1 is of rank p and X_2 is of rank q, i.e. model A is a subset of model B. The difference in the observed deviance

$$D(y, \widehat{\mu}_A) - D(y, \widehat{\mu}_B) = 2 \log \frac{L(\widehat{\mu}_B; y)}{L(\widehat{\mu}_A; y)},$$

is the usual likelihood ratio test for the hypothesis H_0: $\beta_2 = 0$; the null distribution is approximately χ^2 with q degrees of freedom, equal to the difference in degrees of freedom of $D(y, \widehat{\mu}_A)$ and $D(y, \widehat{\mu}_B)$. Assuming $\phi = 1$, this use of deviance is asymptotically valid, regardless of whether the individual deviances are χ^2 or not.

Example 6.11: For the logistic regression analysis of the surgical data (Example 6.3), it is common to report an analysis of deviance table

Model	Deviance	df	Change	df
Constant	51.796	39	–	–
Constant + Age	46.000	38	5.796	1

The change in deviance (5.796 with 1 degree of freedom) indicates that the effect of age is significant. For comparison, we have shown before that $\widehat{\beta}_1 = 0.160$ with standard error equal to 0.072, so the Wald test gives $(0.160/0.072)^2 = 4.94$, comparable to the change in deviance. □

Example 6.12: The analysis of deviance of the accident data in Example 6.6 can be summarized as follows.

Model	Deviance	df	Change	df
Constant	58.589	15	–	–
Constant + Treatment	50.863	14	7.726	1

The χ^2 test for treatment effect is 7.726 with one degree of freedom, indicating a strong treatment effect (P-value= 0.005). □

Scaled deviance

Deviance is defined using $\phi = 1$. If $\phi \neq 1$ then the change in deviance no longer matches the likelihood ratio statistic, but we only need to divide it by ϕ to make it valid. The quantity $D(y, \mu)/\phi$ is called the *scaled deviance.* To compare models A and B above we would use

$$\frac{D(y, \widehat{\mu}_A) - D(y, \widehat{\mu}_B)}{\phi},$$

which is approximately χ^2 with q degrees of freedom. If ϕ is unknown it is common practice simply to plug in an estimated value.

Normal models

Suppose y_i is $N(\mu_i, \sigma^2)$, independently over $i = 1, \ldots, n$. The observed deviance is still

$$D(y, \widehat{\mu}) = \sum_i (y_i - \widehat{\mu}_i)^2.$$

If the error variance σ^2 is known externally, the scaled deviance can be used to test whether the model for μ is acceptable. Assuming $\mu = X\beta$, where X is of rank p,

$$D(y, \widehat{\mu})/\sigma^2 \sim \chi^2_{n-p}.$$

More often than not σ^2 is unknown, in which case D cannot work as a goodness-of-fit statistic. From the normal theory linear models, to compare two models A versus B we use the scaled deviance

$$\frac{D(y, \widehat{\mu}_A) - D(y, \widehat{\mu}_B)}{\widehat{\sigma}^2} \sim \chi^2_{\mathrm{df}_A - \mathrm{df}_B}$$

where the error variance $\widehat{\sigma}^2$ is usually estimated from the larger model B, and df_A and df_B are the degrees of freedom of $D(y, \widehat{\mu}_A)$ and $D(y, \widehat{\mu}_B)$, respectively. Note that, under the normal assumption, we also have an exact F-distribution for the change in scaled deviance.

Deviance as a measure of lack of fit

Under some conditions we can use the deviance for a goodness-of-fit test: a large deviance indicates a poor fit, which can happen for one or both of the following reasons:

- the mean model is not adequate; for example, there should be more predictors in the model
- there is overdispersion, i.e. the assumption of $\phi = 1$ is not tenable.

Uncovering the reasons for the lack of fit is not always straightforward. Some subject matter knowledge about the model or a careful residual analysis might be required. If we can attribute the lack of fit to overdispersion, model comparison should be based on scaled deviances.

We have derived in Example 6.9 for binomial data

$$D(y,\widehat{\mu}) = 2\sum_i \left\{ y_i \log \frac{y_i}{\widehat{\mu}_i} + (n_i - y_i)\log \frac{n_i - y_i}{n_i - \widehat{\mu}_i} \right\}.$$

We can think of the data as an $n \times 2$ contingency table with y_i's and $(n_i - y_i)$'s as the observed (O) frequencies, and $\widehat{\mu}_i$'s and $(n_i - \widehat{\mu}_i)$'s the expected (E) frequencies. Thus, we can recognize

$$D = 2\sum O \log \frac{O}{E},$$

which is approximately the same as Pearson's χ^2 goodness-of-fit statistic (see Theorem 9.9)

$$\chi^2 = \sum \frac{(O-E)^2}{E}.$$

So if the expected frequencies are large enough the deviance may be used as a measure of lack of fit. The same reasoning applies to counting data generally.

In the extreme case $n_i \equiv 1$, defining $0 \log 0 \equiv 0$, we get

$$D(y,\widehat{\mu}) = -2\log L(\widehat{\mu}; y),$$

which is not meaningful as a measure of goodness-of-fit. Here $D(y,\widehat{\mu})$ is used only for model comparisons. Checking the adequacy of the model takes more work, for example by splitting the data into several groups.

Example 6.13: For the analysis of surgery data in Example 6.11, $n_i = 1$, so the deviance value $D = 46.0$ with 38 degrees of freedom is not meaningful as a measure of lack of fit.

Example 6.14: Deviance also works as a measure of lack of fit in Poisson regression, provided the means $\widehat{\mu}_i$'s are large enough. For the analysis of accident data in Example 6.12, the deviance of 50.863 with 14 degrees of freedom indicates a lack of fit for a model that only contains the treatment effect. It can be verified that adding location as a categorical variable into the model (adding 7 parameters) would give a final deviance of 16.28 with 7 degrees of freedom. This is a significant improvement on the model fit, though the deviance is still borderline significant (P-value=0.03).

Estimating dispersion parameter ϕ

From Section 4.9 the approximate log-likelihood contribution from a single observation y_i is

$$\log L_i \approx -\frac{1}{2}\log\{2\pi\phi v(y_i)\} - \frac{1}{2\phi} D(y_i, \mu_i). \tag{6.3}$$

The formula is exact if y_i is $N(\mu_i, \sigma^2)$. Nelder and Pregibon (1987) call it the extended quasi-likelihood. The approximation is sensible if the likelihood based on y_i is reasonably regular.

Given independent data $y_1, \ldots, y_n$, for any ϕ the estimate of μ_i is the minimizer of the total deviance

$$\sum_i D(y_i, \mu_i).$$

Therefore the profile log-likelihood of ϕ is

$$\log L(\phi) \approx \sum_i \left\{ -\frac{1}{2} \log\{2\pi\phi v(y_i)\} - \frac{1}{2\phi} D(y_i, \widehat{\mu}_i) \right\},$$

and the approximate MLE of ϕ is the average deviance

$$\widehat{\phi} = \frac{1}{n} \sum_i D(y_i, \widehat{\mu}_i).$$

Likelihood inference on ϕ is available using the profile likelihood. In practice it is common to use a bias-corrected estimate

$$\widehat{\phi} = \frac{1}{n-p} \sum_i D(y_i, \widehat{\mu}_i).$$

where $n - p$ is the degrees of freedom of the deviance. For example, in Example 6.14 above, assuming Poisson model with overdispersion for the final model, we can estimate ϕ by $\widehat{\phi} = 16.28/7 = 2.3$.

If the likelihood approximation is doubtful, we can use the method of moments estimate. Since $Ey_i = \mu_i$ and $\text{var}(y_i) = \phi v(\mu_i)$, we have

$$\text{var}\left(\frac{y_i - \mu_i}{\sqrt{v(\mu_i)}} \right) = \phi,$$

suggesting

$$\widehat{\phi} = \frac{1}{n-p} \sum_i \frac{(y_i - \widehat{\mu}_i)^2}{v(\widehat{\mu}_i)}$$

as a sensible estimate of ϕ.

Profile deviance plot

There is a variety of software programs to perform GLM, and most would report the deviance of a model. Although there is usually no option to output the profile likelihood for a particular parameter of interest, it is quite easy to generate one. This is useful when the normality of the Wald statistic is in doubt. GLM programs generally allow an offset term, a known part of the mean model, so we can compute the deviance over a range of fixed values of the parameter of interest.

For example, for the surgical data in Example 6.3:

- fix β_1 and run

$$\text{logit } p_i = \beta_0 + \text{offset}(\beta_1 \text{Age})$$

and record the value of the deviance as $D(\beta_1)$. (In a general regression model, there will be other predictors in the model.)

- repeat this over a range of reasonable values of β_1 around $\widehat{\beta}_1$
- up to an additive constant, the deviance and the profile likelihood are related by

$$D(\beta_1)/\phi = -2 \log L(\beta_1).$$

(The dispersion parameter $\phi = 1$ in the standard logistic regression. If it is unknown then we use the estimated value.) Likelihood-based CIs can be read off the deviance plot:

$$\text{CI} = \left\{ \beta_1;\ \frac{D(\beta_1) - D(\widehat{\beta}_1)}{\phi} < \chi^2_{1,(1-\alpha)} \right\}.$$

It is convenient to set the minimum of the deviance plot to zero, which is equivalent to setting the maximum of the likelihood plot to one. We can gauge the validity of the Wald statistic by how close the deviance plot is approximated by a quadratic function around $\widehat{\beta}_1$. See Figure 6.6.

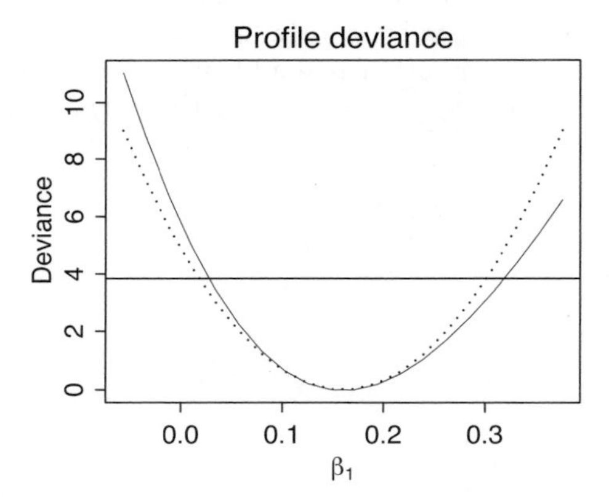

Figure 6.6: *Profile deviance of the slope parameter β_1 (solid line) and its normal approximation (dotted line).*

Example 6.15: A survey is conducted of a number of companies to find out whether they are planning to use internet trading facilities (internet=1 if yes). The following table shows the breakdown of the companies by whether they are located in a big city (city=1), and whether they serve only the domestic market (domestic=1). We want to establish if differences exist among various types of companies.

City	Domestic	Internet 1	0
0	0	0	3
0	1	3	4
1	0	50	2
1	1	27	14

Consider the full model

$$\text{logit } p_i = \beta_0 + \beta_1 \text{City} + \beta_2 \text{Domestic} + \beta_3 \text{City} \times \text{Dom.}$$

Standard logistic regression programs produce the following output:

	Binomial		Binary	
Effects	$\widehat{\beta}$	se	$\widehat{\beta}$	se
Constant	−20.9	12,111	−7.6	15.4
City	20.6	12,111	7.3	15.4
Domestic	24.1	12,111	10.8	15.4
City × Dom.	−23.1	12,111	−9.8	15.4

The 'Binomial' columns show the output using data given in the table as binomial outcomes, while the 'Binary' columns show the output when the outcome was set at zero-one value. In both cases the interaction effect is not significant. That the two results are so different indicates something suspicious.

The analysis of deviance for the additive model

$$\text{logit } p_i = \beta_0 + \beta_1 \text{City} + \beta_2 \text{Domestic}$$

yields $D = 8.44$, highly significant at 1 degree of freedom. The deviance of the additive model provides a test for the term City×Dom., so $D = 8.44$ indicates a significant interaction effect, inconsistent with the output table. The Wald statistics computed from the table are either $z^2 = (-23.1/12,111)^2 = 0.000004$ or $z^2 = (-9.8/15.4)^2 = 0.40$.

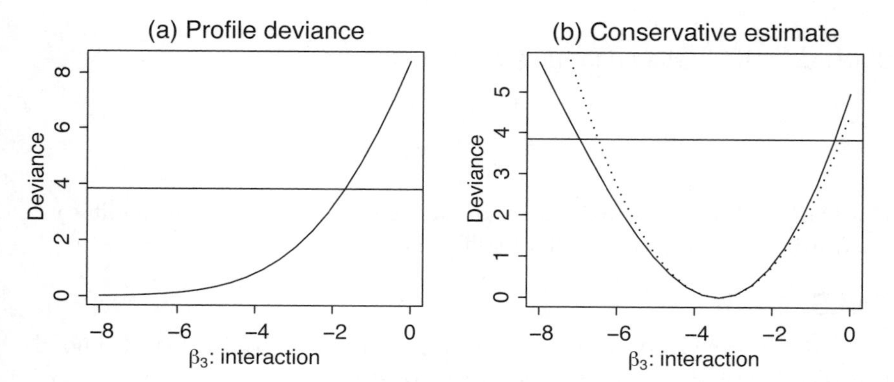

Figure 6.7: *(a) Profile deviance for interaction effect β_3. (b) A conservative estimate of β_3 is obtained by changing the zero outcome to one in the first row of the data. The profile deviance (solid) in this case is well approximated by a quadratic (dotted).*

The Wald statistic fails here since the MLE $\widehat{\beta}_3 = -\infty$; this is due to the zero-level outcome in one of the categories. See Figure 6.7 for the profile deviance of

β_3. This means that the quadratic approximation for the log-likelihood is off, and the Wald statistic or the standard error term is meaningless. This problem will always occur in logistic regression with categorical predictors where one category has zero outcome. That $\widehat{\beta}_3$ is highly significant is ascertained by considering a conservative analysis where the zero outcome in the first line of the data is replaced by one. Figure 6.7 shows that even in this small sample the log-likelihood is well approximated by a quadratic. □

6.7 Iterative weighted least squares

Numerical algorithms to find the MLEs and standard errors are crucial for routine applications of GLM. It turns out that there is one general algorithm, called the iterative weighted least squares (IWLS), that works reliably for GLM. There are several ways to derive IWLS (Section 14.2), but one that is relevant now is via the Newton–Raphson procedure.

Newton–Raphson procedure

This is a general procedure to solve $g(x) = 0$. We start with an initial estimate x^0, then linearize $g(x)$ around x^0, and set it to zero:

$$g(x) \approx g(x^0) + g'(x^0)(x - x^0) = 0.$$

The solution of the linear equation provides an update formula

$$x^1 = x^0 - g(x^0)/g'(x^0).$$

For maximum likelihood estimation we want to solve the score equation

$$S(\beta) = 0.$$

Starting with β^0, the updating formula is

$$\begin{aligned} \beta^1 &= \beta^0 - \{S'(\beta^0)\}^{-1}S(\beta^0) \\ &= \beta^0 + \{I(\beta^0)\}^{-1}S(\beta^0). \end{aligned}$$

Note that a linear approximation of the score function is equivalent to a quadratic approximation of the log-likelihood.

IWLS

Applying this to GLM estimation problems, we start with the log-likelihood contribution of an observation y_i of the form

$$\log L(\theta_i, \phi) = \frac{y_i\theta_i - A(\theta_i)}{\phi} + c(y_i, \phi),$$

and assume that the observations are independent. We consider a regression model $h(\mu_i) = x_i'\beta$, and try to get the estimate of β at a fixed value of ϕ. The score equation is

$$S(\beta) = \phi^{-1} \sum_i \frac{\partial \theta_i}{\partial \beta} \{y_i - A'(\theta_i)\},$$

and the Fisher information is

$$I(\beta) = \phi^{-1} \sum_i \left[-\frac{\partial^2 \theta_i}{\partial \beta \partial \beta'} \{y_i - A'(\theta_i)\} + \frac{\partial \theta_i}{\partial \beta} \frac{\partial \theta_i}{\partial \beta'} A''(\theta_i) \right], \tag{6.4}$$

which in general can be complicated.

Since $A'(\theta_i) = \mu_i$, we have

$$\begin{aligned} A''(\theta_i) &= \partial \mu_i / \partial \theta_i = v_i \\ \frac{\partial \theta_i}{\partial \beta} &= \frac{\partial \theta_i}{\partial \mu_i} \frac{\partial \mu_i}{\partial h} \frac{\partial h}{\partial \beta} \\ &= v_i^{-1} \frac{\partial \mu_i}{\partial h} x_i, \end{aligned}$$

so the second term of $I(\beta)$ is

$$\sum_i \left\{ \left(\frac{\partial h}{\partial \mu_i} \right)^2 \phi v_i \right\}^{-1} x_i x_i' \equiv U.$$

Similarly

$$S(\beta) = \sum_i \left\{ \left(\frac{\partial h}{\partial \mu_i} \right)^2 \phi v_i \right\}^{-1} x_i \frac{\partial h}{\partial \mu_i} (y_i - \mu_i).$$

A major simplification of the Newton–Raphson algorithm occurs when we use the canonical-link function $\theta_i = h(\mu_i) = x_i'\beta$, from which

$$\begin{aligned} \frac{\partial \theta_i}{\partial \beta} &= x_i \\ \frac{\partial^2 \theta_i}{\partial \beta \partial \beta'} &= 0, \end{aligned}$$

so

$$I(\beta) = U$$

and the Newton–Raphson update is

$$\beta^1 = \beta^0 + U^{-1} S(\beta^0).$$

With the canonical link we also have an interesting relationship $\partial \mu_i / \partial h = \partial \mu_i / \partial \theta_i = v_i$, or $\partial h / \partial \mu_i = v_i^{-1}$.

Now let X be the design matrix of predictor variables, Σ a diagonal matrix with elements

$$\Sigma_{ii} = \left(\frac{\partial h}{\partial \mu_i}\right)^2 \phi v_i,$$

so $U = (X'\Sigma^{-1}X)$ and

$$S(\beta) = X'\Sigma^{-1}\frac{\partial h}{\partial \mu}(y - \mu),$$

where $\frac{\partial h}{\partial \mu}(y - \mu)$ is a vector of $\frac{\partial h}{\partial \mu_i}(y_i - \mu_i)$. We can now re-express the update formula as

$$\begin{aligned}\beta^1 &= \beta^0 + (X'\Sigma^{-1}X)^{-1}X'\Sigma^{-1}\frac{\partial h}{\partial \mu}(y - \mu)\\ &= (X'\Sigma^{-1}X)^{-1}X'\Sigma^{-1}\{X\beta^0 + \frac{\partial h}{\partial \mu}(y - \mu)\}\\ &\equiv (X'\Sigma^{-1}X)^{-1}X'\Sigma^{-1}Y, \qquad (6.5)\end{aligned}$$

where Y is a vector of

$$Y_i = x_i'\beta^0 + \frac{\partial h}{\partial \mu_i}(y_i - \mu_i) \qquad (6.6)$$

and all unknown parameters are evaluated at the current values. Note that ϕ cancels out in the formula to compute β^1. In GLM terminology Y is called the *working vector*. The iteration continues by first recomputing μ, Y and Σ. So, (6.5) and (6.6) are the key formulae in IWLS.

These formulae can be connected to a quadratic approximation of the log-likelihood. In effect, starting with β^0, the exponential family log-likelihood is approximated by

$$-\frac{1}{2}\log|\Sigma| - \frac{1}{2}(Y - X\beta)'\Sigma^{-1}(Y - X\beta) \qquad (6.7)$$

with Y and Σ defined above.

At convergence, we can evaluate the standard errors for the estimates from inverse of

$$I(\widehat{\beta}) = (X'\Sigma^{-1}X),$$

where the variance matrix Σ is evaluated using the estimates $\widehat{\beta}$ and $\widehat{\phi}$.

For general link functions we can actually still view the IWLS algorithm as a Newton–Raphson algorithm with the so-called Fisher scoring, i.e. by using the expected Fisher information

$$\mathcal{I}(\beta) = EI(\beta)$$

instead of the observed information $I(\beta)$. Since $Ey_i = A'(\theta_i)$, we get from (6.4)

$$\mathcal{I}(\beta) = U,$$

so the IWLS algorithm stays as it is.

Example 6.16: Suppose y_i is Poisson with mean μ_i (dispersion parameter $\phi = 1$), and we specify a log-linear model

$$h(\mu_i) = \log \mu_i = x_i'\beta.$$

From these assumptions, $\text{var}(y_i) = v_i = \mu_i$, and $\partial h/\partial \mu_i = 1/\mu_i$, so the algorithm proceeds as follows. Start with β^0, then iterate the following until convergence:

- compute

$$\begin{aligned} \mu_i^0 &= e^{x_i'\beta^0} \\ Y_i &= x_i'\beta^0 + (y_i - \mu_i^0)/\mu_i^0 \\ \Sigma_{ii} &= (1/\mu_i^0)^2\mu_i^0 = 1/\mu_i^0 \end{aligned}$$

- update $\beta^1 = (X'\Sigma^{-1}X)^{-1}X'\Sigma^{-1}Y$.

The starting value β^0 can be computed, for example, from the ordinary least-squares estimate of β in the model $\log(y_i + 0.5) = x_i'\beta$. Alternatively, we can start will all the β-coffecients set to zero, except for the constant term. As a numerical exercise, the reader can now verify the output summaries given in Section 6.3.

Example 6.17: When a noncanonical-link function is used, the IWLS is still the algorithm of choice to compute the parameter estimates. In practice, however, it is common to then compute the standard errors of the estimate using the expected Fisher information. To see that there is something at issue, the first term in the observed Fisher information in (6.4) contains a general formula

$$\frac{\partial^2\theta_i}{\partial\beta\partial\beta'} = \frac{\partial^2\theta_i}{\partial\mu_i^2}\left(\frac{\partial\mu_i}{\partial h}\right)^2 x_i x_i' + \frac{\partial\theta_i}{\partial\mu_i}\frac{\partial^2\mu_i}{\partial h^2}x_i x_i'.$$

In the Poisson model, we have the canonical parameter

$$\theta_i = \log(\mu_i),$$

so if we use, for example, the identity link

$$h(\mu_i) = \mu_i = x_i'\beta$$

we obtain

$$\frac{\partial\mu_i}{\partial h} = 1 \quad \text{and} \quad \frac{\partial^2\mu_i}{\partial h^2} = 0,$$

and

$$I(\beta) = \phi^{-1}\sum_i \left\{\frac{1}{\mu_i^2}(y_i - \mu_i)x_i x_i' + \frac{1}{\mu_i}x_i x_i'\right\}.$$

In contrast, the expected Fisher information is

$$\mathcal{I}(\beta) = \phi^{-1}\sum_i \frac{1}{\mu_i}x_i x_i'.$$

Standard errors derived from these can be quite different if μ_i's are not too large. We will discuss in Section 9.6 that, for the purpose of inference, $I(\beta)$ is better than $\mathcal{I}(\beta)$. □

6.8 Box–Cox transformation family

The use of transformation is discussed in Section 4.10 as a way of extending the normal model for positive-valued continuous data. It is assumed that there is $\lambda \neq 0$ such that a transformation of the observed data y according to

$$y_\lambda = \frac{y^\lambda - 1}{\lambda}$$

has a normal model $N(\mu, \sigma^2)$. The value $\lambda = 0$ is defined to represent the log-transformation. Extension to a regression model is clear: we can specify that the transformed value $y_{\lambda i}$ follows a linear model in terms of some predictors x_i, i.e.

$$EY_{\lambda i} = \mu_i = x_i'\beta$$

and $\text{var}(Y_{\lambda i}) = \sigma^2$.

The log-likelihood contribution of a single observation y_i is

$$\log L(\lambda, \beta, \sigma^2) = -\frac{1}{2}\log\sigma^2 - \frac{(y_{\lambda i} - x_i'\beta)^2}{2\sigma^2} + (\lambda - 1)\log y_i.$$

At each value of λ, the estimation of the other parameters follows exactly the usual normal-based regression analysis of the transformed data. By defining the design matrix X and vector Y_λ appropriately

$$\widehat{\beta}(\lambda) = (X'X)^{-1}X'Y_\lambda$$

and, setting $\widehat{\mu}_i(\lambda) = x_i'\widehat{\beta}(\lambda)$,

$$\widehat{\sigma}^2(\lambda) = \frac{1}{n}\sum_i \{y_{\lambda i} - \widehat{\mu}_i(\lambda)\}^2.$$

The appropriate value of λ can be found from the profile log-likelihood of λ

$$\log L(\lambda) = -\frac{n}{2}\log\widehat{\sigma}^2(\lambda) - \frac{n}{2} + (\lambda - 1)\sum_i \log y_i.$$

In practice we use simple values of λ so that there is no serious problem of interpretation.

Example 6.18: Sulphur dioxide is one of the major air pollutants; it is released into the air by gas, oil or coal burning, and, on chronic exposure, it can cause respiratory diseases. The dataset in Table 6.7 (from Sokal and Rohlf 1981) was collected in 41 US cities in 1969–1971. The outcome variable y is the sulphur dioxide content of air (micrograms/m^3), and the predictor variable x is the number of manufacturing enterprises employing 20 or more workers. Since there is no indication that the industry is limited to those relying on oil or coal for energy, we expect to see a lot of noise in the relationship; see Figure 6.8(a).

We first log-transform the x-axis to show a clearer pattern. From Figure 6.8(b) it seems obvious that we should consider a quadratic model

$$y_i = \beta_0 + \beta_1 \log x_i + \beta_2 \log^2 x_i + e_i.$$

(We will treat this as an empirical model and not try to interpret the relationship. Note also that β_1 and β_2 cannot be interpreted separately; for that we need, at least, to centre the predictor variable before analysis.)

City	x	y	City	x	y
1	35	31	22	361	28
2	44	46	23	368	24
3	46	11	24	379	29
4	80	36	25	381	14
5	91	13	26	391	11
6	96	31	27	412	56
7	104	17	28	434	29
8	125	8	29	453	12
9	136	14	30	454	17
10	137	28	31	462	23
11	181	14	32	569	16
12	197	26	33	625	47
13	204	9	34	641	9
14	207	10	35	699	29
15	21	10	36	721	10
16	266	26	37	775	56
17	275	18	38	1007	65
18	291	30	39	1064	35
19	337	10	40	1692	69
20	343	94	41	3344	110
21	347	61			

Table 6.7: *Pollution data from 41 US cities (Sokal and Rohlf 1981). The variable x is the number of manufacturing enterprises employing 20 or more workers, and y is the sulphur dioxide content of air (micrograms/m^3).*

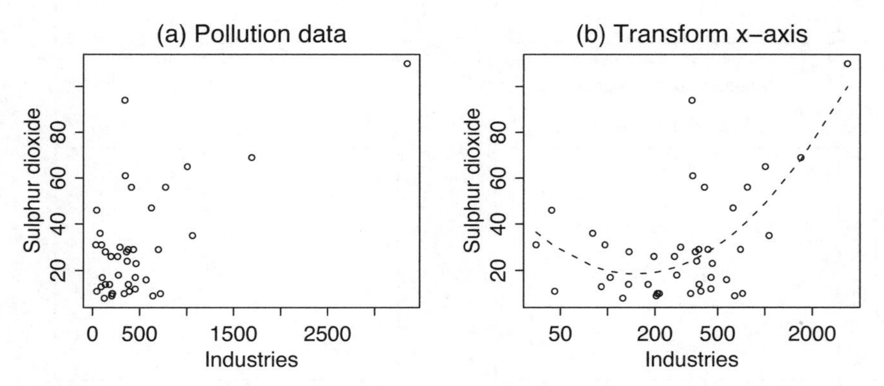

Figure 6.8: *(a) It is hard to see any relationship in the original scale. (b) Putting industry on a log-scale shows a clearer pattern. The dashed line is a quadratic fit.*

Assuming a normal model on e_i we get the following summary table for the model estimates.

Effect	Parameter	Estimate	se
Intercept	β_0	231.44	68.37
$\log x$	β_1	−84.97	24.36
$\log^2 x$	β_2	8.47	2.15

The estimated residual variance is $\widehat{\sigma}^2 = 18.3^2$, and, as expected, the quadratic model is highly significant. The residual plot in Figure 6.9(a) indicates some nonnormality, so it is a question whether y is the right scale for analysis.

Now consider the family of Box–Cox transforms so that

$$y_{\lambda i} = \beta_0 + \beta_1 \log x_i + \beta_2 \log^2 x_i + e_i.$$

The profile likelihood of λ in Figure 6.9(b) shows that we should use $\lambda = 0$ or log transformation. From the log-transformed data we obtain the following summary table:

Effect	Parameter	Estimate	se
Intercept	β_0	8.14	2.29
$\log x$	β_1	−2.08	0.81
$\log^2 x$	β_2	0.20	0.072

The estimated residual variance is $\widehat{\sigma}^2 = 0.61^2$. The normal plot in Figure 6.9(d) shows better behaved residuals. □

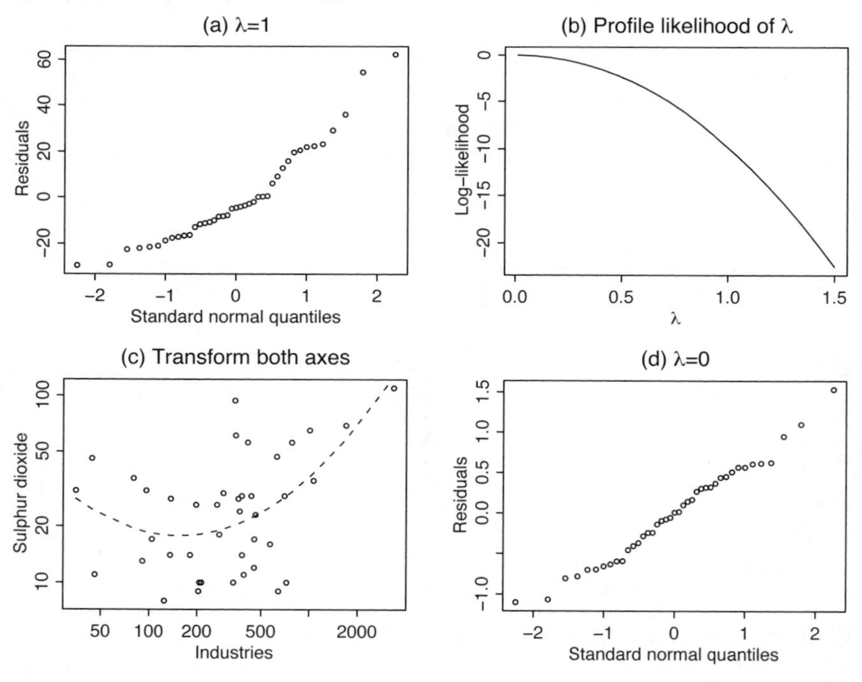

Figure 6.9: *(c) Normal plot of the residuals from Figure 6.8(b). (b) The profile likelihood of λ indicates we should use $\lambda = 0$ or a log-transform on the outcome. (c) The data in a log-log scale with the quadratic fit. (d) Normal plot of the residuals from (c).*

Transform the mean or the observation?

The exponential and Box–Cox transformation families are the two main approaches in dealing with nonlinear transformations. Their common objective is to arrive at a sensible linear model. In the former we apply a link function $h(\cdot)$ on the mean parameter μ such that

$$h(\mu) = x'\beta,$$

so the model can be aptly called a *parameter–transform model.* With Box–Cox models we apply the transformation $g(\cdot)$ on the observations y so that

$$Eg(y) = x'\beta,$$

where $g(\cdot)$ belongs to a certain class of functions. Such a model is called an *observation–transform model.* We have actually used both on the same dataset: see the analysis of plant competition data in Sections 6.4 and 6.5.

The main advantage of the parameter–transform model is that the distribution of the data is not affected by the transformation; this may make the analysis easier to interpret, especially if the result is used for prediction. See also the discussion in Section 4.10. When used empirically to describe relationships both models are on an equal footing. 'Let the data decide' would be the best approach. We can use the AIC (Section 3.5) for such a purpose, but generally it can be a difficult question with no definitive answer.

A joint approach is possible. For example, we might consider the Box–Cox transformation as a family of link functions

$$\frac{\mu_i^\lambda - 1}{\lambda} = x_i'\beta.$$

The parameter λ gives the link function an extra flexibility in possible shapes. Additionally, we can compare the likelihood of different link functions such as identity, inverse or log links by comparing the likelihood of different λ values.

6.9 Location-scale regression models

Example 6.19: The stack-loss dataset in Table 6.8 has been analysed by many statisticians. Brownlee (1965), the source of the data, Daniel and Wood (1971) and Draper and Smith (1981) used the classical regression model. Denby and Mallows (1977), Li (1985) and Lange *et al.* (1989) applied the robust regression approach. The data were recorded from 21 days of operation of a chemical plant to oxidize ammonia NH_3 into nitric acid HNO_3. The variables are

x_1 = air flow, which measures the rate of operation.

x_2 = cooling temperature of the coils in the absorbing tower of HNO_3.

x_3 = concentration of HNO_3 in the absorbing liquid (coded as $10\times$(original data-50)).

y = the 'stack loss', which is the percentage loss of NH_3 ($\times 10$).

Figure 6.10(a) show the relationship between stack loss and air flow. A linear fit of stack loss on air flow produces heavy-tailed residuals, as shown in Figure 6.10(b). This indicates the need to consider a heavy-tailed model such as the Cauchy. □

Day	Flow	Temp.	Concen.	Loss
1	80	27	89	42
2	80	27	88	37
3	75	25	90	37
4	62	24	87	28
5	62	22	87	18
6	62	23	87	18
7	62	24	93	19
8	62	24	93	20
9	58	23	87	15
10	58	18	80	14
11	58	18	89	14
12	58	17	88	13
13	58	18	82	11
14	58	19	93	12
15	50	18	89	8
16	50	18	86	7
17	50	19	72	8
18	50	19	79	8
19	50	20	80	9
20	56	20	82	15
21	70	20	91	15

Table 6.8: *Stack-loss data from Brownlee (1965).*

We can model the outcome y_i to have a location μ_i and scale σ, and a regression model

$$\mu_i = x_i'\beta.$$

The standardized variable $z_i = (y_i - \mu_i)/\sigma$ is assumed to have a known density $f_0(z)$. We will consider the Cauchy model

$$f_0(z_i) = \frac{1}{\pi(1+z_i^2)},$$

so the likelihood contribution of observation y_i is

$$L_i(\beta,\sigma) = \sigma^{-1}\left\{1 + \frac{(y_i - x_i'\beta)^2}{\sigma^2}\right\}^{-1}$$

and, assuming independence, the total likelihood is

$$L(\beta,\sigma) = \prod_i L_i(\beta,\sigma).$$

All estimates and profile likelihood computations for this model must be done numerically; in Section 12.6 we show how to use the IWLS to

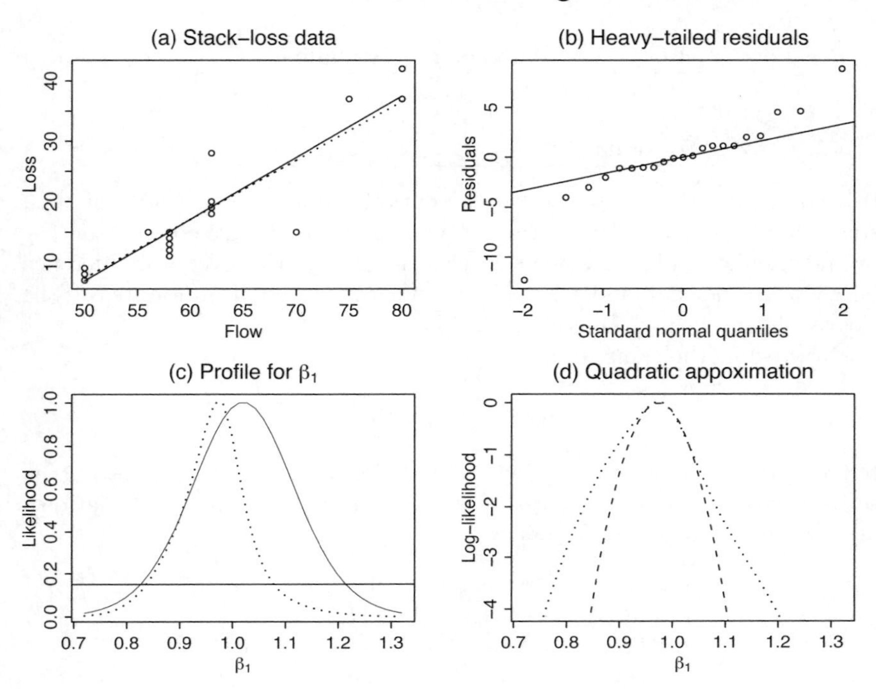

Figure 6.10: *(a) Relationship between stack loss and air flow, and the linear fits using normal error (solid line) and Cauchy error (dotted). (b) Normal plot of the residuals from the normal model. (c) The profile likelihood of β_1 using normal error (solid) and Cauchy error (dotted). (d) Poor quadratic approximation (dashed) of the profile likelihood of β_1 using Cauchy error (dotted).*

compute β. For example, if $\beta = (\beta_0, \beta_1)$, to get the profile likelihood for a scalar β_1, we simply compute

$$L(\beta_1) = \max_{\{\beta_0,\sigma\}} L(\beta_0, \beta_1, \sigma)$$

for each fixed β_1 over a range of values. Such a profile likelihood is important, since experience with Cauchy models indicates that the quadratic approximation does not usually hold.

Analysis of air flow and stack loss

For the data in Figure 6.10(a) we first show the summary of the normal regression model

$$y_i = \beta_0 + \beta_1 x_{1i} + e_i$$

in the following table:

Effect	Parameter	Estimate	se
Intercept	β_0	-44.13	6.11
Air flow	β_1	1.02	0.10
Residual	σ	4.1	

It is no surprise that there is a strong relationship between air flow and loss, since air flow is a measure of rate of operation. For future comparison the profile likelihood of β_1 is shown in Figure 6.10(c). However, the inference based on the normal model is doubtful, since the residuals in Figure 6.10(b) clearly show a heavy-tailed distribution.

Now we fit the same model

$$y_i = \beta_0 + \beta_1 x_{1i} + e_i$$

but e_i's are assumed to iid Cauchy with location zero and scale σ (the scale parameter does not have the usual meaning as standard deviation). We obtain the following summary:

Effect	Parameter	Estimate	se
Intercept	β_0	-41.37	
Air flow	β_1	0.97	0.045
Residual	σ	1.28	

The standard error of $\widehat{\beta}_1$ is computed numerically from the observed profile likelihood. (The standard errors for $\widehat{\beta}_0$ and σ are not computed, since they are not relevant.)

There is little difference in the estimates compared with those from the normal model. However, Figure 6.10(c) shows that the Cauchy model leads to a more precise likelihood. This gain in efficiency is the reward for using a better model for the errors. Figure 6.10(d) shows a poor quadratic approximation of the log profile likelihood of β_1 from the Cauchy model. This means that the standard error quantity reported in the table (0.045) is not meaningful.

To select between the normal or the Cauchy model, we can use the AIC defined in Section 3.5:

$$\text{AIC} = -2 \log L(\widehat{\theta}) + 2p,$$

where $\widehat{\theta}$ is the MLE of the model parameters; the number of parameter p equals 3 for both models. (Note that all the constant terms in the density function must be included in the computation of the maximized likelihood in the AIC formula.) The AIC is 160.26 for the normal model and 115.49 for the Cauchy model, so the Cauchy model is preferable.

As an alternative to the Cauchy model, we can fit a t-distribution to the error term and vary the degrees of freedom k. This family includes the Cauchy at $k = 1$ and the normal at large k, so model selection can be based on testing the parameter k (Exercise 6.27).

Analysis of cooling temperature and stack loss

Analysis of the cooling temperature reveals a surprising aspect of the data. Figure 6.11(a) shows the relationship between stack loss and cooling temperature. Using the same methodology as before, we first perform a normal-based regression model, giving the following output:

Effect	Parameter	Estimate	se
Intercept	β_0	−41.91	7.61
Temperature	β_1	2.82	0.36
Residual	σ	5.04	

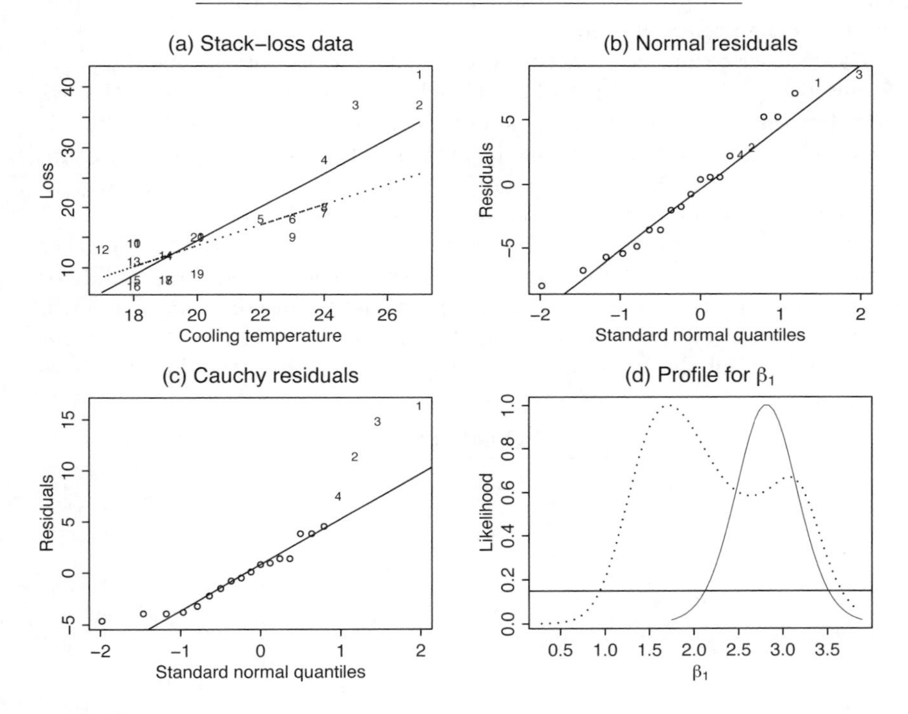

Figure 6.11: *(a) Relationship between stack loss and cooling temperature, and the linear fits using normal error (solid) and Cauchy error (dotted); the printed value is the operating day. (b) Normal plot of the residuals from the normal model. (c) Normal plot of the residuals from the Cauchy model. (d) The profile likelihood of β_1 using normal error (solid) and Cauchy error (dotted).*

The normal plot in Figure 6.11(b) shows that the residuals are reasonably normal. By comparison, the Cauchy-based model gives

Effect	Parameter	Estimate	se
Intercept	β_0	−20.69	
Temperature	β_1	1.72	0.50
Residual	σ	2.71	

Now we have a genuine disagreement in the estimation of β_1. As shown in Figure 6.11(a) the normal fit follows all of the data; the Cauchy fit only follows the bulk of the data and allows some large errors. The normal QQ-plot of the Cauchy errors in Figure 6.11(c) indicates a heavy right-tail. The profile likelihood of β_1 in Figure 6.11(d) is bimodal, implying that the standard error (0.50) for $\widehat{\beta}_1$ is meaningless. The likelihood interval at 15% cutoff, which does not have a CI interpretation, is wide, indicating a large uncertainty in the estimate.

One of the two models is obviously wrong. The AIC is 131.4 for the normal model, and 141.4 for the Cauchy model, pointing to the normal model as the preferred model.

The observations were actually taken as a time series. As shown in Figure 6.11(c) large residuals of the Cauchy model are associated with days 1 to 4 of the operation. If there is a transient state, then we can include it as an effect in a multiple regression; then the effect of cooling temperature would be closer to the Cauchy model. Perhaps the most satisfying way to check the model is to collect more data at high cooling temperature.

As another interpretation, the discrepancy might indicate that neither model is a good fit. Assuming that all the measurements are valid, or that there is no transient state in the operation, Figure 6.12 shows a much better fit achieved by a (normal) quadratic model; its AIC is 120.5, better than the AIC of the linear model. Verifying the fit and the AIC is left as an exercise.

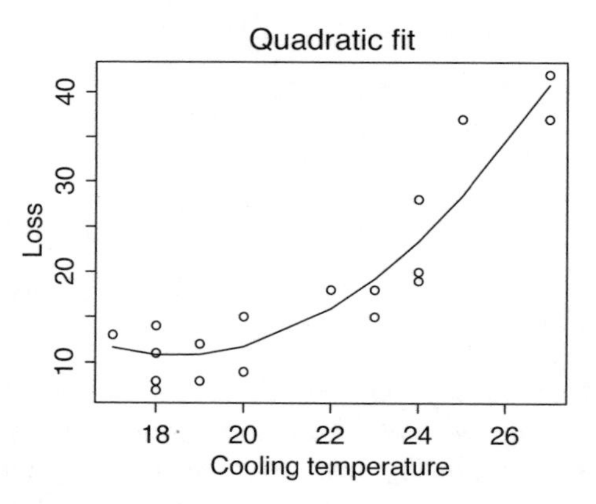

Figure 6.12: *Quadratic fit of stack-loss as a function of cooling temperature.*

6.10 Exercises

Exercise 6.1: Verify the MLEs of β and σ^2 in Section 6.1. You can use some matrix calculus, or verify that

$$\| Y - X\beta \|^2 = \| Y - X\widehat{\beta} \|^2 + \| X\widehat{\beta} - X\beta \|^2$$

where $\| a \|^2 = a'a$ for any vector a.

Exercise 6.2: Derive in general the profile likelihood for a scalar slope parameter β_1 in a normal linear model, and describe how it should be computed. Use the Hanford data in Example 6.2 and verify the likelihoods in Figure 6.2.

Exercise 6.3: For the Hanford data example in Example 6.2, check the quadratic approximation of the profile likelihood of β_1, and verify the plot in Figure 6.13.

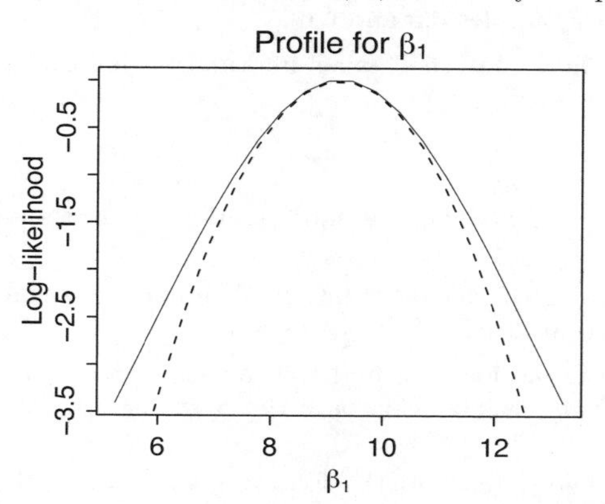

Figure 6.13: *Quadratic approximation (dashed line) of the profile likelihood of β_1 for the Hanford data.*

Exercise 6.4: The following dataset shows the weight (in pounds) of a lamb over a period of time (in weeks).

Time	0	1	2	3	4	6	9	12
Weight	12.6	15.3	26.2	27.9	32.7	34.4	42.3	39.1

Fit a sensible model that captures the growth curve and find the profile likelihood for the growth rate parameter.

Exercise 6.5: Singh *et al.* (1992) reported the length y (in cm) of plants germinating from seeds irradiated with a dose x of gamma radiation.

x	0	10	20	30	40	50	60	70	80	90	100	110
y	8.85	9.40	9.18	8.70	7.53	6.43	5.85	4.73	3.98	3.50	3.10	2.80

Plot the data and fit a nonlinear model

$$y_i = \frac{\beta_0}{1 + \exp\{-\beta_1(x_i - \mu)\}} + e_i,$$

and report the estimates and their standard errors. Compute the profile likelihood of β_1. Check the normality of the residuals.

Exercise 6.6: Verify the regression output (including the standard errors) and the likelihood plots for Examples 6.3 and 6.4.

Exercise 6.7: Show that the Fisher information of the vector parameter β in the logistic regression is of the form

$$I(\widehat{\beta}) = \sum_i \widehat{\theta}_i(1 - \widehat{\theta}_i)x_i x_i'$$

where the vector $x_i \equiv (1, \text{Age}_i)$.

Exercise 6.8: Group the subjects in Example 6.3 into five age groups; the i'th group has n_i subjects and y_i deaths. Use the binomial model to derive the likelihood for a logistic regression, and compare the results with the full data analysis. What is the advantage of the grouped data version?

Exercise 6.9: Verify the Poisson regression estimates and the plots given for the data analysis in Examples 6.5 and 6.6.

Exercise 6.10: Show that the Fisher information for the Poisson log-linear model is

$$I(\widehat{\beta}) = \sum_i \widehat{\theta}_i x_i x_i'$$

where $x_i \equiv (1, \text{Age}_i)$. Verify the standard errors given in the analysis of claims data.

Exercise 6.11: Using the accident data in Example 6.6, suggest ways to check the Poisson assumption in a Poisson regression.

Exercise 6.12: Fit the location and treatment effects in model (6.2) for the accident data in Example 6.6. Compare the inference on the treatment effect when location is not in the model.

Exercise 6.13: The dataset in the following table was collected in a study to compare the number of naevi (pronounced neeVYE), a raised skin lesion of darker colour similar to a mole, among children with spina bifida and normal controls. The controls were match in terms of sex and age. The main hypothesis is that spina bifida is associated with more occurrence of naevi.

Pair	Sex	Age	Case	Cont.	Pair	Sex	Age	Case	Cont.
1	f	16	5	6	22	m	17	27	6
2	f	5	0	3	23	m	10	11	3
3	m	10	15	15	24	f	12	17	1
4	m	6	2	1	25	m	8	3	8
5	f	12	11	7	26	f	11	16	4
6	f	18	22	6	27	m	15	22	3
7	m	11	15	4	28	m	4	0	0
8	m	16	10	73	29	f	11	31	52
9	m	14	29	4	30	f	7	3	17
10	m	10	13	3	31	m	10	18	6
11	f	8	8	14	32	f	8	4	3
12	f	8	6	0	33	f	11	10	0
13	f	5	0	1	34	f	12	5	52
14	m	7	5	10	35	f	15	63	5
15	m	8	7	12	36	f	10	0	4
16	m	17	30	52	37	f	16	47	11
17	f	12	31	2	38	f	8	20	1
18	f	18	19	10	39	f	5	10	5
19	m	3	1	0	40	m	19	20	8
20	f	11	8	3	41	f	4	2	1
21	f	9	7	0	42	m	5	0	3

Table 6.9: *The naevi data from 42 pairs of spina bifida cases and their matched controls. The column under 'Case' gives the number of naevi for the cases.*

(a) Investigate the relationship between the number of naevi and age and sex among the controls, and separately among the cases.

(b) Let y_{i1} and y_{i2} be the number of naevi for the i'th case and control, respectively. Assume y_{ij} is Poisson with mean λ_{ij}, which includes the effects of pair i and other covariates. We have shown that, conditional on the sum $n_i \equiv y_{i1} + y_{i2}$, y_{i1} is binomial with paramaters n_i and

$$\pi_i = \frac{\theta_i}{\theta_i + 1},$$

where $\theta_i = \lambda_{i1}/\lambda_{i2}$. Explain how the conditioning removes the pair effect. Show that the logistic regression based on binomial data (n_i, y_{i1}) is equivalent to a log-linear model for rate ratio θ_i.

(c) Fit the simplest model $\log \theta_i = \beta_0$, and interpret the test of $\beta_0 = 0$.

(d) Fit the model

$$\log \theta_i = \beta_0 + \beta_1 \text{Sex}_i + \beta_2 \text{Age}_i,$$

and interpret the results.

(e) In each case of (a), (c) and (d) above, check the goodness of fit of the model. State the overall conclusion for the study.

Exercise 6.14: The data in the following table (Fairley 1977) are the monthly accident counts on Route 2, a major highway entering Boston, Massachusetts, from the west. The data for the last three months of 1972 are not available.

Year	1	2	3	4	5	6	7	8	9	10	11	12
1970	52	37	49	29	31	32	28	34	32	39	50	63
1971	35	22	17	27	34	23	42	30	36	65	48	40
1972	33	26	31	25	23	20	25	20	36			

(a) Using the Poisson regression with log link, fit an additive model with year and month effects, and describe the result in plain language.

(b) Use the model in (a) to predict the last three months of 1972.

(c) Check the goodness of fit of the model in (a).

Exercise 6.15: The following table (from Simon 1985) shows the occurrences of rare words in James Joyce's *Ullyses*, for example there are 16,432 different words that occur *exactly once* in the whole book. It is also known that a total of 29,899 *different* words were used in the book.

Number of occurrences	Number of words
1	16,432
2	4776
3	2194
4	1285
5	906
6	637
7	483
8	371
9	298
10	222

(a) A naive model for the word frequencies is like this: a word in Joyce's vocabulary will appear x number of times according to a simple Poisson model with mean λ. Estimate λ from the data and assess the adequacy of the model. (Hint: in deriving the likelihood for this model note that the total of 29,899 words must occur at least *once*. So, obviously words that did not occur are not observed here.)

(b) Now consider a slightly more complex model: the number of words that occur exactly k times is Poisson with mean λ_k, where

$$\log \lambda_k = \beta_0 + \beta_1 \log(k+1).$$

Use the IWLS algorithm to estimate the parameters and assess the adequacy of the model. Is it really better than the other model? Plot the data and the model fits from both models.

(c) Using models (a) and (b), compare the estimates of the number of words that Joyce knew.

Exercise 6.16: Verify the regression outputs and plots for the competition data given in Section 6.4.

Exercise 6.17: Show that the Fisher information for the parameter $\beta = (\beta_0, \beta_1)$ in the exponential model where $1/\mu_i = \beta_0 + \beta_1 \log d_i$ is

$$I(\widehat{\beta}) = \sum_i \widehat{\mu}_i^2 x_i x_i',$$

by defining $x_i' \equiv (1, \log d_i)$.

Exercise 6.18: Estimate the dispersion parameter ϕ in the Poisson regression example in Section 6.3. Perform an approximate test of H_0: $\phi = 1$.

Exercise 6.19: Discuss the application of the general exponential family model for logistic regression with zero-one outcomes. Use the surgical data in Section 6.2 as an example.

Exercise 6.20: Verify the Fisher information matrix

$$I(\widehat{\beta}) = \phi^{-1}(XWX)$$

given in Section 6.5.

Exercise 6.21: Figure 6.14 is a log-log plot of the plant competition data, showing we can try a model

$$\log y_i = \beta_0 + \beta_1 \log d_i + e_i,$$

where β_1 is negative.

(a) Estimate the linear regression above, report a summary result and check the normality of the residuals.

(b) As a comparison run an exponential family model as given in Section 6.5, but using a log-link function

$$\log \mu_i = \beta_0 + \beta_1 \log d_i.$$

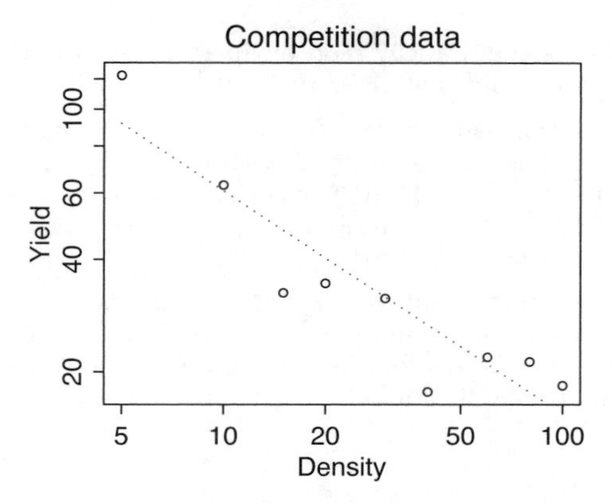

Figure 6.14: *Competition data in a log-log plot, showing an underlying linear relationship.*

(c) For the link function in part (b) show that the Fisher information is given by

$$I(\widehat{\beta}) = \phi^{-1}X'WX,$$

by defining the design matrix X appropriately and the weight matrix is $W = \text{diag}\{y_i/\widehat{\mu}_i\}$.

(d) Compare the fits of these two models and the previous models using inverse link. Explain the advantages and disadvantages of each model.

Exercise 6.22: Using the canonical and some sensible noncanonical links, derive the IWLS algorithm for the binomial and exponential outcome data.

Exercise 6.23: Implement the IWLS for the logistic and Poisson regression models in Examples 6.3 and 6.5.

Exercise 6.24: The inverse Gaussian distribution provides a framework for a regression model where the variance is a cubic function of the mean. Specifically, the density is

$$f(y) = \left(\frac{\lambda}{2\pi y^3}\right)^{1/2} \exp\left\{-\frac{\lambda}{2\mu^2}\frac{(y-\mu)^2}{y}\right\}, \qquad y > 0$$

and it has mean μ and variance μ^3/λ.

(a) Given outcome values $y_1, ..., y_n$ and covariates $x_1, ..., x_n$, suppose we model y_i as $\text{IG}(\mu_i, \lambda)$, where $\mu_i = Ey_i = h(x_i'\beta)$. Write down the log-likelihood function for the parameters β and λ, and identify the canonical link function.

(b) Describe the iteratively weighted least squares (IWLS) procedure to get the estimate for β using the canonical link function.

(c) Find the MLE of the dispersion parameter λ.

(d) Give the formula for the deviance in this case and describe what the deviance may be used for. Can we use it as a measure of goodness of fit of the current model?

Exercise 6.25: Verify all of the computations given in Section 6.9.

Exercise 6.26: Check the 'Cauchy plot' of the residual from the Cauchy-based regression of stack loss on air flow in Section 6.9.

Exercise 6.27: Fit the regression of stack loss on air flow in Section 6.9 assuming the error term has a t-distribution with unknown degrees of freedom k. The density is given in Section 4.11. Perform the likelihood computation at fixed k, and try several values of k, so you obtain a profile likelihood of k. Report the profile likelihood of β_1 when k is fixed at the MLE.

Exercise 6.28: Perform a multiple regression analysis of the stack-loss data with both air flow and cooling temperature as predictors in the model. Obtain the individual profile likelihoods for the slope parameters. Check the quadratic approximation of the profile likelihoods.

7

Evidence and the likelihood principle*

There is a strong consensus about the utility of likelihood for modelling, but its direct use for inference is controversial. This is not surprising: most of the consensus in statistics is associated with modelling, and most of the controversies with inference. While we are able to specify and estimate very complex models, statisticians still cannot agree on how to interpret the CIs (see Section 5.10). The persistent controversies indicate that the issues are not simple.

7.1 Ideal inference machine?

Assuming a correct model, the likelihood function $L(\theta)$ is an 'inference machine': every inference we need about θ can be derived from $L(\theta)$; and, in principle, once a model is constructed, we can proceed fairly automatically. The development of a model is the only statistical step. Finding the likelihood and quantities for inference is merely computational. The advent of computers and computational methods have allowed us (i) to concentrate more on the modelling aspect and (ii) to compromise less on the model complexity.

Can we rely on the likelihood alone? If yes, the likelihood forms an ideal inference machine. Unfortunately we cannot answer it categorically. While the likelihood is appealing as an objective quantity that is totally determined by the observed data, are the observed data all that are relevant from an experiment? How about the sampling scheme or the way we collect the data?

These fundamental issues remain controversial in statistical inference, and are related to the so-called *likelihood principle*. An informal statement of the (strong) likelihood principle is that two datasets that produce equal (proportional) likelihoods should lead to the same conclusion. The principle has far-reaching implications in statistical inference. For example, the ubiquitous P-value violates the likelihood principle: it is possible to find two datasets (from two experiments) producing the same likelihood, but different P-values (see Example 7.1).

Most statisticians do not accept the seeming implication of the likelihood principle that we base conclusions on the likelihood alone. There are serious reservations:

- It is possible to generate a biased dataset, e.g. by a sequential sampling scheme (Section 7.5), but the likelihood function completely ignores the sampling scheme. This can lead to spurious evidence that does not stand up under repeated experiments.
- The computation of likelihood requires a probability model. While the likelihood contains all the information about the parameter of the model, we may not have full conviction in the model. The likelihood cannot convey uncertainty about the model it is based on, while the correctness of the conclusion might depend on the correctness of the model.
- There is a fundamental difficulty in dealing with multi-parameter models. Without frequentist consideration a joint likelihood inference of all parameters can have a very poor repeated sampling property (Section 3.5). Generally, if the parameter space is 'too large' relative to the available data the likelihood function can produce spurious results. This problem can be called 'parameter snooping' as opposed to 'data snooping'. In 'parameter snooping' the observed data are fixed but we keep enlarging the parameter space to find a model that best explains the data. The result is a spurious model that overfits the data, see the example in Section 7.7, but the likelihood on its own cannot inform us about the problem.

7.2 Sufficiency and the likelihood principles

The so-called *sufficiency principle* states that

> all sufficient statistics based on data x for a given model $p_\theta(x)$ should lead to the same conclusion.

This seems reasonable: since any sufficient statistic contains all of the information about θ, different choices of statistic should carry the same information; then, from the same information one should reach the same conclusion. Since all sufficient statistics lead to the same likelihood function, as a consequence we have the *weak likelihood principle* that

> any set of observations from a given model $p_\theta(x)$ with the same likelihood should lead to the same conclusion.

Thus the sufficiency principle is equivalent to the weak likelihood principle.

Now let us restate the idea above in terms of *evidence*. A sufficient statistic summarizes all the evidence (about the parameter) from an experiment. Then it seems reasonable that the likelihood principle should say that *the likelihood function contains all the evidence about θ*. This feels somewhat stronger. To state that the humble $\overline{x}$ is in a one-to-one map with the likelihood function is fine from the frequentist point of view. However, we still feel intuitively that the likelihood function contains more than just

$\overline{x}$. We know that $\overline{x}$ is not meaningful without the original sampling model $p_\theta(x)$, e.g. we cannot get a measure of uncertainty from $\overline{x}$ alone, so we still need $p_\theta(x)$. But the likelihood function *contains more*: it captures the model in its computation and, seemingly, we can derive inferences from it.

If the likelihood function contains all the evidence, after obtaining the likelihood, *can we throw away the model?* The strong version of the likelihood principle says: yes, the likelihood is *it,* do throw the model away, since it is no longer relevant as evidence. It is a dramatic proposition, but it is not as crazy as it sounds.

We have seen before that data from the binomial and negative binomial experiments can produce the same likelihood. If they do produce the same likelihood, can we say they carry the same evidence about θ, even if sampling properties of the likelihood functions are different? The strong likelihood principle says yes. Note the crucial difference: the weak likelihood principle states that different outcomes from *the same experiment* having the same likelihood carry the same evidence. The strong likelihood principle allows the outcomes to come from different experiments with different sampling schemes. So, according to the strong likelihood principle, evidence about θ does not depend on the sampling scheme.

If the likelihood functions carry the same evidence, shouldn't they lead to the same conclusion about θ? It is difficult to say no, but unfortunately the answer is controversial. If we think that we make conclusions based on evidence alone, then the same likelihood should lead to the same conclusion. But in statistics we deal with uncertainties, not just evidence, so there could be other nondata considerations that may affect our conclusions.

Example 7.1: To see that there is a real and fundamental issue involved, consider closely the binomial versus the negative binomial case. Suppose we obtain $x = 8$ out of $n = 10$ trials in a binomial experiment, and $x = 8$ successes in a negative binomial experiment where we had planned to stop when we get two failures. The likelihood functions of the success probability θ from these observations are the same

$$L(\theta) = \text{constant} \times \theta^8(1-\theta)^2,$$

so the strong likelihood principle would assert that the two experiments have the same evidence about θ.

Now consider the test of hypothesis H_0: $\theta = 0.5$ versus H_1: $\theta > 0.5$. The one-sided P-value from the first experiment is

$$\begin{aligned} p_1 &= P(X \geq 8|\theta = 0.5) \\ &= \sum_{x=8}^{10} \binom{10}{x} 0.5^{10} \\ &= 0.055, \end{aligned}$$

so we commonly say 'there is not enough evidence to reject H_0 at the 5% level'. From the second experiment we have

$$p_2 = P(X \geq 8|\theta = 0.5)$$

$$= \sum_{x=8}^{\infty}(x+1)0.5^{x+2}$$
$$= 0.020,$$

leading to rejection of H_0 at the 5% level, i.e. 'we do have enough evidence' to reject H_0. Hence the standard significance testing is in conflict with the strong likelihood principle. □

The frequentists' first reaction is to suspect and to reject the strong likelihood principle, while Bayesians generally accept the principle, since it is a consequence of their axiomatics. However, even the frequentists cannot simply ignore the principle, since Birnbaum (1962) showed that it is a consequence of the sufficiency and conditionality principles, two seemingly reasonable principles.

7.3 Conditionality principle and ancillarity

Informally, the conditionality principle asserts that only the observed data matter, or information about θ should only depend on the experiment performed. To appreciate that this is sensible, consider an example which is a variant of Cox (1958) or Berger and Wolpert (1988).

Example 7.2: A certain substance can be sent for analysis to either Lab B or Lab C. A coin is tossed to decide the location, and suppose it is decided that we use Lab C. Now, when we evaluate the lab result, do we account for the coin toss? To be specific, suppose Lab B measures with normal error with variance $\sigma_B^2 = 1$, and similarly Lab C has $\sigma_C^2 = 4$. We receive a report that $x = 3$ and we want to test H_0: $\mu = 0$.

If we account for the coin toss in the overall randomness, the outcome X has a normal mixture

$$0.5N(\mu, 1) + 0.5N(\mu, 4)$$

and the one-sided P-value is

$$p_1 = 0.5P(Z > 3) + 0.5P(Z > 3/2) = 0.034,$$

where Z has the standard normal distribution. But, knowing that Lab C produced the actual measurement we obtain a one-sided P-value

$$p_2 = P(Z > 3/2) = 0.067.$$

We say p_2 is a conditional P-value, i.e. it is conditional on the result of the coin toss, while p_1 is unconditional. □

Which P-value is more meaningful? Even frequentists would be tempted to say that p_2 is more meaningful as evidence. This is the essence of the conditionality principle. As remarked earlier the sufficiency principle is also reasonable, so Birnbaum's result about the likelihood principle deserves a closer look.

The last example is not as contrived as it seems, since in practice there are many 'nontechnical' contingencies in real experiments that resemble a coin toss, for example funding, patience, industrial strikes, etc. Theoretically, the same issue always arises when there is an *ancillary statistic*, a function of the data whose distribution is free of the unknown parameter. For example, the sample size in most experiments is typically an ancillary information. If $x_1, \ldots, x_n$ are an iid sample from $N(\mu, 1)$, then any difference $x_i - x_j$ is ancillary.

The idea of conditionality is that our inference should be made conditional on the observed value of the ancillary statistic. This follows from the likelihood principle: if our data are (x, a), where a is ancillary, then the likelihood of the observed data is

$$L(\theta) = p_\theta(x, a) = p(a)p_\theta(x|a).$$

This means the part that matters in the likelihood is only the conditional model $p_\theta(x|a)$, rather than the marginal model $p_\theta(x)$.

7.4 Birnbaum's theorem

Following Berger and Wolpert (1988) we will describe Birnbaum's theorem for the discrete case. Barnard *et al.* (1962) gave a very closely related development. First recall what we mean by an experiment E as a collection $\{X, \theta, p_\theta(x)\}$. From an experiment we obtain evidence about θ; this is a function of E and the observed data x, denoted by $\text{Ev}(E, x)$. This does not need to be specified exactly, in principle anything sensible will do. For example, we may use the likelihood function itself; or, for an iid sample $x_1, \ldots, x_n$ from $N(\mu, \sigma^2)$, with σ^2 known, we may define $\text{Ev}(E, x) \equiv (\overline{x}, \sigma^2/n)$.

We first state the sufficiency principle formally in terms of evidence. Suppose we perform an experiment E, and $T(X)$ is a sufficient statistic for θ.

Definition 7.1 SUFFICIENCY PRINCIPLE: *If x and y are sample data from E such that $T(x) = T(y)$ then*

$$Ev(E, x) = Ev(E, y).$$

The idea of a mixture experiment as in Example 7.2 is important in the formal definition of the conditionality principle. Suppose there are two experiments $E_1 = \{X_1, \theta, p_{1,\theta}(x_1)\}$ and $E_2 = \{X_2, \theta, p_{2,\theta}(x_2)\}$; only the parameter θ need to be common in the two experiments. Now consider a mixture experiment E^*, where a random index $J = 1$ or 2 is first generated with probability 0.5 each, and E_J is then performed. Formally $E^* = \{X^*, \theta, p^*_\theta(x^*)\}$, where $X^* = (J, X_J)$ and $p^*_\theta(j, x_j) = 0.5 p_{j,\theta}(x_j)$.

Definition 7.2 CONDITIONALITY PRINCIPLE: *The evidence from a mixture experiment is equal to the evidence from the experiment performed, that is*

$$Ev(E^*, x^*) = Ev(E_j, x_j).$$

To define the likelihood principle in terms of evidence consider two experiments E_1 and E_2 again.

Definition 7.3 STRONG LIKELIHOOD PRINCIPLE: *Suppose x_1 and x_2 are data from E_1 and E_2 respectively such that the likelihood functions are proportional, namely*

$$L_1(\theta) = cL_2(\theta),$$

for some constant c, i.e. $p_{1,\theta}(x_1) = cp_{2,\theta}(x_2)$ for all θ. Then the evidence is the same:

$$Ev(E_1, x_1) = Ev(E_2, x_2).$$

Theorem 7.1 (Birnbaum's theorem) *The sufficiency and conditionality principles together are equivalent to the strong likelihood principle.*

Proof: We first show that the sufficiency and conditionality principles together imply the strong likelihood principle. Let us start with the premise of the strong likelihood principle that data x_1 and x_2 have proportional likelihoods. Now consider the mixed experiment E^* as defined for the conditionality principle, and recall that on observing any (j, x_j) we have

$$\text{Ev}\{E^*, (j, x_j)\} = \text{Ev}(E_j, x_j). \tag{7.1}$$

For the mixed experiment E^* with random outcome (J, X_J) we define

$$T(J, X_J) \equiv \begin{cases} (1, x_1) & \text{if } J = 2, X_2 = x_2 \\ (J, X_J) & \text{otherwise.} \end{cases}$$

Note that there is some data reduction in $T(J, X_J)$ since if we observe $T = t = (1, x_1)$ we do not know if we have performed $(1, x_1)$ or $(2, x_2)$; in fact this is the only data reduction. The essence of the proof is to show that such a reduction does not result in any loss of information; that is to show that $T(J, X_J)$ is sufficient for θ. To satisfy the definition of sufficient statistic we find the conditional probabilities

$$P\{X^* = (j, x_j) | T = t \neq (1, x_1)\} = \begin{cases} 1 & \text{if } (j, x_j) = t \\ 0 & \text{otherwise,} \end{cases}$$

and

$$\begin{aligned} P\{X^* = (1, x_1) | T = t = (1, x_1)\} &= 1 - P(X^* = (2, x_2) | T = t = (1, x_1)) \\ &= \frac{0.5p_{1,\theta}(x_1)}{0.5p_{1,\theta}(x_1) + 0.5p_{2,\theta}(x_2)} \\ &= \frac{c}{c+1} \end{aligned}$$

and see that they are independent of θ. The sufficiency principle implies

$$\text{Ev}\{E^*, (1, x_1)\} = \text{Ev}\{E^*, (2, x_2)\}. \tag{7.2}$$

Now it is clear that (7.1) and (7.2) imply $\text{Ev}(E_1, x_1) = \text{Ev}(E_2, x_2)$. This proves the first part of the theorem.

To prove the converse, the likelihood of θ based on (j, x_j) in experiment E^* is

$$L^*(\theta) = 0.5 p_{j,\theta}(x_j),$$

so it is proportional to the likelihood of θ based on x_j in experiment E_j. The strong likelihood principle implies

$$\text{Ev}(E^*, (j, x_j)) = \text{Ev}(E_j, x_j),$$

which is the conditionality principle. In Section 3.2 we have shown that if $T(X)$ is sufficient and $T(x) = T(y)$, then x and y have proportional likelihood functions, so by the strong likelihood principle

$$\text{Ev}(E, x) = \text{Ev}(E, y),$$

which is the sufficiency principle. □

A simple corollary of Birnbaum's theorem is the requirement that $\text{Ev}(E, x)$ *depend on* E *and* x *only through the likelihood function.* We can see that by defining a new experiment E^Y where we only record

$$Y = \begin{cases} 1 & \text{if } X = x \\ 0 & \text{otherwise,} \end{cases}$$

then $P(Y = 1) = p_\theta(x)$, so

$$\text{Ev}(E, x) = \text{Ev}(E^Y, 1),$$

but E^Y depends only on $p_\theta(x) = L(\theta)$.

The corollary has fundamental implications on statistical inference. For example, the standard P-value is generally not a function of the likelihood; hence if we adopt the strong likelihood principle, we have to admit that P-value is not evidence from the data. The key idea in the computation of P-value is the notion of more extreme values in the sample space other than the observed data; such a notion is in conflict with the likelihood principle. Recall the binomial–negative binomial experiments in Example 7.1; it is a special case of sequential experiments, an important branch of statistics where the likelihood and traditional frequentist inference do not match.

7.5 Sequential experiments and stopping rule

Without loss of generality, suppose independent observations are taken one at a time, where after each new observation we make a decision about taking another independent observation. In many sequential experiments, the decision to carry on is typically of this form:

> at time m and given $x_1, \ldots, x_m$, continue to obtain x_{m+1} with probability $h_m(x_1, \ldots, x_m)$; that is, the decision depends only on the past data, but not on the unobserved x_{m+1}.

For example, in the negative binomial experiment, we continue a Bernoulli trial until the current number of successes is equal to a pre-specified value r. In this experiment x_i is zero or one, and

$$h_m(x_1, \ldots, x_m) = \begin{cases} 1 & \text{if } \sum_{i=1}^m x_i < r \\ 0 & \text{if } \sum_{i=1}^m x_i = r. \end{cases}$$

Given a model $p_\theta(x)$, and observations $x_1, \ldots, x_n$, and assuming that the decision function $h_i(\cdot)$ does not depend on the unknown parameter, the likelihood function is

$$\begin{aligned} L(\theta) &= P(\text{deciding to observe } x_1)p_\theta(x_1) \times \\ &\quad P(\text{deciding to observe } x_2|x_1)p_\theta(x_2) \times \\ &\quad \ldots \times \\ &\quad P(\text{deciding to observe } x_n|x_1, \ldots, x_{n-1})p_\theta(x_n) \times \\ &\quad P(\text{stop at } x_n|x_1, \ldots, x_n) \\ &= h_0 p_\theta(x_1) \times \\ &\quad h_1(x_1)p_\theta(x_2) \times \\ &\quad \ldots \times \\ &\quad h_{n-1}(x_1, \ldots, x_{n-1})p_\theta(x_n) \times \\ &\quad (1 - h_n(x_1, \ldots, x_n)) \\ &= \text{constant} \times \prod_i p_\theta(x_i). \end{aligned}$$

So, we arrive at a remarkable conclusion that the likelihood function ignores the stopping rule altogether. A strict adherence to the strong likelihood principle in this case implies that *the evidence from such an experiment does not depend on the stopping rule.* That is, if we want to find out what the data say, we can ignore the stopping rule in the analysis. This is convenient if we stop a study for some unrelated reason (e.g. a power cut), but it is also the case even if *we decide to stop because 'we are ahead' or the 'data look good'.*

To see that this is not a trivial issue, consider the problem of sequential trials from the frequentist point of view. In industrial experiments involving expensive units or destructive testing, for which sequential testing was originally conceived (Wald 1947), it makes sense to proceed sequentially to try to minimize cost. In clinical trials sequential testing is adopted for an added ethical reason: if a drug or a procedure is harmful then we want to stop a trial early to avoid harming the study subjects, while if a drug is beneficial then it should be made readily available.

Example 7.3: Suppose we observe $x_1, x_2, \ldots$ sequentially and independently from $N(\theta, 1)$, where we are interested to test H_0: $\theta = 0$ versus H_1: $\theta > 0$. (This setup includes the group sequential testing since we can think of x_i as representing the average of, say, 100 observations.) A simple sequential procedure is to test at each step $n = 1, 2, \ldots$

$$\sqrt{n}\overline{x}_n > k,$$

where $\overline{x}_n$ is the current sample mean and k is a fixed critical value, and reject H_0 the first time the test is significant. There is an immediate problem: under H_0 the test will be significant for some n with probability one! This is a consequence of Kolmogorov's law of iterated logarithm that $\sqrt{n}|\overline{x}_n|$ 'grows' like $\sqrt{2\log\log n}$, so eventually it will cross any fixed boundary. This means the true type I error probability of the test is 100%, not a very desirable property.

In practice we may not want to wait too long for a conclusion, which can happen if the true θ is near zero. Typically we plan to have, say, a maximum of $N = 4$ tests. Using the same rule above the type I error probability can be computed as

$$\sum_{n=1}^{N=4} p_n$$

where p_n can be interpreted as the current α-level defined as

$$p_n = P(\sqrt{n}\overline{x}_n > k, \text{and } \sqrt{n-j}\overline{x}_{n-j} < k, \text{for } j = 1, \ldots, n-1).$$

While this looks complicated, it can be easily found using the Monte Carlo technique. For example, for $N = 4$ and $k = 1.65$ we obtain (Exercise 7.1)

$$\begin{aligned} p_1 &= 0.050 \\ p_2 &= 0.033 \\ p_3 &= 0.018 \\ p_4 &= 0.010, \end{aligned}$$

so the overall type I error probability is $\alpha = \sum p_i = 0.111$. To obtain a sequential test with 5% level we can set the critical value to $k = 2.13$. In sequential analysis terminology we 'spend' our α-level according to a prescription of (p_1, p_2, p_3, p_4). We can make it hard or easy to stop early by controlling the relative sizes of p_1 to p_4. □

The main consideration above has been the frequentist concern over the type I error probability. That this concern is not universally accepted can be seen from the following example (Berger and Berry 1987).

Example 7.4: A scientist has $n = 100$ iid observations assumed coming from $N(\theta, 1)$ and wants to test H_0: $\theta = 0$ versus H_1: $\theta \neq 0$. The observed average is $\overline{x}_{100} = 0.2$, so the standardized test statistic is $z = \sqrt{n}|\overline{x}_{100} - 0| = 2.0$. 'A careless classical statistician might simply conclude that there is significant evidence against H_0 at the 0.05 level. But a more careful one will ask the scientist, "Why did you cease experimentation after 100 observations?" If the scientist replies, "I just decided to take a batch of 100 observations", there would seem to be no problem, and few statisticians would pursue the issue. But there is another important question that should be asked (from the classical perspective), namely: "What would you have done had the first 100 observations not yielded significance?"

To see the reasons for this question, suppose the scientist replies: "I would then have taken another batch of 100 observations." The scientist was implicitly considering a procedure of the form:

(a) take 100 observations

(b) if $\sqrt{100}|\overline{x}_{100}| \geq k$ then stop and reject H_0;

(c) if $\sqrt{100}|\overline{x}_{100}| < k$ then take another 100 observations and reject H_0 if $\sqrt{200}|\overline{x}_{200}| \geq k$.

For this procedure to have level $\alpha = 0.05$, k must be chosen to be $k = 2.18$ (Exercise 7.2). For the observed $z = 2.0$ at step (b), the scientist could not actually conclude significance, and hence would have to take the next 100 observations!' What is obviously disturbing is that the 'significance' of an observed P-value depends on the thoughts or intention of the scientist rather than the data alone.

This example can be elaborated to create further paradoxes. Suppose the scientist got another 100 observations and the final $z = \sqrt{200}|\overline{x}_{200}| = 2.1 < 2.18$, which is not significant at $\alpha = 0.05$. The 'proper P-value' that takes the half-way analysis into account is

$$P(\sqrt{100}|\overline{x}_{100}| > 2.18) + P(\sqrt{100}|\overline{x}_{100}| < 2.18, \sqrt{200}|\overline{x}_{200}| > 2.1) = 0.055.$$

If the full dataset is published and another scientist analyses it, the latter will get a significant result, with P-value

$$P(\sqrt{200}|\overline{x}_{200}| > 2.1) = 0.036.$$

So the same data produce different 'evidence'. □

That we can ignore the optional stopping is a liberating implication of the likelihood principle. It simplifies our analysis and, regardless of our intention, we would report the same evidence. Adherence to the strong likelihood principle, however, creates a puzzle of its own: consider the following example from Armitage (1961).

Example 7.5: Fix a value k in advance and take $x_1, x_2, \ldots$ randomly from $N(\theta, 1)$ until

$$z_n = \sqrt{n}\overline{x}_n \geq k.$$

Think of z_n as a time series with index n and the stopping rule as a boundary crossing. Figure 7.1 shows a sample realization of z_n, where the boundary of $k = 2$ is crossed at $n = 48$. This means that only $x_1, \ldots, x_{48}$ are available for analysis.

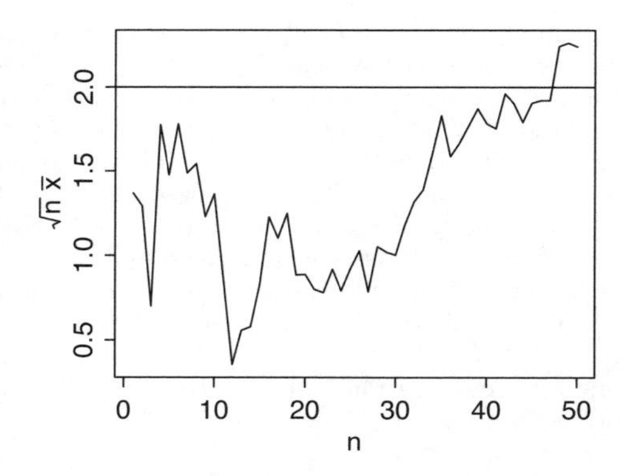

Figure 7.1: *A sample realization of* $\sqrt{n}\overline{x}_n$ *computed from* $x_1, x_2, \ldots$ *for* $n = 1, \ldots, 50$. *The boundary is crossed at* $n = 48$.

Since the stopping rule is a function of the observed data alone, the likelihood of θ based on the data $x_1, \ldots, x_n$ is

$$L(\theta) = \text{constant} \times e^{-\frac{n}{2}(\overline{x}_n - \theta)^2}.$$

In particular $\widehat{\theta} = \overline{x}_n$ as before. The likelihood of H_0: $\theta = 0$ is

$$\frac{L(0)}{L(\widehat{\theta})} = e^{-n\overline{x}^2/2} \leq e^{-k^2/2}.$$

What is puzzling here is that the evidence about H_0 appears to be determined in advance by the choice of k; for example, using $k = 2$, the likelihood of $\theta = 0$ is less than 15%. Here the data have been 'rigged', but the computation of evidence via the likelihood does not take that into account. □

It must be emphasized that even if the likelihood is not affected by the stopping rule, the likelihood principle *does not* say 'so, go ahead and use the standard frequentist-style likelihood ratio test for inference'. The principle only states something about evidence, but not any particular course of action.

Suppose we decide to stop if $\sqrt{n}\overline{x}_n \geq 2$, and we observe $x_1 = 2.5$ and stop. What can we say about θ given that x_1 is a sample from $N(\theta, 1)$? To illustrate frequentist calculations, suppose $\theta = 0$; given that we stop at $n = 1$, the average of X_1 is

$$E(X_1 | X_1 > 2) = 2.4.$$

This seems unacceptable: if we stop early then there is the potential of a large positive bias. However, if θ is much larger than $k = 2$, then there is very little bias; for example, if $x_1 = 50$ then we know that there should be very little bias due to stopping.

Stopping at a large n is evidence that θ is small. In this case the sample mean $\overline{x}$ will be near $k/\sqrt{n}$, which is small if k is moderate. So the likelihood, which is concentrated at the sample mean, is not going to be very biased. Also, as the sample size n goes to infinity $L(\theta = 0) > L(\theta_1)$, meaning that $\theta = 0$ has better support, for any *fixed* $\theta_1 > 0$.

The preceding discussion suggests that the sample size n carries information about θ: a small n is evidence of a large θ, and a large n a small θ. This is true, but unlike in the negative binomial case, in the normal case n is not sufficient so inference based on n alone will be inefficient.

Sequential analysis such as those in industrial quality control applications is 'made' for a true frequentist testing. In this case there is no difficulty in accepting the relevance of long-term operating characteristics such as false alarm rate, etc. Later applications such as sequential clinical trials are justified more on ethical rather than on efficiency grounds; here we find that a full-blown inference including estimation and confidence procedure meets inherent logical difficulties.

Even a frequentist estimation procedure can lead to a logical quandary: Exercise 8.11 shows a simple example where the best unbiased estimate from a two-stage experiment ignores the second-stage data. Frequentist inference from a stopped experiment can be found, for example, in Whitehead (1986), Rosner and Tsiatis (1988), or Whitehead *et al.* (2000). Note, however, that any analysis that takes the sampling scheme into account will violate the likelihood principle and, as illustrated by Example 7.4, any such violation will make the analysis or its conclusions open to paradoxes.

A common practice in sequential clinical trials is to use a formal sequential testing procedure during the trial (see Example 7.3), but then to ignore the stopping rule when analysing the data for further scientific publications. This seems a sensible compromise. Adopting a sequential testing procedure would prevent experimenters from stopping too soon unless the evidence is very strong. Ignoring the stopping rule when presenting the evidence avoids having the intention of the experimenters interfere with the interpretation of the data (Example 7.4).

Informative stopping

There is a subtle point in the analysis of a sequential trial that has a general relevance. In the above analysis, the likelihood ignores the optional stopping if $h_m(x_1, \ldots, x_m)$, the probability of stopping at time m, is a function of observed data alone. We can pretend that the values $x_1, x_2, \ldots$ were already observed, but when we decide to stop at n, then we simply drop/delete $x_{n+1}, x_{n+2}, \ldots$, and keep only $x_1, \ldots, x_n$.

Imagine a scheme where we take one observation at a time, note the last value and drop/delete it if it is too large, and stop the experiment. In this case the probability of stopping is also a function of the unknown parameter and the likelihood will not ignore it. So, the general rule is that dropping an observation because of its value will always affect the likelihood, but dropping it with a probability determined by other available data does not affect the likelihood from the available data.

For example, suppose $x_1, \ldots, x_n$ are an iid sample, and we drop all values except the maximum. The likelihood based on the maximum value is quite different from the likelihood based on a single random value x_i (Example 2.4). Similarly, other processing of the data prior to analysis may or may not have an effect on the likelihood from the final dataset.

Thus the idea of optional stopping is also related to analysis of missing data. If the missing mechanism is a function of available data, the likelihood ignores it; if it is also a function of the missing data the likelihood will take it into account.

7.6 Multiplicity

The frequentist concern in the sequential experiment example is a special case of the problem of 'multiplicity', which is pervasive in applied science or applied statistics. If we get a significant result after performing many tests or 'repeated looks' at the data, how should we account for these in our conclusion?

In its mild form, multiplicity is known as a 'multiple comparison problem' (Miller 1981; Hochberg and Tamhane 1987). If we are comparing many things, some of them are bound to be significant purely by chance: under the null hypothesis, if each test has a 5% level then for every 20 tests we expect one significant result, and even if such a result is unexpected it is usually 'easy' to supply a 'scientific explanation'.

The standard F-test protects the α-level against this problem, as does the Bonferroni adjustment or Tukey's procedure. At what point do we stop protecting our α-level against multiplicity? Within a single analysis of variance table? Within an experiment? Or within the lifetime of a scientist? In addition, the use of P-value as a measure of evidence presents problems. There will be paradoxes similar to Example 7.4: the 'intention' of the experimenter in planning how many tests to perform becomes relevant in evaluating the 'significance' of a revealed pattern.

In its extreme form multiplicity carries a pejorative term such as 'data snooping' or 'fishing expedition' or 'data dredging', etc. The problem is obvious: if we look long enough around a large dataset we are going to see some statistically significant patterns, even when in truth no such pattern exists; this means the true α-level of the procedure is near 100%. The strength of evidence as reported by the likelihood or standard P-value becomes meaningless, since it usually does not take into account the unorganized nature of the data snooping.

The likelihood might not be affected by data snooping. For example, after seeing the result of one variable, it may occur to us to test another variable. It is obvious that the evidence about the first variable should not be affected by our intention about the second variable. In situations where we ignore data after seeing them (e.g. they are not 'significant'), data snooping can be more complicated than a sequential trial, and the likelihood will be affected.

As with the optional stopping it is liberating not to worry about multiplicity or various decisions at data pre-processing. However, we learn from the sequential trial that, even when the likelihood is not affected by the optional stopping, inference from the final data can be elusive.

Inference after model selection

Model selection is a bread-and-butter statistical activity that is accompanied by the problem of multiplicity. It can be formal as in a stepwise or best-subset variable selection procedure. Or it can be informal and undocumented, where we make decisions about inclusion/exclusion criteria for subjects of analysis, definition of categories or groups for comparison, etc. The formal process is clearly guided by the data to arrive at simple but significant results, so it is well known and intuitive that model selection can produce spurious results. The question is how to account for the selection process in the reporting of evidence in the final model. This is not a question of model selection itself, which can be done for example using the AIC, but of the uncertain inference associated with the parameters of the chosen model.

The common practice is in fact to ignore the selection process: P-values are typically reported as if the final model is what has been contemplated initially. While convenient, such P-values can be spurious. Moreover, generally we cannot invoke the likelihood principle to justify the evidence in the final model.

There are general techniques we can use to measure uncertainty due to model selection. For example, we can use the cross-validation approach (Section 5.2): split the dataset into two parts, then use one part ('the training set') to develop the best model, and the other ('the validation set') to validate the model or measure its uncertainty. For example, the training set may indicate that 'social-economic level' is an important predictor of 'health status' in the best-subset model; then we find the P-value of the variable in the best subset model based on an analysis of the validation set.

The method is not feasible if the dataset is small. In this case it is common practice to use the whole dataset as the training set. If a scientifically interesting model is found after an extensive model selection, then it is probably best to collect more data to get an external validation. This is consistent with the conservative tendency in the scientific community to believe only results that have been validated by several experiments.

7.7 Questioning the likelihood principle

It is common to hear some statisticians declare that they 'reject' the (strong) likelihood principle. How can they reject a correct theorem? It is important here to distinguish between the formal and informal statements of the likelihood principle. The formal statement is that

> two datasets (regardless of experimental source) with the same likelihood carry the same evidence, so that evidence is in the likelihood alone – assuming the model is correct, of course.

The informal likelihood principle states that

> two datasets (regardless of experimental source) with the same likelihood should lead to the same conclusion.

These are two very different statements, with the latter being the commonly rejected one. What we do (e.g. drawing conclusions or other actions) given the likelihood is outside the boundary of the formal likelihood principle.

Lying in the gap between the statements is a fundamental question whether our conclusion should depend only on the observed data. If yes, then it should follow from the likelihood alone. *Most statisticians from all persuasions in fact reject this.* Bayesians use the prior information formally to arrive at conclusions, while frequentists care very much about how we arrive at the data. The informal principle is also stronger than the Fisherian view that both likelihood and probability inference are possible, and that likelihood is weaker than probability.

Interpretation of likelihood ratio

An important aspect of Fisherian inference is that, when needed as a measure of support, the likelihood ratio can be interpreted subjectively. For example, when considering hypotheses A versus B, a likelihood ratio of 209 in favour of B means B is preferred over A, and 209 measures the strength of preference. Now, if in another experiment we report a likelihood ratio of 21 between two competing hypotheses, can we say that the first experiment shows stronger evidence? This is a question of calibration: can we interpret the likelihood ratio as is, or does it require another measure to calibrate it?

The sequential experiment in Example 7.5 shows that the likelihood cannot tell if a dataset has been 'rigged'. We now come back to the experiment in Section 2.4 to show that the likelihood also cannot tell that the parameter space may be too large, or that an observed pattern is spurious, and we have overfitted the data.

Imagine taking a card at random from a deck of $N = 209$ well-shuffled cards and consider the following two hypotheses:

H_0 : the deck contains N different cards labelled as 1 to N.
H_2 : the deck contains N similar cards labelled as, say, 2.

Suppose card with label 2 is obtained; then the likelihood ratio of the two hypotheses is

$$\frac{L(H_2)}{L(H_0)} = N = 209,$$

that is, H_2 is $N = 209$ times more likely than H_0, so the evidence indicates that H_2 should be preferred over H_0.

There is nothing unusual in the above interpretation that a likelihood ratio of 209 gives strong evidence for H_2. Now suppose the card experiment is conducted *without* any hypothesis in mind. One card is taken at random, and a card with label 5 is obtained; *then* one sets the hypothesis

H_5 : the deck contains N similar cards labelled as 5.

Then $L(H_5)/L(H_0) = N = 209$, that is H_5 is 209 times more likely than H_0. Some may find this disturbing: H_k for any obtained label $k > 0$ is always more likely than H_0.

The difference with the previous interpretation, that the hypothesis H_5 is set after seeing the data (label 5), is actually *not* important. We may consider $N+1$ hypotheses in advance: H_k for $k = 0, \ldots, N$. Then, for any observed label k we have $L(H_k)/L(H_0) = N$. This is a potent reminder that the likelihood ratio compares the relative merits of two hypotheses in light of the data; it does not provide an absolute support for or against a particular hypothesis on its own. That $L(H_5)/L(H_0) = N = 209$ does not mean H_5 in itself is a reasonable hypothesis.

One may find the original interpretation above more acceptable; there is an implicit suggestion, before we collect the data (pick a card), that H_0 and H_2 are the only possible hypotheses because they are explicitly stated;

hence observing label 2 'feels' like evidence for H_2. This 'suggestion' is not based on data, but on prior information.

To make the last argument more specific and to clarify the way the data enters our reasoning, suppose we conduct an extra layer of experiment by first choosing H_0 or H_2 at random so that

$$P(H_0) = P(H_2) = 1/2.$$

Then, on observing a card with label 2,

$$\begin{aligned} P(H_0|2) &= \frac{P(H_0)\,P(2|H_0)}{P(2)} \\ &= \frac{1/2 \times 1/N}{1/2 \times 1/N + 1/2} \\ &= 1/(N+1) \end{aligned}$$

and $P(H_2|2) = N/(N+1)$. Here it is true a posteriori that H_2 has more chance to be true. However, it is not the case if a priori $P(H_2)$ is tiny, e.g. a lot smaller than $1/N$. Note that the ratio of the posterior probability

$$\begin{aligned} \frac{P(H_2|2)}{P(H_0|2)} &= \frac{P(H_2)}{P(H_0)} \times \frac{P(2|H_2)}{P(2|H_0)} \\ &= \frac{P(H_2)}{P(H_0)} \times \frac{L(H_2)}{L(H_0)}, \end{aligned}$$

so the data enter the posterior comparison only through the likelihood ratio.

Our sense of surprise in viewing an experimental result is a function of both our prior belief and likelihood. This is how we can explain why the interpretation of $L(H_5)/L(H_0) = N$ above feels 'surprising': we are mixing up evidence in the data (which is strong) with a prior belief (which is weak).

If H_5 is truly considered after seeing the data, then it is a spurious hypothesis; including it in the list of possible hypotheses enlarges the 'size of the parameter space'. This example shows that some prior knowledge can be important to avoid over-interpretation of likelihood. The knowledge is used mainly to limit consideration of sensible hypotheses. This leans towards the Bayesian attitude, though we may not want to invest formally in our prior.

A Fisherian attitude here is closer to a frequentist's: if some evidence is not likely to be repeatable, then it is spurious and the uncertainty is not adequately measured by the likelihood alone. What we know theoretically is that if the parameter space is large the likelihood ratio tends to be large, so to avoid spurious results we have to adapt our assessment of likelihood according to the size of the parameter space (Section 3.5).

Empirical distribution function

The above example is not far fetched. In statistics we regularly use the empirical distribution function (EDF), which is the (nonparametric) MLE in the class of all possible distributions (discrete and continuous). For example, given data $x_1 = 1.2$, the EDF is a degenerate distribution at x_1; this has higher likelihood than any other distribution. (To avoid the technical problem of comparing the discrete and continuous models, simply assume that measurements have finite precision, so 1.2 means $1.2 \pm \epsilon$ for a small ϵ.) Given n distinct sample points, the EDF puts a weight $1/n$ at each point (see Section 15.1).

It is not necessary to 'believe' that our sample actually comes from the EDF to accept that the EDF is a reasonable estimate of the distribution function.

Example from Fraser et al.

Fraser *et al.* (1984) give an example where a pure likelihood inference seems to be in conflict with the repeated sampling principle. Suppose $\theta \in \{1, 2, \dots\}$ and

$$p_\theta(x) = 1/3 \text{ for } x = \begin{cases} 1, 2, 3 & \text{if } \theta = 1 \\ \theta/2, 2\theta, 2\theta + 1 & \text{if } \theta \text{ is even} \\ (\theta - 1)/2, 2\theta, 2\theta + 1 & \text{if } \theta \text{ is odd.} \end{cases}$$

This probability model is best shown in a graph: see Figure 7.2. The nonzero probabilities, all with value 1/3, are on the circles. From the

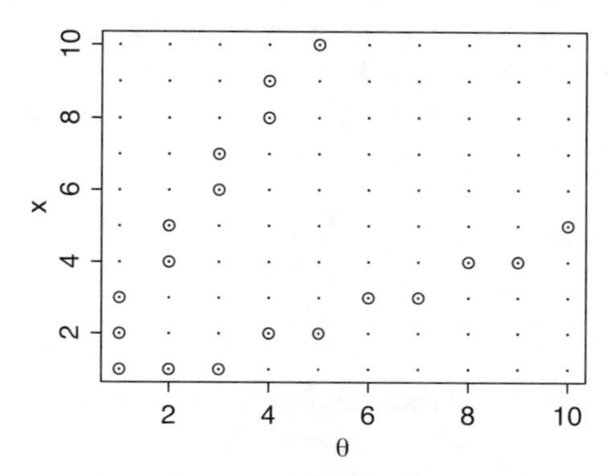

Figure 7.2: *Probability model from Fraser* et al. *(1984): the probability is 1/3 on the circles and zero otherwise.*

graph it is obvious that, for any observed x, the likelihood function is flat on the three possible values of θ:

$$L(\theta) = 1/3 \text{ for } \theta = \begin{cases} 1, 2, 3 & \text{if } x = 1 \\ x/2, 2x, 2x + 1 & \text{if } x \text{ is even} \\ (x - 1)/2, 2x, 2x + 1 & \text{if } x \text{ is odd.} \end{cases}$$

So from the likelihood function alone we cannot prefer one value of θ over the other.

Now consider using the first, middle and last values as an estimate of θ and call them $\widehat{\theta}_1$, $\widehat{\theta}_2$ and $\widehat{\theta}_3$. Then, the probability of hitting the true value is

$$
\begin{aligned}
P_\theta(\widehat{\theta}_1 = \theta) &= \begin{cases} P_\theta(X = \{1,2,3\}) = 1 & \text{if } \theta = 1 \\ P_\theta(X = \{2\theta, 2\theta+1\}) = 2/3 & \text{otherwise} \end{cases} \\
P_\theta(\widehat{\theta}_2 = \theta) &= P_\theta(X = \theta/2) = \begin{cases} 1/3 & \text{if } \theta \text{ is even} \\ 0 & \text{otherwise} \end{cases} \\
P_\theta(\widehat{\theta}_3 = \theta) &= P_\theta(X = (\theta-1)/2) = \begin{cases} 1/3 & \text{if } \theta \text{ is odd, but not 1} \\ 0 & \text{otherwise.} \end{cases}
\end{aligned}
$$

This computation suggests that we should use $\widehat{\theta}_1$.

The probability of hitting the true value is not evidence in the data, it is a consequence of repeated sampling from the assumed probability model. In this example evidence in the data from a single experiment cannot tell us which choice of θ is best, but the experimental setup tells us it is better *in the long run* to use $\widehat{\theta}_1$. If the experiment is to be repeated exactly, say by different participants, where θ is fixed in advance and it is desirable to have maximum hitting probability, then $\widehat{\theta}_1$ is better than $\widehat{\theta}_2$ or $\widehat{\theta}_3$.

Goldstein and Howard (1991) modified the example to become more contradictory. Let θ be a nonnegative integer. Given θ, 99 balls are marked with θ and put in a bag together with 99^2 balls marked with $99^2\theta + 1, \ldots, 99^2(\theta+1)$. For example, if $\theta = 0$, 99 balls are marked with 0, and 99^2 balls with $1, \ldots, 99^2$. One picks a ball at random from the bag and notes the mark. If it is $X = k$, then θ is either $\widehat{\theta}_1 = [(k-1)/99^2]$ or $\widehat{\theta}_2 = k$. The likelihood of $\widehat{\theta}_2$ is 99 times the likelihood of $\widehat{\theta}_1$, so $\widehat{\theta}_2$ is the MLE of θ, while the hitting probabilities are

$$
\begin{aligned}
P(\widehat{\theta}_1 = \theta) &= \begin{cases} 1 & \text{if } \theta = 0 \\ 0.99 & \text{if } \theta > 0 \end{cases} \\
P(\widehat{\theta}_2 = \theta) &= P(X = \theta) = 0.01,
\end{aligned}
$$

which indicates that $\widehat{\theta}_1$ is a better choice 'in the long run'.

The long-run interpretation requires that θ is fixed, and the repeat experiments are performed without any accumulation of data (otherwise θ is known exactly once we observe two distinct marks). Such a model is appropriate, for example, if the experiment is to be repeated by different people who act independently. Without such restriction we can end up with a paradox: on observing, say, $k = 55$, θ is either 0 or 55; if we restrict θ to these two values, then the MLE has a superior long-run property.

It should be emphasized that Fraser *et al.*'s example is not a counterexample of the likelihood principle. It only highlights the fundamental fact that the likelihood on its own is silent regarding the long-term properties.

Conditionality

A serious question on the formal likelihood principle is directed at the conditionality principle: there can be more information from a mixed experiment. That is, in the previous notation,

$$\mathrm{Ev}\{E^*, (j, x_j)\} \geq \mathrm{Ev}(E_j, x_j).$$

Under this assumption the likelihood principle does not follow. Here is a famous example using a finite sampling experiment (Godambe and Thompson 1976; Helland 1995).

Suppose there are N units in a population each having, on a certain measurement, values $\mu_1, \ldots, \mu_N$. We take a sample of size n without replacement from the population and are interested in the mean parameter

$$\theta = \frac{1}{N} \sum_1^N \mu_i.$$

The full parameter space is $(\mu_1, \ldots, \mu_N, \theta)$. Enumerate all K possible samples of size n, i.e.

$$K = \binom{N}{n}$$

and let E_k be the experiment of measuring the n units in sample k. Now it is obvious that E_k does not contain evidence about θ. However, a *mixed experiment* which chooses one sample k at random with probability $1/K$ would produce the usual simple random sample and have some information about θ.

The problem with this example is that the parameter θ is not well defined within each experiment E_k, so the setup lies outside the formal definition of the strong likelihood principle.

7.8 Exercises

Exercise 7.1: Verify the current α-levels shown in Example 7.3.

Exercise 7.2: Verify the critical value $k = 2.18$ stated in Example 7.4.

8

Score function and Fisher information

Given data x and probability model $p_\theta(x)$ in Chapter 2 we have defined the likelihood function as

$$L(\theta) = p_\theta(x)$$

and the first derivative of the log-likelihood as

$$S(\theta) = \frac{\partial}{\partial\theta} \log L(\theta).$$

As a function of θ we call $S(\theta)$ the *score function*, while as a random variable for fixed θ we call $S(\theta)$ the *score statistic*.

The score statistic turns out to have rich frequentist properties. In likelihood theory we use the frequentist analytical tools to operate or manipulate likelihood quantities, to establish needed characteristics, and to derive approximations.

8.1 Sampling variation of score function

The sampling distribution of the score function shows what we should expect as the data vary from sample to sample. We will study this through a series of specific models.

Normal model

Let $x_1, \ldots, x_n$ be an iid sample from $N(\theta, \sigma^2)$ with σ^2 known. Then the log-likelihood and the score functions are

$$\begin{aligned}
\log L(\theta) &= -\frac{n}{2}\log\sigma^2 - \frac{1}{2\sigma^2}\sum_i (x_i - \theta)^2 \\
S(\theta) &= \frac{\partial}{\partial\theta}\log L(\theta) \\
&= \frac{1}{\sigma^2}\sum_i (x_i - \theta) \\
&= \frac{n}{\sigma^2}(\overline{x} - \theta).
\end{aligned}$$

Figure 8.1(a) shows 20 score functions, each based on an iid sample of size $n = 10$ from $N(4, 1)$. We have noted before that the score function of the normal mean is exactly linear. At the true parameter $\theta = 4$ the score varies around zero.

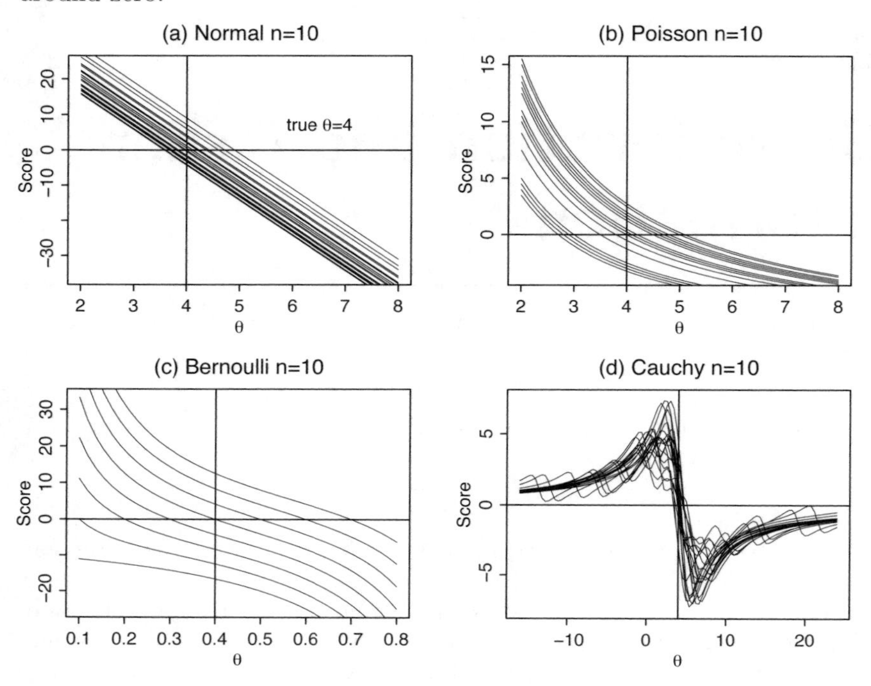

Figure 8.1: *Sampling variation of score functions for different models.*

Poisson model

Let $x_1, \ldots, x_n$ be an iid sample from Poisson(θ). The log-likelihood and the score function are

$$\begin{aligned}\log L(\theta) &= -n\theta + \sum x_i \log\theta \\ S(\theta) &= -n + \frac{\sum x_i}{\theta} \\ &= \frac{n}{\theta}(\overline{x} - \theta).\end{aligned}$$

Figure 8.1(b) shows 20 score functions based on 20 independent samples of size 10 from a Poisson distribution with mean 4. Here the score function is only approximately linear. At the true parameter ($\theta = 4$) the score function also varies around zero.

Binomial model

Let $x_1, \ldots, x_n$ be an iid sample from binomial(N, θ). The log-likelihood and the score functions are

$$\begin{aligned}\log L(\theta) &= \sum_{i=1}^{n} \{x_i \log\theta + (N - x_i)\log(1-\theta)\} \\ S(\theta) &= \sum_{i=1}^{n} \left\{ \frac{x_i}{\theta} - \frac{N - x_i}{1-\theta} \right\} \\ &= \frac{n(\overline{x} - N\theta)}{\theta(1-\theta)}.\end{aligned}$$

Figure 8.1(c) shows 20 score functions based on 20 independent samples of size $n = 10$ from binomial$(10, 0.4)$.

Cauchy model

Let $x_1, \ldots, x_n$ be an iid sample from Cauchy(θ) with density

$$p_\theta(x) = \{\pi(1 + (x-\theta)^2)\}^{-1},$$

where θ is the location parameter. The log-likelihood and the score functions are

$$\begin{aligned}L(\theta) &= -\sum_i \log\{(1 + (x_i - \theta)^2)\} \\ S(\theta) &= \sum_i \frac{2(x_i - \theta)}{1 + (x_i - \theta)^2}.\end{aligned}$$

Figure 8.1(d) shows 20 score functions based on independent samples of size 10 from Cauchy$(\theta = 4)$. The score function is quite irregular compared with the previous three examples. The Cauchy model can be considered as a representative of problems with complicated likelihood.

8.2 The mean of $S(\theta)$

We have seen in all cases above that, *at the true parameter* the score statistic varies around zero. For normal data

$$S(\theta) = \frac{n}{\sigma^2}(\overline{x} - \theta)$$

and for Poisson data

$$S(\theta) = \frac{n}{\theta}(\overline{x} - \theta).$$

In both cases $E_\theta S(\theta) = 0$. The subscript θ on the expected value is important: it communicates that θ is the true parameter that generates the data. The result is true in general:

Theorem 8.1 *Assuming regularity conditions so that we can take the derivative under the integral sign, we have*

$$E_\theta S(\theta) = 0.$$

Proof: Without loss of generality, we consider the continuous case:

$$\begin{aligned}
E_\theta S(\theta) &= \int S(\theta) p_\theta(x) dx \\
&= \int \left\{ \frac{\partial}{\partial\theta} \log L(\theta) \right\} p_\theta(x) dx \\
&= \int \frac{\frac{\partial}{\partial\theta} L(\theta)}{L(\theta)} p_\theta(x) dx \\
&= \int \frac{\partial}{\partial\theta} L(\theta) dx \\
&= \frac{\partial}{\partial\theta} \int p_\theta(x) dx = 0. \quad \square
\end{aligned}$$

The general conditions for taking the derivative under the integral sign are rather technical (e.g. Apostol 1974, page 283). For example, θ cannot be a boundary parameter, and, as a function of x in the neighbourhood of the true parameter θ, $|\partial L(\theta; x)/\partial\theta| \leq g(x)$ where $\int g(x)dx$ is finite. Essentially, this requires the model $p_\theta(x)$ to be 'smooth enough' as a function of θ. It can be shown that the exponential family models or other models commonly used in practice, including the Cauchy model, satisfy the conditions.

8.3 The variance of $S(\theta)$

We have defined in Section 2.5 the *observed* Fisher information as minus the second derivative of the log-likelihood function evaluated at the MLE $\widehat{\theta}$. Its generic form is

$$I(\theta) = -\frac{\partial^2}{\partial\theta^2} \log L(\theta) = -\frac{\partial}{\partial\theta} S(\theta).$$

This is minus the slope of the score function. This quantity varies from sample to sample; see Figure 8.1. Now we define the average or *expected Fisher information* as

$$\mathcal{I}(\theta) \equiv E_\theta I(\theta).$$

The expected value is taken at the *fixed and true* value of θ; 'true' in the sense that the data are generated at that value of θ.

There are notable qualitative differences between the expected and the observed Fisher information. The expected information is meaningful as a function of θ across the admissible values of θ, but $I(\theta)$ is only meaningful in the neighbourhood of $\widehat{\theta}$. More importantly, as an observed likelihood quantity the observed information applies to a single dataset; it is better to think of it as a single value, or a single statistic, rather than as a function of θ. In contrast, the 'expected information' is an average quantity over all possible datasets generated at a true value of the parameter. It is not immediately obvious whether $\mathcal{I}(\theta)$ is a relevant measure of information for a particular dataset.

As a function of θ, the expected information tells how 'hard' it is to estimate θ: parameters with greater information can be estimated more easily, requiring less sample to achieve a required precision. It might seem surprising then to arrive at a theorem that the expected information is equal to the variance of the score statistic. The proof is left as an exercise.

Theorem 8.2 *Assuming regularity conditions so that we can take two derivatives under the integral sign, we have*

$$var_\theta S(\theta) = \mathcal{I}(\theta).$$

Since $E_\theta S(\theta) = 0$, the theorem is equivalent to stating that

$$E_\theta \left\{ \frac{\partial}{\partial\theta} \log L(\theta) \right\}^2 = -E_\theta \left\{ \frac{\partial^2}{\partial\theta^2} \log L(\theta) \right\}.$$

The regularity conditions are satisfied by the exponential family or other commonly used models. In our discussions we will assume that this theorem holds.

Normal model

Let $x_1, \ldots, x_n$ be an iid sample from $N(\theta, \sigma^2)$ with σ^2 known. We have

$$\begin{aligned} \text{var}_\theta S(\theta) &= \text{var}\left\{ \frac{n}{\sigma^2}(\overline{x} - \theta) \right\} \\ &= \left(\frac{n}{\sigma^2} \right)^2 \frac{\sigma^2}{n} = \frac{n}{\sigma^2} \\ I(\theta) &= -\frac{\partial}{\partial\theta} S(\theta) = \frac{n}{\sigma^2}, \end{aligned}$$

so $\text{var}_\theta S(\theta) = E_\theta I(\theta)$. In this case $\mathcal{I}(\theta) = I(\theta)$, a happy coincidence in any exponential family model with canonical parameter θ (Exercise 8.3). We have noted before that $\text{var}(\widehat{\theta}) = 1/I(\widehat{\theta})$, so larger $I(\widehat{\theta})$ implies more precise information about θ.

Poisson model

For $x_1, \ldots, x_n$ an iid sample from Poisson(θ) we have

$$\begin{aligned} \text{var}_\theta S(\theta) &= \text{var}\left\{ \frac{n}{\theta}(\overline{x} - \theta) \right\} = \frac{n}{\theta} \\ I(\theta) &= -\frac{\partial}{\partial\theta} S(\theta) = \frac{n\overline{x}}{\theta^2} \\ \mathcal{I}(\theta) &= \frac{n}{\theta}. \end{aligned}$$

Now $\mathcal{I}(\theta) \neq I(\theta)$, but at $\theta = \widehat{\theta} = \overline{x}$ we have $\mathcal{I}(\widehat{\theta}) = I(\widehat{\theta})$. This is true generally for the full exponential family (Exercise 8.3). It means we can estimate the variance of the score statistic by either $\mathcal{I}(\widehat{\theta})$ or $I(\widehat{\theta})$. This is a remarkable result, since $I(\widehat{\theta})$ is a feature of the observed likelihood function, while the variance of the score statistic is a frequentist quantity.

Cauchy model

Let $x_1, \ldots, x_n$ be an iid sample from Cauchy(θ) with density $p(x) = \{\pi(1+(x-\theta)^2)\}^{-1}$. Here we have

$$
\begin{aligned}
I(\theta) &= -\sum_i \frac{2\{(x_i-\theta)^2-1\}}{\{(x_i-\theta)^2+1\}^2} \\
\mathcal{I}(\theta) &= -2nE_\theta\left\{\frac{(X_1-\theta)^2-1}{\{(X_1-\theta)^2+1\}^2}\right\} = \frac{n}{2} \\
\text{var}_\theta S(\theta) &= n\ \text{var}\left\{\frac{2(X_1-\theta)}{(X_1-\theta)^2+1}\right\} \\
&= \frac{n}{2}.
\end{aligned}
$$

So $\text{var}_\theta S(\theta) = \mathcal{I}(\theta)$, but now $\mathcal{I}(\widehat{\theta}) \neq I(\widehat{\theta})$, a common occurrence in complex cases. Both quantities are related to the precision of $\widehat{\theta}$, so when they are different, there is a genuine question: which is a better or more relevant measure of information? For sample size $n = 20$, there is 90% probability for $I(\widehat{\theta})$ to fall between 5.4 and 16.3, while $\mathcal{I}(\widehat{\theta}) = 10$. The quantity $I(\widehat{\theta})$ is 'closer to the data', as it describes the curvature of the observed likelihood, while $\mathcal{I}(\theta)$ is an average curvature that may be quite far from the observed data. Efron and Hinkley (1978) made this idea more rigorous; we will come back to this isssue in Section 9.6.

Censored data example

Let $t_1, \ldots, t_n$ be an iid sample from the exponential distribution with mean $1/\theta$, which are *censored* at fixed time c. This means an event time is observed only if it is less than c. So, the observed data are (y_i, δ_i), for $i = 1, \ldots, n$, where

$$y_i = t_i \text{ and } \delta_i = 1, \text{ if } t_i \leq c,$$

and

$$y_i = c \text{ and } \delta_i = 0, \text{ if } t_i > c.$$

The variable δ_i is an event indicator, which is Bernoulli with probability $(1 - e^{-\theta c})$. Using

$$
\begin{aligned}
p_\theta(t) &= \theta e^{-\theta t} \\
P_\theta(t) &= P(T > t) = e^{-\theta t},
\end{aligned}
$$

the contribution of (y_i, δ_i) to the likelihood is

$$L_i(\theta) = \{p_\theta(y_i)\}^{\delta_i}\{P_\theta(y_i)\}^{1-\delta_i}.$$

So,

$$\log L(\theta) = \sum_i \{\delta_i \log p_\theta(y_i) + (1-\delta_i)\log P_\theta(y_i)\}$$

$$
\begin{aligned}
&= \sum \delta_i \log\theta - \theta \sum y_i. \\
S(\theta) &= \frac{\sum \delta_i}{\theta} - \sum y_i \\
I(\theta) &= \frac{\sum \delta_i}{\theta^2} \\
\mathcal{I}(\theta) &= \frac{\sum E\delta_i}{\theta^2} \\
&= \frac{n(1-e^{-\theta c})}{\theta^2}
\end{aligned}
$$

and $\widehat{\theta} = \sum y_i / \sum \delta_i$. The sum $\sum \delta_i$ is the number of uncensored events.

So in general $\mathcal{I}(\widehat{\theta}) \neq I(\widehat{\theta})$, but there is a sense here that $I(\widehat{\theta})$ is the proper amount of information. For example, if there is no censored value

$$I(\widehat{\theta}) = n/\widehat{\theta}^2,$$

but

$$\mathcal{I}(\widehat{\theta}) = n(1-e^{-\widehat{\theta}c})/\widehat{\theta}^2 < I(\widehat{\theta}).$$

The quantity $n(1-e^{-\theta c})$, the expected number of events, is not relevant when the observed number is known. The dependence on c, an arbitrary censoring parameter that depends on the study design, makes $\mathcal{I}(\widehat{\theta})$ undesirable. The standard practice of survival analysis always uses the observed information $I(\widehat{\theta})$.

8.4 Properties of expected Fisher information

Additive property for independent data

Let $\mathcal{I}_x(\theta)$ be the Fisher information on θ based on data X and $\mathcal{I}_y(\theta)$ based on Y. If X and Y are independent, then the total information contained in X and Y is $\mathcal{I}_x(\theta) + \mathcal{I}_y(\theta)$. In particular if $x_1, \ldots, x_n$ are an iid sample from $p_\theta(x)$, then the information contained in the sample is $n\mathcal{I}_{x_1}(\theta)$.

Information for location parameter

If $p_\theta(x) = f(x-\theta)$ for some standard density $f(\cdot)$, then $\mathcal{I}(\theta)$ is a constant free of θ. This means that any location θ can be estimated with the same precision. To prove it:

$$
\begin{aligned}
\mathcal{I}(\theta) &= \int S(\theta)^2 p_\theta(x)dx \\
&= \int \left\{ \frac{f'(x-\theta)}{f(x-\theta)} \right\}^2 f(x-\theta)dx \\
&= \int \left\{ \frac{f'(u)}{f(u)} \right\}^2 f(u)du.
\end{aligned}
$$

Information for scale parameter

We call θ a scale parameter if

$$p_\theta(x) = \frac{1}{\theta} f\left(\frac{x}{\theta}\right),$$

for some standard density $f(\cdot)$. The family $p_\theta(x)$ is called a scale family; for example, $N(0, \theta^2)$ is a scale family. In this case we can show that

$$\mathcal{I}(\theta) = \text{constant}/\theta^2,$$

implying that it is easier to estimate a small θ than a large one. The proof is left as an exercise.

Transformation of parameter

Let $\psi = g(\theta)$ for some function $g(\cdot)$. The score function of ψ is

$$\begin{aligned} S^*(\psi) &= \frac{\partial}{\partial \psi} \log L(\theta) \\ &= \frac{\partial \theta}{\partial \psi} \frac{\partial}{\partial \theta} \log L(\theta) \\ &= \frac{\partial \theta}{\partial \psi} S(\theta), \end{aligned}$$

so the expected information on ψ is

$$\begin{aligned} \mathcal{I}^*(\psi) &= \text{var} S^*(\psi) \\ &= \left(\frac{\partial \theta}{\partial \psi}\right)^2 \mathcal{I}(\theta) \\ &= \frac{\mathcal{I}(\theta)}{(\partial \psi / \partial \theta)^2}. \end{aligned}$$

For example, based on a sample of size one from Poisson(θ) we have

$$\begin{aligned} \mathcal{I}(\theta) &= \frac{1}{\theta} \\ \mathcal{I}^*(\log \theta) &= \frac{1/\theta}{(1/\theta)^2} = \theta \\ \mathcal{I}^*(\sqrt{\theta}) &= 4. \end{aligned}$$

Therefore, in the log-scale, it is easier to estimate a large parameter value than a small one. In generalized linear modelling of Poisson data this means that inference in the log-scale is easier if we have data with large means. The parameter $\sqrt{\theta}$ behaves like a location parameter.

8.5 Cramér–Rao lower bound

We estimate the normal mean μ by $T(x) = \overline{x}$. Is this the best possible? What do we mean by 'best'? These are questions of optimality, which drove much of classical estimation theory. The standard way to express 'best' is to say that among unbiased estimates $\overline{x}$ has the smallest variance, implying that it is the most precise. The restriction to unbiasedness is quite arbitrary, but it is necessary in order to develop some theory. Furthermore, for the variance to be meaningful the sampling distribution must be normal or close to normal.

In the normal model we know that

$$\text{var}(\overline{X}) = \frac{\sigma^2}{n}.$$

Is it possible to get another estimate with lower variance? The Cramér–Rao lower bound theorem addresses this. It states the best we can achieve in terms of variability in estimating a parameter $g(\theta)$. If the bound is achieved we know that we cannot do better.

Theorem 8.3 *Let $E_\theta T(X) = g(\theta)$ and $\mathcal{I}(\theta)$ be the Fisher information for θ based on X. Assuming regularity conditions we have*

$$var_\theta\{T(X)\} \geq \frac{\{g'(\theta)\}^2}{\mathcal{I}(\theta)}.$$

In particular, if $E_\theta T = \theta$, then

$$var_\theta\{T(X)\} \geq \frac{1}{\mathcal{I}(\theta)}.$$

The value $\{g'(\theta)\}^2/\mathcal{I}(\theta)$ is called the Cramér–Rao lower bound (CRLB). The proof, to be given later, does not illuminate why such a result should be true.

Example 8.1: Let $x_1, \ldots, x_n$ be an iid sample from Poisson(θ). We have shown before that $\mathcal{I}(\theta) = n/\theta$, so to estimate θ by an unbiased estimate T, we must have

$$\text{var}(T) \geq \frac{1}{\mathcal{I}(\theta)} = \frac{\theta}{n}.$$

Since $\text{var}(\overline{X}) = \theta/n$ we conclude that $\overline{X}$ is the best unbiased estimate. Theoretically, it will be satisfying to be able to say that this is a unique best estimate, but that would require the concept of *completeness*, which we defer until Section 8.6. □

Example 8.2: Let $x_1, \ldots, x_n$ be an iid sample from Poisson(θ) and we would like to estimate $g(\theta) = P(X_1 = 0) = e^{-\theta}$. If T is unbiased for $g(\theta)$, then

$$\text{var}(T) \geq \frac{(-e^{-\theta})^2}{n/\theta} = \frac{\theta e^{-2\theta}}{n}.$$

The theorem does not provide guidance to finding an estimate that achieves the bound, or even to say whether such an estimate exists. If we use the MLE $T(x) = e^{-\overline{x}}$, then, using the Delta method,

$$\text{var}\left(e^{-\overline{X}}\right) \approx \frac{\theta e^{-2\theta}}{n} = \text{CRLB},$$

so the MLE approximately achieves the CRLB. The best unbiased estimate in this case is given by (see Example 8.9)

$$T(x) = \left(1 - \frac{1}{n}\right)^{n\overline{x}}.$$

To see the connection with the MLE, we know from calculus that $(1-1/n)^n \to e^{-1}$ as n gets large. This estimate does not achieve the CRLB; from the proof of the CRLB theorem we shall see that only T that is linearly related to the score statistic will achieve the CRLB exactly. □

Example 8.3: Let $x_1, \ldots, x_n$ be an iid sample from Poisson(θ) and we would like to estimate $g(\theta) = \theta^2$. If T is unbiased for θ^2, then according to the theorem

$$\text{var}(T) \geq \frac{4\theta^3}{n}.$$

What is the best unbiased T? First guess $T = \overline{X}^2$, which has

$$E\overline{X}^2 = \text{var}(\overline{X}) + (E\overline{X})^2 = \frac{\theta}{n} + \theta^2,$$

so $T = \overline{X}^2 - \overline{X}/n$ is unbiased for θ^2. In fact, using the method given in the next section, we can show that T is the best unbiased estimate for θ^2, but

$$\begin{aligned} \text{var}(T) &= E(\overline{X}^2 - \overline{X}/n)^2 - (\theta^2)^2 \\ &= E\overline{X}^4 - \frac{2}{n}E\overline{X}^3 + \frac{1}{n^2}E\overline{X}^2 - \theta^4 \\ &= \frac{4\theta^3}{n} + \frac{2\theta^2}{n^2} > \text{CRLB}. \end{aligned}$$

This generally occurs in small or fixed sample cases. □

Proof of the CRLB theorem

From the covariance inequality, for any two random variables S and T we have

$$\text{var}(T) \geq \frac{|\text{cov}(S,T)|^2}{\text{var}(S)},$$

with equality if and only if S and T are perfectly correlated. Choose S to be the score statistic $S(\theta)$, so we only need to show that $\text{cov}\{S(\theta), T\} = g'(\theta)$. We showed before that $ES(\theta) = 0$, so

$$\text{cov}\{S(\theta), T\} = E\{S(\theta)T(x)\}$$

$$
\begin{aligned}
&= \int S(\theta)T(x)p_\theta(x)dx \\
&= \int \frac{\frac{\partial}{\partial\theta}p_\theta(x)}{p_\theta(x)}T(x)p_\theta(x)dx \\
&= \int \frac{\partial}{\partial\theta}p_\theta(x)T(x)dx \\
&= \frac{\partial}{\partial\theta}\int T(x)p_\theta(x)dx \\
&= g'(\theta).
\end{aligned}
$$

Connection with the exponential family

From the proof above, the only statistics $T(x)$ that can achieve CRLB are those which are linearly related to the score statistic. In such cases, for some functions $u(\theta)$ and $v(\theta)$,

$$\frac{\partial}{\partial\theta}\log p_\theta(x) = u(\theta)T(x) + v(\theta).$$

This implies that, if there exists a statistic that achieves the CRLB, the model must be in the exponential family, with log-density of the form

$$\log p_\theta(x) = \eta(\theta)T(x) - A(\theta) + c(x).$$

This also means that, up to a linear transformation, the CRLB is only achievable for one function $g(\theta)$.

Example 8.4: For the normal, Poisson and binomial models, the sample mean is the natural statistic, so it achieves the CRLB for estimation of the mean. No other statistics that are not linearly related to the sample mean can achieve the CRLB. □

Example 8.5: Suppose $x_1, \ldots, x_n$ are a sample from the Cauchy distribution with location θ. Since it is not in the exponential family there is no statistic that can achieve the CRLB. □

8.6 Minimum variance unbiased estimation*

If the CRLB is not achieved, is there still a 'best' estimate? Among unbiased estimates, suppose we define the best estimate as the one with minimum variance. Establishing such a minimum variance unbiased estimate (MVUE) requires a new concept called *completeness*. (The MVUE theory is a class of theory of statistics that investigates how much can be achieved theoretically given a certain criterion of optimality. It is quite different from the likelihood approach followed in this book, where data modelling is the primary consideration.)

Definition 8.1 *A sufficient statistic T is complete if, for any function $g(T)$,*

$$E_\theta g(T) = 0$$

for all θ implies

$$g(t) = 0.$$

Example 8.6: Suppose x is a sample from the binomial(n, θ). Then x is a complete sufficient statistic for θ. Sufficiency is trivial, since x is the whole data. Now, for any function $g(x)$,

$$\begin{aligned} E_\theta g(X) &= \sum_{x=0}^{n} g(x) \binom{n}{x} \theta^x (1-\theta)^{n-x} \\ &\equiv \sum_{x=0}^{n} c(x)\theta^x, \end{aligned}$$

for some function $c(x)$ not involving θ. Hence $E_\theta g(X)$ is a polynomial of maximum order n. Therefore, the condition $E_\theta g(X) = 0$ for all θ implies $c(x) = 0$, or $g(x) = 0$ for all x. □

Example 8.7: Let $x_1, \ldots, x_n$ be a sample from a one-parameter exponential family, with log-density of the form

$$\log p_\theta(x_i) = \eta(\theta)t_1(x_i) - A(\theta) + c(x).$$

Then $T = \sum_i t_1(x_i)$ is a complete sufficient statistic for θ. By the factorization theorem T is clearly sufficient. The density of T is also of the exponential family form

$$\log p_\theta(t) = \eta(\theta)t - nA(\theta) + c^*(t)$$

so, for any function $g(T)$,

$$E_\theta g(T) = \int g(t) e^{\eta(\theta)t - nA(\theta) + c^*(t)}$$

and $E_\theta g(T) = 0$ for all θ means

$$\int g(t) e^{c^*(t)} e^{\eta(\theta)t} = 0$$

for all $\eta(\theta)$. From the theory of Laplace transform,

$$g(t) e^{h^*(t)} = 0$$

but $e^{h^*(t)} > 0$, so $g(t) = 0$. □

Example 8.8: Extending the result to the p-parameter exponential family, where $\theta \in R^p$, follows the same route. Specifically, if $x_1, \ldots, x_n$ are an iid sample from a distribution with log-density

$$\log p_\theta(x) = \sum_{k=1}^{p} \eta_k(\theta)t_k(x) - A(\theta) + c(x),$$

then $\{\sum_i t_1(x_i), \ldots, \sum_i t_p(x_i)\}$ is a complete sufficient statistic. The only condition in the Laplace transform theory is that the set $\{(\eta_1, \ldots, \eta_p), \text{for all } \theta\}$ contains an open set in R^p. This is exactly the requirement of a full exponential family.

If $x_1, \ldots, x_n$ are an iid sample from $N(\mu, \sigma^2)$, then $(\sum_i x_i, \sum_i x_i^2)$ is complete sufficient for $\theta = (\mu, \sigma^2)$. But, if, additionally, $\sigma = \mu$ then the set $\{(\eta_1, \eta_2), \text{for all } \theta\}$ is a curve in R^2. In this case $(\sum_i x_i, \sum_i x_i^2)$ is not complete sufficient, though it is still minimal sufficient. □

Completeness implies unique estimates

In connection with estimation, completeness implies uniqueness: if T is a complete sufficient statistic for θ, then $h(T)$ is a unique unbiased estimate of $E_\theta\{h(T)\}$. To see this, suppose $u(T)$ is another unbiased estimate of $E_\theta\{h(T)\}$. Then,

$$E_\theta\{h(T) - u(T)\} = 0$$

for all θ, which, by completeness of T, implies $h(t) = u(t)$.

Completeness implies minimal sufficiency

Completeness implies minimal sufficiency, but not vice versa. Let U be the minimal sufficient statistic and T be the complete sufficient statistic for θ. From minimality, U is a function of T:

$$U = h(T)$$

and, since

$$E_\theta\{E(T|U) - T\} = 0,$$

completeness of T implies

$$T - E(T|U) = 0$$

or $T = g(U)$. This means U is a one-to-one function of T, so T is minimal sufficient.

Construction of MVUE

Lehmann and Scheffé (1950) established that, in a model that admits a complete sufficient statistic we can always construct the MVUE. This is regardless of the CRLB. The construction of the estimate follows the so-called Rao–Blackwell step:

Theorem 8.4 *Suppose $U(x)$ is any unbiased estimate of $g(\theta)$, and $T(x)$ is a complete sufficient statistic for θ. Then a new estimate*

$$S(x) = E(U|T)$$

is the unique MVUE of $g(\theta)$.

Proof: The new estimate $S(x)$ is a proper statistic, i.e. it does not depend on the unknown θ, since T is sufficient. It is also unbiased, since

$$E_\theta S = E_\theta\{E(U|T)\} = E_\theta U = g(\theta).$$

From the variance equality

$$\text{var}(U) = E\{\text{var}(U|T)\} + \text{var}\{E(U|T)\},$$

we immediately get

$$\text{var}(U) \geq \text{var}(S).$$

So S has smaller variance among any unbiased estimate U. Completeness of T implies that S is a unique function of T that is unbiased for $g(\theta)$. □

Example 8.9: Suppose $x_1, \ldots, x_n$ are an iid sample from the Poisson distribution with mean θ and we are interested to estimate $g(\theta) = P(X_1 = 0) = e^{-\theta}$. Since the Poisson model is in the exponential family, $T = \sum_i x_i$ is a complete sufficient statistic for θ. Let

$$U = I(x_1 = 0),$$

so $E_\theta U = P(X_1 = 0)$, or U is an unbiased estimate of $g(\theta)$. The MVUE of $g(\theta)$ is

$$\begin{aligned} S &= E(U|T) \\ &= P(X_1 = 0 | \sum_i x_i) \\ &= \left(1 - \frac{1}{n}\right)^{\sum_i x_i} \end{aligned}$$

since the conditional distribution of X_1 given $\sum_i x_i$ is binomial with $n = \sum_i x_i$ and $p = 1/n$. □

8.7 Multiparameter CRLB

In practice, we are more likely to have many parameters, some of which may be a nuisance. Is there any effect of the extra parameters on the CRLB of the parameter of interest? This question is meaningful since the CRLB is usually achieved by the MLE in large samples. This section is closely related to Section 3.3 on multiparameter (observed) Fisher information, but for completeness we repeat some of the notations.

Let $\theta = (\theta_1, \ldots, \theta_p)$. The score function is now a gradient vector

$$S(\theta) = \frac{\partial}{\partial\theta} \log L(\theta) = \begin{pmatrix} \frac{\partial}{\partial\theta_1} \log L(\theta) \\ \vdots \\ \frac{\partial}{\partial\theta_p} \log L(\theta) \end{pmatrix}.$$

The observed Fisher information is minus the Hessian of the log-likelihood function

$$I(\theta) = -\frac{\partial^2}{\partial\theta\partial\theta'} \log L(\theta).$$

The *expected* Fisher information is

$$\mathcal{I}(\theta) = E_\theta I(\theta).$$

Using similar methods as in the scalar case, assuming regularity conditions, we can show that

$$\begin{aligned} E_\theta S(\theta) &= 0 \\ \text{var}_\theta\{S(\theta)\} &= \mathcal{I}(\theta). \end{aligned}$$

The information matrix $\mathcal{I}(\theta)$ is now a $p \times p$ variance matrix, which means it is a nonnegative definite matrix.

Example 8.10: Let $x_1, \ldots, x_n$ be an iid sample from $N(\mu, \sigma^2)$ and let $\theta = (\mu, \sigma^2)$. Then we have the following:

$$
\begin{aligned}
\log L(\theta) &= -\frac{n}{2}\log\sigma^2 - \frac{1}{2\sigma^2}\sum_i (x_i - \mu)^2 \\
S(\theta) &= \begin{pmatrix} \frac{\partial}{\partial\mu}\log L(\theta) \\ \frac{\partial}{\partial\sigma^2}\log L(\theta) \end{pmatrix} = \begin{pmatrix} \frac{n}{\sigma^2}(\overline{x} - \mu) \\ -\frac{n}{2\sigma^2} + \frac{\sum(x_i-\mu)^2}{2\sigma^4} \end{pmatrix} \\
I(\theta) &= \begin{pmatrix} \frac{n}{\sigma^2} & \frac{n}{\sigma^4}(\overline{x} - \mu) \\ \frac{n}{\sigma^4}(\overline{x} - \mu) & -\frac{n}{2\sigma^4} + \frac{\sum(x_i-\mu)^2}{\sigma^6} \end{pmatrix} \\
\mathcal{I}(\theta) &= \begin{pmatrix} \frac{n}{\sigma^2} & 0 \\ 0 & \frac{n}{2\sigma^4} \end{pmatrix}. \; \Box
\end{aligned}
$$

We now state the multiparameter version of the CRLB theorem:

Theorem 8.5 *Let $T(X)$ be a scalar function, $E_\theta T = g(\theta)$ and $\mathcal{I}(\theta)$ the expected Fisher information for θ based on data X. Then*

$$var_\theta(T) \geq \alpha' \mathcal{I}(\theta)^{-1}\alpha,$$

where $\alpha = \frac{\partial}{\partial\theta}g(\theta)$.

Proof: The proof relies on an extension of the covariance inequality involving a scalar random variable T and a vector of random variables S:

$$\text{var}(T) \geq \text{cov}(S,T)'\{\text{var}(S)\}^{-1}\text{cov}(S,T).$$

and showing that $\text{cov}\{S(\theta), T\} = \frac{\partial}{\partial\theta}g(\theta)$. The proof of these statements is left as an exercise. $\Box$

Example 8.11: Let $g(\theta) = a'\theta$ for some known vector a. Then for any unbiased estimate T we have

$$\text{var}(T) \geq a'\mathcal{I}(\theta)^{-1}a.$$

In particular, let $g(\theta) = \theta_1$ (or any other θ_i), which is obtained using $a = (1, 0, \ldots, 0)$, so

$$\text{var}(T) \geq [\mathcal{I}(\theta)^{-1}]_{11}, \tag{8.1}$$

where $[\mathcal{I}(\theta)^{-1}]_{11}$ is the (1,1) element of the inverse Fisher information, which is sometimes denoted by $\mathcal{I}^{11}(\theta)$. This is the bound for the variance if θ_1 is the parameter of interest, and we do not know the value of the other parameters.

For comparison, if θ_1 is the only unknown parameter then any unbiased estimate T of θ_1 satisfies

$$\text{var}(T) \geq \frac{1}{\mathcal{I}_{11}(\theta)}. \tag{8.2}$$

To see which bound is larger, partition the information matrix $\mathcal{I}(\theta)$ and its inverse $\mathcal{I}(\theta)^{-1}$ as

$$\mathcal{I}(\theta) = \begin{pmatrix} \mathcal{I}_{11} & \mathcal{I}_{12} \\ \mathcal{I}_{21} & \mathcal{I}_{22} \end{pmatrix}$$

and

$$\mathcal{I}(\theta)^{-1} = \begin{pmatrix} \mathcal{I}^{11} & \mathcal{I}^{12} \\ \mathcal{I}^{21} & \mathcal{I}^{22} \end{pmatrix}.$$

Using some matrix algebra we can show (Exercise 8.10)

$$(\mathcal{I}^{11})^{-1} = \mathcal{I}_{11} - \mathcal{I}_{12}\mathcal{I}_{22}^{-1}\mathcal{I}_{21}. \tag{8.3}$$

Since $\mathcal{I}_{22}$ is a variance matrix and $\mathcal{I}_{12} = \mathcal{I}'_{21}$, the quadratic form is nonnegative, so

$$(\mathcal{I}^{11})^{-1} \leq \mathcal{I}_{11},$$

and the bound in (8.2) is smaller than in (8.1). This bound is generally achieved by the MLE in large samples, so this result has an important statistical modelling implication: there is a cost in having to estimate extra parameters.

We can also interpret $(\mathcal{I}^{11})^{-1}$ as the Fisher information on θ_1 when the other parameters are unknown. It makes statistical sense that the information is less than if other parameters are known. In modelling we consider extra parameters to explain the data better; for example, $N(\theta, \sigma^2)$ is likely to fit better than $N(\theta, 1)$. The reward for better fitting is a reduced bias, but our discussion here warns that this must be balanced against the increase in variability.

From (8.3) the cost in terms of increased variability is (asymptotically) zero if $\mathcal{I}_{12} = 0$. This happens in the normal example above. The 'asymptotic' qualifier is ever present if we are thinking of CRLB as a variance quantity. From the normal example we also know that the zero-cost benefit does not apply in small samples. □

8.8 Exercises

Exercise 8.1: Prove Theorem 8.2.

Exercise 8.2: Let $x_1, \ldots, x_n$ be an iid sample from the following distributions. In each case, find the score statistics and the Fisher information.

(a) Gamma with density

$$p(x) = \frac{1}{\Gamma(\alpha)}\lambda^\alpha x^{\alpha-1}e^{-\lambda x}, \quad x > 0,$$

first assuming one parameter is known, then assuming both parameters are unknown.

(b) Weibull with distribution function

$$F(x) = 1 - e^{-(\lambda x)^\alpha}, \quad x > 0,$$

first assuming one parameter is known, then assuming both parameters are unknown. The parameter α is the shape parameter.

(c) Beta with density

$$p(x) = \frac{\Gamma(\alpha+\beta)}{\Gamma(\alpha)\Gamma(\beta)}x^{\alpha-1}(1-x)^{\beta-1}, \quad 0 < x < 1,$$

first assuming one parameter is known, then assuming both parameters are unknown.

Exercise 8.3: Show that for the general exponential family model with log-density of the form

$$\log p_\theta(x) = t(x)\eta(\theta) - A(\theta) + c(x)$$

we have

$$\mathcal{I}(\widehat{\theta}) = I(\widehat{\theta}),$$

where $\widehat{\theta}$ is the MLE of θ. If θ is the canonical parameter, then $\mathcal{I}(\theta) = I(\theta)$.

Exercise 8.4: Suppose $y_1, \ldots, y_n$ are an iid sample from $N(\theta, \sigma^2)$ with known σ^2. Find the CRLB for unbiased estimates of the following parameters:

(a) $P(Y_1 < 2)$.

(b) $P(-2 < Y_1 < 2)$.

(c) $p_\theta(2)$, the density function at $y = 2$.

Exercise 8.5: Repeat the previous exercise for unknown σ^2.

Exercise 8.6: Suppose $y_1, \ldots, y_n$ are an iid sample from $N(0, \sigma^2)$. Find the CRLB for estimating σ. Is it achievable by some statistic?

Exercise 8.7: Suppose $y_1, \ldots, y_n$ are an iid sample from Poisson(θ). Based on $y_1, \ldots, y_n$, find the CRLB for unbiased estimates of the following parameters:

(a) $P(Y_1 = 1)$.

(b) $P(Y_1 \leq 1)$.

Find the MLE of the parameters above, and compute the bias and variance of the MLE. Compare the variance of the MLE with the CRLB.

Exercise 8.8: Suppose $y_1, \ldots, y_n$ are an iid sample from a Weibull distribution (Exercise 8.2) with the shape parameter known. What is the CRLB for estimates of the median? Repeat the exercise for both parameters unknown.

Exercise 8.9: If $x_1, \ldots, x_n$ are an iid sample from $N(\theta, \theta^2)$, then show that $(\sum x_i, \sum x_1^2)$ is minimal sufficient but not complete. (Hint: give an example of a non-zero statistic that has mean zero.) Give another example showing a minimal sufficient statistic that is not complete.

Exercise 8.10: Prove the matrix equality given by (8.3).

Exercise 8.11: To study the toxicity of a certain chemical a two-stage experiment was conducted. In the first stage, seven rats were injected with the chemical and monitored. If none died then another batch of seven rats was injected, otherwise the experiment was stopped. Say S is the number of survivors from the first batch, and T from the second batch. Let θ be the probability of survival, so S is binomial$(7, \theta)$, and $P(T = 0|S < 7) = 1$, and T is binomial$(7, \theta)$ if $S = 7$.

(a) Show that the joint probability distribution of S and T is

$$p_\theta(s,t) = \binom{7}{s}\binom{7}{t}\theta^{s+t}(1-\theta)^{7-s-t+7I(s+t\geq 7)}$$

for $t = 0$ if $0 \leq s < 7$, and for $0 \leq t \leq 7$ if $s = 7$, and $I(s + t \geq 7)$ is an indicator function.

(b) Find and interpret the maximum likelihood of θ.

(c) Show that $S + T$ is a complete sufficient statistic.

(d) Show that $S/7$ is the MVUE for θ. (This means the best unbiased estimate uses data only from the first batch!)

9

Large-sample results

While likelihood-based quantities can be computed at any sample size, small-sample frequentist calibration depends on specific models such as the normal, Poisson, binomial, etc. (Chapter 5). A general technique for complex models is provided by the large-sample theory. As the price for this generality, the theory is only approximate. The results in this chapter provide theoretical justifications for the previous claims regarding the sampling properties of likelihood ratio and the MLE. In modelling, we use the large-sample theory to suggest approximate likelihoods.

9.1 Background results

What we need to know for our large-sample theory is mostly captured by the behaviour of the sample mean $\overline{X}$ as the sample size gets large. The theorems listed in this section are sufficient for most of the standard likelihood theory. Further grounding is needed, however, if we want to prove theoretical extensions or nonstandard results. It is important to recognize two types of results:

- first-order results that capture the magnitude of an estimate. The basis for these results is the law of large numbers, and in particular we would use the concept of convergence in probability.
- second- or higher-order results that deal with the variability or the distribution of an estimate. We rely mainly on the central limit theorem to establish them.

Law of large numbers

As a first-order behaviour, we expect intuitively that $\overline{X}$ will become close to the true mean μ. This can be shown easily as follows. Let $X_1, \ldots, X_n$ be an iid sample from a population with mean μ and variance σ^2. By Chebyshev's inequality, for any $\epsilon > 0$,

$$\begin{aligned} P(|\overline{X} - \mu| > \epsilon) &\leq \frac{\text{var}(\overline{X})}{\epsilon^2} \\ &= \frac{\sigma^2}{\epsilon^2 n} \to 0 \end{aligned}$$

as n goes to infinity. (It helps to think of ϵ as a small but fixed positive number.) Equivalently,

$$P(|\overline{X} - \mu| < \epsilon) \to 1,$$

that is, with high probability, with large enough n, we expect $\overline{X}$ to be within ϵ of μ. We say that $\overline{X}$ converges to μ *in probability* or

$$\overline{X} \xrightarrow{p} \mu.$$

This result is called the *weak law of large numbers* (WLLN). It is far from the best result, which is known as the *strong law of large numbers* (SLLN); it states

$$P(|\overline{X} - \mu| \to 0) = 1$$

as long as μ exists. This means $\overline{X}$ is guaranteed (i.e. with probability one) to converge to μ in the usual numerical sense; hence we are correct in thinking that an observed $\overline{x}$ is numerically close to the true μ; this is not guaranteed by the convergence in probability. This mode of convergence is called *almost sure convergence* and we write

$$\overline{X} \xrightarrow{a.s.} \mu.$$

We may interpret the WLLN as a frequentist theorem: if we repeat an experiment a large number of times, each time computing the sample mean, then a large proportion of the sample means is close to the true mean. The statement 'the sample mean is close to the true mean' does not apply to a particular realization. In contrast, the SLLN deals with what happens to the result of a single realization of the data.

The SLLN is one of Kolmogorov's celebrated theorems, and is standard fare in any course on advanced probability theory (e.g. Chung 1974). The techniques needed to establish almost sure convergence are beyond the scope of this text, so we will rely on the convergence in probability.

Central limit theorem

For statistical inference the first-order property given by the law of large numbers is not enough. We need to know the second-order property, the variability of $\overline{X}$ around μ. This is given by the *central limit theorem* (CLT): if $X_1, \ldots, X_n$ are an iid sample from a population with mean μ and variance σ^2 then

$$\sqrt{n}(\overline{X} - \mu) \to N(0, \sigma^2).$$

It is striking that only μ and σ^2 matter in the asymptotic distribution of $\overline{X}$; other features of the parent distribution, such as skewness or discreteness, do not matter. These features do, however, determine how fast the true distribution of $\overline{X}$ converges to the normal distribution. Figure 9.1 shows the normal QQ-plots of $\overline{X}$ simulated from the standard uniform, standard exponential and Poisson(3) distributions; note the effect of skewness and discreteness.

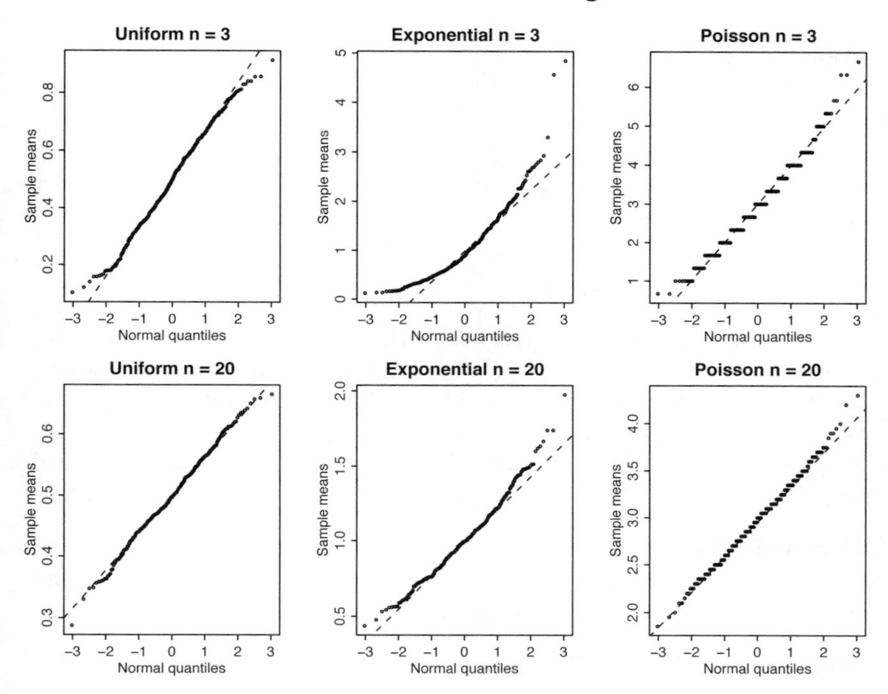

Figure 9.1: *First row: the normal QQ-plots of simulated* $\overline{X}$ *from the standard uniform, standard exponential and Poisson(3) distributions. In each case 400 samples of size* $n = 3$ *were generated. Second row: the corresponding QQ-plots for* $n = 20$.

The CLT is the most important example of *convergence in distribution* or *convergence in law.* Suppose X_n has distribution F_n, and X has distribution F; we say X_n converges to X in distribution or

$$X_n \xrightarrow{d} X$$

if $F_n(x) \to F(x)$ for all x such that $F(x)$ is continuous. If X is degenerate, convergence in distribution is equivalent to the convergence in probability.

Example 9.1: To see why we need the last condition, let X_n be a random variate with distribution $N(0, 1/n)$ and X is a degenerate distribution at zero. As we expect, $X_n \xrightarrow{d} X$. In this case $F_n(x) \to F(x)$ at every $x \neq 0$, but at $x = 0$ we have $F_n(x) = 0.5 \neq F(0) = 1$. □

The CLT states that $S_n = \sum X_i$ is approximately normal with mean ES_n and variance $\text{var}(S_n)$. We expect this to be true also for non-iid observations as long as no individual X_i dominates in the summation. There are many non-iid versions of the CLT of varying generalities (Chung 1974, Chapter 7), we will state only the one due to Liapounov. Suppose $X_1, \ldots, X_n$ are an independent sample; let μ_i and σ_i^2 be the mean and variance of X_i, and

$$\gamma_i = E|X_i - \mu_i|^3.$$

If

$$\frac{\sum_i \gamma_i}{(\sum_i \sigma_i^2)^{3/2}} \to 0$$

as n goes to infinity, then

$$\frac{S_n - ES_n}{\{\text{var}(S_n)\}^{1/2}} = \frac{\sum_i (X_i - \mu_i)}{(\sum_i \sigma_i^2)^{1/2}} \xrightarrow{d} N(0,1).$$

The condition of the theorem guarantees that no individual X_i dominates in the summation S_n. This is usually satisfied in statistical applications where the observations have comparable weights or variability. It is trivially satisfied in the iid case, since $\sum \gamma_i = n\gamma_1$ and $\sum \sigma_i^2 = n\sigma_1^2$. More generally, it is satisfied if $\gamma_i < M < \infty$ and $\sigma_i^2 > m > 0$, i.e. the individual third moment is finite and the individual variance is bounded away from zero.

Other results

The following results are intuitive, although the proofs are actually quite technical (e.g. Serfling 1980, Chapter 1).

Theorem 9.1 (Slutsky) *If $A_n \xrightarrow{p} a$, $B_n \xrightarrow{p} b$ and $X_n \xrightarrow{d} X$ then*

$$A_n X_n + B_n \xrightarrow{d} aX + b.$$

It is worth noting that there is no statement of dependency between A_n, B_n and X_n.

In applications we often consider the transformation of parameters. If $\overline{x} \xrightarrow{p} \mu$ then we expect that $e^{-\overline{x}} \xrightarrow{p} e^{-\mu}$. Such a result is covered under the following 'continuous mapping theorems'.

Theorem 9.2 *If $X_n \xrightarrow{p} X$ and $g(x)$ is a continuous function then*

$$g(X_n) \xrightarrow{p} g(X).$$

Theorem 9.3 *If $X_n \xrightarrow{d} X$ and $g(x)$ is a continuous function then*

$$g(X_n) \xrightarrow{d} g(X).$$

Using these theorems we can show that, from an iid sample,

$$s^2 = \frac{1}{n-1} \sum_i (x_i - \overline{x})^2 \xrightarrow{p} \sigma^2,$$

so the sample standard deviation $s \xrightarrow{p} \sigma$, and

$$\frac{\sqrt{n}(\overline{X} - \mu)}{s} \xrightarrow{d} N(0,1).$$

Furthermore,

$$\frac{n(\overline{X} - \mu)^2}{s^2} \xrightarrow{d} \chi_1^2.$$

9.2 Distribution of the score statistic

The results in this section form the basis for most of the large-sample likelihood theory, and for the so-called *score tests* or *locally most powerful tests.* We will assume that our data $x_1, \ldots, x_n$ are an iid sample from $p_\theta(x)$ and θ is not a boundary parameter; the independent but not identical case follows from Liapounov's CLT. Thus,

$$\begin{aligned}
\log L(\theta) &= \sum_i \log p_\theta(x_i) \\
S(\theta) &= \sum_i \frac{\partial}{\partial\theta} \log p_\theta(x_i) \\
I(\theta) &= -\sum_i \frac{\partial^2}{\partial\theta^2} \log p_\theta(x_i) \\
\mathcal{I}(\theta) &= E_\theta I(\theta).
\end{aligned}$$

Let

$$y_i \equiv \frac{\partial}{\partial\theta} \log p_\theta(x_i),$$

the individual score statistic from each x_i. In view of Theorems 8.1 and 8.2, $y_1, \ldots, y_n$ are an iid sample with mean $Ey_1 = 0$ and variance

$$\text{var}(y_1) \equiv \mathcal{I}_1(\theta),$$

so immediately by the CLT we get

$$\sqrt{n}(\overline{y} - 0) \stackrel{d}{\to} N\{0, \mathcal{I}_1(\theta)\},$$

or

$$\frac{S(\theta)}{\sqrt{n}} \stackrel{d}{\to} N\{0, \mathcal{I}_1(\theta)\}.$$

Since $\mathcal{I}(\theta) = n\mathcal{I}_1(\theta)$, as the sample size gets large, we have approximately

$$\{\mathcal{I}(\theta)\}^{-1/2} S(\theta) \sim N(0,1), \tag{9.1}$$

or, informally, $S(\theta) \sim N\{0, \mathcal{I}(\theta)\}$.

These arguments work whether θ is a scalar or a vector parameter. In the vector case $S(\theta)$ is a vector and $\mathcal{I}(\theta)$ is a matrix.

Example 9.2: Let $x_1, \ldots, x_n$ be an iid sample from Poisson(θ) with $n = 10$ and $\theta = 4$. Figure 9.2(a) shows the score function for a single sample, and Figure 9.2(b) shows the functions from 25 samples. Figure 9.2(c) shows the distribution of $S(\theta)$ at the true $\theta = 4$ over repeated Monte Carlo samples. The superimposed normal curve is the density of $N\{0, \mathcal{I}(\theta)\}$, where $\mathcal{I}(\theta) = n/\theta = 2.5$. Figure 9.2(d) is shown to illustrate the variability of $I(\theta = 4)$. □

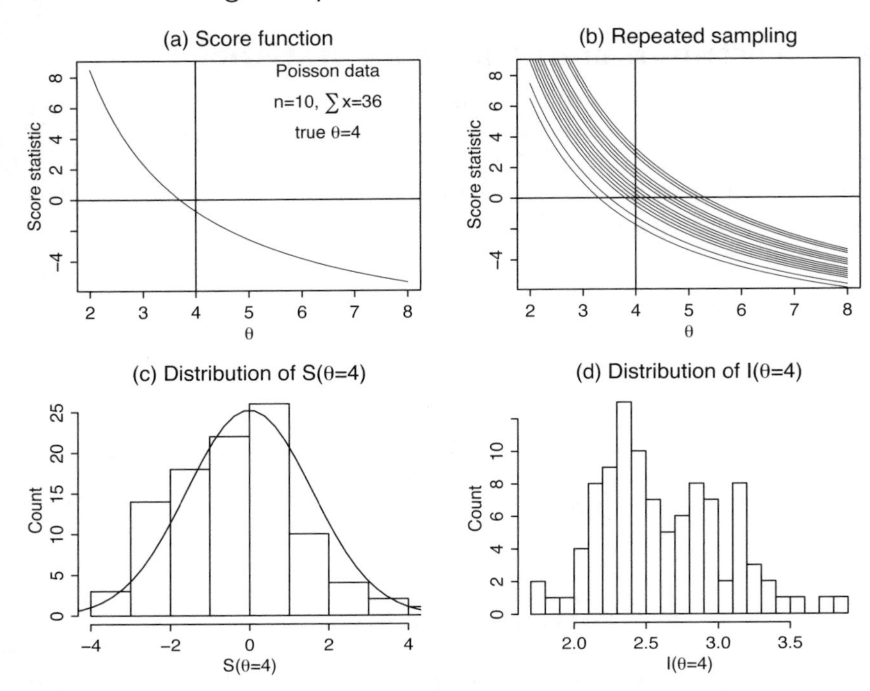

Figure 9.2: *Distribution of the score statistic in the Poisson case. (a) Score function from a single realization. (b) Score function from 25 realizations. (c) Histogram of $S(\theta = 4)$ and its normal approximation. (d) Histogram of $I(\theta = 4)$.*

Score tests

From classical perspectives, we can immediately use the distribution theory for testing

$$H_0: \theta = \theta_0 \quad \text{versus} \quad H_1: \theta \neq \theta_0.$$

H_0 is rejected if $|S(\theta_0)|$ is 'too large'. According to the theory, we may compute the standardized z-statistic

$$z = \frac{S(\theta_0)}{\sqrt{\mathcal{I}(\theta_0)}}$$

and

$$\text{P-value} = P(|Z| > |z|),$$

where Z has the standard normal distribution. This test is called the *score test*, or *Rao's test*, or the locally most powerful test.

From Figure 9.2(a) the test is intuitively clear: a large $|S(\theta_0)|$ means θ_0 is far from the MLE $\widehat{\theta}$. The normal approximation is sensible if $S(\theta)$ is close to linear in the neighbourhood of $\widehat{\theta} \pm 2/\sqrt{\mathcal{I}(\widehat{\theta})}$. The linearity of $S(\theta)$ is equivalent to a quadratic log-likelihood.

Example 9.3: Let $X_1, \ldots, X_n$ be an iid sample from $N(\theta, \sigma^2)$ with σ^2 known. Then

$$\begin{aligned} S(\theta) &= \frac{n}{\sigma^2}(\overline{x} - \theta) \\ \mathcal{I}(\theta) &= \frac{n}{\sigma^2}. \end{aligned}$$

The score test rejects H_0: $\theta = \theta_0$ if

$$|z| = \left| \frac{\overline{x} - \theta}{\sqrt{\sigma^2/n}} \right| > |z_{\alpha/2}|,$$

which is the standard z-test. □

Example 9.4: Let $x_1, \ldots, x_{10}$ be an iid sample from Poisson(θ). Suppose we observe $\overline{x} = 3.5$ and we would like to test H_0: $\theta = 5$ versus H_1: $\theta \neq 5$. In Section 8.1 we obtain

$$\begin{aligned} S(\theta) &= \frac{n}{\theta}(\overline{x} - \theta) \\ \mathcal{I}(\theta) &= \frac{n}{\theta}, \end{aligned}$$

so the score test yields

$$\begin{aligned} z &= \frac{S(\theta_0)}{\sqrt{\mathcal{I}(\theta_0)}} \\ &= \frac{3.5 - 5}{\sqrt{5/10}} = 2.12 \end{aligned}$$

which is significant at the 5% level. □

Using observed Fisher information

The observed information version of the previous result can be developed easily. By the WLLN

$$\frac{1}{n} \sum_i \frac{\partial^2}{\partial \theta^2} \log p_\theta(x_i) \xrightarrow{p} E_\theta \frac{\partial^2}{\partial \theta^2} \log p_\theta(X_1),$$

or

$$\frac{I(\theta)}{n} \xrightarrow{p} \mathcal{I}_1(\theta),$$

where $I(\theta)$ is the observed Fisher information. So, by Slutsky's theorem

$$\{I(\theta)\}^{-1/2} S(\theta) \xrightarrow{d} N(0, 1), \tag{9.2}$$

which is to be compared with (9.1). Now, in most models $\mathcal{I}(\theta) \neq I(\theta)$ and at the moment it is not clear which is better. The other option is to use $I(\widehat{\theta})$, with a disadvantage that we have to find the MLE $\widehat{\theta}$. We will compare these three quantities in Section 9.7.

Example 9.4: continued. In the Poisson example above we have

$$\begin{aligned} z_1 = \{\mathcal{I}(\theta_0)\}^{-1/2} S(\theta_0) &= \frac{\sqrt{n}(\overline{x} - \theta_0)}{\sqrt{\theta_0}} \\ I(\theta_0) &= \frac{n\overline{x}}{\theta_0^2} \\ z_2 = \{I(\theta_0)\}^{-1/2} S(\theta_0) &= \frac{\sqrt{n}(\overline{x} - \theta_0)}{\sqrt{\overline{x}}} \\ I(\widehat{\theta}) &= \frac{n}{\overline{x}} \\ z_3 = \{I(\widehat{\theta})\}^{-1/2} S(\theta_0) &= \frac{\sqrt{n}(\overline{x} - \theta_0)}{\sqrt{\theta_0^2/\overline{x}}} \end{aligned}$$

Under the null hypothesis, all these statistics are equivalent. The statistic z_1 is commonly used, but z_2 and z_3 are rather unusual. To put it in a slightly different form:

$$z_1^2 = \frac{(n\overline{x} - n\theta_0)^2}{n\theta_0} \equiv \frac{(O-E)^2}{E},$$

where 'O' and 'E' are the usual notation for 'observed' and 'expected' frequencies, and

$$z_2^2 = \frac{(n\overline{x} - n\theta_0)^2}{n\overline{x}} \equiv \frac{(O-E)^2}{O}. \;\square$$

9.3 Consistency of MLE for scalar θ

Given an estimation procedure it is reasonable to require that it produces a 'good' estimate if the experiment is large enough, and a 'better' estimate as the experiment becomes larger. One simple requirement is as follows. Suppose θ_0 is the true parameter, and ϵ is a small positive value. For any choice of ϵ, by making the experiment large enough, can we guarantee (with large probability) that the estimate $\widehat{\theta}$ will fall within ϵ of θ_0? If yes, we say that $\widehat{\theta}$ is *consistent*. Put it more simply, $\widehat{\theta}$ is consistent for θ_0 if $\widehat{\theta} \xrightarrow{p} \theta_0$. This is a frequentist requirement: if we repeat the large experiment many times then a large proportion of the resulting $\widehat{\theta}$ will be within ϵ of θ_0.

Before we state and prove the main theorem it is useful to mention *Jensen's inequality* involving convex functions. By definition $g(x)$ is convex on an interval (a, b) if for any two points x_1, x_2 in the interval, and any $0 < \alpha < 1$,

$$\alpha g(x_1) + (1-\alpha)g(x_2) \geq g(\alpha x_1 + (1-\alpha)x_2).$$

It is strictly convex if the inequality is strict. This condition is clear graphically: $g(x)$ lies under the line connecting points $(x_1, g(x_1))$ and $(x_2, g(x_2))$. Statistically, the (weighted) average of the function value is greater than the value of the function at some weighted average.

If $g(x)$ is convex and $(x_0, g(x_0))$ is any point on the function, there exists a straight line with some slope m:

$$L(x) = g(x_0) + m(x - x_0)$$

passing through $(x_0, g(x_0))$ such that $g(x) \geq L(x)$. If $g(x)$ is differentiable we can simply choose $L(x)$ to be the tangent line. If $g(x)$ is twice differentiable, it is convex if $g''(x) \geq 0$, and strictly convex if $g''(x) > 0$.

Theorem 9.4 (Jensen's inequality) *If X has the mean EX and $g(x)$ is convex then*

$$E\{g(X)\} \geq g(EX),$$

with strict inequality if $g(x)$ is strictly convex and X is nondegenerate.

Proof: Since $g(x)$ is convex, there is a straight line passing through $\{EX, g(EX)\}$ such that

$$g(x) \geq g(EX) + m(x - EX).$$

We finish the proof by taking an expected value of the inequality. □

Using Jensen's inequality we can immediately claim the following for nondegenerate X:

$$E(X^2) \quad > \quad (EX)^2,$$

and if additionally $X > 0$:

$$\begin{aligned} E(1/X) &> 1/E(X) \\ E(-\log X) &> -\log EX. \end{aligned}$$

The last inequality provides a proof of the *information inequality.* The term 'information' here refers to the so-called Kullback–Leibler information (Section 13.2), not Fisher information.

Theorem 9.5 (Information inequality) *If $f(x)$ and $g(x)$ are two densities, then*

$$E_g \log \frac{g(X)}{f(X)} \geq 0,$$

where E_g means the expected value is taken assuming X has density $g(x)$. The inequality is strict unless $f(x) = g(x)$.

One way to interpret Theorem 9.5 is that

$$E_g \log g(X) \geq E_g \log f(X),$$

which means the log-likelihood of the 'true' model tends to be larger than the log-likelihood of a 'wrong' model. This, in fact, provides an intuition for the consistency of the MLE.

For our main theorem assume that the support of $p_\theta(x)$ does not depend on θ, and $p_{\theta_1}(x) \neq p_{\theta_2}(x)$ if $\theta_1 \neq \theta_2$. If the likelihood has several maxima, we will call each local maximizer a potential MLE $\widehat{\theta}$.

Theorem 9.6 *Let $x_1, \ldots, x_n$ be an iid sample from $p_{\theta_0}(x)$, and assume that $p_\theta(x)$ is a continuous function of θ. Then as $n \to \infty$* there exists, *with probability tending to one, a consistent sequence of MLE $\widehat{\theta}$.*

Proof: For any fixed $\epsilon > 0$ we need to show that there is a potential MLE $\widehat{\theta}$ in the interval $(\theta_0 - \epsilon, \theta_0 + \epsilon)$. This is true if we can show that

$$\begin{aligned} L(\theta_0) &> L(\theta_0 - \epsilon) \\ L(\theta_0) &> L(\theta_0 + \epsilon), \end{aligned}$$

with probability tending to one as $n \to \infty$. The first inequality above follows from

$$\begin{aligned} \frac{1}{n} \log \frac{L(\theta_0)}{L(\theta_0 - \epsilon)} &= \frac{1}{n} \sum_i \log \frac{p_{\theta_0}(x)}{p_{\theta_0 - \epsilon}(x)} \\ &\stackrel{p}{\to} E_{\theta_0} \log \frac{p_{\theta_0}(X_1)}{p_{\theta_0 - \epsilon}(X_1)} \\ &> 0 \end{aligned}$$

using the WLLN and the information inequality. The second inequality is proved the same way. □

The essence of the proof is that, as we enlarge the sample, the true parameter θ_0 becomes more likely than any pre-specified point in its local neighbourhood. A global result that captures this property is that, for any $\theta \neq \theta_0$ and any constant c (think of large c),

$$P_{\theta_0} \left(\frac{L(\theta)}{L(\theta_0)} \geq c \right) \leq \frac{1}{c}.$$

This means that when the likelihood function becomes concentrated, it is unlikely that we will find an estimate far from the true value; see the proof following equation (5.1).

The condition and the proof of the consistency result are simple, but the conclusion is far from the best possible result. It is only an 'existence theorem': if the likelihood contains several maxima the theorem does not say which one is the consistent estimate; if we define the global maximizer as the MLE, then the theorem *does not guarantee* that the MLE is consistent. Conditions that guarantee the consistency of the MLE are given, for example, in Wald (1949).

The method of proof also does not work for a vector parameter $\theta \in R^p$. A lot more assumptions are required in this case, though it will not discussed further here; see, for example, Lehmann (1983, Chapter 6).

However, Theorem 9.6 does guarantee that if the MLE is unique for all n, or if it is unique as $n \to \infty$, then it is consistent. As discussed in Section 4.9, this is true in the full exponential family models. Furthermore, the proof can be modified slightly to argue that in the world of finite precision and finite parameter space, the MLE is consistent. This is because we can guarantee, with probability tending to one, that θ_0 becomes more likely than any other point in a finite set.

9.4 Distribution of MLE and the Wald statistic

Consistency properties are not enough for statistical inference. The results in this section provide methods of inference under very general conditions. As the price for the generality, the method is only approximate. The approximation is typically accurate in large samples, but in small to medium samples its performance varies. The difficulty in practice is in knowing when the sample is large enough. From the previous discussions we can check whether the log-likelihood is nearly quadratic, or the score statistic is nearly linear.

Theorem 9.7 *Let $x_1, \ldots, x_n$ be an iid sample from $p_{\theta_0}(x)$, and assume that the MLE $\widehat{\theta}$ is consistent. Then, under some regularity conditions,*

$$\sqrt{n}(\widehat{\theta} - \theta_0) \to N(0, 1/\mathcal{I}_1(\theta_0)),$$

where $\mathcal{I}_1(\theta_0)$ is the Fisher information from a single observation.

First recall the CRLB (Section 8.5): if $ET = \theta$ then

$$\text{var}(T) \geq \frac{1}{\mathcal{I}(\theta)}.$$

The theorem states that $\widehat{\theta}$ is approximately normal with mean θ_0 and variance

$$\text{var}(\widehat{\theta}) = \frac{1}{n\mathcal{I}_1(\theta_0)} = \frac{1}{\mathcal{I}(\theta_0)},$$

which means that asymptotically $\widehat{\theta}$ achieves the CRLB, or it is asymptotically the best estimate.

For a complete list of standard 'regularity conditions' see Lehmann (1983, Chapter 6). Essentially the conditions ensure that

- θ is not a boundary parameter (otherwise the likelihood cannot be regular);
- the Fisher information is positive and bounded (otherwise it cannot be a variance);
- we can take (up to third) derivatives of $\int p_\theta(x)dx$ under the integral sign;
- simple algebra up to a second-order expansion of the log-likelihood is sufficient and valid.

Proof: A linear approximation of the score function $S(\theta)$ around θ_0 gives

$$S(\theta) \approx S(\theta_0) - I(\theta_0)(\theta - \theta_0)$$

and since $S(\widehat{\theta}) = 0$, we have

$$\sqrt{n}(\widehat{\theta} - \theta_0) \approx \{I(\theta_0)/n\}^{-1} S(\theta_0)/\sqrt{n}.$$

The result follows using Slutsky's theorem, since

$$I(\theta_0)/n \xrightarrow{p} \mathcal{I}_1(\theta_0)$$

and

$$S(\theta_0)/\sqrt{n} \xrightarrow{d} N\{0, \mathcal{I}_1(\theta_0)\}. \ \square$$

We can then show that all of the following are true:

$$\begin{aligned}
\sqrt{\mathcal{I}(\theta_0)}(\widehat{\theta} - \theta_0) &\rightarrow N(0,1) \\
\sqrt{I(\theta_0)}(\widehat{\theta} - \theta_0) &\rightarrow N(0,1) \\
\sqrt{\mathcal{I}(\widehat{\theta})}(\widehat{\theta} - \theta_0) &\rightarrow N(0,1) \\
\sqrt{I(\widehat{\theta})}(\widehat{\theta} - \theta_0) &\rightarrow N(0,1).
\end{aligned}$$

The last two forms are the most practical, and informally we say

$$\begin{aligned}
\widehat{\theta} &\sim N(\theta_0, 1/\mathcal{I}(\widehat{\theta})) \\
\widehat{\theta} &\sim N(\theta_0, 1/I(\widehat{\theta})).
\end{aligned}$$

In the full exponential family these two versions are identical. In more complex cases where $\mathcal{I}(\widehat{\theta}) \neq I(\widehat{\theta})$, the use of $I(\widehat{\theta})$ is preferable (Section 9.6).

Wald tests and intervals

We have now proved our previous claim in Section 2.7 that the approximate standard error of the MLE in regular cases is

$$\text{se}(\widehat{\theta}) = I^{-1/2}(\widehat{\theta}).$$

The MLE distribution theory can be used for testing H_0: $\theta = \theta_0$ versus H_1: $\theta \neq \theta_0$ using the Wald statistic

$$z = \frac{\widehat{\theta} - \theta_0}{\text{se}(\widehat{\theta})}$$

or

$$\chi^2 = \frac{(\widehat{\theta} - \theta_0)^2}{\text{se}^2(\widehat{\theta})},$$

which, under H_0, are distributed as $N(0,1)$ and χ^2_1 respectively.

Secondly, the result can be used to get the approximate $100(1-\alpha)\%$ CI formula

$$\widehat{\theta} \pm z_{\alpha/2}\,\mathrm{se}(\widehat{\theta}).$$

This is the same as the approximate likelihood interval, based on the quadratic approximation, at cutoff equal to

$$\exp\left\{-z_{\alpha/2}^2/2\right\}.$$

Example 9.5: For the aspirin data in Section 4.7 we have

$$\begin{aligned}
\log L(\theta) &= 139\log\theta - 378\log(\theta+1)\\
S(\theta) &= \frac{139}{\theta} - \frac{378}{\theta+1}\\
I(\theta) &= \frac{139}{\theta^2} - \frac{378}{(\theta+1)^2},
\end{aligned}$$

so

$$\begin{aligned}
\widehat{\theta} &= 139/(378-139) = 0.58\\
I(\widehat{\theta}) &= \frac{139}{(139/239)^2} - \frac{378}{(1+139/239)^2} = 259.8287\\
\mathrm{var}(\widehat{\theta}) &= 1/I(\widehat{\theta}) = 0.003849.
\end{aligned}$$

Note that $\mathrm{var}(\widehat{\theta})$ is equal to that found using the Delta method. So

$$\mathrm{se}(\widehat{\theta}) = \sqrt{0.003849} = 0.062,$$

and the Wald 95% CI for θ is

$$0.46 < \theta < 0.70. \ \square$$

9.5 Distribution of likelihood ratio statistic

In the development of the distribution theory of $S(\theta)$ and $\widehat{\theta}$ we do not refer to the likelihood function itself. We now show that the previous results are equivalent to a quadratic approximation of the log-likelihood function. However, as described in Section 2.9, we do get something more from the likelihood function. The approximate method in this section is the basis for frequentist calibration of the likelihood function.

Using second-order expansion around $\widehat{\theta}$

$$\begin{aligned}
\log L(\theta) &\approx \log L(\widehat{\theta}) + S(\widehat{\theta})(\theta-\widehat{\theta}) - \frac{1}{2}I(\widehat{\theta})(\theta-\widehat{\theta})^2\\
&= \log L(\widehat{\theta}) - \frac{1}{2}I(\widehat{\theta})(\theta-\widehat{\theta})^2. \qquad (9.3)
\end{aligned}$$

This means

$$L(\theta) \approx \text{constant} \times \exp\left\{-\frac{1}{2}I(\widehat{\theta})(\theta-\widehat{\theta})^2\right\},$$

which is the likelihood based on a single observation $\widehat{\theta}$ taken from $N(\theta, 1/I(\widehat{\theta}))$.

From (9.3) we get

$$\begin{aligned} W &\equiv 2\log\frac{L(\widehat{\theta})}{L(\theta)} \\ &= I(\widehat{\theta})(\widehat{\theta}-\theta)^2 \stackrel{d}{\to} \chi^2_1, \end{aligned}$$

where W is Wilk's likelihood ratio statistic. The distribution theory may be used to get an approximate P-value for testing $H_0: \theta = \theta_0$ versus $H_1: \theta \neq \theta_0$. Specifically, on observing a normalized likelihood

$$\frac{L(\theta_0)}{L(\widehat{\theta})} = r$$

we compute $w = -2\log r$, and

$$\text{P-value} = P(W \geq w),$$

where W has a χ^2_1 distribution. The approximate connection between likelihood and P-value follows the pattern under the normal model in Example 5.6.

From the distribution theory we can also set an approximate $100(1-\alpha)\%$ CI for θ as

$$\text{CI} = \left\{\theta;\ 2\log\frac{L(\widehat{\theta})}{L(\theta)} < \chi^2_{1,(1-\alpha)}\right\}.$$

For example, an approximate 95% CI is

$$\begin{aligned} \text{CI} &= \left\{\theta;\ 2\log\frac{L(\widehat{\theta})}{L(\theta)} < 3.84\right\} \\ &= \{\theta; L(\theta) > 0.15 \times L(\widehat{\theta})\}. \end{aligned}$$

This is the likelihood interval at 15% cutoff. So we have established that, in general, the confidence level of a likelihood interval at cutoff α is approximately

$$P(W \leq -2\log\alpha).$$

9.6 Observed versus expected information*

According to our theory the following statistics are asymptotically equivalent

$$\begin{aligned} W_1 = \mathcal{I}(\widehat{\theta})(\widehat{\theta}-\theta)^2 &\sim \chi^2_1 \\ W_2 = I(\widehat{\theta})(\widehat{\theta}-\theta)^2 &\sim \chi^2_1 \\ W = 2\log\frac{L(\widehat{\theta})}{L(\theta)} &\sim \chi^2_1. \end{aligned}$$

It was emphasized previously that W_1 and W_2 are sensible only if the likelihood is reasonably regular. The only difference between W_1 and W_2

is in the use of the observed versus expected Fisher information. In the full exponential family $\mathcal{I}(\widehat{\theta}) = I(\widehat{\theta})$, so $W_1 = W_2$ (see Exercise 8.3). If $\mathcal{I}(\widehat{\theta}) \neq I(\widehat{\theta})$, how should we choose between W_1 and W_2, and how do they compare with W? Efron and Hinkley (1978) discussed these questions in detail and showed that $I(\widehat{\theta})$ is better than $\mathcal{I}(\widehat{\theta})$. Overall, W is preferred.

If θ is a location parameter then $I(\widehat{\theta})$ does not carry information about where the true θ_0 is. However, $I(\widehat{\theta})$ tells us something about precision, so potentially different probability statements can be made conditional or unconditional on $I(\widehat{\theta})$. This means $I(\widehat{\theta})$ is relevant information. We call all those properties together *ancillarity*; for example, the sample size in an experiment is typically ancillary information. From the discussion in Section 5.10 it makes sense to require our inference to be conditional on the the observed ancillary statistic. This will make the inferential statements (such as confidence level or significance level) relevant for the data at hand.

As a specific example, we will describe the simulation study of the Cauchy location parameter (Efron and Hinkley 1978). Let $x_1, \ldots, x_n$ be a sample from the Cauchy distribution with density

$$p_\theta(x) = \frac{1}{\pi\{1 + (x - \theta)^2\}}.$$

From routine calculations

$$\begin{aligned} S(\theta) &= \sum_i \frac{2(x_i - \theta)}{1 + (x_i - \theta)^2} \\ I(\theta) &= -\sum_i \frac{2\{(x_i - \theta)^2 - 1\}}{\{(x_i - \theta)^2 + 1\}^2} \\ \mathcal{I}(\theta) &= E_\theta I(\theta) = \frac{n}{2}. \end{aligned}$$

For $n = 20$, we generate $x_1, \ldots, x_{20}$ from the Cauchy distribution with location $\theta_0 = 0$. From each realization we estimate the MLE $\widehat{\theta}$, and compute W_1, W_2 and W. This is repeated 2000 times. The 2000 values of W_1 can be checked against the χ^2_1 distribution. However, it is more convenient to compute the signed root of all of the statistics and check them against the standard normal distribution. Thus, under H_0: $\theta = \theta_0$ and asymptotically

$$\begin{aligned} \text{sign}(\widehat{\theta} - \theta_0)\sqrt{W_1} = \sqrt{\mathcal{I}(\widehat{\theta})}(\widehat{\theta} - \theta_0) &\sim N(0,1) \\ \text{sign}(\widehat{\theta} - \theta_0)\sqrt{W_2} = \sqrt{I(\widehat{\theta})}(\widehat{\theta} - \theta_0) &\sim N(0,1) \\ \text{sign}(\widehat{\theta} - \theta_0)\sqrt{W} &\sim N(0,1), \end{aligned}$$

where 'sign(·)' is +1 if the value in the bracket is positive, and −1 otherwise.

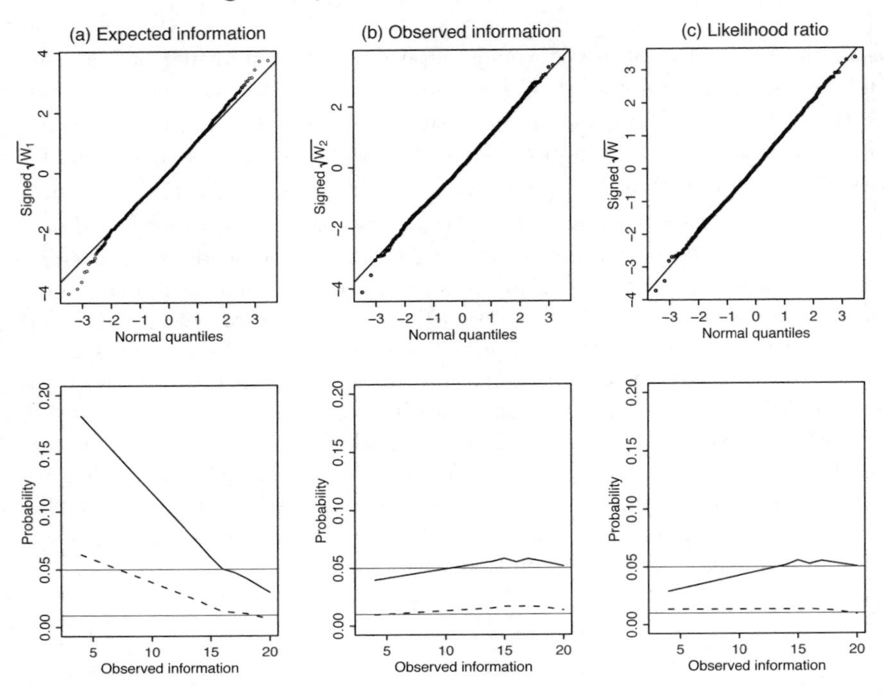

Figure 9.3: *First row: (a) Wald statistic normalized using the expected Fisher information* $\mathcal{I}(\widehat{\theta})$. *(b) Wald statistic normalized using* $I(\widehat{\theta})$. *(c) Likelihood ratio statistic. Second row: the corresponding estimated conditional probability* $P\{W_1 > 3.84|I(\widehat{\theta})\}$, *in solid line, and* $P\{W_1 > 6.63|I(\widehat{\theta})\}$, *in dashed line.*

The normal plots in the first row of Figure 9.3 show that these three statistics are reasonably normal; the tail of W is slightly better behaved than the others. These plots indicate that the problem is quite regular, so here we are not concerned with the regularity issue. We can also check that, conditional on $I(\widehat{\theta})$, the distribution is also quite normal. We can do this by grouping together realizations where $I(\widehat{\theta})$ falls in a small interval.

The first plot in the second row of Figure 9.3 shows the estimated conditional probability

$$P\{W_1 > 3.84|I(\widehat{\theta})\}$$

as a function of $I(\widehat{\theta})$. Unconditionally the probability is 0.05. We compute the estimate using a scatterplot smoother for the paired data

$$\{I(\widehat{\theta}), \text{Ind}(W_1 > 3.84)\}$$

where $\text{Ind}(W_1 > 3.84)$ is the indicator function, taking value one if the condition inside the bracket is true and zero otherwise. A similar curve (dashed line) can be computed for

$$P\{W_1 > 6.63 | I(\widehat{\theta})\},$$

which unconditionally is 0.01. The plot shows that the distribution of W_1 varies with $I(\widehat{\theta})$, so statements based on W_1 are open to the criticisms discussed in Section 5.10. For example, a reported confidence level is not relevant for the data at hand, since it is different if we make it conditional on an observed $I(\widehat{\theta})$.

The other plots in the second row of Figure 9.3 show that the distributions of W_2 and W are relatively constant across $I(\widehat{\theta})$, so inferential statements based on them are safe from the previous criticisms.

9.7 Proper variance of the score statistic*

The same issue also occurs with the score test. The following score tests for H_0: $\theta = \theta_0$ are asymptotically equivalent:

$$\begin{aligned} z_1 &= \frac{S(\theta_0)}{\sqrt{\mathcal{I}(\theta_0)}} \\ z_2 &= \frac{S(\theta_0)}{\sqrt{I(\theta_0)}} \\ z_3 &= \frac{S(\theta_0)}{\sqrt{I(\widehat{\theta})}}. \end{aligned}$$

Which should be the preferred formula, especially if the denominators are very different?

Using the same Cauchy simulation setup as in the previous section, but increasing the sample size to 30, at each realization we can compute z_1, z_2, z_3 and $I(\widehat{\theta})$. Under the null hypothesis these are all supposed to be standard normal. The first row of Figure 9.4 shows that all three test statistics are reasonably normal.

The second row shows the conditional probabilities of the three statistics given $I(\widehat{\theta})$. The score statistic z_1 has the poorest conditional property, while z_2 is not as good as z_3. This means $I(\widehat{\theta})$ is actually the best variance quantity for $S(\theta_0)$. Recall the discussion in Section 8.3 that we should think of the observed Fisher information as a single quantity $I(\widehat{\theta})$. The quantity $I(\theta_0)$ in complicated models, as in the Cauchy example, is not even guaranteed to be positive.

9.8 Higher-order approximation: magic formula*

What if we do not believe in the normal approximation? From the standard theory we have, approximately,

$$\widehat{\theta} \sim N\{\theta, I(\widehat{\theta})^{-1}\}$$

so the approximate density of $\widehat{\theta}$ is

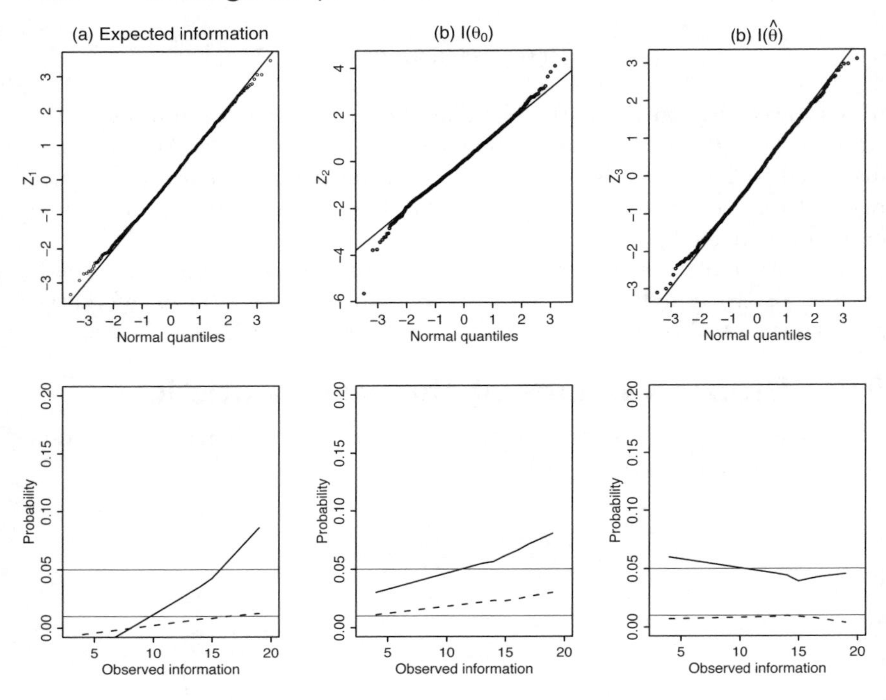

Figure 9.4: *First row: (a) Score statistic normalized using the expected Fisher information* $\mathcal{I}(\theta_0)$. *(b) Score statistic normalized using* $I(\theta_0)$. *(c) Score statistic normalized using* $I(\widehat{\theta})$. *Second row: the corresponding estimated conditional probability* $P\{|Z_1| > 1.96|I(\widehat{\theta})\}$, *in solid line, and* $P\{|Z_1| > 2.57|I(\widehat{\theta})\}$, *in dashed line.*

$$p_\theta(\widehat{\theta}) \approx (2\pi)^{-1/2}|I(\widehat{\theta})|^{1/2}\exp\left\{-\frac{I(\widehat{\theta})}{2}(\widehat{\theta}-\theta)^2\right\}. \tag{9.4}$$

We have also shown the quadratic approximation

$$\log\frac{L(\theta)}{L(\widehat{\theta})} \approx -\frac{I(\widehat{\theta})}{2}(\widehat{\theta}-\theta)^2,$$

so we have another approximate density

$$p_\theta(\widehat{\theta}) \approx (2\pi)^{-1/2}|I(\widehat{\theta})|^{1/2}\frac{L(\theta)}{L(\widehat{\theta})}. \tag{9.5}$$

We will refer to this as the likelihood-based p-formula, which turns out to be much more accurate than the normal-based formula (9.4). Intuitively, if $p_\theta(\widehat{\theta})$ were available then we would obtain $L(\theta)$ from it directly, but $L(\theta)$ is an exact likelihood, so (9.5) must be a good density for $\widehat{\theta}$.

Even though we are using a likelihood ratio it should be understood that (9.5) is a formula for a sampling density: θ is fixed and $\widehat{\theta}$ varies. The *observed* likelihood function is not enough to derive $p_\theta(\widehat{\theta})$ over $\widehat{\theta}$; for such a purpose the original model for the data is still required.

Recall that the p-formula (9.5) has been used in Section 4.9 to develop an approximate likelihood of the exponential dispersion model.

Example 9.6: Let $x_1, \ldots, x_n$ be an iid sample from $N(\theta, \sigma^2)$ with σ^2 known. Here we know that $\widehat{\theta} = \overline{X}$ is $N(\theta, \sigma^2/n)$. To use formula (9.5) we need

$$\begin{aligned}
\log L(\theta) &= -\frac{1}{2\sigma^2}\left\{\sum_i (x_i - \overline{x})^2 + n(\overline{x} - \theta)^2\right\} \\
\log L(\widehat{\theta}) &= -\frac{1}{2\sigma^2}\sum_i (x_i - \overline{x})^2 \\
I(\widehat{\theta}) &= n/\sigma^2,
\end{aligned}$$

so

$$p_\theta(\overline{x}) \approx (2\pi)^{-1/2}|n/\sigma^2|^{1/2}\exp\left\{-\frac{n}{2\sigma^2}(\overline{x} - \theta)^2\right\},$$

exactly the density of the normal distribution $N(\theta, \sigma^2/n)$. □

Example 9.7: Let y be Poisson with mean θ. The MLE of θ is $\widehat{\theta} = y$, and the Fisher information is $I(\widehat{\theta}) = 1/\widehat{\theta} = 1/y$. So, the p-formula (9.5) is

$$\begin{aligned}
p_\theta(y) &\approx (2\pi)^{-1/2}(1/y)^{1/2}\frac{e^{-\theta}\theta^y/y!}{e^{-y}y^y/y!} \\
&= \frac{e^{-\theta}\theta^y}{(2\pi y)^{1/2}e^{-y}y^y},
\end{aligned}$$

so in effect we have approximated the Poisson probability by replacing $y!$ with its Stirling's approximation. The approximation is excellent for $y > 3$, but not so good for $y \leq 3$. Nelder and Pregibon (1987) suggested a simple modification of the denominator to

$$(2\pi(y + 1/6))^{1/2}e^{-y}y^y,$$

which works remarkably well for all $y \geq 0$. □

The p-formula can be improved by a generic normalizing constant to make the density integrate to one. The formula

$$p^*_\theta(\widehat{\theta}) = c(\theta)(2\pi)^{-1/2}|I(\widehat{\theta})|^{1/2}\frac{L(\theta)}{L(\widehat{\theta})} \tag{9.6}$$

is called Barndorff-Nielsen's (1983) p^*-formula. As we would expect in many cases $c(\theta)$ is very nearly one; in fact, $c(\theta) \approx 1 + B(\theta)/n$, where $B(\theta)$ is bounded over n. If difficult to derive analytically, $c(\theta)$ can be computed numerically. In many examples the approximation is so good that the p^*-formula is called a 'magic formula' (Efron 1998).

In the exponential family formula (9.6) coincides with the so-called *saddlepoint approximation.* Durbin (1980) shows that

$$p_\theta(\widehat{\theta}) = c(\theta)(2\pi)^{-1/2}|I(\widehat{\theta})|^{1/2}\frac{L(\theta)}{L(\widehat{\theta})}\{1+O(n^{-3/2})\},$$

where we can think of the error term $O(n^{-3/2}) = bn^{-3/2}$ for some bounded b. This is a highly accurate approximation; by comparison, the standard rate of asymptotic approximation using (9.4) is only of order $O(n^{-1/2})$.

In the full exponential family the MLE $\widehat{\theta}$ is sufficient (Section 4.9), so the likelihood ratio $L(\theta)/L(\widehat{\theta})$ depends on the data x only through $\widehat{\theta}$. To make it explicit, we may write

$$L(\theta) \equiv L(\theta; x) = L(\theta; \widehat{\theta}),$$

so (9.5) or (9.6) are not ambiguous.

If $\widehat{\theta}$ is not sufficient the formulae are ambiguous. It turns out, however, that it still provides an approximate conditional density of $\widehat{\theta}$ given some ancillary statistic $a(x)$. Suppose there is a one-to-one function of the data to $\{\widehat{\theta}, a(x)\}$, so that the likelihood

$$L(\theta) \equiv L(\theta; x) = L(\theta; \widehat{\theta}, a(x)),$$

where the dependence of the likelihood on the data is made explicit. Then

$$p_\theta(\widehat{\theta}|a) \approx (2\pi)^{-1/2}|I(\widehat{\theta})|^{1/2}\frac{L(\theta; \widehat{\theta}, a)}{L(\widehat{\theta}; \widehat{\theta}, a)}.$$

In particular, and this is its most important application, it gives an exact conditional distribution in the location family; see the subsection below.

Exponential family models

Example 9.8: Let $x_1, \ldots, x_n$ be an iid sample from gamma(μ, β), where $EX_i = \mu$ and β is the shape parameter, and the density is given by

$$p_\theta(x) = \frac{1}{\Gamma(\beta)}\left(\frac{\beta}{\mu}\right)^\beta x^{\beta-1}e^{-\beta x/\mu}.$$

For simplicity assume that μ is known; note, however, that formula (9.6) also applies in multivariate settings. First we get

$$\log L(\beta) = -n\log\Gamma(\beta) + n\beta\log\frac{\beta}{\mu} + (\beta-1)\sum_i \log x_i - \frac{\beta}{\mu}\sum_i x_i,$$

so

$$\frac{\partial \log L(\beta)}{\partial\beta} = -n\psi(\beta) + n\left(\log\frac{\beta}{\mu}+1\right) + \sum_i \log x_i - \frac{1}{\mu}\sum_i x_i,$$

where $\psi(\beta) = \partial \log \Gamma(\beta)/\partial\beta$. The MLE $\widehat{\beta}$ satisfies

$$-n\psi(\widehat{\beta}) + n \log \widehat{\beta} = -\sum_i \log x_i + \frac{1}{\mu}\sum_i x_i + n\log\mu - n,$$

and

$$\begin{aligned}\log\frac{L(\beta)}{L(\widehat{\beta})} &= -n\log\Gamma(\beta) + n\log\Gamma(\widehat{\beta}) + n\left(\beta\log\frac{\beta}{\mu} - \widehat{\beta}\log\frac{\widehat{\beta}}{\mu}\right) \\ &\quad +(\beta-\widehat{\beta})\sum_i \log x_i - (\beta - \widehat{\beta})\frac{1}{\mu}\sum_i x_i \\ &= -n\log\Gamma(\beta) + n\log\Gamma(\widehat{\beta}) + n\{(\beta-\widehat{\beta})\psi(\widehat{\beta}) - \beta\log\widehat{\beta} + \widehat{\beta}\}.\end{aligned}$$

On taking another derivative of the log-likelihood we obtain the observed Fisher information

$$I(\widehat{\beta}) = n\{\psi'(\widehat{\beta}) - 1/\widehat{\beta}\}.$$

The approximate density of $\widehat{\beta}$ is then given by

$$p_\beta(\widehat{\beta}) \approx \text{constant} \times |I(\widehat{\beta})|^{1/2}\frac{L(\beta)}{L(\widehat{\beta})}. \tag{9.7}$$

To show how close the approximation is, we simulate data $x_1, \ldots, x_{10}$ iid from gamma($\mu = 1, \beta = 1$), which is equal to the exponential distribution. The parameter μ is assumed known. For each dataset we compute $\widehat{\beta}$ by solving

$$-n\psi(\widehat{\beta}) + n\log\widehat{\beta} = -\sum_i \log x_i + \sum_i x_i - 10.$$

This is repeated 500 times, so we have a sample of size 500 from the true distribution of $\widehat{\beta}$. Figure 9.5(a) shows that $\widehat{\beta}$ is far from normal. Figure 9.5(b) shows the histogram of the 500 values and the approximate density from (9.7). □

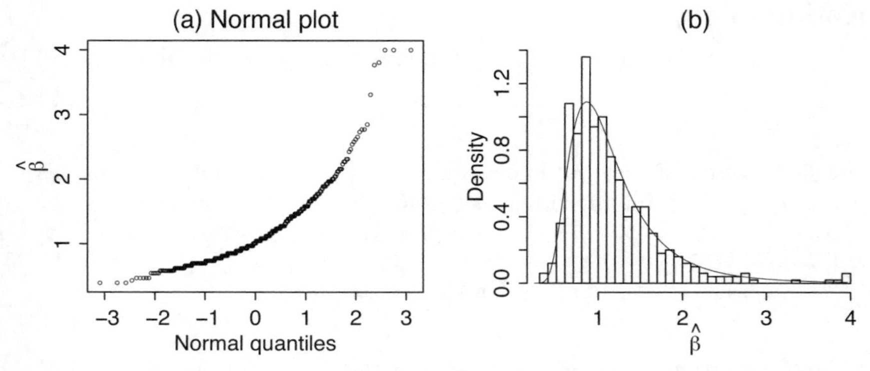

Figure 9.5: *(a) The normal plot of $\widehat{\beta}$ from simulation. (b) The histogram of $\widehat{\beta}$ and the approximate density using (9.7).*

Example 9.9: A key result in saddlepoint approximation theory is an improved formula for the distribution of the sample mean. Let $x_1, \ldots, x_n$ be an iid sample from some distribution with density $e^{h(x)}$ and moment generating function $m(\theta) = Ee^{\theta X}$. Let $K(\theta) \equiv \log m(\theta)$, the cumulant generating function of X. Then

$$p_\theta(x) \equiv e^{\theta x - K(\theta) + h(x)}$$

defines an exponential family with parameter θ, called the exponential tilting of X (Section 4.9). Given $x_1, \ldots, x_n$, the MLE $\widehat{\theta}$ is the solution of the 'saddlepoint equation'

$$K'(\widehat{\theta}) = \overline{x}$$

and the Fisher information is $I(\widehat{\theta}) = nK''(\widehat{\theta})$. From (9.6), the approximate density of $\widehat{\theta}$ is

$$p^*(\widehat{\theta}) = \text{constant} \times |K''(\widehat{\theta})|^{1/2} \exp\left[(\theta - \widehat{\theta})\sum x_i - n\{K(\theta) - K(\widehat{\theta})\}\right],$$

and

$$\begin{aligned} p^*(\overline{x}) &= p^*(\widehat{\theta}) \left|\frac{\partial \widehat{\theta}}{\partial \overline{x}}\right| \\ &= \text{constant} \times |K''(\widehat{\theta})|^{-1/2} \exp\left[(\theta - \widehat{\theta})\sum x_i - n\{K(\theta) - K(\widehat{\theta})\}\right]. \end{aligned}$$

At $\theta = 0$, we have the original distribution of the data, and the saddlepoint formula for the distribution of the sample mean:

$$p^*(\overline{x}) = \text{constant} \times |K''(\widehat{\theta})|^{-1/2} \exp[n\{K(\widehat{\theta}) - \widehat{\theta}\overline{x}\}]. \quad (9.8)$$

If there is no explicit formula for $\widehat{\theta}$ in terms of $\overline{x}$, then we need to solve the saddlepoint equation numerically at each value of $\overline{x}$. If we approximate $K(\theta)$ around zero by a quadratic function, then formula (9.8) reduces to the CLT (Exercise 9.2). □

Location family

Let $x_1, \ldots, x_n$ be an iid sample from the location family with density

$$p_\theta(x) = f_0(x - \theta),$$

where $f_0(\cdot)$ is an arbitrary but known density. Without further assumptions on the density, the minimal sufficient statistic for θ is the whole set of order statistics $\{x_{(1)}, \ldots, x_{(n)}\}$. That is, we need the whole set to compute the likelihood

$$L(\theta) = \prod_i f_0(x_{(i)} - \theta).$$

The distribution of $(x_i - \theta)$ is free of θ, so is the distribution of $(x_{(i)} - \theta)$ for each i, and, consequently, that of any spacing $a_i = x_{(i)} - x_{(i-1)}$. The relevant ancillary statistic $a(x)$ in this case is the set of $(n-1)$ spacings $a_2, \ldots, a_n$.

Let $\widehat{\theta}$ be the MLE of θ. There is a one-to-one map between $\{x_{(1)}, \ldots, x_{(n)}\}$ and $(\widehat{\theta}, a)$, and the Jacobian of the transformation is equal to one (Exercise 9.7). The joint density of $x_{(1)}, \ldots, x_{(n)}$ is

$$c \prod_i f_0(x_{(i)} - \theta).$$

Each $x_{(i)} - \theta$ is expressible in terms of $(a, \widehat{\theta} - \theta)$, so the joint density of $(a, \widehat{\theta})$ is

$$c \prod_i f_i(a, \widehat{\theta} - \theta),$$

for some functions $f_i(\cdot)$ not depending on θ. Therefore, the conditional distribution of $\widehat{\theta}$ given a is in the location family

$$p_\theta(\widehat{\theta}|a) = p_a(\widehat{\theta} - \theta),$$

for some (potentially complicated) function $p_a(\cdot)$. This means the original density of x can be decomposed as

$$p_\theta(x) = c p_a(\widehat{\theta} - \theta) g(a)$$

where $g(a)$ is the density of a, which is free of θ. Writing $p_\theta(x) = L(\theta)$, we obtain

$$\frac{p_a(\widehat{\theta} - \theta)}{p_a(0)} = \frac{L(\theta)}{L(\widehat{\theta})}$$

or

$$p_\theta(\widehat{\theta}|a) = p_a(\widehat{\theta} - \theta) = c(a) \frac{L(\theta)}{L(\widehat{\theta})},$$

where it is understood that $L(\theta) = L(\theta; \widehat{\theta}, a)$, so the formula is exactly in the form of (9.6).

If $\widehat{\theta}$ is symmetric around θ then the estimated conditional density of $\widehat{\theta}$ is given by the likelihood function itself (normalized to integrate to one). This is a remarkable simplification, since the likelihood function $L(\theta)$ is easy to compute. There is an immediate frequentist implication: likelihood intervals have an exact coverage probability given by the area under the likelihood curve. A simple example is given by the normal and Cauchy models. Another way to state the result is that the confidence density of θ matches the likelihood function; see Example 5.9 in Section 5.6.

Improved approximation of P-value

One major use of the normal approximation is to provide the P-value for testing H_0: $\theta = \theta_0$. Specifically, on observing $\widehat{\theta}$, we compute

$$w = 2 \log \frac{L(\widehat{\theta})}{L(\theta_0)}$$

and approximate the P-value by

$$\text{P-value} = P(\chi_1^2 \geq w).$$

To get a one-sided P-value we can also compute the signed root of the likelihood ratio test

$$r = \text{sign}(\widehat{\theta} - \theta_0)\sqrt{w}$$

and compute the left-side P-value by

$$p_r = P(Z < r),$$

where Z is the standard normal variate. Alternatively, we can compute the Wald statistic

$$z = |I(\widehat{\theta})|^{1/2}(\widehat{\theta} - \theta_0)$$

and evaluate the left-side P-value by

$$p_z = P(Z < z).$$

If r and z are very different then it is an indication that the normal approximation is poor, and both P-values are inappropriate. The saddlepoint approximation leads to an improved formula. (Note that once a P-value is defined we can compute the associated confidence distribution for a more complete inference; see Section 5.6).

For the exponential models of the form

$$p_\theta(x) = e^{\theta t(x) - A(\theta) + h(x)}$$

there are tail probability formulae based on (9.6). In particular, the left-side P-value is

$$p^* = P(Z < r^*) \tag{9.9}$$

where

$$r^* = r + \frac{1}{r}\log\frac{z}{r}$$

or, by expanding the normal probability around r, we can also use

$$p^* = P(Z < r) + \phi(r)\left(\frac{1}{r} - \frac{1}{z}\right), \tag{9.10}$$

where $\phi(\cdot)$ is the standard normal density function. Note that the parameter θ must be the canonical parameter, and the formulae are not invariant under transformation.

Example 9.10: Let $x_1, \ldots, x_n$ be an iid sample from the exponential distribution with mean $1/\theta$, so the density is

$$p_\theta(x) = \theta e^{-\theta x}.$$

Routine algebra yields

$$\log L(\theta) = n\log\theta - \theta\sum_i x_i$$

$$\begin{aligned} \widehat{\theta} &= \frac{1}{\overline{x}} \\ I(\widehat{\theta}) &= \frac{n}{\widehat{\theta}^2}, \end{aligned}$$

therefore

$$\begin{aligned} w &= 2\log\frac{L(\widehat{\theta})}{L(\theta)} = 2n\log\frac{\widehat{\theta}}{\theta} - 2(\widehat{\theta}-\theta)\sum_i x_i \\ r &= \operatorname{sign}(\widehat{\theta}-\theta)\sqrt{w} \\ z &= |I(\widehat{\theta})|^{1/2}(\widehat{\theta}-\theta) = \frac{\sqrt{n}(\widehat{\theta}-\theta)}{\widehat{\theta}}. \end{aligned}$$

Here we actually know the exact distribution of $\widehat{\theta}$, since $\sum_i X_i$ has a gamma distribution, so the exact P-value can be compared.

To be specific, suppose we want to test $H_0 : \theta = 1$ based on a sample of size $n = 5$. The computed P-values are left-side or right-side, depending on whether $\widehat{\theta}$ falls on the left or the right side of $\theta_0 = 1$. (In this example the saddlepoint formula (9.6) actually gives an exact density.)

$\widehat{\theta}$	w	r	p_r (%)	z	p_z (%)	r^*	p^* (%)	Exact (%)
0.35	8.07	−2.84	0.22	−4.15	0.00	−2.97	0.15	0.15
0.55	2.20	−1.48	6.89	−1.83	3.37	−1.63	5.21	5.21
2.50	3.16	1.78	3.77	1.34	8.99	1.62	5.26	5.27
4.00	6.26	2.52	0.58	1.68	4.68	0.36	0.91	0.91

Figure 9.6 shows the corresponding log-likelihood functions, and their normal approximations, for the four values of $\widehat{\theta}$ in the table. □

Bartlett correction

The saddlepoint improvement in the distribution of the MLE is closely related to the Bartlett correction for the distribution of the likelihood ratio statistic. Asymptotically, under the null hypothesis, W is χ_1^2, so $EW \approx 1$. Following the general bias formula, the expected value of the statistic is of the form

$$EW = 1 + \frac{b(\theta)}{n} + O(n^{-2})$$

for some function $b(\theta)$. Bartlett (1953) suggested a simple correction

$$W_b = \frac{W}{1 + b(\theta)/n}$$

and treated W_b as a χ_1^2 variate. The Bartlett correction factor $1 + b(\theta)/n$ is connected to the unspecified normalizing constant in (9.6). A rigorous proof is given by Barndorff-Nielsen and Cox (1984).

We can estimate the Bartlett factor using the Delta method or, in complicated cases, using the Monte Carlo technique discussed in Section 5.2. Here we want to estimate

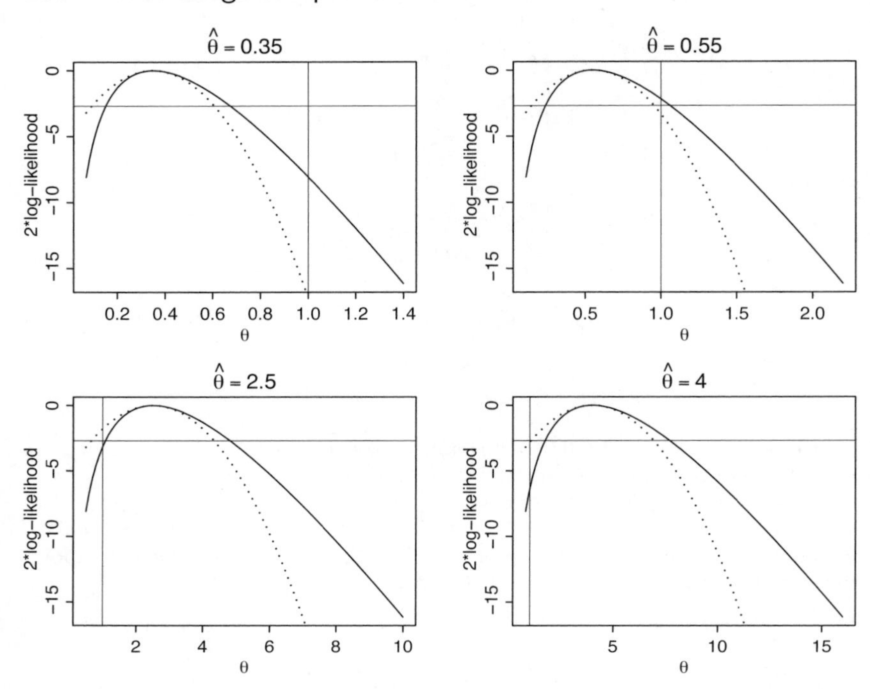

Figure 9.6: *The log-likelihood functions (solid curves) and normal approximations (dotted curves) associated with the four different values of $\widehat{\theta}$ in the table. The horizontal line is the 5% limit for the two-sided P-value based on the likelihood ratio statistic W.*

$$E_\theta W = 2E_\theta \log \frac{L(\widehat{\theta})}{L(\theta)}$$

where $\widehat{\theta}$ is the random estimate, and θ is fixed at the null hypothesis. So,

1. generate a new dataset x^* from the parametric model $p_\theta(x)$,
2. compute $\widehat{\theta}^*$ and the corresponding likelihood ratio statistic w^* from the data x^*,
3. repeat steps 1 and 2 a large number of times and simply take the average of w^* as an estimate of $E_\theta W$.

We can actually check from the w^*'s whether the distribution of W is well approximated by the χ^2 distribution.

9.9 Multiparameter case: $\theta \in R^p$

The asymptotic theory for the multiparameter case follows the previous development with very little change in the notation. The score statistic $S(\theta)$

is now a vector and the Fisher information $I(\theta)$ a matrix. It is convenient to define the so-called *square root* of any covariance matrix A

$$A^{1/2} \equiv \Gamma \Lambda^{1/2} \Gamma'$$

where Γ is the matrix of eigenvectors of A and Λ is a diagonal matrix of the corresponding eigenvalues. We shall use this concept of square-root matrix only in an abstract way, i.e. in practice we seldom need to compute one. It is a useful concept since the square-root matrix can be treated like the usual square root for scalar quantities. Let us denote the identity matrix of size p by 1_p; then we can show

$$\begin{aligned} A^{1/2}A^{1/2} &= A \\ (A^{1/2})^{-1} &= (A^{-1})^{1/2} = A^{-1/2} \\ A^{1/2}A^{-1/2} &= 1_p, \end{aligned}$$

and if random vector X is multivariate $N(\mu, \Sigma)$ then $\Sigma^{-1/2}(X - \mu)$ is $N(0, 1_p)$.

Basic results

Let $x_1, \ldots, x_n$ be an iid sample from $p_\theta(x)$, where $\theta \in R^p$. Under similar regularity conditions, the following results are direct generalizations of those for scalar parameters. All results can be used to test $H_0 : \theta = \theta_0$. The Wald statistic is particularly convenient to test individual parameters, while, in the current setting, the score and the likelihood ratio statistics test all parameters jointly. The same issue of expected versus observed Fisher information also arises with the same resolution: it is generally better to use the observed Fisher information.

Score statistic

For the score statistic, the basis of the score test or Rao's test, we have

$$\begin{aligned} n^{-1/2}S(\theta) &\stackrel{d}{\to} N\{0, \mathcal{I}_1(\theta)\} \\ \mathcal{I}(\theta)^{-1/2}S(\theta) &\stackrel{d}{\to} N(0, 1_p) \\ I(\widehat{\theta})^{-1/2}S(\theta) &\stackrel{d}{\to} N(0, 1_p) \\ S(\theta)'I(\widehat{\theta})^{-1}S(\theta) &\stackrel{d}{\to} \chi^2_p. \end{aligned}$$

From our discussion in Section 9.7, we can informally write

$$S(\theta) \sim N\{0, I(\widehat{\theta})\}.$$

Wald statistic

The asymptotic distribution of the MLE $\widehat{\theta}$ is given by the following equivalent results:

$$\sqrt{n}(\widehat{\theta} - \theta) \stackrel{d}{\to} N(0, \mathcal{I}_1(\theta)^{-1})$$

$$\begin{aligned}
\mathcal{I}(\theta)^{1/2}(\widehat{\theta} - \theta) &\stackrel{d}{\to} N(0, 1_p) \\
I(\widehat{\theta})^{1/2}(\widehat{\theta} - \theta) &\stackrel{d}{\to} N(0, 1_p) \\
(\widehat{\theta} - \theta)' I(\widehat{\theta})(\widehat{\theta} - \theta) &\stackrel{d}{\to} \chi^2_p.
\end{aligned}$$

In practice, we would use

$$\widehat{\theta} \sim N(\theta, I(\widehat{\theta})^{-1}).$$

The standard error of $\widehat{\theta}_i$ is given by the estimated standard deviation

$$\mathrm{se}(\widehat{\theta}_i) = \sqrt{I^{ii}},$$

where I^{ii} is the i'th diagonal term of $I(\widehat{\theta})^{-1}$. A test of an individual parameter $H_0 : \theta_i = \theta_{i0}$ is given by the Wald statistic

$$z_i = \frac{\widehat{\theta}_i - \theta_{i0}}{\mathrm{se}(\widehat{\theta}_i)},$$

which has approximately a standard normal distribution as its null distribution.

Likelihood ratio statistic

The asymptotic behaviour of $L(\widehat{\theta})$ is governed by Wilk's likelihood ratio statistic:

$$\begin{aligned}
W &= 2 \log \frac{L(\widehat{\theta})}{L(\theta)} \\
&\approx (\widehat{\theta} - \theta)' I(\widehat{\theta})(\widehat{\theta} - \theta) \stackrel{d}{\to} \chi^2_p.
\end{aligned}$$

This result gives the connection between the normalized likelihood and P-value, as well as the confidence and likelihood intervals. For example, a normalized likelihood

$$r = \frac{L(\theta)}{L(\widehat{\theta})}$$

is associated with the likelihood ratio statistic

$$w = -2 \log r$$

and

$$\text{P-value} = P(W \geq w),$$

where W is χ^2_p.

The asymptotic distribution theory also gives an approximate $100(1-\alpha)\%$ likelihood-based confidence region:

$$\mathrm{CR} = \left\{ \theta ; 2 \log \frac{L(\widehat{\theta})}{L(\theta)} < \chi^2_{p,(1-\alpha)} \right\}$$

$$= \left\{ \theta; \frac{L(\theta)}{L(\widehat{\theta})} > e^{-\frac{1}{2}\chi^2_{p,(1-\alpha)}} \right\}.$$

This type of region is unlikely to be useful for $p > 2$ because of the display problem. The case of $p = 2$ is particularly simple, since $100\alpha\%$ likelihood cutoff has an approximate $100(1-\alpha)\%$ confidence level. This is true since

$$\exp\left\{ -\frac{1}{2}\chi^2_p(1-\alpha) \right\} = \alpha,$$

so the contour $\{\theta; L(\theta) = \alpha L(\widehat{\theta})\}$ defines an approximate $100(1-\alpha)\%$ confidence region.

9.10 Examples

Logistic regression

Recall our logistic regression example (Example 6.3) where the outcome y_i is surgical mortality and the predictor x_{1i} is the age of the i'th patient. We assume that y_i is Bernoulli with parameter p_i, where

$$p_i = \frac{e^{\beta_0 + \beta_1 \text{ Age}_i}}{1 + e^{\beta_0 + \beta_1 \text{ Age}_i}}$$

or

$$\text{logit } p_i = \beta_0 + \beta_1 \text{ Age}_i.$$

Denoting $\beta = (\beta_0, \beta_1)'$ and $x_i = (1, x_{1i})'$ we obtain

$$\begin{aligned} \log L(\beta) &= \sum_i \{y_i x_i'\beta - \log(1 + e^{x_i'\beta})\} \\ S(\beta) &= \sum_i (y_i - p_i) x_i \\ I(\beta) &= \sum_i p_i(1-p_i) x_i x_i'. \end{aligned}$$

Using the IWLS algorithm discussed in Section 6.7, we get $\widehat{\beta}_0 = -0.723$ and $\widehat{\beta}_1 = 0.160$. To get the standard errors for these estimates we can verify that

$$I(\widehat{\beta}) = \begin{pmatrix} 7.843175 & 9.365821 \\ 9.365821 & 202.474502 \end{pmatrix}$$

so, the estimated covariance matrix of $\widehat{\beta}$ is

$$I(\widehat{\beta})^{-1} = \begin{pmatrix} 0.367^2 & -0.006242531 \\ -0.006242531 & 0.0723^2 \end{pmatrix}$$

and the standard errors are 0.367 and 0.072. The Wald statistic for age effect is

$$z = 0.160/0.072 = 2.22.$$

Poisson regression

In the Poisson regression example (Section 6.3) the outcome y_i is the number of claims, and the predictor x_{1i} is the age of client i. Assuming y_i is Poisson with mean μ_i and

$$\mu_i = e^{\beta_0 + \beta_1 \text{ Age}_i} \equiv e^{x_i'\beta},$$

we can derive the following:

$$\begin{aligned}
\log L(\beta) &= \sum_i (-e^{x_i'\beta} + y_i x_i'\beta) \\
S(\beta) &= \sum_i (y_i - \mu_i) x_i \\
I(\beta) &= \sum_i \mu_i x_i x_i'.
\end{aligned}$$

We can verify that $\widehat{\beta} = (0.43, 0.066)$ and

$$I(\widehat{\beta}) = \begin{pmatrix} 57.17 & 103.67 \\ 103.67 & 1650.75 \end{pmatrix}$$

so we can summarize the analysis in the following table

Effect	Parameter	Estimate	se	z
Intercept	β_0	0.43	0.14	
Age	β_1	0.066	0.026	2.54

Previous likelihood analysis shows that the profile likelihood of β_1 is reasonably regular, so the Wald test can be used safely.

One-way random effects

Table 9.1 (from Fears et al. 1996) shows the estrone measurements from five menopausal women, where 16 measurements were taken from each woman. The questions of interest include the variability between the women and reliability of the measurements. The data are plotted in Figure 9.7(a).

It is natural to model persons as random, so

$$y_{ij} = \mu + a_i + e_{ij}$$

where

$$\begin{aligned}
y_{ij} &= 10 \times \log_{10} \text{ of estrone measurements,} \\
a_i &= \text{person effect, for } i = 1, \ldots, N = 5, \\
e_{ij} &= \text{residual effect, for } j = 1, \ldots, n = 16.
\end{aligned}$$

We assume that a_i's are iid $N(0, \sigma_a^2)$, e_{ij}'s are iid $N(0, \sigma^2)$ and they are independent. The standard analysis of variance table from the dataset is

$i=1$	2	3	4	5
23	25	38	14	46
23	33	38	16	36
22	27	41	15	30
20	27	38	19	29
25	30	38	20	36
22	28	32	22	31
27	24	38	16	30
25	22	42	19	32
22	26	35	17	32
22	30	40	18	31
23	30	41	20	30
23	29	37	18	32
27	29	28	12	25
19	37	36	17	29
23	24	30	15	31
18	28	37	13	32

Table 9.1: *Estrone measurements from five menopausal women; there were 16 measurements taken from each woman from Fears* et al. *(1996).*

Source	df	SS	MS
Person	4	SSA	28.32
Error	75	SSE	0.325

The standard F-test for H_0: $\sigma_a^2 = 0$ gives

$$F = 28.32/0.325 = 87.0,$$

with 4 and 75 degrees of freedom; this is highly significant as we expect from the plot. Now we will show that the MLE of σ_a^2 is $\widehat{\sigma}_a^2 = 1.395$ with standard error 0.895, so the Wald test gives

$$z = 1.395/0.895 = 1.56,$$

which is not at all significant. What is wrong? To get a clear explanation we need to analyse the likelihood function.

Measurements within a person are correlated according to

$$\text{cov}(y_{ij}, y_{ik}) = \sigma_a^2.$$

So, $y_i = (y_{i1}, \ldots, y_{in})'$ is multivariate normal with mean μ and variance

$$S = \sigma^2 I_n + \sigma_a^2 J_n \tag{9.11}$$

where I_n is an $n \times n$ identity matrix and J_n is an $n \times n$ matrix of ones. The likelihood of $\theta = (\mu, \sigma^2, \sigma_a^2)$ is

$$L(\theta) = -\frac{N}{2}\log|S| - \frac{1}{2}\sum_i (y_i - \mu)' S^{-1} (y_i - \mu).$$

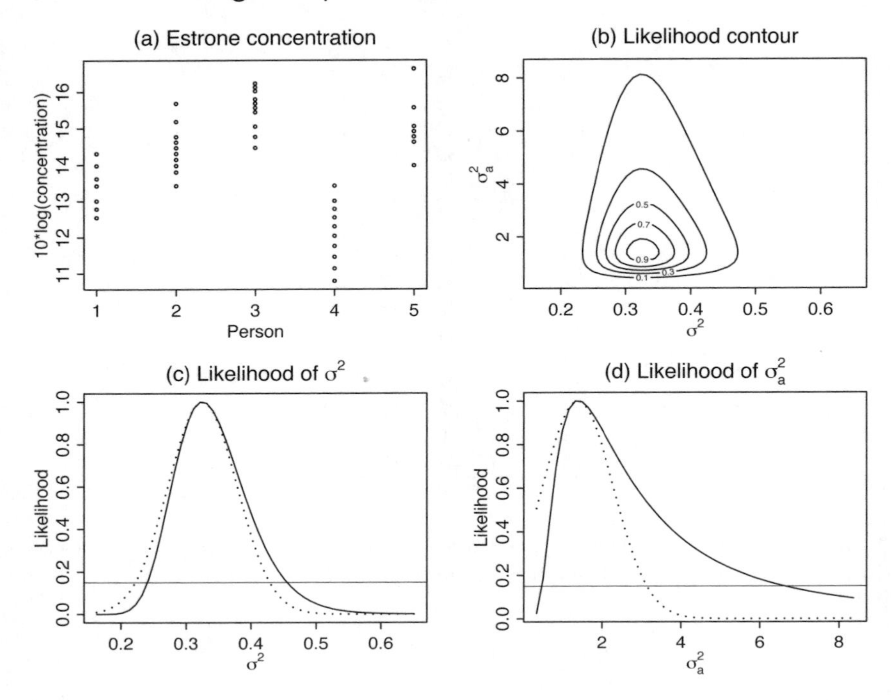

Figure 9.7: *Likelihood analysis of a one-way random effects experiment. (a) This plot shows a significant person-to-person variability. (b) Joint likelihood of* (σ^2, σ_a^2)*. (c) The profile likelihood of* σ^2 *is well approximated by the normal. (d) Poor normal approximation of the profile likelihood of* σ_a^2*.*

To simplify the terms in the likelihood, we need some matrix algebra results (Rao 1973, page 67) that

$$\begin{aligned} |S| &= \sigma^{2(n-1)}(\sigma^2 + n\sigma_a^2) \\ S^{-1} &= \frac{I_n}{\sigma^2} - \frac{\sigma_a^2}{\sigma^2(\sigma^2 + n\sigma_a^2)} J_n. \end{aligned}$$

We can then compute a profile likelihood of (σ^2, σ_a^2) found by maximizing over μ for a fixed value of (σ^2, σ_a^2). We can show that

$$\begin{aligned} \widehat{\mu}(\sigma^2, \sigma_a^2) &= \frac{\sum_i \mathbf{1}' S^{-1} y_i}{\sum_i \mathbf{1}' S^{-1} \mathbf{1}} \\ &= \sum_{ij} y_{ij}/Nn = \overline{y}. \end{aligned}$$

We now define the following (corrected) total, person and error sum-of-squares

$$\begin{aligned}
\text{SST} &= \sum_{ij}(y_{ij} - \overline{y})^2 \\
\text{SSA} &= \sum_i\{\sum_j(y_{ij} - \overline{y})\}^2/n \\
\text{SSE} &= \text{SST} - \text{SSA}.
\end{aligned}$$

Then the profile likelihood can be shown to be

$$\log L(\sigma^2, \sigma_a^2) = -\frac{N}{2}\{(n-1)\log\sigma^2 + \log(\sigma^2 + n\sigma_a^2)\} - \frac{1}{2}\left\{\frac{\text{SSE}}{\sigma^2} + \frac{\text{SSA}}{\sigma^2 + n\sigma_a^2}\right\}.$$

Figure 9.7(b) shows that the contour of the profile likelihood is far from quadratic. From the likelihood we obtain the MLEs

$$\begin{aligned}
\widehat{\sigma}^2 &= \frac{\text{SSE}}{N(n-1)} \\
\widehat{\sigma}_a^2 &= (\text{SSA}/N - \widehat{\sigma}^2)/n.
\end{aligned}$$

From the ANOVA table we can verify $\widehat{\sigma}^2 = 0.325$ and $\widehat{\sigma}_a^2 = 1.395$ as stated earlier.

Taking the second derivatives produces (Exercise 9.8)

$$I(\sigma^2, \sigma_a^2) = \begin{pmatrix} \frac{N(n-1)}{2\sigma^4} + \frac{N}{2(\sigma^2+n\sigma_a^2)^2} & \frac{Nn}{2(\sigma^2+n\sigma_a^2)^2} \\ \frac{Nn}{2(\sigma^2+n\sigma_a^2)^2} & \frac{Nn^2}{2(\sigma^2+n\sigma_a^2)^2} \end{pmatrix}, \tag{9.12}$$

and the observed Fisher information

$$I(\widehat{\sigma}^2, \widehat{\sigma}_a^2) = \begin{pmatrix} 354.08 & 0.0779 \\ 0.0779 & 1.2472 \end{pmatrix},$$

from which we get the asymptotic variance matrix of $(\widehat{\sigma}^2, \widehat{\sigma}_a^2)$

$$I(\widehat{\sigma}^2, \widehat{\sigma}_a^2)^{-1} = \begin{pmatrix} 0.053^2 & 0.0 \\ 0.0 & 0.895^2 \end{pmatrix}$$

and standard errors $\text{se}(\widehat{\sigma}^2) = 0.053$ and $\text{se}(\widehat{\sigma}_a^2) = 0.895$.

Figures 9.7(c)–(d) compare the approximate normal likelihoods for σ^2 and σ_a^2 versus the profile likelihood. The normal likelihood is based on the asymptotic theory that, approximately, $\widehat{\sigma}^2 \sim N(\sigma, 0.053^2)$ and $\widehat{\sigma}_a^2 \sim N(\sigma_a^2, 0.895^2)$. It is obvious now that the normal approximation for $\widehat{\sigma}_a^2$ is inappropriate. As an exercise it can be verified that the likelihood of $\log\sigma_a$ is reasonably regular. The likelihood-based 95% CIs are $0.24 < \sigma^2 < 0.45$ and $0.51 < \sigma_a^2 < 6.57$.

The reliability of the measurement can be expressed as the correlation of y_{ij} and y_{ik} for $j \neq k$, i.e. the similarity of different measurements from the same person. The measurement is reliable if the correlation is high. From the model

$$
\begin{aligned}
\text{cor}(y_{ij}, y_{ik}) &= \frac{\text{cov}(y_{ij}, y_{ik})}{\{\text{var } y_{ij} \ \text{var } y_{ik}\}^{1/2}} \\
&= \frac{\text{cov}(\mu + a_i + e_{ij}, \mu + a_i + e_{ik})}{\sigma^2 + \sigma_a^2} \\
&= \frac{\sigma_a^2}{\sigma^2 + \sigma_a^2}. \qquad (9.13)
\end{aligned}
$$

This quantity is also called the intraclass correlation. Its estimate for the above data is

$$\frac{\widehat{\sigma}_a^2}{\widehat{\sigma}^2 + \widehat{\sigma}_a^2} = 0.81.$$

Finding the profile likelihood for the intraclass correlation is left as an exercise.

9.11 Nuisance parameters

While convenient for dealing with individual parameters in a multiparameter setting, the Wald statistic has serious weaknesses. In particular, parameter transformation has a great impact on Wald-based inference, so the choice of parameterization becomes unduly important. In contrast, because of its invariance property, the likelihood ratio test is safer to use (Section 2.9). We now develop a likelihood ratio theory for some parameters while treating the others as nuisance parameters.

The theory we develop is also useful for situations where we want to test a hypothesis that is not easily parameterized. For example,

- goodness-of-fit tests
- test of independence for multiway tables.

We follow the general method of profile likelihood to remove the nuisance parameters. Let $\theta = (\theta_1, \theta_2) \in R^p$, where $\theta_1 \in R^q$ is the parameter of interest and $\theta_2 \in R^r$ is the nuisance parameter, so $p = q + r$. Given the likelihood $L(\theta_1, \theta_2)$ we compute the profile likelihood as

$$
\begin{aligned}
L(\theta_1) &\equiv \max_{\theta_2} L(\theta_1, \theta_2) \\
&\equiv L(\theta_1, \widehat{\theta}_2(\theta_1)),
\end{aligned}
$$

where $\widehat{\theta}_2(\theta_1)$ is the MLE of θ_2 at a fixed value of θ_1.

The theory indicates that we can treat $L(\theta_1)$ as if it is a true likelihood; in particular, the profile likelihood ratio follows the usual asymptotic theory:

$$W = 2 \log \frac{L(\widehat{\theta}_1)}{L(\theta_1)} \xrightarrow{d} \chi_q^2 = \chi_{p-r}^2. \qquad (9.14)$$

Here is another way of looking at the profile likelihood ratio from the point of view of testing H_0: $\theta_1 = \theta_{10}$. This is useful to deal with hypotheses that are not easily parameterized. By definition,

$$\begin{aligned} L(\theta_{10}) &= \max_{\theta_2, \theta_1 = \theta_{10}} L(\theta_1, \theta_2) \\ &= \max_{H_0} L(\theta) \\ L(\widehat{\theta}_1) &= \max_{\theta_1}\{\max_{\theta_2} L(\theta_1, \theta_2)\} \\ &= \max_{\theta} L(\theta). \end{aligned}$$

Therefore,

$$W = 2 \log \frac{\max L(\theta), \text{ no restriction on } \theta}{\max L(\theta), \ \theta \in H_0}.$$

A large value of W means H_0 has a small likelihood, or there are other values with higher support, so we should reject H_0.

How large is 'large' will be determined by the sampling distribution of W. We can interpret p and r as

$$\begin{aligned} p &= \text{dimension of the whole parameter space } \theta \\ &= \text{the total number of free parameters} \\ &= \text{total degrees of freedom of the parameter space} \\ r &= \text{dimension of the parameter space under } H_0 \\ &= \text{the number of free parameters under } H_0 \\ &= \text{degrees of freedom of the model under } H_0. \end{aligned}$$

Hence the degree of freedom in (9.14) is the change in the dimension of the parameter space from the whole space to the one under H_0.

Before we prove the general asymptotic result, it is important to note that in some applications it is possible to get an exact distribution for W. Many normal-based classical tests, such as the t-test or F-test, are exact likelihood ratio tests.

Example 9.11: Let $x_1, \ldots, x_n$ be an iid sample from $N(\mu, \sigma^2)$ with σ^2 unknown and we are interested in testing H_0: $\mu = \mu_0$ versus H_1: $\mu \neq \mu_0$. Under H_0 the MLE of σ^2 is

$$\widehat{\sigma}^2 = \frac{1}{n}\sum_i (x_i - \mu_0)^2.$$

Up to a constant term,

$$\begin{aligned} \max_{H_0} L(\theta) &= \left\{\frac{1}{n}\sum_i (x_i - \mu_0)^2\right\}^{-n/2} \\ \max L(\theta) &= \left\{\frac{1}{n}\sum_i (x_i - \overline{x})^2\right\}^{-n/2} \end{aligned}$$

and

$$\begin{aligned} W &= n\log\frac{\sum_i(x_i-\mu_0)^2}{\sum_i(x_i-\overline{x})^2} \\ &= n\log\frac{\sum_i(x_i-\overline{x})^2+n(\overline{x}-\mu_0)^2}{\sum_i(x_i-\overline{x})^2} \\ &= n\log\left(1+\frac{t^2}{n-1}\right), \end{aligned}$$

where $t=\sqrt{n}(\overline{x}-\mu_0)/s)$ and s^2 is the sample variance. Now, W is monotone increasing in t^2, so we reject H_0 for large values of t^2 or $|t|$. This is the usual t-test. A critical value or a P-value can be determined from the t_{n-1}-distribution. □

Having an exact distribution for a likelihood ratio statistic is a fortunate coincidence; generally there is a reliance on a large-sample approximation given in the following theorem. Its proof will also provide a justification of the claim made in Section 3.4 about the curvature of the profile likelihood.

Theorem 9.8 *Assuming regularity conditions, under* H_0*:* $\theta_1=\theta_{10}$

$$W=2\log\frac{\max L(\theta)}{\max_{H_0}L(\theta)}\to\chi^2_{p-r}.$$

Proof: Let $\widehat{\theta}=(\widehat{\theta}_1,\widehat{\theta}_2)$ be the unrestricted MLE and $\widehat{\theta}_0=(\theta_{10},\widehat{\theta}_{20})$ be the MLE under H_0. Let the true parameter be $\theta_0=(\theta_{10},\theta_2)$. We want to show that under H_0

$$W=2\log\frac{L(\widehat{\theta})}{L(\widehat{\theta}_0)}\to\chi^2_{p-r}.$$

The difficult step is to find the adjusted estimate $\widehat{\theta}_{20}$ in terms of $\widehat{\theta}$. From our basic results we have, approximately,

$$\begin{pmatrix}\widehat{\theta}_1-\theta_1\\ \widehat{\theta}_2-\theta_2\end{pmatrix}\sim N\left\{0,I(\widehat{\theta})^{-1}\equiv\begin{pmatrix}I^{11} & I^{12}\\ I^{21} & I^{22}\end{pmatrix}\right\}.$$

The problem can be stated more transparently as follows. Suppose we observe

$$\begin{pmatrix}x\\ y\end{pmatrix}\sim N\left\{\begin{pmatrix}\mu_x\\ \mu_y\end{pmatrix},\begin{pmatrix}\sigma_{xx} & \sigma_{xy}\\ \sigma_{yx} & \sigma_{yy}\end{pmatrix}\right\},$$

where the variance matrix is assumed known. If (μ_x,μ_y) are both unknown then we obtain the MLE $\widehat{\mu}_y=y$. But what if μ_x is known? Intuitively, if x and y are correlated then x will contribute some information about μ_y. In fact, from the standard normal theory we have the conditional distribution

$$y|x\sim N(\mu_y+\sigma_{yx}\sigma_{xx}^{-1}(x-\mu_x),\sigma_{yy.x}\equiv\sigma_{yy}-\sigma_{yx}\sigma_{xx}^{-1}\sigma_{xy}).$$

So, given x, y and μ_x, the MLE of μ_y is

$$\widehat{\mu}_y = y - \sigma_{yx}\sigma_{xx}^{-1}(x - \mu_x).$$

So, equivalently, given $\theta_1 = \theta_{10}$, $\widehat{\theta}_1$ and $\widehat{\theta}_2$, we get

$$\widehat{\theta}_{20} = \widehat{\theta}_2 - I^{21}(I^{11})^{-1}(\widehat{\theta}_1 - \theta_{10}).$$

A simple manipulation of the partitioned matrix gives a simpler form

$$\widehat{\theta}_{20} = \widehat{\theta}_2 + I_{22}^{-1}I_{21}(\widehat{\theta}_1 - \theta_{10}),$$

using the partition

$$I = \begin{pmatrix} I_{11} & I_{12} \\ I_{21} & I_{22} \end{pmatrix}$$

for the Fisher information.

Previously we have used the quadratic approximation

$$2\log\frac{L(\widehat{\theta})}{L(\theta)} \approx (\widehat{\theta} - \theta)'I(\widehat{\theta})(\widehat{\theta} - \theta),$$

so under $H_0 : \theta_1 = \theta_{10}$, and assuming $I(\widehat{\theta}) = I(\widehat{\theta}_0)$

$$\begin{aligned} 2\log\frac{L(\widehat{\theta})}{L(\widehat{\theta}_0)} &= 2\log\frac{L(\widehat{\theta})}{L(\theta_0)} - 2\log\frac{L(\widehat{\theta}_0)}{L(\theta_0)} \\ &\approx (\widehat{\theta} - \theta_0)'I(\widehat{\theta})(\widehat{\theta} - \theta_0) - (\widehat{\theta}_0 - \theta_0)'I(\widehat{\theta})(\widehat{\theta}_0 - \theta_0). \end{aligned}$$

Since $(\widehat{\theta}_0 - \theta_0) = (0, \widehat{\theta}_{20} - \theta_2)$ and

$$\widehat{\theta}_{20} - \theta_2 = \widehat{\theta}_2 - \theta_2 + I_{22}^{-1}I_{21}(\widehat{\theta}_1 - \theta_{10})$$

then

$$\begin{aligned} (\widehat{\theta}_0 - \theta_0)'I(\widehat{\theta})(\widehat{\theta}_0 - \theta_0) &= (\widehat{\theta}_{20} - \theta_2)'I_{22}(\widehat{\theta}_{20} - \theta_2) \\ &= \begin{pmatrix} \widehat{\theta}_1 - \theta_{10} \\ \widehat{\theta}_2 - \theta_2 \end{pmatrix}' \begin{pmatrix} I_{12}I_{22}^{-1}I_{21} & I_{12} \\ I_{21} & I_{22} \end{pmatrix} \begin{pmatrix} \widehat{\theta}_1 - \theta_{10} \\ \widehat{\theta}_2 - \theta_2 \end{pmatrix}. \end{aligned}$$

Collecting all the terms, we get

$$\begin{aligned} 2\log\frac{L(\widehat{\theta})}{L(\widehat{\theta}_0)} &\approx (\widehat{\theta}_1 - \theta_{10})'(I_{11} - I_{12}I_{22}^{-1}I_{21})(\widehat{\theta}_1 - \theta_{10}) \\ &= (\widehat{\theta}_1 - \theta_{10})'(I^{11})^{-1}(\widehat{\theta}_1 - \theta_{10}), \end{aligned}$$

so the profile likelihood ratio is again asymptotically equivalent to the Wald test on θ_1. This quadratic approximation shows that the curvature of the profile likelihood is given by $(I^{11})^{-1}$, the claim made in Section 3.4. From the asymptotics of the MLE $\widehat{\theta}$ we have

$$\widehat{\theta}_1 - \theta_{10} \stackrel{d}{\rightarrow} N(0, I^{11}),$$

so we arrive at

$$2\log\frac{L(\widehat{\theta})}{L(\widehat{\theta}_0)} \stackrel{d}{\rightarrow} \chi^2_{p-r}. \ \Box$$

9.12 χ^2 goodness-of-fit tests

One major use of the likelihood ratio test is in the test of hypotheses involving categorical data, including the goodness-of-fit tests. We will first consider the simple case where there is no nuisance parameter.

Example 9.12: Are birthdays uniformly distributed throughout the year? Here is the monthly breakdown of birthdays of 307 students in a first-year statistics class. Almost all students have the same age. □

Month	1	2	3	4	5	6	7	8	9	10	11	12
No.	28	18	30	30	32	21	30	25	30	20	22	21

Suppose $N_1, \ldots, N_K$ are multinomial with total size n and probability $\theta = (p_1, \ldots, p_K)$, with $\sum N_i = n$ and $\sum p_i = 1$. We want to test a null hypothesis that birthdays are uniformly distributed. There is no explicit parameter of interest; it is a lot easier to express the problem using the likelihood ratio test than using the Wald test. Specifically,

$$H_0 : p_i = p_{i0} = \frac{\text{No. of days in month } i}{365}$$

versus H_1: $p_i \neq p_{i0}$ for some i. If there is no restriction on the parameters, we get the MLEs $\widehat{p}_i = n_i/n$, so the likelihood ratio test is simply

$$\begin{aligned} 2\log\frac{L(\widehat{\theta})}{L(\theta)} &= 2\sum_i n_i \log\frac{n_i}{np_{i0}} \\ &\equiv 2\sum O\log\frac{O}{E}, \end{aligned}$$

where 'O' stands for the observed frequencies and 'E' the expected frequencies under H_0. W is in fact numerically close to the more commonly used Pearson's χ^2 statistic

$$\chi^2 = \sum \frac{(O-E)^2}{E}.$$

Theorem 9.9 *If the expected frequencies E are large enough in every cell then, under H_0,*

$$2\sum O\log\frac{O}{E} \approx \sum\frac{(O-E)^2}{E}.$$

Proof: Consider a second-order expansion of $\log x$ around 1

$$\log x \approx (x-1) - \frac{1}{2}(x-1)^2.$$

Under H_0 we expect $O/E \approx 1$, so we apply the second-order expansion on $\log O/E$ and finish the algebra. □

Example 9.12: continued. For the birthday data we can verify that $W = 9.47$ (and the corresponding Pearson's $\chi^2 = 9.35$), which is not significant at $12 - 1 = 11$ degrees of freedom. Therefore, there is no evidence of nonuniform birth pattern. Note, however, that a test with high degrees of freedom is not very desirable in practice, since it has low power against specific alternatives. In this case, grouping the months into three-month (3 degree-of-freedom test) and six-month intervals (1 degree-of-freedom test) does not reveal any significant nonuniformity (Exercise 9.13). □

Nuisance parameters

As usual, the more important case is when there are nuisance parameters. For example:

- Checking the distributional assumption of the residual after model fitting. To test if the errors are normally distributed, the regression model parameters are treated as nuisance parameters.
- Testing the independence in a 2-way table: the marginal distributions are nuisance parameters.

Suppose $n_1, \ldots, n_K$ are multinomial with parameters n and $(p_1, \ldots, p_K)$. We want to test

$$H_0: \; p_i = p_i(\theta_0),$$

i.e. p_i's follow a parametric form with $\dim(\theta_0) = r$, versus H_1: p_i is arbitrary, satisfying only $\sum p_i = 1$. Here the parameter θ_0 is the nuisance parameter. The likelihood ratio test is

$$\begin{aligned} W &= 2\sum O \log \frac{O}{E} \\ &= 2\sum n_i \log \frac{n_i}{np_i(\widehat{\theta}_0)}, \end{aligned}$$

where $\widehat{\theta}_0$ is the MLE of θ_0 *based on data* $n_1, \ldots, n_K$. (This point is important if the group data are based on grouping continuous data, in which case there is a temptation to use $\widehat{\theta}$ based on the original data.) According to our theory, under the null hypothesis, W is χ^2 with $K - 1 - r$ degrees of freedom.

Example 9.13: One of the most common applications of the χ^2 test is in testing the independence of two characteristics; for example, eye versus hair colour. The data are usually presented in a two-way contingency table. Consider a table with cell frequencies n_{ij} for $i = 1, \ldots, I$ and $j = 1, \ldots, J$, and corresponding probabilities p_{ij}, such that $\sum_{ij} p_{ij} = 1$. The log-likelihood of the parameter $\theta \equiv \{p_{ij}\}$ given the observed data n_{ij} is

$$L(\theta) = \sum n_{ij} \log p_{ij}.$$

Under the null hypothesis of independence between the row and column characteristics: $p_{ij} = r_i c_j$, where r_i is the true proportion of the i'th row characteristic and c_j is the true proportion of the j'th column characteristic. The free parameter under the null hypothesis is $\theta_0 = (r_1, \ldots, r_I, c_1, \ldots, c_J)$, satisfying the constraint $\sum_i r_i = 1$ and $\sum_j c_j = 1$. Under independence we obtain

$$p_{ij}(\widehat{\theta}_0) = \frac{n_{i.}n_{.j}}{n_{..}^2},$$

where the row total $n_{i.} \equiv \sum_j n_{ij}$, the column total $n_{.j} \equiv \sum_i n_{ij}$ and the grand total $n_{..} \equiv \sum_{ij} n_{ij}$. The test of independence is

$$\begin{aligned} W &= 2\sum_{ij} n_{ij} \log \frac{n_{ij}}{n_{..}p_{ij}(\widehat{\theta}_0)} \\ &\approx \sum_{ij} \frac{\{n_{ij} - n_{..}p_{ij}(\widehat{\theta}_0)\}^2}{n_{..}p_{ij}(\widehat{\theta}_0)}. \end{aligned}$$

Since $\dim(\theta_0) = I + J - 2$, the degrees of freedom of the test is $IJ - 1 - I - J + 2 = (I-1)(J-1)$. □

9.13 Exercises

Exercise 9.1: The noncentral hypergeometric probability is defined as

$$P(X = x) = \frac{\binom{m}{x}\binom{n}{t-x} e^{\theta x}}{\sum_{s=0}^{t} \binom{m}{s}\binom{n}{t-s} e^{\theta s}},$$

for $x = 0, \ldots, t$, where m, n and t are known constants. At $\theta = 0$ we have the (central) hypergeometric model, where $P(X = x)$ is the probability of getting x black balls in a random sample of t balls without replacement from an urn with m black and n white balls. Show that the score test for testing H_0: $\theta = 0$ is of the form

$$z = \frac{x - \mu_0}{\sigma_0}$$

where μ_0 and σ_0 are the mean and variance of the (central) hypergeometric distribution.

Exercise 9.2: As stated in Example 9.9, show that if we approximate $K(\widehat{\theta})$ around zero by a quadratic function

$$K(\widehat{\theta}) \approx K(0) + K'(0)\widehat{\theta} + \frac{1}{2}K''(0)\widehat{\theta}^2$$

then we obtain the standard central limit theorem from the saddlepoint formula.

Exercise 9.3: The saddlepoint approximation of the distribution of the sample mean in Example 9.9 can be used as a theoretical alternative of the bootstrap computation. Define the empirical cumulant generating function

$$K(\theta) = \log\left(\frac{1}{n}\sum_i e^{\theta x_i}\right).$$

For the following observations

50	44	102	72	22	39	3	15	197	188	79	88
46	5	5	36	22	139	210	97	30	23	13	14

compare the bootstrap distribution of the sample mean with the saddlepoint approximation (9.8) based on the empirical cumulant function.

Exercise 9.4: Suppose $y_1, \ldots, y_n$ are an iid sample from the inverse Gaussian distribution with density

$$p(y) = \left(\frac{\lambda}{2\pi y^3}\right)^{1/2} \exp\left\{-\frac{\lambda}{2\mu^2}\frac{(y-\mu)^2}{y}\right\}, \qquad y > 0.$$

(a) Assuming λ is known, find the saddlepoint approximation of the density of the MLE of μ.

(b) Assuming μ is known find the saddlepoint approximation of the density of the MLE of λ.

(c) For $\mu = 1$, $\lambda = 1$, and $n = 10$, show how good is the approximation in (b) by performing a Monte Carlo simulation similar to the one in Example 9.8.

Exercise 9.5: Let $x_1, \ldots, x_n$ be an iid sample from $N(\mu, \sigma^2)$ with μ known. Give the approximate density of the sample variance using formula (9.6).

Exercise 9.6: Let $x_1, \ldots, x_n$ be an iid sample from gamma(μ, β), with known β. Derive the approximate density for μ using formula (9.6) and show that the formula is exact.

Exercise 9.7: For the location family in Section 9.8 show that the Jacobian in the transformation from $x_{(1)}, \ldots, x_{(n)}$ to $(a, \widehat{\theta})$ is equal to one. (Hint: first transform the data to $(x_{(1)}, a)$, then transform $(x_{(1)}, a)$ to $(\widehat{\theta}, a)$, so the Jacobian is $|dx_{(1)}/d\widehat{\theta}|$.)

Exercise 9.8: Verify the Fisher information (9.12) for the variance components in one-way random effects.

Exercise 9.9: For the variance matrix (9.11) verify that its inverse is

$$S^{-1} = \frac{I_n}{\sigma^2} - \frac{\sigma_a^2}{\sigma^2(\sigma^2 + n\sigma_a^2)} J_n.$$

To find its determinant one needs to get the eigenvalues of S. Treat this as an exercise only if you are familiar enough with matrix algebra.

Exercise 9.10: For the random effects example in Section 9.10 show that the likelihood of $\log \sigma_a$ is reasonably regular. Find the normal approximation of the distribution of $\log \widehat{\sigma}_a$.

Exercise 9.11: Compute the profile likelihood for the intraclass correlation (9.13) based on the data in Table 9.1. Comment on the regularity of the likelihood.

Exercise 9.12: Complete the detail of the proof of Theorem 9.9 that

$$2\sum O \log \frac{O}{E} \approx \sum \frac{(O-E)^2}{E}.$$

Exercise 9.13: Test the uniformity of the birthdays in Example 9.12 by splitting the data into three-month and six-month intervals. Compute both W and Pearson's χ^2 statistics.

10

Dealing with nuisance parameters

Nuisance parameters create most of the complications in likelihood theory. They appear on the scene as a natural consequence of our effort to use 'bigger and better' models: while some parameters are of interest, others are only required to complete the model. The issue is important since nuisance parameters can have a dramatic impact on the inference for the parameters of interest. Even if we are interested in all of the parameters in a model, our inability to view multidimensional likelihood forces us to see individual parameters in isolation. While viewing one, the other parameters are a nuisance.

We have used the idea of profile likelihood as a general method to eliminate nuisance parameters. The generality comes with a price, namely the potential for bias (even in large samples) and overly optimistic precision. For example, the MLE of the normal variance is

$$\widehat{\sigma}^2 = \frac{1}{n}\sum_i (x_i - \overline{x})^2.$$

Since $\sum_i (x_i - \overline{x})^2/\sigma^2$ is χ^2_{n-1},

$$E\widehat{\sigma}^2 = \frac{n-1}{n}\sigma^2.$$

For $n = 2$ this is a severe underestimate. Furthermore, the profile likelihood of the variance

$$\log L(\sigma^2) = -\frac{n}{2}\log\sigma^2 - \frac{1}{2\sigma^2}\sum_i (x_i - \overline{x})^2 \tag{10.1}$$

is the same as the likelihood of σ^2 if the mean μ is known at $\overline{x}$. This means we are not 'paying' the price for not knowing μ. Hence, bias is only a symptom of a potentially more serious problem. The bias itself can be traced from the score function

$$S(\sigma^2) = \frac{\partial}{\partial\sigma^2}\log L(\sigma^2) = -\frac{n}{2\sigma^2} + \frac{\sum_i (x_i - \overline{x})^2}{2\sigma^4}.$$

This yields $E_{\sigma^2} S(\sigma^2) = -1/(2\sigma^2) \neq 0$, not satisfying the usual zero-mean property of a true score statistic. We can also show that the variance of the score statistic does not match the expected Fisher information:

$$\text{var}_{\sigma^2}\{S(\sigma^2)\} = \frac{n-1}{2\sigma^4} \neq E_{\sigma^2} I(\sigma^2) = \frac{n-2}{2\sigma^4}.$$

If there are more mean parameters to estimate, as in analysis of variance problems, the mismatch is worse.

In regular problems, bias is small relative to standard error, and it goes away as the sample gets large. That is typically the case when the number of nuisance parameters is small relative to the sample size. There is a genuine concern, however, when bias does not disappear as the sample size gets large, or bias is large relative to standard error, resulting in an inconsistent estimation. This usually occurs when the number of nuisance parameters is of the same order of magnitude as the sample size, technically known as 'infinitely many nuisance parameters'.

The main theoretical methods to eliminate nuisance parameters are via conditioning or marginalizing. Unlike the profile likelihood, the resulting conditional or marginal likelihoods are a true likelihood, based on the probability of observed quantities. These methods typically correct the profile likelihood in terms of the bias in the MLE, or the overly optimistic precision level, or both. If an exact method is not available, we use approximate conditional or marginal likelihoods based on a modification of the profile likelihood.

The simplest method to deal with nuisance parameters is to replace the unknowns by their estimates. This is especially useful when other methods are either not available or too complicated. The resulting likelihood will be called the *estimated likelihood.* For example, the likelihood (10.1) is an estimated likelihood. The main problem with the estimated likelihood is that it does not take into account the extra uncertainty due to the unknown nuisance parameters.

10.1 Inconsistent likelihood estimates

Neyman and Scott (1948) demonstrated that the profile likelihood can be severely biased even as the sample size gets large. This is a common ocurrence if there are 'infinitely many' nuisance parameters.

Example 10.1: Consider a highly stratified dataset below where y_{i1} and y_{i2} are an iid sample from $N(\mu_i, \sigma^2)$, for $i = 1, \ldots, N$, and they are all independent over index i. The parameter of interest is σ^2. The total number of unknown parameters is $N+1$ and the number of observations is $2N$. To convince ourselves of the bias, and appreciate the corrected procedure, we simulate data from the model as shown in Table 10.1. The advantage of simulated data is that we know the true $\sigma^2 = 1$ and μ_i's, so we can show the 'true' likelihood of σ^2. To make the bias visible, the number of strata N should be large enough; $N = 20$ is sufficient in this case.

Letting $\theta = (\mu_1, \ldots, \mu_N, \sigma^2)$, the full likelihood is

i	μ_i	y_{i1}	y_{i2}	$\overline{y}_i$
1	0.88	−0.31	−0.51	−0.41
2	2.51	3.20	3.57	3.38
3	1.74	1.70	2.81	2.26
4	−6.74	−6.67	−5.37	−6.02
5	1.42	3.12	3.74	3.43
6	−3.34	−3.15	−3.27	−3.21
7	−2.72	−2.11	−3.10	−2.60
8	6.89	6.88	6.53	6.70
9	0.67	0.81	−2.70	−0.94
10	−4.18	−4.25	−3.64	−3.94
11	8.43	8.58	7.08	7.83
12	0.15	1.88	−1.15	0.36
13	3.89	5.18	4.31	4.74
14	4.52	3.81	5.86	4.84
15	−4.05	−4.68	−4.55	−4.62
16	−6.95	−6.58	−7.52	−7.05
17	−2.61	−2.16	−1.13	−1.64
18	−6.52	−7.99	−7.53	−7.76
19	−6.06	−6.48	−6.13	−6.30
20	0.92	0.89	1.33	1.11

Table 10.1: *Simulated highly stratified data: y_{i1} and y_{i2} are iid $N(\mu_i, \sigma^2)$.*

$$\log L(\theta) = -N \log \sigma^2 - \frac{1}{2\sigma^2} \sum_{i=1}^{N} \sum_{j=1}^{2} (y_{ij} - \mu_i)^2.$$

Assuming μ_i's are known, the full likelihood is the 'true' likelihood of σ^2; this serves as the gold standard in our analysis. The true likelihood is shown as the dotted curve in Figure 10.1.

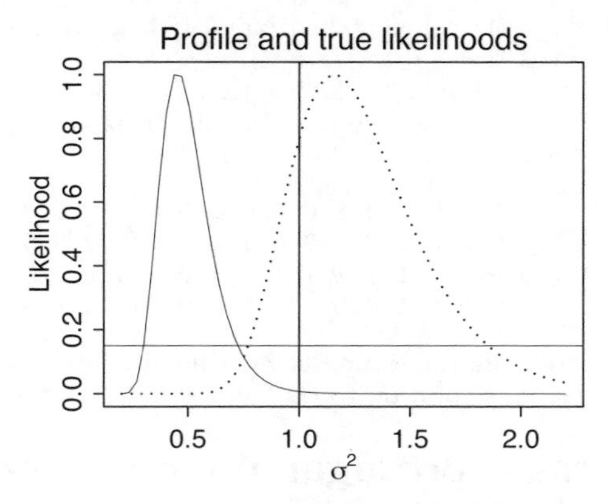

Figure 10.1: *Simulation from highly stratified data: the profile likelihood (solid line) is compared with the true likelihood (dotted line).*

To compute the profile likelihood of σ^2, at each σ^2 we can show that $\widehat{\mu}_i = \overline{y}_i$. We denote the residual sum of squares $\text{RSS} = \sum_i \sum_j (y_{ij} - \overline{y}_i)^2$; hence the profile likelihood of σ^2 is

$$\begin{aligned}\log L(\sigma^2) &= \max_{\mu_1,\ldots,\mu_N} \log L(\mu_1,\ldots,\mu_N,\sigma^2)\\ &= -N\log\sigma^2 - \frac{\text{RSS}}{2\sigma^2}\end{aligned}$$

and the MLE is

$$\widehat{\sigma}^2 = \frac{\text{RSS}}{2N}.$$

From the data we can verify that RSS $= 18.086$ and $\widehat{\sigma}^2 = 0.452$. Figure 10.1 shows the profile likelihood function; also, using the profile likelihood, the true value $\sigma^2 = 1$ only has 0.7% likelihood, clearly unexpected in a regular problem.

Note that what we have is simply a one-way analysis of variance model, where the residual degrees of freedom is equal to N. It is clear that the RSS is $\sigma^2\chi^2_N$, so $E\widehat{\sigma}^2 = \sigma^2/2$ for any N. Furthermore, $\text{var}(\widehat{\sigma}^2) = \sigma^4/(2N)$, so

$$\widehat{\sigma}^2 \xrightarrow{p} \sigma^2/2,$$

or the estimate is not consistent. □

Example 10.2: A similar problem can also occur with a regression parameter. Suppose y_{ij} are binomial outcomes with parameters n_{ij} and p_{ij}, following a logistic model

$$\text{logit } p_{ij} = \beta_0 + S_i + \tau_j,$$

where S_i's are the strata effects, for $i = 1,\ldots,I$, and τ_j's are the treatment effects, for $j = 0, 1$. For identifiability assume that $\tau_0 = 0$; the parameter of interest is treatment contrast τ_1. This model assumes that the treatment effect is the same across strata.

If n_{ij} is small and I is large, we have highly stratified data. For example, a stratum may represent a subject, and treatments are assigned within each subject; or, a stratum may represent a family, and the treatments are assigned to the specific members of the family. The standard MLE of τ_1 is seriously biased, and inference based on the ordinary profile likelihood is questionable. In the extreme case where $n_{ij} \equiv 1$ (e.g. matched pairs with Bernoulli outcomes), the MLE $\widehat{\tau}_1 \to 2\tau_1$ (Breslow 1981).

The following y_{ij}'s are simulated data with $\tau_1 = 1$ and some random S_i's. The first 50 values (first two rows, representing 50 strata) come from treatment $j = 0$, and the second from $j = 1$:

```
0 0 1 0 1 0 1 0 0 0 1 0 0 0 0 0 1 0 0 0 0 0 0 0 0
0 0 0 1 0 0 0 0 0 0 0 0 0 1 0 1 0 1 0 0 0 0 0 0 0
0 1 0 1 0 0 1 0 1 1 1 1 1 0 1 0 1 0 1 1 0 1 1 1 0
0 1 1 1 1 0 0 1 1 0 0 1 1 1 0 0 1 0 0 1 0 0 0 0 1
```

Fitting the logistic regression, the estimated treatment effect $\widehat{\tau}_1 = 3.05$ (se=0.70) indicates a serious bias. The solution to this problem is given in Section 10.5. □

10.2 Ideal case: orthogonal parameters

An ideal situation occurs if we have data x depending on a model $p_\theta(x)$ and y on $p_\eta(y)$, where x and y are independent, and there is no logical connection between θ and η. The joint likelihood of (θ, η) is

$$L(\theta,\eta) = p_\theta(x)p_\eta(y)$$

$$= L(\theta; x)L(\eta; y).$$

There can be no argument that the true likelihood of θ should be

$$L(\theta) = L(\theta; x),$$

since y does not carry any information about θ.

Also ideal is the situation where we can factorize the likelihood

$$L(\theta, \eta) = L_1(\theta)L_2(\eta),$$

where we do not care how the data enter $L_1(\cdot)$ and $L_2(\cdot)$. It is clear that the information on θ is captured by $L_1(\theta)$. When such a factorization exists θ and η are called *orthogonal parameters.*

Example 10.3: In the traffic deaths example in Section 4.6 we assume that the number of deaths x and y are independent Poisson with parameters λ_x and λ_y. The joint likelihood function is

$$L(\lambda_x, \lambda_y) = e^{-(\lambda_x+\lambda_y)}\lambda_x^x\lambda_y^y.$$

Assuming the parameter of interest is $\theta = \lambda_y/\lambda_x$, now let the nuisance parameter be $\eta = \lambda_x + \lambda_y$. So

$$\begin{aligned} L(\theta, \eta) &= \left(\frac{\theta}{1+\theta}\right)^y \left(\frac{1}{1+\theta}\right)^x \eta^{x+y}e^{-\eta} \\ &\equiv L_1(\theta)L_2(\eta), \end{aligned}$$

where

$$L_1(\theta) \equiv \left(\frac{\theta}{1+\theta}\right)^y \left(\frac{1}{1+\theta}\right)^x.$$

As shown before $L_1(\theta)$ is also the profile likelihood of θ. This is generally true: if there exists an orthogonal parameter for θ then, without having to specify η, the profile likelihood computation would automatically provide $L_1(\theta)$. □

Often we do not achieve the ideal case, but only

$$L(\theta, \eta) = L_1(\theta)L_2(\theta, \eta),$$

with the additional argument that $L_2(\theta, \eta)$ contains little information about θ, or $L_1(\theta)$ captures most of the information about θ.

Example 10.4: Suppose $x_1, \ldots, x_n$ are an iid sample from $N(\mu, \sigma^2)$ with both parameters unknown. It is well known that the sample mean $\overline{x}$ and the sample variance

$$s^2 = \frac{1}{n-1}\sum_i (x_i - \overline{x})^2$$

are independent. However, $\overline{x}$ is $N(\mu, \sigma^2/n)$ and $(n-1)s^2$ is $\sigma^2\chi^2_{n-1}$, so the parameters do not separate cleanly. In likelihood terms we can write with obvious notation

$$L(\mu, \sigma^2) = L(\mu, \sigma^2; \overline{x})L(\sigma^2; s^2).$$

If we are interested in σ^2, and μ is unknown, we can ponder whether there is information in $\overline{x}$ about σ^2. In repeated sampling terms, yes there is, but it is intuitive that the observed $\overline{x}$ itself does not carry any information about the variance. This means that we can ignore $\overline{x}$, and concentrate our likelihood based on s^2

$$\log L(\sigma^2) = -\frac{n-1}{2}\log\sigma^2 - \frac{(n-1)s^2}{2\sigma^2},$$

now free of the unknown parameter μ. Such a likelihood is called a marginal likelihood. □

10.3 Marginal and conditional likelihoods

As a general method, consider a transformation of the data x to (v, w) such that either the marginal distribution of v or the conditional distribution of v given w depends only on the parameter of interest θ. Let the total parameter be (θ, η). In the first case

$$\begin{aligned} L(\theta,\eta) &= p_{\theta,\eta}(v,w) \\ &= p_\theta(v)p_{\theta,\eta}(w|v) \\ &\equiv L_1(\theta)L_2(\theta,\eta), \end{aligned}$$

so the *marginal likelihood* of θ is defined as

$$L_1(\theta) = p_\theta(v).$$

In the second case

$$\begin{aligned} L(\theta,\eta) &= p_\theta(v|w)p_{\theta,\eta}(w) \\ &\equiv L_1(\theta)L_2(\theta,\eta), \end{aligned}$$

where the *conditional likelihood* is defined as

$$L_1(\theta) = p_\theta(v|w).$$

The question of which one is applicable has to be decided on a case-by-case basis. If v and w are independent the two likelihood functions coincide.

In 1922 Fisher used a two-stage maximum likelihood estimation to estimate the error variance in one-way classification problems; a similar argument was used in 1915 for the correlation coefficient, but then the likelihood terminology was not explicit. Suppose $(\widehat{\theta}, \widehat{\eta})$ is the usual MLE of (θ, η). If the distribution of $\widehat{\theta}$ depends only on θ, then a second-stage estimation of θ should be based on $p_\theta(\widehat{\theta})$. This corresponds to a marginal likelihood approach. To see intuitively why the second-stage estimate has less bias, suppose $\widehat{\theta}$ is normal with mean $\theta + b(\theta)/n$. Up to a first-order approximation, the second-stage estimate is $\widehat{\theta} - b(\widehat{\theta})/n$, i.e. it is a bias-corrected estimate.

The marginal or conditional likelihoods are useful if

- $p_\theta(v)$ or $p_\theta(v|w)$ are simpler than the original model $p_{\theta,\eta}(x)$.
- Not much information is lost by ignoring $L_2(\theta, \eta)$.

- The use of full likelihood is inconsistent.

The second condition is usually argued informally on an intuitive basis. Under the last condition the use of marginal or conditional likelihood is essential.

When available, these likelihoods are true likelihoods in the sense that they correspond to a probability of the observed data; this is their main advantage over profile likelihood. However, the problem is that it is not always obvious how to transform the data to arrive at a model that is free of the nuisance parameter.

Example 10.1: continued. To get an unbiased inference for σ^2, consider the following transformations:

$$\begin{aligned} v_i &= (y_{i1} - y_{i2})/\sqrt{2} \\ w_i &= (y_{i1} + y_{i2})/\sqrt{2}. \end{aligned}$$

Clearly v_i's are iid $N(0, \sigma^2)$, and w_i's are iid $N(\mu_i\sqrt{2}, \sigma^2)$. The likelihood of σ^2 based on v_i's is a marginal likelihood, given by

$$L_v(\sigma^2) = \left(\frac{1}{\sqrt{2\pi\sigma^2}}\right)^N \exp\left(-\frac{1}{2\sigma^2}\sum_{i=1}^N v_i^2\right).$$

Since v_i and w_i are independent, in this case it is also a conditional likelihood. Figure 10.2 shows that the marginal likelihood corrects both the bias and over-precision of the profile likelihood. In fact, the MLE from the marginal likelihood is

$$\widehat{\sigma}^2 = \frac{1}{N}\sum_{i=1}^N v_i^2 = \frac{\text{RSS}}{N},$$

the same as the unbiased estimator from the analysis of variance. □

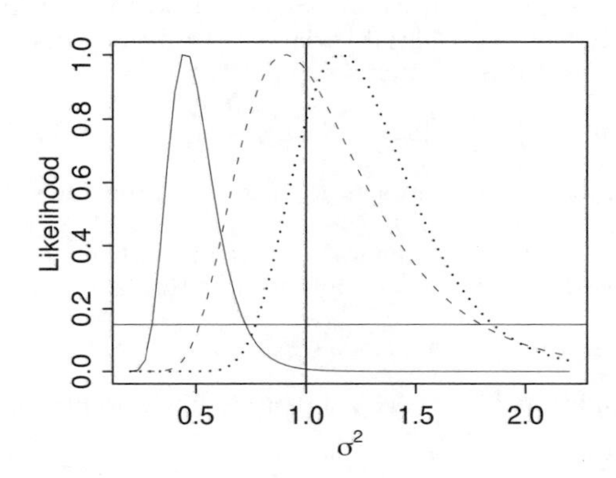

Figure 10.2: *The marginal likelihood (dashed line) corrects the bias of the profile likelihood (solid), with the 'true' likelihood (dotted) shown as comparison.*

Example 10.5: Conditional likelihood is generally available if both the parameter of interest and the nuisance parameter are the natural parameters of an exponential family model. Suppose x is in the $(q+r)$-parameter exponential family with log-density

$$\log p_{\theta,\eta}(x) = \theta' t_1(x) + \eta' t_2(x) - A(\theta,\eta) + c(x),$$

where θ is a q-vector of parameters of interest, and η is an r-vector of nuisance parameters. The marginal log-density of $t_1(x)$ is of the form

$$\log p_{\theta,\eta}(t_1) = \theta' t_1(x) - A(\theta,\eta) + c_1(t_1,\eta),$$

which involves both θ and η. But the conditional density of t_1 given t_2 depends only on θ, according to

$$\log p_\theta(t_1|t_2) = \theta' t_1(x) - A_1(\theta, t_2) + h_1(t_1, t_2),$$

for some (potentially complicated) functions $A_1(\cdot)$ and $h_1(\cdot)$. A simple approximation of the conditional likelihood using the likelihood-based p-formula is given in Section 10.6. □

Example 10.6: Let $y_1, \ldots, y_n$ be independent exponential outcomes with mean $\mu_1, \ldots, \mu_n$, where

$$\frac{1}{\mu_i} = \beta_0 + \beta_1 x_i$$

and x_i is a known predictor. The log-likelihood is

$$\begin{aligned}\log L(\beta_0,\beta_1) &= -\sum_i (\log \mu_i + y_i/\mu_i)\\ &= \sum_i \log(\beta_0 + \beta_1 x_i) - \beta_0 \sum_i y_i - \beta_1 \sum_i y_i x_i,\end{aligned}$$

so we can get a conditional likelihood of β_1. This result can be extended to several predictors. □

Example 10.7: Even in the exponential family, parameters of interest can appear in a form that cannot be isolated using conditioning or marginalizing. Let y_1 and y_2 be independent exponential variates with mean η and $\theta\eta$ respectively; the parameter of interest θ is the mean ratio. Here

$$\log p(y_1, y_2) = -\log\theta - 2\log\eta - y_1/\eta - y_2/(\theta\eta).$$

The parameter of interest is not a natural parameter, and the conditional distribution of y_2 given y_1 is not free of η.

The same problem occurs in general regression with noncanonical link: let $y_1, \ldots, y_n$ be independent exponential outcomes with mean $\mu_1, \ldots, \mu_n$, where

$$\log \mu_i = \beta_0 + \beta_1 x_i.$$

An approximate conditional inference using a modified profile likelihood is given in Section 10.6. □

Information loss

In Example 10.1 it is natural to ask if we lose information by ignoring $\mu_1, \ldots, \mu_N$. Without further assumptions about μ_i's, it seems intuitively clear that little information is lost, but it is not easy to quantify the amount of loss. The upper bound on the information loss is reached when μ_i's are known: using the original data, the expected Fisher information on σ^2 is N/σ^4, and using v_i's alone the Fisher information is $N/(2\sigma^4)$, so the loss is 50%. (Note, however, the quadratic approximation of the log-likelihood of variance parameters is usually poor, except the sample is quite large. This means a comparison based on the Fisher information is meaningful only in large samples.)

Further analysis can be made assuming μ_i's are an iid sample from $N(\mu, \sigma_\mu^2)$, where σ_μ^2 is known. This is now a random effects model, where σ_μ^2 is the variance component parameter for the strata variable i. One can compare the Fisher information of σ^2 under this assumption. Using the result in Section 9.10, the expected Fisher information on σ^2 is

$$\mathcal{I}(\sigma^2) = \frac{N}{2\sigma^4} + \frac{N}{2(\sigma^2 + 2\sigma_\mu^2)^2}.$$

Compared with the Fisher information we get from the marginal likelihood, the proportion of information loss is

$$\frac{1}{1 + \left(1 + 2\frac{\sigma_\mu^2}{\sigma^2}\right)^2}.$$

The ratio σ_μ^2/σ^2 measures the variability between strata relative to within-strata variability; if it is large the information loss is small. For example, if $\sigma^2 = \sigma_\mu^2 = 1$ there is a 10% loss. If there are no strata effects ($\sigma_\mu^2 = 0$) we get the upper bound of 50% loss.

10.4 Comparing Poisson means

Traffic deaths example

In the traffic deaths example (Section 4.6) we assume that the number of deaths x and y are independent Poisson with parameters λ_x and λ_y. The conditional distribution of y given the sum $x + y$ is binomial with parameters $n = x + y$ and probability

$$\pi = \frac{\lambda_y}{\lambda_x + \lambda_y}.$$

Assuming the parameter of interest is $\theta = \lambda_y/\lambda_x$ we have

$$\pi = \frac{\theta}{1 + \theta},$$

which is free of nuisance parameters. The total $x + y$ intuitively carries little or no information about the ratio parameter θ, so on observing y, the conditional likelihood of θ is

$$L(\theta) \equiv \left(\frac{\theta}{1+\theta}\right)^y \left(\frac{1}{1+\theta}\right)^x$$

as we have seen before using the orthogonal parameter or profile likelihood arguments.

This example may be used to illustrate another fundamental difference between the profile and the conditional/marginal likelihoods: the profile likelihood is totally determined by the probability of the observed data, while the latter is affected by the sampling scheme or a contrived rearrangement of the sample space. In the comparison of two Poisson means, the conditional argument produces the binomial distribution as the basis for the likelihood provided we have a standard sample. If parts of the sample space are censored, then the conditional distribution is affected, even if the observed data are not censored.

To be specific, suppose x values greater than five cannot be observed exactly, and in such a case only '$x \geq 5$' is reported; the underlying variate X is assumed Poisson with mean λ_x. Suppose we observe $x = 3$ and $y = 7$, i.e. the data are actually observed exactly; the joint likelihood of (λ_x, λ_y) is

$$L(\lambda_x, \lambda_y) = e^{-\lambda_x}\lambda_x^3 e^{-\lambda_y}\lambda_y^7,$$

and the profile likelihood of θ is

$$L(\theta) = \left(\frac{\theta}{1+\theta}\right)^7 \left(\frac{1}{1+\theta}\right)^3$$

as before. However, the conditional likelihood is no longer available, since the probability $P(X+Y=10)$ and the conditional probability $P(Y = 7|X+Y=10)$ cannot be computed.

Aspirin data example

Let us go back to the aspirin data example in Section 1.1. We assume that the number of heart attacks in the aspirin group x_a is binomial(n_a, θ_a) and that in the placebo group x_p is binomial(n_p, θ_p). We observed $x_a = 139$ from a total $n_a = 11,037$ subjects, and $x_p = 239$ from a total of $n_p = 11,034$. The parameter of interest is $\theta = \theta_a/\theta_p$.

Consider a one-to-one transformation of the data (x_a, x_p) to $(x_a, x_a + x_p)$ and the parameter (θ_a, θ_p) to (θ, θ_p). The likelihood function based on $(x_a, x_a + x_p)$ is

$$\begin{aligned} L(\theta, \theta_p) &= p_{\theta,\theta_p}(x_a, x_a + x_p) \\ &= p_\theta(x_a|x_a + x_p)p_{\theta,\theta_p}(x_a + x_p). \end{aligned}$$

Intuitively, $x_a + x_p$ does not contain much information about θ, so inference about θ can be based on the first term in the likelihood function.

Since θ_a and θ_p are small, we consider the useful approximation that x_a and x_p are Poisson with parameters $n_a\theta_a$ and $n_p\theta_p$, respectively. Therefore, conditionally on $x_a + x_p = t$, x_a is binomial with parameters t and π, where

$$\pi = \frac{n_a\theta_a}{n_a\theta_a + n_p\theta_p} = \frac{n_a\theta}{n_a\theta + n_p}.$$

So the likelihood function of θ based on $X_a|X_a + X_p$ is

$$\begin{aligned} L(\theta) &= P(X_a = x_a|X_a + X_p = x_a + x_p) \\ &= \text{constant} \times \left(\frac{n_a\theta}{n_a\theta + n_p}\right)^{x_a} \left(1 - \frac{n_a\theta}{n_a\theta + n_p}\right)^{x_p}. \end{aligned}$$

This is exactly the profile likelihood of θ we derived previously. See Section 4.7 for exact numerical results and plots of the likelihood function.

10.5 Comparing proportions

Section 4.3 used the profile likelihood for the comparison of two binomial proportions. We now show the conditional likelihood solution.

Suppose we want to compare the proportion of a certain characteristic in two groups. Let x be the number of cases where the characteristic is present in the first group; assume x is binomial $B(m, \pi_x)$; independently we observe y as $B(n, \pi_y)$. We present the data as

	Group 1	Group 2	total
present	x	y	t
absent	$m - x$	$n - y$	u
total	m	n	$m + n$

As the parameter of interest, just as before, we consider the log odds-ratio θ defined by

$$\theta = \log \frac{\pi_x/(1 - \pi_x)}{\pi_y/(1 - \pi_y)}.$$

In terms of θ the hypothesis of interest H_0: $\pi_x = \pi_y$ is equivalent to H_0: $\theta = 0$.

Now we make the following transformation: (x, y) to $(x, x + y)$. The conditional probability of $X = x$ given $X + Y = t$ is

$$P(X = x|X + Y = t) = \frac{P(X = x, X + Y = t)}{P(X + Y = t)}.$$

The numerator is equal to

$$\begin{aligned} & \binom{m}{x}\binom{n}{t-x} \pi_x^x (1 - \pi_x)^{m-x} \pi_y^{t-x} (1 - \pi_y)^{n-t+x} \\ = & \binom{m}{x}\binom{n}{t-x} \left(\frac{\pi_x/(1 - \pi_x)}{\pi_y/(1 - \pi_y)}\right)^x \left(\frac{\pi_y}{1 - \pi_y}\right)^t \end{aligned}$$

$$\times(1-\pi_x)^m(1-\pi_y)^n$$

$$= \binom{m}{x}\binom{n}{t-x} e^{\theta x}\left(\frac{\pi_y}{1-\pi_y}\right)^t (1-\pi_x)^m(1-\pi_y)^n,$$

so the conditional probability is

$$P(X=x|X+Y=t) = \frac{\binom{m}{x}\binom{n}{t-x} e^{\theta x}}{\sum_{s=0}^t \binom{m}{s}\binom{n}{t-s} e^{\theta s}},$$

which is independent of the nuisance parameter. This conditional model is known as the noncentral hypergeometric probability, which is in the exponential family with θ being the canonical parameter. At $\theta = 0$ we obtain the standard hypergeometric probability, which forms the basis of Fisher's exact test (Section 4.3).

The total number of cases $t = x + y$ intuitively does not carry much information about the odds-ratio parameter, so we can use the conditional distribution of X given $X + Y = t$ as the basis for likelihood. Specifically, on observing $X = x$ the conditional likelihood of θ is

$$L(\theta) = \frac{\binom{m}{x}\binom{n}{t-x} e^{\theta x}}{\sum_{s=0}^t \binom{m}{s}\binom{n}{t-s} e^{\theta s}}. \tag{10.2}$$

While the conditional and profile likelihoods are not the same, they are numerically very close, even for small datasets. Figures 10.3(a) and (b) compare the conditional likelihood (dotted line) based on formula (10.2) with the profile likelihood (solid), using the genetic data in Example 4.4.

Series of 2×2 tables

Example 10.2 states the problem of bias in highly stratified binomial data. We can think of the data as a series of 2×2 tables, where each stratum contributes one table. There is a rich body of applications in epidemiology associated with this data structure (see e.g. Breslow and Day 1980).

With so many nuisance parameters for strata effects, bias in the standard MLE of the common odds ratio can accumulate and dominate variability. One solution of the problem is to condition on the margin of each table, so stratum i contributes a likelihood $L_i(\theta)$ given by (10.2) with the corresponding m_i, n_i, t_i and x_i. Conditioning eliminates the strata effects in the original logistic model. The total log-likelihood from all the tables is the sum of individual log-likelihoods:

$$\log L(\theta) = \sum_i \log L_i(\theta).$$

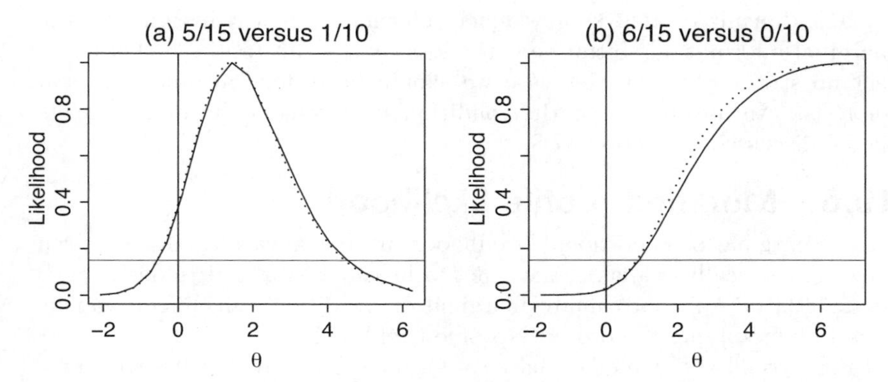

Figure 10.3: *Comparing two binomial proportions: conditional likelihood is very close to profile likelihood. (a) Profile likelihood (solid line) and conditional likelihood (dotted) for the genetic data in Example 4.4. (b) Same as (a) for an extreme case where* $\widehat{\theta} = \infty$.

The score test of H_0: $\theta = 0$ derived from this conditional log-likelihood is known as Mantel–Haenzel test. From Exercise 9.1 it is of the form

$$z = \frac{\sum_i (x_i - \mu_i)}{\{\sum \sigma_i^2\}^{1/2}},$$

where μ_i and σ_i^2 are the mean and variance of the standard hypergeometric random variable x_i:

$$\mu_i = \frac{m_i t_i}{m_i + n_i}, \quad \sigma_i^2 = \frac{m_i n_i t_i (m_i + n_i - t_i)}{(m_i + n_i)^2 (m_i + n_i - 1)}.$$

In the extreme case $m = n = 1$ for each table (e.g. matched-pairs with Bernoulli outcomes), only tables with discordant entries $(x = 1, y = 0)$ or $(x = 0, y = 1)$ contribute to the likelihood. We can think of each pair as a new success–failure outcome (i.e. 'success' $= (x = 1, y = 0)$, 'failure' $= (x = 0, y = 1)$). The null hypothesis of odds ratio equal to one is simply a test of binomial proportion equal to 0.5; this is known as McNemar's test.

For the data in Example 10.2 there are 28 discordant pairs, of which 23 (=82%) belong to treatment $j = 1$. The bias-corrected estimate of the treatment contrast parameter τ_1 is

$$\widehat{\tau}_1 = \log \frac{0.82}{0.18} = 1.52.$$

Using the formula in Example 2.18 the standard error is

$$\text{se}(\widehat{\tau}_1) = \left\{ \frac{1}{23} + \frac{1}{5} \right\}^{0.5} = 0.49.$$

Additionally, the likelihood function of τ_1 can be computed based on the binomial probability.

It is difficult to analyse how much information is lost by ignoring the marginal information. Intuitively, the loss can be substantial if there is in fact no strata effect; in this case we should have done an unconditional analysis. An alternative to the conditional analysis is the mixed effects model discussed in Section 17.8.

10.6 Modified profile likelihood*

Exact marginal or conditional likelihoods are not always available. Even when theoretically available, the exact form may be difficult to derive (Example 10.6). An approximate marginal or conditional likelihood can be found by modifying the ordinary profile likelihood.

First recall the likelihood-based p-formula from Section 9.8 that provides an approximate density of $\widehat{\theta}$:

$$p_\theta(\widehat{\theta}) \approx (2\pi)^{-1/2}|I(\widehat{\theta})|^{1/2}\frac{L(\theta)}{L(\widehat{\theta})}.$$

A better approximation is possible via Barndorff-Nielsen's p^*-formula, which sets a normalizing constant $c(\theta)$ such that the density integrates to one. However, it is less convenient for likelihood approximation since $c(\theta)$ is rarely available. We will use the simpler p-formula, since the normalizing constant is free of θ.

In a multiparameter setting let $(\widehat{\theta}, \widehat{\eta})$ be the MLE of (θ, η); then we have the approximate density

$$p(\widehat{\theta}, \widehat{\eta}) \approx c|I(\widehat{\theta}, \widehat{\eta})|^{1/2}\frac{L(\theta, \eta)}{L(\widehat{\theta}, \widehat{\eta})}$$

where $c = (2\pi)^{p/2}$, and p is the dimensionality of (θ, η). Throughout this section the constant c is free of θ.

The profile likelihood appears naturally in the approximate marginal distribution of $\widehat{\eta}$. In fact, this is the theoretical basis of the construction of modified profile likelihood. Let $\widehat{\eta}_\theta$ be the MLE of η at a fixed value of θ, and $I(\widehat{\eta}_\theta)$ the corresponding observed Fisher information. The profile likelihood of θ is

$$L_p(\theta) = L(\theta, \widehat{\eta}_\theta).$$

The marginal density of $\widehat{\eta}$ is

$$\begin{aligned} p(\widehat{\eta}) &= p(\widehat{\eta}_\theta)\left|\frac{\partial\widehat{\eta}_\theta}{\partial\widehat{\eta}}\right| \\ &\approx c|I(\widehat{\eta}_\theta)|^{1/2}\frac{L(\theta, \eta)}{L(\theta, \widehat{\eta}_\theta)}\left|\frac{\partial\widehat{\eta}_\theta}{\partial\widehat{\eta}}\right|. \end{aligned} \tag{10.3}$$

The conditional distribution of $\widehat{\theta}$ given $\widehat{\eta}$ is

$$p(\widehat{\theta}|\widehat{\eta}) = \frac{p(\widehat{\theta}, \widehat{\eta})}{p(\widehat{\eta})}$$

$$\approx \quad c|I(\widehat{\eta}_\theta)|^{-1/2}\frac{L(\theta,\widehat{\eta}_\theta)}{L(\widehat{\theta},\widehat{\eta})}\left|\frac{\partial\widehat{\eta}}{\partial\widehat{\eta}_\theta}\right|,$$

where we have used the p-formula on both the numerator and the denominator. Hence, the approximate conditional log-likelihood of θ is

$$\begin{aligned}\log L_m(\theta) &= \log L(\theta,\widehat{\eta}_\theta) - \frac{1}{2}\log|I(\widehat{\eta}_\theta)| + \log\left|\frac{\partial\widehat{\eta}}{\partial\widehat{\eta}_\theta}\right| \\ &= \log L_p(\theta) - \frac{1}{2}\log|I(\widehat{\eta}_\theta)| + \log\left|\frac{\partial\widehat{\eta}}{\partial\widehat{\eta}_\theta}\right|. \qquad (10.4)\end{aligned}$$

$L_m(\theta)$ is the required modified profile likelihood. We can arrive at the same formula using a marginal distribution of $\widehat{\theta}$.

The quantity $\frac{1}{2}\log|I(\widehat{\eta}_\theta)|$ can be interpreted as a penalty term, which subtracts from the profile log-likelihood the 'undeserved' information on the nuisance parameter η. The Jacobian term $|\partial\widehat{\eta}/\partial\widehat{\eta}_\theta|$ works as an 'invariance-preserving' quantity, which keeps the modified profile likelihood invariant with respect to transformations of the nuisance parameter. Being a difficult quantity to evaluate, it is a major theoretical hurdle preventing a routine application of (10.4).

The modified profile likelihood formula (10.4) applies in the general setting where $(\widehat{\theta},\widehat{\eta})$ is not sufficient. Suppose there is a one-to-one function of the data x to $(\widehat{\theta},\widehat{\eta},a(x))$, where $a(x)$ is ancillary. To make explicit the dependence on the data, we can write

$$L(\theta,\eta) \equiv L(\theta,\eta;x) = L(\theta,\eta;\widehat{\theta},\widehat{\eta},a).$$

For fixed θ, the MLE $\widehat{\eta}_\theta$ satisfies

$$\frac{\partial}{\partial\widehat{\eta}_\theta}\log L(\theta,\widehat{\eta}_\theta;\widehat{\theta},\widehat{\eta},a) = 0.$$

Taking the derivative with respect to $\widehat{\eta}_\theta$ we obtain

$$\frac{\partial^2}{\partial\widehat{\eta}_\theta^2}\log L(\theta,\widehat{\eta}_\theta;\widehat{\theta},\widehat{\eta},a) + \frac{\partial^2}{\partial\widehat{\eta}_\theta\partial\widehat{\eta}}\log L(\theta,\widehat{\eta}_\theta;\widehat{\theta},\widehat{\eta},a)\frac{\partial\widehat{\eta}}{\partial\widehat{\eta}_\theta} = 0,$$

so

$$\left|\frac{\partial\widehat{\eta}}{\partial\widehat{\eta}_\theta}\right| = \frac{|I(\widehat{\eta}_\theta)|}{\left|\frac{\partial^2}{\partial\widehat{\eta}_\theta\partial\widehat{\eta}}\log L(\theta,\widehat{\eta}_\theta;\widehat{\theta},\widehat{\eta},a)\right|}. \qquad (10.5)$$

The denominator is potentially difficult to get, since we may not have an explicit dependence of the likelihood on $(\widehat{\theta},\widehat{\eta})$.

In lucky situations we might have $\widehat{\eta}_\theta = \widehat{\eta}$, implying $|\partial\widehat{\eta}/\partial\widehat{\eta}_\theta| = 1$, and the last term of (10.4) vanishes. If θ is scalar it is possible to set the nuisance parameter η such that $|\partial\widehat{\eta}/\partial\widehat{\eta}_\theta| \approx 1$ (Cox and Reid 1987). This is achieved by choosing η so that

$$E\frac{\partial^2}{\partial\theta\partial\eta}\log L(\theta,\eta) = 0. \qquad (10.6)$$

Such parameters are called 'information orthogonal'.

It is interesting to compare these results with Bayesian formulae as it suggests how to compute modified likelihoods using Bayesian computational methods. For scalar parameters, the quadratic approximation

$$\log \frac{L(\theta)}{L(\widehat{\theta})} \approx -\frac{1}{2} I(\widehat{\theta})(\theta - \widehat{\theta})^2$$

implies

$$\begin{aligned} \int L(\theta) d\theta &\approx L(\widehat{\theta}) \int e^{-\frac{1}{2} I(\widehat{\theta})(\theta-\widehat{\theta})^2} d\theta \\ &= L(\widehat{\theta})(2\pi)^{1/2} |I(\widehat{\theta})|^{-1/2}. \end{aligned}$$

This is known as Laplace's integral approximation; it is highly accurate if $\log L(\theta)$ is well approximated by a quadratic. For a two-parameter model, we immediately have, for fixed θ, the integrated likelihood

$$L_{int}(\theta) \equiv \int L(\theta, \eta) d\eta \approx c L(\theta, \widehat{\eta}_\theta) |I(\widehat{\eta}_\theta)|^{-1/2},$$

where c is free of θ, so

$$\log L_{int}(\theta) \approx \log L_p(\theta) - \frac{1}{2} \log |I(\widehat{\eta}_\theta)|,$$

exactly the modified profile likelihood in the case of orthogonal parameters.

Example 10.8: Suppose x is in the $(q+r)$-parameter exponential family with log-density

$$\log p_{\theta,\eta}(x) = \theta' t_1(x) + \eta' t_2(x) - A(\theta, \eta) + c(x),$$

where θ is a q-vector of parameters of interest, and η is an r-vector of nuisance parameters. We know that the conditional distribution of t_1 given t_2 is free of η (Example 10.5), but the explicit form can be complicated. The modified profile likelihood provides an explicit, but approximate, conditional likelihood.

The information term is

$$I(\widehat{\eta}_\theta) = \frac{\partial^2}{\partial \widehat{\eta}_\theta^2} A(\theta, \widehat{\eta}_\theta) \equiv A''(\theta, \widehat{\eta}_\theta).$$

To get the Jacobian term, first note that MLE $(\widehat{\theta}, \widehat{\eta})$ satisfies

$$\begin{aligned} t_1 &= \frac{\partial}{\partial \widehat{\theta}} A(\widehat{\theta}, \widehat{\eta}) \\ t_2 &= \frac{\partial}{\partial \widehat{\eta}} A(\widehat{\theta}, \widehat{\eta}) \equiv A'(\widehat{\theta}, \widehat{\eta}). \end{aligned}$$

At fixed value of θ we have

$$t_2 - A'(\theta, \widehat{\eta}_\theta) = 0,$$

so

$$A'(\widehat{\theta},\widehat{\eta}) - A'(\theta,\widehat{\eta}_\theta) = 0,$$

and taking the derivative with respect to $\widehat{\eta}_\theta$, we obtain

$$A''(\widehat{\theta},\widehat{\eta})\frac{\partial\widehat{\eta}}{\partial\widehat{\eta}_\theta} - A''(\theta,\widehat{\eta}_\theta) = 0$$

or

$$\frac{\partial\widehat{\eta}}{\partial\widehat{\eta}_\theta} = \frac{A''(\theta,\widehat{\eta}_\theta)}{A''(\widehat{\theta},\widehat{\eta})}.$$

Hence, up to a constant term, the modified profile likelihood is

$$\log L_m(\theta) = \log L(\theta,\widehat{\eta}_\theta) + \frac{1}{2}\log A''(\theta,\widehat{\eta}_\theta).$$

Ignoring the Jacobian term would have led to a minus sign rather than the correct plus sign on the right-hand side. □

Example 10.9: We observe the following

```
-5.3 -4.5 -1.0 -0.7  3.7  3.9  4.2  5.5  6.8  7.4  9.3
```

and assume that they are an iid sample from $N(\mu,\sigma^2)$ with both parameters unknown. The log-likelihood is

$$\log L(\mu,\sigma^2) = -\frac{n}{2}\log\sigma^2 - \frac{1}{2\sigma^2}\sum_i (x_i-\mu)^2$$

where $n = 11$.

(a) We first find the modified profile likelihood of μ. Given μ the MLE of σ^2 is

$$\begin{aligned}
\widehat{\sigma}^2_\mu &= \frac{1}{n}\sum_i (x_i-\mu)^2 \\
&= \frac{1}{n}\sum_i (x_i-\overline{x})^2 + (\overline{x}-\mu)^2 \\
&= \widehat{\sigma}^2 + (\overline{x}-\mu)^2.
\end{aligned}$$

Immediately,

$$\log L(\mu,\widehat{\sigma}^2_\mu) = -\frac{n}{2}\log\widehat{\sigma}^2_\mu - \frac{n}{2},$$

and

$$I(\widehat{\sigma}^2_\mu) = \frac{n}{2\widehat{\sigma}^4_\mu},$$

and

$$\frac{\partial\widehat{\sigma}^2_\mu}{\partial\widehat{\sigma}^2} = 1.$$

Up to a constant term, the modified profile likelihood is

$$\begin{aligned}
\log L_m(\mu) &= -\frac{n}{2}\log\widehat{\sigma}^2_\mu - \frac{1}{2}\log I(\widehat{\sigma}^2_\mu) \\
&= -\frac{n-2}{2}\log\widehat{\sigma}^2_\mu,
\end{aligned}$$

the same as the profile likelihood based on $(n-2)$ observations.

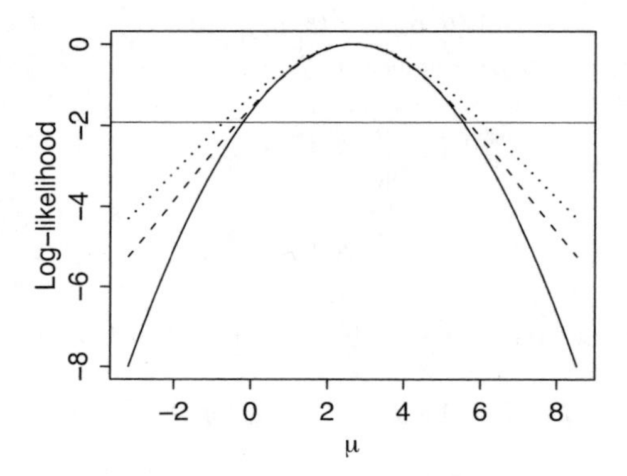

Figure 10.4: *Log-likelihoods of μ: normal approximation (solid), ordinary profile (dashed) and modified profile (dotted).*

Figure 10.4 shows the log-likelihoods of μ based on the normal approximation, profile and modified profile. The normal approximation is equivalent to the quadratic approximation of the profile likelihood. All likelihoods are maximized in the same location, but they have varying precision. The exact 95% CI for μ is the t-interval

$$\overline{x} \pm t_{9,0.025} s/\sqrt{n},$$

producing $-0.60 < \mu < 5.93$. The normal-based interval $-0.21 < \mu < 5.54$ is too narrow; the profile and modified profile likelihood intervals at 15% cutoff are $-0.35 < \mu < 5.67$ and $-0.73 < \mu < 6.06$. In this case the modified likelihood gives the closest likelihood-based interval to the exact interval.

(b) Now suppose σ^2 is the parameter of interest. At each σ^2 we obtain

$$\widehat{\mu}_{\sigma^2} = \overline{x} = \widehat{\mu},$$

so $\partial\widehat{\mu}/\partial\widehat{\mu}_{\sigma^2} = 1$. The Fisher information is

$$I(\widehat{\mu}_{\sigma^2}) = \frac{n}{\sigma^2},$$

so the modified profile likelihood is

$$\log L_m(\sigma^2) = -\frac{n-1}{2}\log\sigma^2 - \frac{(n-1)s^2}{2\sigma^2},$$

exactly the same as the marginal likelihood based on s^2 shown in Example 10.4. □

Example 10.10: With highly stratified data, a modification of the profile likelihood will have a dramatic impact. Suppose y_{ij}'s are independent normal data with mean μ and variance σ_i^2, where $i = 1, \ldots, I$ and $j = 1, \ldots, n_i$. Given μ, the MLEs of the variance parameters are

$$\widehat{\sigma}^2_{i,\mu} = \frac{1}{n_i}\sum_j (y_{ij} - \mu)^2.$$

The profile likelihood of μ is

$$\log L_p(\mu) = -\sum_i \frac{n_i}{2} \log \widehat{\sigma}^2_{i,\mu}.$$

Using the same derivations as in the previous example, the modified profile likelihood is

$$\log L_m(\mu) = -\sum_i \frac{n_i - 2}{2} \log \widehat{\sigma}^2_{i,\mu}.$$

Both likelihoods will produce the same estimate of μ, but different levels of precision. □

Example 10.11: Suppose $y_1, \ldots, y_n$ are independent normal outcomes with means $\mu_1, \ldots, \mu_n$ and common variance σ^2, where

$$\mu_i = x_i'\beta,$$

and x_i is a vector of p predictors. For fixed β the MLE of σ^2 is

$$\begin{aligned}
\widehat{\sigma}^2_\beta &= \frac{1}{n}\sum_i (y_i - x_i'\beta)^2 \\
&= \frac{1}{n}\sum_i (y_i - x_i'\widehat{\beta})^2 + \frac{1}{n}\sum_i (x_i'\widehat{\beta} - x_i'\beta)^2 \\
&= \widehat{\sigma}^2 + \frac{1}{n}\sum_i (x_i'\widehat{\beta} - x_i'\beta)^2,
\end{aligned}$$

which implies $\partial\widehat{\sigma}^2_\beta / \partial\widehat{\sigma}^2 = 1$. The required Fisher information is $I(\widehat{\sigma}^2_\beta) = n/(2\widehat{\sigma}^4_\beta)$ so the modified profile likelihood for β is

$$\log L_m(\beta) = -\frac{n-2}{2} \log \widehat{\sigma}^2_\beta.$$

There is no simple formula for the modified profile likelihood of the individual coefficients.

Using similar derivations as in the one-sample case, the modified profile likelihood of σ^2 is

$$\log L_m(\sigma^2) = -\frac{n-p}{2} \log \sigma^2 - \frac{1}{2\sigma^2} \sum_i (y_i - x_i'\widehat{\beta})^2.$$

This leads not only to the $(n-p)$-divisor for the estimate of σ^2

$$s^2 = \frac{1}{n-p} \sum_i (y_i - x_i'\widehat{\beta})^2,$$

but also to a better likelihood-based inference. In Exercise 10.8 this is extended to the general dispersion parameter. □

Example 10.12: The variance estimation in Example 10.11 is a special case of the general variance components estimation. Suppose an array of outcomes y is normal with mean μ and variance V, where

$$\mu = X\beta$$

for known design matrix X, and $V \equiv V(\theta)$. Let θ be the parameter of interest; it is known as the variance component parameter. The overall likelihood is

$$\log L(\beta, \theta) = -\frac{1}{2} \log |V| - \frac{1}{2}(y - X\beta)'V^{-1}(y - X\beta).$$

Given θ, the MLE of β is the usual weighted least-squares estimate

$$\widehat{\beta}_\theta = (X'V^{-1}X)^{-1}X'V^{-1}y.$$

The profile likelihood of θ is

$$\log L_p(\theta) = -\frac{1}{2} \log |V| - \frac{1}{2}(y - X\widehat{\beta}_\theta)'V^{-1}(y - X\widehat{\beta}_\theta).$$

The observed Fisher information is

$$I(\widehat{\beta}_\theta) = X'V^{-1}X.$$

There is no general formula for $\partial\widehat{\beta}/\partial\widehat{\beta}_\theta$, but we can check that

$$E\left\{\frac{\partial^2}{\partial\beta\partial\theta_i} \log L(\beta, \theta)\right\} = E\left\{X'V^{-1}\frac{\partial V}{\partial\theta_i}V^{-1}(Y - X\beta)\right\} = 0$$

for any θ_i, so β and θ are information orthogonal, and the Jacobian $|\partial\widehat{\beta}/\partial\widehat{\beta}_\theta| \approx 1$. Hence the modified profile likelihood is

$$\log L_m(\theta) = \log L_p(\theta) - \frac{1}{2} \log |X'V^{-1}X|.$$

This matches exactly the so-called restricted maximum likelihood (REML), derived by Patterson and Thompson (1971) and Harville (1974) using the marginal distribution of the error term $y - X\widehat{\beta}_\theta$. See also Harville (1977) for further discussion on normal-based variance component estimation. □

10.7 Estimated likelihood

Suppose the total parameter space is (θ, η), where θ is the parameter of interest. Let $\widehat{\eta}$ be an estimate of η; it can be any reasonable estimate, and in particular it does not have to be an MLE. The estimated likelihood of θ is

$$L_e(\theta) = L(\theta, \widehat{\eta}).$$

Not to be confused with the profile likelihood $L(\theta, \widehat{\eta}_\theta)$, the estimate $\widehat{\eta}$ here is to be estimated free from the parameter of interest θ. Some authors (e.g. Gong and Samaniego 1981) use the term 'pseudo' likelihood, but we will keep the descriptive name 'estimated'.

Example 10.13: Suppose $x_1, \ldots, x_n$ are an iid sample from $N(\mu, \sigma^2)$ with both parameters unknown. Using the sample variance

$$s^2 = \frac{1}{n-1} \sum_i (x_i - \overline{x})^2$$

as a sensible estimate of σ^2, the estimated likelihood of μ is

$$\begin{aligned} L_e(\mu) &= \text{constant} \times \exp\left\{-\frac{1}{2s^2}\sum_i (x_i-\mu)^2\right\} \\ &= \text{constant} \times \exp\left\{-\frac{1}{2}n(\overline{x}-\mu)^2/s^2\right\}. \end{aligned}$$

This is exactly the likelihood based on an iid sample from $N(\mu, s^2)$, i.e. we assume that σ^2 is known at the observed value s^2. □

The estimated likelihood does not account for the extra uncertainty due to the nuisance parameter. For the normal variance parameter and the highly stratified data in Example 10.1 the estimated and profile likelihoods are the same. Inference from the estimated likelihood typically relies on the asymptotic distribution of the estimate $\widehat{\theta}$, the solution of

$$S(\theta, \widehat{\eta}) = \frac{\partial}{\partial\theta} L(\theta, \widehat{\eta}) = 0,$$

accounting for the extra variability from estimating η. The following theorem is due to Gong and Samaniego (1981).

Theorem 10.1 *Assume similar regularity conditions as stated in Section 9.4, and let (θ_0, η_0) be the true parameters.*

(a) If $\widehat{\eta}$ is consistent, then there exists a consistent sequence of solution $\widehat{\theta}$.

(b) If $n^{-1/2}S(\theta_0, \eta_0)$ and $\sqrt{n}(\widehat{\eta}-\eta_0)$ are asymptotically normal with mean zero and covariance matrix

$$\begin{pmatrix} \sigma_{11} & \sigma_{12} \\ \sigma_{12} & \sigma_{22} \end{pmatrix}$$

then $\sqrt{n}(\hat{\theta}-\theta_0)$ is asymptotically normal with mean zero and variance

$$\sigma^2 = \mathcal{I}_{11}^{-1} + \mathcal{I}_{11}^{-2}\mathcal{I}_{12}(\sigma_{22}\mathcal{I}_{12} - 2\sigma_{12}).$$

(c) If the estimate $\widehat{\eta}$ is asymptotically equivalent to the MLE of η_0, then $\sigma_{12} = 0$, $\sigma_{22} = (\mathcal{I}_{22} - \mathcal{I}_{21}\mathcal{I}_{11}^{-1}\mathcal{I}_{12})^{-1}$ and $\sigma^2 = (\mathcal{I}_{11} - \mathcal{I}_{12}\mathcal{I}_{22}^{-1}\mathcal{I}_{21})^{-1}$, so that $\widehat{\theta}$ is asymptotically equivalent to the MLE.

The simplest case occurs if we use the MLE $\widehat{\eta}$ and $\mathcal{I}_{12} = 0$ (θ and η are information orthogonal). Asymptotically we can treat $L(\theta, \widehat{\eta})$ like a standard likelihood; this is a compromise solution in more general cases. In the normal example above, suppose we replace the unknown σ^2 by its MLE $\widehat{\sigma}^2$. Then, inference on μ is the same as if σ^2 is known at $\widehat{\sigma}^2$. In general, if we use an estimate other than the MLE for $\widehat{\eta}$, the covariance σ_{12} is the most difficult quantity to evaluate.

10.8 Exercises

Exercise 10.1: Prove the statement in Example 10.3 that if there exists an orthogonal parameter for θ then, without having to specify η, the profile likelihood computation would automatically provide the likelihood factor $L_1(\theta)$.

Exercise 10.2: Verify the conditional analysis of the data in Example 10.2 as given in Section 10.5. Reanalyse the full data assuming there is no strata effect and compare the results. Discuss the advantages and disadvantages of the conditional analysis.

Exercise 10.3: For the stratified data in Exercise 4.9, compare the profile likelihood of the *common* odds ratio with the conditional likelihood given in Section 10.5. Report also the Mantel-Haenzel test of the odds ratio.

Exercise 10.4: Show that the modified profile likelihood (10.4) is invariant with respect to transformation of the nuisance parameter. For example, define a new nuisance parameter $\psi = g(\eta)$ and show that the modified profile likelihood for θ stays the same, up to a constant term. Note the role of the Jacobian term as an invariance preserver.

Exercise 10.5: Suppose y_1 and y_2 are independent exponentials with mean η and $\theta\eta$; the parameter of interest θ is the mean ratio.

(a) Express the likelihood $L(\theta, \eta)$ as a function of both the parameters and the MLEs:

$$\begin{aligned} L(\theta,\eta) &\equiv L(\theta,\eta;\widehat{\theta},\widehat{\eta}) \\ &= -\log\theta - 2\log\eta - \frac{\widehat{\eta}(\theta+\widehat{\theta})}{\theta\eta}. \end{aligned}$$

(b) Derive the observed Fisher information

$$I(\widehat{\eta}_\theta) = \frac{8\theta^2}{\widehat{\eta}^2(\theta+\widehat{\theta})^2}.$$

(c) Find $\widehat{\eta}_\theta$ in terms of $\widehat{\eta}$ and show that

$$\frac{\partial\widehat{\eta}}{\partial\widehat{\eta}_\theta} = \frac{2\theta}{\theta+\widehat{\theta}}.$$

(d) Show that the modified profile likelihood of θ is the same as the ordinary profile likelihood.

Exercise 10.6: Suppose $y_1, \ldots, y_n$ are independent exponential outcomes with mean $\mu_1, \ldots, \mu_n$, where

$$\log\mu_i = \beta_0 + \beta_1 x_i,$$

and $\sum_i x_i = 0$. Verify the following results:

(a) β_0 and β_1 are information orthogonal.

(b) The profile likelihood of β_1 is

$$\log L_p(\beta_1) = -n\log(\sum_i y_i e^{-\beta_1 x_i}).$$

(c) The Fisher information on $\widehat{\beta}_{0,\beta_1}$ is

$$I(\widehat{\beta}_{0,\beta_1}) = n,$$

so the modified profile likelihood of β_1 is the same as the ordinary profile likelihood.

Exercise 10.7: Let $y_1, \ldots, y_n$ be an iid sample from the gamma distribution with mean θ and shape parameter η. The likelihood of the parameters is

$$\log L(\theta, \eta) = n\eta \log \frac{\eta}{\theta} - n \log \Gamma(\eta) + \eta \sum \log y_i - \frac{\eta}{\theta} \sum_i y_i.$$

Let $D(\eta) = \partial \log \Gamma(\eta)/\partial \eta$. Verify the following results:

(a) We can express the likelihood in terms of the MLEs by using

$$\sum_i \log y_i = n \log \frac{\widehat{\eta}}{\widehat{\theta}} - nD(\widehat{\eta})$$
$$\sum y_i = n\widehat{\theta}.$$

(b) The required Fisher information is

$$I(\widehat{\eta}_\theta) = n\{D'(\widehat{\eta}_\theta) - 1/\widehat{\eta}_\theta\}.$$

(c) The Jacobian term is

$$\frac{\partial \widehat{\eta}}{\partial \widehat{\eta}_\theta} = \frac{I(\widehat{\eta}_\theta)}{I(\widehat{\eta})},$$

so the modified profile likelihood of θ is

$$\log L_m(\theta) = \log L(\theta, \widehat{\eta}_\theta) + \frac{1}{2} \log I(\widehat{\eta}_\theta).$$

Exercise 10.8: Assuming the exponential dispersion model, the log-likelihood contribution from a single observation y_i is

$$\log L(\theta_i, \phi; y_i) = \{y_i\theta_i - A(\theta_i)\}/\phi + c(y_i, \phi).$$

In Sections 4.9 and 6.6 we describe an approximation

$$\log L(\theta_i, \phi; y_i) \approx -\frac{1}{2} \log\{2\pi\phi v(y_i)\} - \frac{1}{2\phi} D(y_i, \mu_i),$$

where

$$D(y_i, \mu_i) = 2 \log \frac{L(y_i, \phi = 1; y_i)}{L(\mu_i, \phi = 1; y_i)},$$

and $L(\mu_i, \phi = 1; y_i)$ is the likelihood of μ_i based on a single observation y_i, assuming $\phi = 1$. Assuming a regression model $h(\mu_i) = x_i'\beta$, where $\beta \in R^p$, show that the modified profile log-likelihood of ϕ is

$$\log L(\phi) \approx -\frac{n-p}{2} \log(2\pi\phi) - \frac{1}{2\phi} \sum_i D(y_i, \widehat{\mu}_i).$$

This justifies the $(n - p)$ divisor for the estimate of ϕ.

Exercise 10.9: Compare the profile and the estimated likelihoods in the two-sample Poisson and binomial examples in Sections 10.4 and 10.5.

Exercise 10.10: Provide a rough proof of Theorem 10.1

11
Complex data structures

The focus in this chapter is on problems with complex data structure, in particular those involving dependence and censoring. Modelling is crucial for simplifying and clarifying the structure. Compromises with the likelihood, for example using the marginal, conditional or estimated likelihood, also become a necessity. With some examples we discuss the simplest thing we can do with the data, with as little modelling as possible. This will provide a safety check for results based on more complex models; similar results would give some assurance that the complex model is not off the mark. It is important to recognize in each application the price and the reward of modelling.

11.1 ARMA models

So far we have considered independent observations, where we only need to model the probability of single outcomes. From independence, the joint probability is simply the product of individual probabilities; this is not true with dependent data. Suppose we observe a time series $x_1, \ldots, x_n$; then, in general, their joint density can be written as

$$p_\theta(x_1, \ldots, x_n) = p_\theta(x_1)\ p_\theta(x_2|x_1) \ldots p_\theta(x_n|x_1, \ldots, x_{n-1}).$$

Difficulty in modelling arises as the list of conditioning variables gets long. Generally, even for simple parametric models, the likelihood of time series models can be complicated. To make the problems tractable we need to assume some special structures. The main objective of standard modelling assumptions is to limit the length of the conditioning list, while capturing the dependence in the series.

A minimal requirement for time series modelling is weak stationarity, meaning that x_t has a constant mean, and the covariance $\text{cov}(x_t, x_{t-k})$ is only a function of lag k. To make this sufficient for likelihood construction, we usually assume a Gaussian model.

A large class of time series models that leads to a tractable likelihood is the class of autoregressive (AR) models. A time series is said to be an AR(p) series if

$$x_t = \theta_0 + \phi_1 x_{t-1} + \cdots + \phi_p x_{t-p} + e_t,$$

where the so-called innovation or driving noise e_t is assumed to be an iid series. The stationarity and correlation structure of the time series are determined solely by the parameters $(\phi_1, \ldots, \phi_p)$. Likelihood analysis of AR models typically assumes e_t's are iid normal.

Example 11.1: An AR(1) model specifies

$$x_t = \theta_0 + \phi_1 x_{t-1} + e_t,$$

where e_t's are iid $N(0, \sigma^2)$. This is equivalent to stating the conditional distribution of x_t given its past is normal with mean $\theta_0 + \phi_1 x_{t-1}$ and variance σ^2. Given $x_1, \ldots, x_n$, the likelihood of the parameter $\theta = (\theta_0, \phi_1, \sigma^2)$ is

$$\begin{aligned} L(\theta) &= p_\theta(x_1) \prod_{t=2}^{n} p_\theta(x_t | x_u, u < t) \\ &= p_\theta(x_1) \prod_{t=2}^{n} p_\theta(x_t | x_{t-1}) \\ &= p_\theta(x_1) \prod_{t=2}^{n} (2\pi\sigma^2)^{-1/2} \exp\left\{ -\frac{1}{2\sigma^2}(x_t - \theta_0 - \phi_1 x_{t-1})^2 \right\} \\ &\equiv L_1(\theta)\, L_2(\theta), \end{aligned}$$

where $L_1(\theta) = p_\theta(x_1)$. The term $L_2(\theta)$ is a conditional likelihood based on the distribution of $x_2, \ldots, x_n$ given x_1. This conditional likelihood is commonly assumed in routine data analysis; it leads to the usual least-squares computations.

How much information is lost by ignoring x_1? Assuming $|\phi_1| < 1$, so the series is stationary with mean μ and variance σ_x^2, we find

$$Ex_t = \theta_0 + \phi_1 E x_{t-1}$$

or $\mu = \theta_0/(1 - \phi_1)$. Iterated expansion of x_{t-1} in terms its past values yields

$$x_t = \mu + \sum_{i=0}^{\infty} \phi_1^i e_{t-i}$$

so

$$\text{var}(x_t) = \sigma^2 \sum_{i=0}^{\infty} \phi_1^{2i}$$

or $\sigma_x^2 = \sigma^2/(1 - \phi_1^2)$. Therefore the likelihood based on x_1 is

$$L_1(\theta) = (2\pi\sigma^2)^{-1/2}(1 - \phi_1^2)^{1/2} \exp\left[-\frac{1 - \phi_1^2}{2\sigma^2}\{x_1 - \theta_0/(1 - \phi_1)\}^2 \right].$$

Hence x_1 creates a nonlinearity in the estimation of ϕ_1. If ϕ_1 is near one, which is the boundary of nonstationarity, the effect of nonlinearity can be substantial.

Figure 11.1 shows the likelihood of ϕ for simulated data with $n = 100$. For simplicity it is assumed $\theta_0 = 0$ and $\sigma^2 = 1$, the true values used in the simulation. For the time series in Figure 11.1(a), the full and conditional likelihoods show little difference. But when $\widehat{\phi}$ is near one, as shown in Figure 11.1(d), the difference is evident. The curvatures of the likelihoods at the maximum are similar, so we cannot measure the effect of x_1 using the Fisher information. □

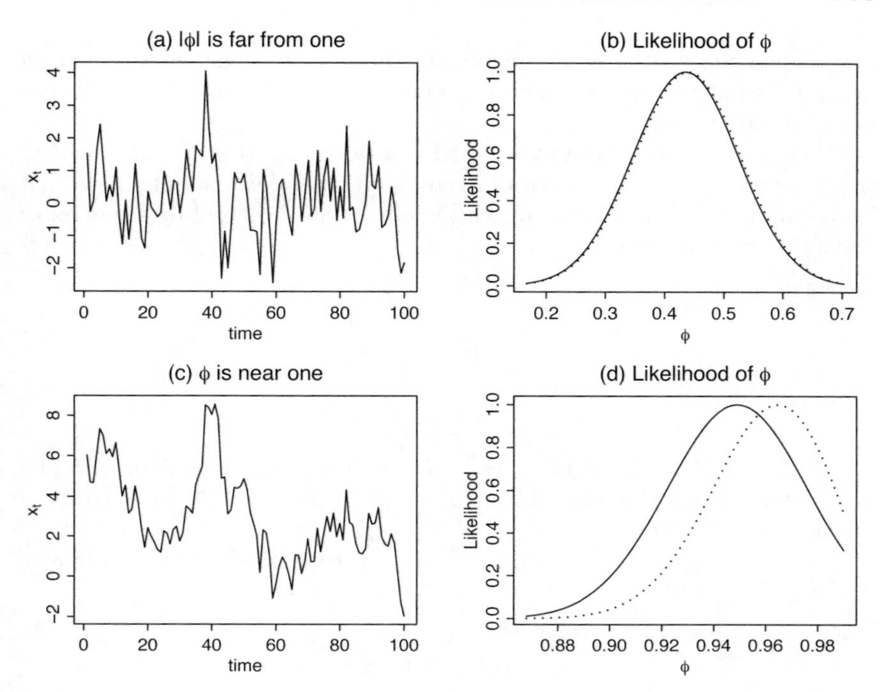

Figure 11.1: *(a) Simulated AR(1) time series, where $\widehat{\phi}$ is far from the boundary of nonstationarity. (b) The conditional (solid line) and full likelihood (dotted) of ϕ based on the time series data in (a). (c) and (d) The same as (a) and (b) for $\widehat{\phi}$ near one.*

A rich class of parametric time series models commonly used in practice is the autoregressive-moving average (ARMA) models (Box *et al.* 1994). A time series x_t that follows an ARMA(p, q) model can be represented as

$$x_t = \theta_0 + \phi_1 x_{t-1} + \cdots + \phi_p x_{t-p} + e_t - \theta_1 e_{t-1} - \cdots - \theta_q e_{t-q},$$

where the e_t's are an iid series. Modelling time series data is more difficult than standard regression analysis, since we are not guided by any meaningful relationship between variables. There are a number of descriptive tools to help us, such as the autocorrelation and partial autocorrelation functions. Given a particular model choice, the derivation of a full Gaussian likelihood is tedious, but there are some fast algorithms based on the so-called state-space methodology (Mélard 1984).

11.2 Markov chains

A time series x_t is a first-order Markov chain if

$$p_\theta(x_t | x_1, \ldots, x_{t-1}) = p_\theta(x_t | x_{t-1}),$$

i.e. x_t depends on the past values only through x_{t-1}, or, conditional on x_{t-1}, x_t is independent of the past values. An AR(1) model is a Markov model of order one.

The simplest but still useful model is a two-state Markov chain, where the data are a dependent series of zeros and ones; for example, $x_t = 1$ if it is raining and zero otherwise. This chain is characterized by a matrix of transition probabilities

		x_t	
		0	1
x_{t-1}	0	θ_{00}	θ_{01}
	1	θ_{10}	θ_{11}

where $\theta_{ij} = P(X_t = j|X_{t-1} = i)$, for i and j equal to 0 or 1. The parameters satisfy the constraints $\theta_{00} + \theta_{01} = 1$ and $\theta_{10} + \theta_{11} = 1$, so there are two free parameters.

On observing time series data $x_1, \ldots, x_n$, the likelihood of the parameter $\theta = (\theta_{00}, \theta_{01}, \theta_{10}, \theta_{11})$ is

$$\begin{aligned} L(\theta) &= p_\theta(x_1) \prod_{t=2}^{n} p(x_t|x_{t-1}) \\ &= p_\theta(x_1) \prod_{t=2}^{n} \theta_{x_{t-1}0}^{1-x_t} \theta_{x_{t-1}1}^{x_t} \\ &= p_\theta(x_1) \prod_{ij} \theta_{ij}^{n_{ij}} \\ &\equiv L_1(\theta) L_2(\theta) \\ &\equiv L_1(\theta) L_{20}(\theta_{01}) L_{21}(\theta_{11}) \end{aligned}$$

where n_{ij} is the number of transitions from state i to state j. For example, if we observe a series

$$0 \;\; 1 \;\; 1 \;\; 0 \;\; 0 \;\; 0$$

then $n_{00} = 2$, $n_{01} = 1$, $n_{10} = 1$ and $n_{11} = 1$. Again, here it is simpler to consider the conditional likelihood given x_1. The first term $L_1(\theta) = p_\theta(x_1)$ can be derived from the stationary distribution of the Markov chain, which requires a specialized theory (Feller 1968, Chapter XV).

The likelihood term $L_2(\theta)$ in effect treats all the pairs of the form (x_{t-1}, x_t) as if they are independent. All pairs with the same x_{t-1} are independent Bernoulli trials with success probability $\theta_{x_{t-1}1}$. Conditional on x_1, the free parameters θ_{01} and θ_{11} are orthogonal parameters, allowing separate likelihood analyses via

$$L_{20}(\theta_{01}) = (1 - \theta_{01})^{n_{00}} \theta_{01}^{n_{01}}$$

and

$$L_{21}(\theta_{11}) = (1-\theta_{11})^{n_{10}}\theta_{11}^{n_{11}}.$$

We would use this, for example, in (logistic) regression modelling of a Markov chain. In its simplest structure the conditional MLEs of the parameters are

$$\widehat{\theta}_{ij} = \frac{n_{ij}}{n_{i0}+n_{i1}}.$$

Example 11.2: Figure 11.2(a) shows the plot of asthma attacks suffered by a child during 110 winter-days. The data (read by row) are the following:

```
0 0 0 0 0 1 0 0 0 0 0 0 0 0 0 0 0 0 0 0 1 1 0
0 0 0 0 1 1 0 0 0 0 1 1 1 1 1 1 1 0 0 0 0 0
0 1 1 0 0 0 0 0 1 1 0 0 0 0 0 0 0 0 0 0 0 0
0 0 0 0 0 0 0 0 0 1 0 0 0 0 0 0 0 0 0 0 0 1
1 0 0 0 0 0 0 0 0 0 0 0 0 0 0 0 0 0 0 0 0 0
```

Assuming a first-order Markov model the observed transition matrix is

		x_t 0	1	Total
x_{t-1}	0	82	8	90
	1	8	11	19

For example, when the child is healthy today the estimated probability of an attack tomorrow is $\widehat{\theta}_{01} = 8/90 = 0.09$ (se $= 0.03$); if there is an attack today, the probability of another attack tomorrow is $\widehat{\theta}_{11} = 11/19 = 0.58$ (se $= 0.11$).

We can check the adequacy of the first-order model by extending the model to a second or higher order, and comparing the likelihoods of the different models; see Section 9.11. Derivation of the likelihood based on the higher-order models is left as an exercise. □

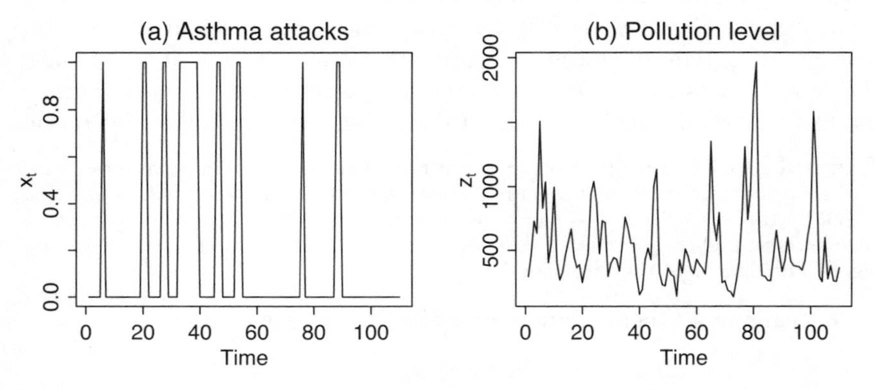

Figure 11.2: *A time series of asthma attacks and a related pollution series.*

Regression analysis

Figure 11.2(b) shows a time series of smoke concentration z_t, measured over the same 110-day period as the asthma series. The values of time series z_t are (read by row)

291	466	730	633	1509	831	1038	403	553	996	400	267	329
466	570	668	446	362	387	246	361	467	940	1041	871	473
732	717	294	396	443	429	336	544	760	672	555	556	298
150	192	428	517	425	1000	1135	322	228	220	360	310	294
138	425	322	512	453	352	317	430	389	357	314	544	1353
720	574	796	246	260	184	172	133	261	406	770	1310	742
976	1701	1965	646	301	295	263	261	450	657	486	333	419
600	415	380	374	370	344	418	617	749	1587	1157	297	253
601	276	380	260	256	363							

The association between asthma attacks and pollution level can be investigated by first-order Markov modelling. For example, we can model

$$\text{logit } \theta_{01t} = \beta_0 + \beta_1 \text{ pollution}_t,$$

so the pollution level modifies the transition probabilities. This model may be analysed using standard logistic regression analysis where the outcome data are pairs of (0,0)'s as failures and (0,1)'s as successes. A simpler model would have been

$$\text{logit } P(x_t = 1) = \beta_0 + \beta_1 \text{ pollution}_t,$$

but in this case the dependence in the asthma series has not been modelled. The actual estimation of the model is left as an exercise.

11.3 Replicated Markov chains

Classical applications of time series, for example in engineering, business or economics, usually involve an analysis of one long time series. New applications in biostatistics have brought short time series data measured on many individuals; they are called repeated measures or longitudinal data. The questions typically focus on comparison between groups of individuals; the dependence structure in the series is a nuisance rather than a feature of interest, and modelling is necessary to attain efficiency or correct inference.

Example 11.3: In a clinical trial of ulcer treatment, 59 patients were randomized to control or treatment groups. These patients were evaluated at baseline (week 0) and at weeks 2, 4, 6 and 8, with symptom severity coded on a six-point scale, where higher is worse. Table 11.1 shows the outcome data. Is there any significant benefit of the treatment? □

In real studies there are potential complications such as

- some follow-up data may be missing,
- length of follow-up differs between subjects, and
- there could be other covariates of interest such as age, sex, etc.

The proper likelihood treatment of these problems is left as Exercises 11.12, 11.13 and 11.15.

The simplest analysis can be obtained by assessing the improvement at the last visit (w8) relative to the baseline (w0):

Control group						Treated group					
No.	w0	w2	w4	w6	w8	No.	w0	w2	w4	w6	w8
1	3	4	4	4	4	1	4	4	3	3	3
2	5	5	5	5	5	2	4	4	3	3	3
3	2	2	2	1	1	3	6	5	5	4	3
4	6	6	5	5	5	4	3	3	3	3	2
5	4	3	3	3	3	5	6	5	5	5	5
6	5	5	5	5	5	6	3	3	3	2	2
7	3	3	3	3	3	7	5	5	6	6	6
8	3	3	3	2	2	8	3	2	2	2	2
9	4	4	4	3	2	9	4	3	2	2	2
10	5	4	4	3	4	10	5	5	5	5	5
11	4	4	4	4	4	11	3	2	2	2	2
12	4	4	4	4	4	12	3	2	2	1	1
13	5	5	5	5	5	13	6	6	6	6	5
14	5	4	4	4	4	14	2	2	1	1	2
15	5	5	5	4	3	15	4	4	4	3	2
16	3	2	2	2	2	16	3	3	3	3	3
17	5	5	6	6	6	17	2	1	1	1	2
18	6	6	6	6	6	18	4	4	4	4	4
19	3	2	2	2	2	19	4	4	4	4	3
20	5	5	5	5	5	20	4	4	3	3	3
21	4	4	4	4	3	21	4	4	4	3	3
22	2	2	2	2	2	22	4	3	2	2	2
23	4	3	3	3	3	23	3	2	2	2	2
24	3	2	2	1	1	24	3	3	3	3	3
25	4	3	3	3	3	25	3	2	2	1	1
26	4	3	3	3	3	26	4	4	3	3	3
27	5	5	5	5	5	27	2	1	1	1	1
28	3	3	3	3	3	28	3	2	3	2	1
29	3	3	3	3	2	29	2	1	1	1	1
30	6	6	6	6	6						

Table 11.1: *Follow-up data on treatment of ulcers.*

Change	Control	Treated	Total
Better	16(53%)	22(76%)	38
No	14	7	21
Total	30	29	59

This table indicates some positive benefit. However, the standard χ^2 statistic for the observed 2×2 table is

$$\chi^2 = \frac{(16 \times 7 - 22 \times 14)^2 59}{30 \times 29 \times 38 \times 21} = 3.29,$$

with 1 degree of freedom; this gives a (one-sided) P-value=0.07, which is not quite significant.

How much has been lost by ignoring most of the data? To include the whole dataset in the analysis we need to consider a more complicated model. Assuming a first-order Markov model

$$P(x_k = j|x_t, t \leq k-1) = P(x_k = j|x_{k-1})$$

would spawn a 6×6 transition matrix with $6 \times (6-1) = 30$ independent parameters *for each group*; using this most general model, it would not be obvious how to compare the treatment versus the placebo groups.

This can be simplified by assuming a patient can only change by one level of severity in a two-week period. The transition probability from state i to state j is given by, for $i = 2, \ldots, 5$,

$$p_{ij} = \begin{cases} p & \text{if } j = i+1 \text{ (worse)} \\ q & \text{if } j = i-1 \text{ (better)} \\ 1-p-q & \text{if } j = i \text{ (same)} \\ 0 & \text{otherwise.} \end{cases}$$

The model is completed by specifying the boundary probabilities

$$p_{1j} = \begin{cases} p & \text{if } j = 2 \\ 1-p & \text{if } j = 1 \\ 0 & \text{otherwise.} \end{cases}$$

$$p_{6j} = \begin{cases} q & \text{if } j = 5 \\ 1-q & \text{if } j = 6 \\ 0 & \text{otherwise.} \end{cases}$$

In view of the model, the data can be conveniently summarized in the following table

Change	Control	Treated	Transition probability
1→ 1	2	11	$1-p$
1→ 2	0	2	p
6→ 6	11	5	$1-q$
6→ 5	1	3	q
other +1	3	2	p
other 0	84	63	$1-p-q$
other −1	19	30	q
Total	120	116	

where 'other +1' means a change of +1 from the states $2, \ldots, 5$. The data satisfy the assumption of no jump of more that one point; extension that allows for a small transition probability of a larger jump is left as an exercise. The likelihood of the parameters p and q for each group can be computed based on the table (Exercise 11.11). The two groups can now be compared, for example, in terms of the parameter q alone or the difference $q - p$.

If we are interested in comparing the rate of improvement we can compare the parameter q; for this we can reduce the data by conditioning on starting at states 2 or worse, so we can ignore the first two lines of the table and combine the rest into a 2×2 table

Change	Control	Treated	Total
Better (-1)	20(17%)	33(32%)	53
No (≥ 0)	98	70	168
Total	118	103	221

which now yields $\chi^2 = 6.87$ with P-value=0.009. The interpretation is different from the first analysis: there is strong evidence that the treatment has a short-term benefit in reducing symptoms.

11.4 Spatial data

Spatial data exhibit dependence like time series data, with the main difference that there is no natural direction in the dependence or time-causality. Spatially dependent data exhibit features such as

- clustering or smooth variation: high values tend to appear near each other; this is indicative of a local positive dependence;
- almost-regular pattern: high or low values tend to appear individually, indicating a negative spatial dependence or a spatial inhibition;
- streaks: positive dependence occurs along particular directions. Long streaks indicate global rather than local dependence.

We should realize, however, that a completely random pattern must show some local clustering. This is because if there is no clustering at all we will end up with a regular pattern. Except in extreme cases, our eyes are not good at judging if a particular clustering is real or spurious.

In many applications it is natural to model the value at a particular point in space in terms of the surrounding values. Modelling of spatial data is greatly simplified if the measurements are done on a regular grid or lattice structure, equivalent to equal-space measurement of time series. However, even for lattice-structured data the analysis is marked by compromises.

The spatial nature of an outcome y ('an outcome' here is a whole image) can be ignored if we model y in terms of spatially varying covariates x. Conditional on x we might model the elements of y as being independent; this assumes any spatial dependence in y is inherited from that in x. For example, the yield y of a plant may be spatially dependent, but the dependence is probably induced by fertility x. With such a model the spatial dependence is assumed to be fully absorbed by x, and it can be left unmodelled. So the technique we discuss in this section is useful if we are interested in the dependence structure itself, or if there is a residual dependence after conditioning on a covariate x.

One-dimensional Ising model

To keep the discussion simple, consider a linear lattice or grid system where we observe an array $y_1, \ldots, y_n$. What is a natural model to describe its probabilistic behaviour? We can still decompose the joint probability as

$$p(y_1, \ldots, y_n) = p(y_1)\ p(y_2|y_1) \ldots p(y_n|y_{n-1}, \ldots),$$

but such a decomposition is no longer natural, since it has an implied left-to-right direction. There are generalizations of the ARMA models to spatial processes. The key idea is that the probabilistic behaviour of the process at a particular location is determined by the nearby values or neighbours.

For example, we might want to specify a first-order model

$$p(y_k|y_1,\ldots,y_{k-1},y_{k+1},\ldots,y_n) = p(y_k|y_{k-1},y_{k+1}).$$

The problem is that, except for the Gaussian case, it is not obvious how to build the likelihood from such a specification. The product of such conditional probabilities for all values $y_1,\ldots,y_n$ is not a true likelihood. In its simplest form, if we observe (y_1,y_2), the product of conditional probabilities is

$$p(y_1|y_2)\,p(y_2|y_1),$$

while the true likelihood is

$$p(y_1)\,p(y_2|y_1).$$

To simplify the problem further, suppose y_i can only take 0–1 values. The famous Ising model in statistical mechanics specifies the joint distribution of $y_1,\ldots,y_n$ as follows. Two locations j and k are called neighbours if they are adjacent, i.e. $|j-k|=1$; for example, the neighbours of location $k=3$ are $j=2$ and $j=4$. Let n_k be the sum of y_j's from neighbours of k. The Ising model specifies that, conditional on the edges y_1 and y_n, the joint probability of $y_2,\ldots,y_{n-1}$ is

$$p(y_2,\ldots,y_{n-1}|y_1,y_n) = \exp\left\{\alpha\sum_{k=2}^{n-1} y_k + \beta\sum_{k=2}^{n-1} y_k n_k - h(\alpha,\beta)\right\},$$

where $h(\alpha,\beta)$ is a normalizing constant. The parameter β measures the local interaction between neighbours. If $\beta=0$ then $y_2,\ldots,y_{n-1}$ are iid Bernoulli with parameter e^α. Positive β implies positive dependence, so that values of y_i's tend to cluster; this will be obvious from the conditional probability below.

It is not easy to compute the true likelihood from the joint probability, since the normalizing factor $h(\alpha,\beta)$ is only defined implicitly. However, we can show (Exercise 11.16) that the conditional probability of y_k given all of the other values satisfies the logistic model

$$P(y_k=1|y_1,\ldots,y_{k-1},y_{k+1},\ldots,y_n) = \frac{\exp(\alpha+\beta n_k)}{1+\exp(\alpha+\beta n_k)},$$

so it is totally determined by the local neighbours. It is tempting to define a likelihood simply as the product of the conditional probabilities. That is in fact the common approach for estimating the Ising model, namely we use

$$L(\alpha, \beta) = \prod_{k=2}^{n-1} p(y_k | y_1, \ldots, y_{k-1}, y_{k+1}, \ldots, y_n),$$

known as *pseudo-likelihood* (Besag 1974, 1975); it is a different likelihood from all we have defined previously, and inference from it should be viewed with care. Besag and Clifford (1989) developed an appropriate Monte Carlo test for the parameters. The practical advantage of the pseudo-likelihood is obvious: parameter estimation can be performed using standard logistic regression packages.

Example 11.4: As a simple illustration of the Ising model, consider analysing the dependence structure in the following spatial data:

0 0 0 1 1 1 0 0 1 0 0 0 1 1 0 1 1 1 0 1 0 0 1 1 0 0 1 1 1 1

The data for the logistic regression can be set up first in terms of data pairs (y_k, n_k) for $k = 2, \ldots, 29$. There are 30 total points in the data, and we have 28 points with two neighbours. Thus $(y_2 = 0, n_2 = 0)$, $(y_3 = 0, n_3 = 1)$, $\ldots$, $(y_{29} = 1, n_{29} = 2)$ and we can group the data into

n_k	$y_k = 0$	$y_k = 1$	Total
0	2	2	4
1	9	9	18
2	2	4	6

Estimating the logistic regression model with n_k as the predictor, the maximum pseudo-likelihood estimates and the standard errors of the parameters (α, β) are given in the following:

Effect	Parameter	Estimate	se
Constant	α	−0.27	0.79
n_k	β	0.38	0.65

The analysis shows that there is no evidence of local dependence in the series. □

Two-dimensional Ising model

Extension to true spatial data with a two-dimensional grid structure (i, j) is as follows. Suppose we observe a two-dimensional array y_{ij} of 0–1 values. The most important step is the definition of neighbourhood structure. For example, we may define the locations (i, j) and (k, l) as 'primary' neighbours if $|i - k| = 1$ and $j = l$, or $|j - l| = 1$ and $i = k$; the 'diagonal' or 'secondary' neighbours are those with $|i - k| = 1$ and $|j - l| = 1$. (Draw these neighbours to get a clear understanding.) Depending on the applications we might treat these different neighbours differently. Let n_{ij} be the sum of y_{ij}'s from the primary neighbours of location (i, j), and m_{ij} be the sum from the diagonal neighbours. Then a general Ising model of the joint distribution of y_{ij} (conditional on the edges) implies a logistic model

$$P(y_{ij} = 1 | \text{other } y_{ij}\text{'s}) = \frac{\exp(\alpha + \beta n_{ij} + \gamma m_{ij})}{1 + \exp(\alpha + \beta n_{ij} + \gamma m_{ij})}.$$

Estimation of the parameters based on the pseudo-likelihood can be done using standard logistic regression packages.

Gaussian models

To describe the Gaussian spatial models it is convenient to first vectorize the data into $y = (y_1, \ldots, y_n)$. The neighbourhood structure can be preserved in an $n \times n$ matrix W with elements $w_{ii} = 0$, and $w_{ij} = 1$ if i and j are neighbours, and zero otherwise. The domain of y can be irregular.

The Gaussian conditional autoregressive (CAR) model specifies that the conditional distribution of y_i given the other values is normal with mean and variance

$$\begin{aligned} E(y_i|y_j,\ j \neq i) &= \mu_i + \sum_j c_{ij}(y_j - \mu_j) \\ \text{var}(y_i|y_j,\ j \neq i) &= \sigma_i^2. \end{aligned}$$

Let $C \equiv [c_{ij}]$ with $c_{ii} = 0$ and $D \equiv \text{diag}[\sigma^2]$. The likelihood can be derived from the unconditional distribution of y, which is $N(\mu, \Sigma)$ with

$$\begin{aligned} \mu &= (\mu_1, \ldots, \mu_n) \\ \Sigma &= (I_n - C)^{-1} D, \end{aligned}$$

provided Σ is a symmetric positive definite matrix; I_n is an $n \times n$ identity matrix. The symmetry is satisfied if $c_{ij}\sigma_j^2 = c_{ji}\sigma_i^2$. For modelling purposes, some simplifying assumptions are needed, such as equal variance and a first-order model, which specifies $c_{ij} = \phi$ if i and j are neighbours and zero otherwise. Some model fitting examples can be found in Ripley (1988).

An important application of spatial data analysis is in smoothing sparse disease maps, where the raw data exhibit too much noise for sensible reading. The input for these maps are count data, which are not covered by the Ising model. However, the pseudo-likelihood approach can be extended for such data (Exercise 11.20). The problem can also be approached using mixed model in Section 18.10. For this purpose it is useful to have a nonstationary process for the underlying smooth function. Besag *et al.* (1991) suggest a nonstationary CAR model defined by the joint distribution of a set of differences. The log-density is

$$\begin{aligned} \log p(y) &= -\frac{N}{2}\log(2\pi\sigma^2) - \frac{1}{2\sigma^2}\sum_{ij} w_{ij}(y_i - y_j)^2 \\ &= -\frac{N}{2}\log(2\pi\sigma^2) - \frac{1}{2\sigma^2} y' R^{-1} y \end{aligned}$$

where $N = \sum w_{ij}$ is the number of neighbour pairs and $\sigma^{-2}R^{-1}$ is the inverse covariance matrix of y. It can be shown that $R^{-1} = [r^{ij}]$, where $r^{ii} = \sum_j w_{ij}$ and $r^{ij} = -w_{ij}$ if $i \neq j$.

11.5 Censored/survival data

In most clinical trials or reliability studies it is not possible to wait for all experimental units to reach their 'end-point'. An end-point in a survival study is the time of death, or appearance of a certain condition; in general it is the event that marks the end of follow-up for a subject. Subjects are said to be censored if they have not reached the end-point when the study is stopped, or are lost during follow-up. In more general settings, censored data are obtained whenever the measurement is not precise; for example, a binomial experiment where (i) the number of successes is known only to be less than a number, or (ii) the data have been grouped.

Example 11.5: Two groups of rats were exposed to carcinogen DBMA, and the number of days to death due to cancer was recorded (Kalbfleisch and Prentice 1980).

$$\begin{array}{ll} \text{Group 1}: & 143, 164, 188, 188, 190, 192, 206, 209, 213, \\ & 216, 220, 227, 230, 234, 246, 265, 304, 216+, 244+ \\ \text{Group 2}: & 142, 156, 163, 198, 205, 232, 232, 233, 233, \\ & 233, 233, 239, 240, 261, 280, 280, 296, 296, \\ & 323, 204+, 344+ \end{array}$$

Values marked with '+' are censored. Is there a significant difference between the two groups? □

In this example four rats were 'censored' at times 216, 244, 204 and 344; those rats were known not to have died of cancer by those times. Possible reasons for censoring are

- deaths due to other causes;
- being alive when the study ends.

The group comparison problem is simple, although the censored data presents a problem. How do we treat these cases? We can

- ignore the censoring information, i.e. we treat all the data as if they are genuine deaths;
- drop the censored cases, so we are dealing with genuine deaths;
- model the censored data properly.

The first two methods can be very biased and misleading if the censoring patterns in the groups differ. The second method is inefficient even if the censoring patterns in the two groups are similar. With a correct model, the last method is potentially the best as it would take into account whatever information is available in the censored data.

The censored data can be written as pairs $(y_1, \delta_1), \ldots, (y_n, \delta_n)$, where δ_i is the *last-known status* or *event indicator*: $\delta_i = 1$ if y_i is a true event time, and zero otherwise. If t_i is the true lifetime of subject i, then $\delta_i = 0$ iff $t_i > y_i$. We would be concerned with modeling the true lifetime t_i rather than the observed y_i, since censoring is usually a nuisance process that does not have any substantive meaning, for example it can be determined by the study design.

Suppose $t_1, \ldots, t_n$ are an iid sample from $p_\theta(t)$. The likelihood contribution of the observation (y_i, δ_i) is

$$L_i(\theta) = P_\theta(T_i > y_i) \quad \text{if } \delta_i = 0$$

or

$$L_i(\theta) = p_\theta(t_i) \quad \text{if } \delta_i = 1.$$

The probability $P_\theta(T_i > y_i)$ is called the survival function of T_i. The overall likelihood is

$$\begin{aligned} L(\theta) &= \prod_{i=1}^{n} L_i(\theta) \\ &= \prod_{i=1}^{n} \{p_\theta(y_i)\}^{\delta_i} \{P_\theta(T_i > y_i)\}^{1-\delta_i}. \end{aligned}$$

As an example, consider a simple exponential model, which is commonly used in survival and reliability studies, defined by

$$\begin{aligned} p_\theta(t) &= \frac{1}{\theta} e^{-t/\theta} \\ P_\theta(T > t) &= e^{-t/\theta}. \end{aligned}$$

In this case

$$L(\theta) = \left(\frac{1}{\theta}\right)^{\sum \delta_i} \exp(-\sum y_i/\theta).$$

Upon taking the derivative of the log-likelihood we get the score function

$$S(\theta) = -\frac{\sum \delta_i}{\theta} + \frac{\sum_i y_i}{\theta^2}$$

and setting it to zero, we get

$$\widehat{\theta} = \frac{\sum y_i}{\sum \delta_i}.$$

Note that $\sum y_i$ is the total observation times including both the censored and uncensored cases, while $\sum \delta_i$ is the number of events. The inverse $1/\widehat{\theta}$ is an estimate of the event rate. This is a commonly used formula in epidemiology, and known as the person-year method. For example, the Steering Committee of the Physicians' Health Study Research Group (1989) reported the rate of heart attacks as 254.8 per 100,000 person-years in the aspirin group, compared with 439.7 in the placebo group; see Section 1.1.

With some algebra the observed Fisher information of θ is

$$I(\widehat{\theta}) = \frac{\sum \delta_i}{\widehat{\theta}^2},$$

so the standard error of $\widehat{\theta}$ is

$$\text{se}(\widehat{\theta}) = \frac{\widehat{\theta}}{(\sum \delta_i)^{1/2}}.$$

Example 11.5: continued. Assume an exponential model for excess lifetime over 100 days (in principle we can make this cutoff value another unknown parameter), so from Group 1 we get $n = 19$, $\sum y_i = 2195$, $\sum \delta_i = 17$ and

$$L(\theta_1) = \left(\frac{1}{\theta_1}\right)^{17} \exp(-2195/\theta_1),$$

which yields $\widehat{\theta}_1 = 2195/17 = 129.1$ (se $= 31.3$). The plot of the likelihood function is given in Figure 11.3(a). Similarly, from Group 2 we have $n = 21$, $\sum y_i = 2923$, $\sum \delta_i = 19$ and

$$L(\theta_2) = \left(\frac{1}{\theta_2}\right)^{19} \exp(-2923/\theta_2),$$

which yields $\widehat{\theta}_2 = 2923/19 = 153.8$ (se $= 35.3$). There is some indication that rats from Group 2 live longer than those from Group 1. The standard error of $\widehat{\theta}_1 - \widehat{\theta}_2$ is

$$\text{se}(\widehat{\theta}_1 - \widehat{\theta}_2) = \{\text{se}(\widehat{\theta}_1)^2 + \text{se}(\widehat{\theta}_2)^2\}^{1/2} = 47.2.$$

The Wald statistic for comparing the mean of Group 1 with the mean of Group 2 is $z = (129.1 - 153.8)/47.2 = -0.53$.

The following table compares the results of the three methods mentioned earlier. The normal model will be described later. □

	Sample mean		t- or Wald
Method	Group 1	Group 2	statistic
Ignore	115.5	139.2	−1.65
Drop cases	113.8	135.5	−1.49
Exp. model	129.1	153.8	−0.53
Normal model	119.1	142.6	−1.56

The result based on the exponential model is the least significant, which of course does not mean it is the best result for this dataset. Such a result is strongly dependent on the chosen model. Is the exponential model a good fit for the data? Figure 11.3(b) shows the QQ-plot of the uncensored observations versus the theoretical exponential distribution, indicating that the exponential model is doubtful. (A proper QQ-plot that takes the censored data into account requires an estimate of the survival function, which is given by the so-called Kaplan–Meier estimate below.)

The mean–variance relationship implied by the exponential model also does not hold: the variance is much smaller than the square of the mean.

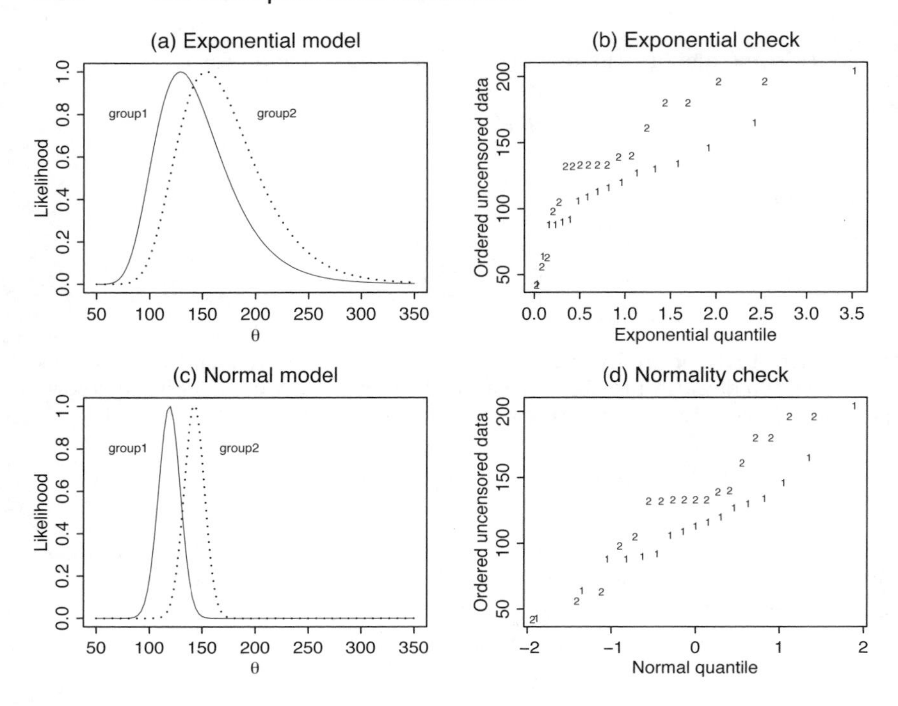

Figure 11.3: *Analysis of the rat data: the first row assumes exponential model and the second row assumes normal model.*

For the uncensored data there is an underdispersion factor of around 0.11. This means that the exponential-based likelihood is too wide. A proper model that takes the dispersion factor into account is given by the general survival regression model in the next section.

Normal model

Suppose t_i's are an iid sample from $N(\mu, \sigma^2)$. The likelihood contribution of an uncensored observation y_i is

$$p_\theta(y_i) = (2\pi\sigma^2)^{-1/2} \exp\left\{-\frac{1}{2\sigma^2}(y_i - \mu)^2\right\}$$

and the contribution of a censored observation is

$$P_\theta(y_i) = 1 - \Phi\left(\frac{y_i - \mu}{\sigma}\right)$$

where $\Phi(\cdot)$ is the standard normal distribution function. The functions are analytically more difficult than those in the exponential model, and there are no explicit formulae.

For the rat data, the likelihood of μ based on the normal model is shown in Figure 11.3(c). To simplify the computations it is assumed that the two groups have equal variance, and the variance is known at the estimated value from the uncensored data. The QQ-plot suggests the normal model seems to be a better fit. A formal comparison can be done using the AIC.

Kaplan–Meier estimate of the survival function

The most commonly used display of survival data is the Kaplan–Meier estimate of the survival function. It is particularly useful for a graphical comparison of several survival functions. Assume for the moment that there is *no censored data* and no tie, and let $t_1 < t_2 < \cdots < t_n$ be the ordered failure times. The empirical distribution function (EDF)

$$F_n(t) = \frac{\text{number of failure times } t_i\text{'s} \le t}{n}$$

is the cumulative proportion of mortality by time t. In Section 15.1 this is shown to the nonparametric MLE of the underlying distribution, therefore the MLE of the survival function is

$$\widehat{S}(t) = 1 - F_n(t).$$

The function $\widehat{S}(t)$ is a step function with a drop of $1/n$ at each failure time, starting at $\widehat{S}(0) \equiv 1$.

We can reexpress $\widehat{S}(t)$ in a form that is extendable to censored data. Let n_i be the number 'at risk' (yet to fail) just prior to failure time t_i. If there is no censoring $n_1 = n$, $n_2 = n - 1$, and $n_i = n - i + 1$. It is easily seen that

$$\widehat{S}(t) = \prod_{t_i \le t} \frac{n_i - 1}{n_i}.$$

For example, for $t_3 \le t < t_4$,

$$\begin{aligned}
\widehat{S}(t) &= \frac{n_1 - 1}{n_1} \times \frac{n_2 - 1}{n_2} \times \frac{n_3 - 1}{n_3} \\
&= \frac{n-1}{n} \times \frac{n-2}{n-1} \times \frac{n-3}{n-2} \\
&= \frac{n-3}{n} = 1 - \frac{3}{n} = 1 - F_n(t).
\end{aligned}$$

If there are ties, we only need a simple modification. Let $t_1, \ldots, t_k$ be the observed failure times, and $d_1, \ldots, d_k$ be the corresponding number of failures. Then

$$\widehat{S}(t) = \prod_{t_i \le t} \frac{n_i - d_i}{n_i}.$$

Exactly the same formula applies for censored data, and it is called Kaplan–Meier's product limit estimate of the survival function. It can be shown that the Kaplan–Meier estimate is the MLE of the survival distribution; see Kalbfleisch and Prentice (1980, page 11), and also Section 11.10.

Information from the censored cases is used in computing the number at risk n_i's. With uncensored data

$$n_i = n_{i-1} - d_{i-1},$$

i.e. the number at risk prior to time t_i is the number at risk prior to the previous failure time t_{i-1} minus the number that fails at time t_{i-1}. With censored data,

$$n_i = n_{i-1} - d_{i-1} - c_{i-1},$$

where c_{i-1} is the number censored between failure times t_{i-1} and t_i. (If there is a tie between failure and censoring times, it is usually assumed that censoring occurs after failure.)

These ideas can be grasped easily using a toy dataset $(y_1, \ldots, y_6) = (3, 4, 6+, 8, 8, 10)$. Here, we have

i	t_i	n_i	d_i	c_i	$\widehat{S}(t)$
0	$t_0 \equiv 0$	6	0	0	1
1	3	6	1	0	$5/6$
2	4	5	1	1	$5/6 \times 4/5 = 4/6$
3	8	3	2	0	$5/6 \times 4/5 \times 1/3 = 2/9$
4	10	1	1	0	0

For larger datasets the computation is tedious, but there are many available softwares. Figure 11.4 shows the Kaplan–Meier estimates of the survival functions of the rat groups in Example 11.5. The plot indicates a survival advantage of group 2 over group 1.

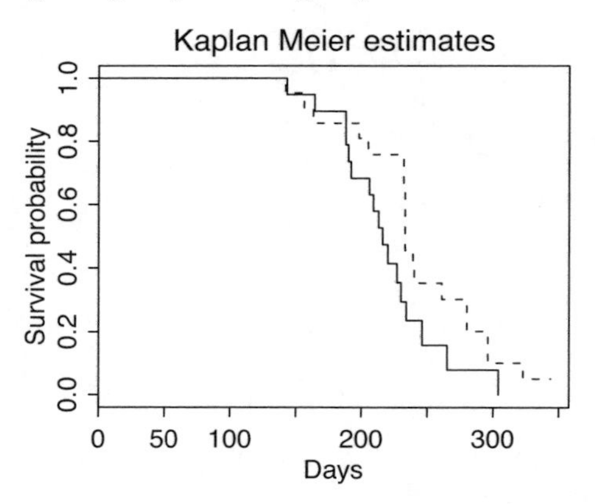

Figure 11.4: *Kaplan–Meier estimates of the survival function of group 1 (solid) and group 2 (dashed) of the rat data in Example 11.5.*

11.6 Survival regression models

In the same spirit as the models we develop in Chapter 6, the previous example can be extended to a general regression framework. Suppose we want to analyse the effect of some characteristics x on survival; for example, x is a group indicator ($x = 0$ or 1 in the rat data example). Using an exponential model

$$t_i \sim \text{Exponential}(\theta_i),$$

where the mean θ_i is a function of the covariates x_i, connected via a link function $h(\cdot)$, such that

$$h(\theta_i) = x_i'\beta.$$

We might consider the identity link

$$\theta_i = x_i'\beta$$

or the log-link

$$\log \theta_i = x_i'\beta.$$

The log-link function is more commonly used since $\theta_i > 0$.

Based on the observed data $(y_1, \delta_1, x_1), \ldots, (y_n, \delta_n, x_n)$ the likelihood function of the regression parameter β can written immediately as

$$L(\beta) = \prod_{i=1}^{n} \{p_{\theta_i}(y_i)\}^{\delta_i} \{P_{\theta_i}(T_i > y_i)\}^{1-\delta_i},$$

where θ_i is a function of β, and

$$p_{\theta_i}(y_i) = \theta_i^{-1} e^{-y_i/\theta_i},$$

and

$$P_{\theta_i}(y_i) = e^{-y_i/\theta_i}.$$

As we have seen before the exponential model specifies a rigid relationship between the mean and the variance. An extension that allows a flexible relationship is essential. To motivate a natural development, note that if T is exponential with mean θ, then

$$\log T = \beta_0 + W$$

where $\beta_0 = \log \theta$ and W has the standard extreme-value distribution with density

$$p(w) = e^w \exp(-e^w)$$

and survival function

$$P(W > w) = \exp(-e^w).$$

A more flexible regression model between a covariate vector x_i and outcome T_i can be developed by assuming

$$\log T_i = \log \theta_i + \sigma W_i$$

where W_i's are iid with standard extreme-value distribution, and σ is a scale parameter. This extension is equivalent to using the Weibull model.

As usual, the parameter θ_i is related to the covariate vector x_i via a link function. For example, using the log-link

$$\log \theta_i = x_i' \beta.$$

This is the so-called *accelerated failure time model*: the parameter e^{β} is interpreted as the multiplicative effect of a unit change in x on average lifetime.

The likelihood can be computed from the Weibull survival function

$$\begin{aligned} P_{\theta_i}(T_i > y_i) &= P\{\log(T_i/\theta_i) > \log(y_i/\theta_i)\} \\ &= P\{W_i > \log(y_i/\theta_i)^{1/\sigma}\} \\ &= \exp\{-(y_i/\theta_i)^{1/\sigma}\} \end{aligned}$$

and density function

$$p_{\theta_i}(y_i) = \sigma^{-1} y_i^{1/\sigma - 1} \theta_i^{-1/\sigma} \exp\{-(y_i/\theta_i)^{1/\sigma}\}.$$

Example 11.6: For the rat data in Example 11.5, suppose we model the mean θ_i as

$$\log \theta_i = \beta_0 + \beta_1 x_i$$

where $x_i = 0$ or 1, for Group 1 or 2, respectively. The following table summarizes the analysis using the exponential model ($\sigma = 1$) and the general model by letting σ be free.

		Exponential		General	
Effect	Parameter	Estimate	se	Estimate	se
Constant	β_0	4.861	0.243	4.873	0.079
Group	β_1	0.175	0.334	0.213	0.108
Scale	σ	1	–	0.32	–

For the exponential model the estimated mean ratio of the two groups is $e^{\widehat{\beta_1}} = 1.19 = 153.8/129.1$ as computed before. The estimated scale $\widehat{\sigma} = 0.32$ gives a dispersion factor $\widehat{\sigma}^2 = 0.10$, as expected from the previous discussion of this example. Since the exponential model ($\sigma = 1$) is a special case of the general model, we can test its adequacy by the likelihood ratio test: find the maximized likelihood under each model, and compute the likelihood ratio statistic $W = 44.8$, which is convincing evidence against the exponential model.

Under the general model, we obtain $z = 0.213/0.108 = 1.98$ for the observed group difference, which is borderline significant. Checking the appropriateness of the quadratic approximation for the profile likelihood of β_1 is left as an exercise. □

11.7 Hazard regression and Cox partial likelihood

The *hazard function* is indispensable in the analysis of censored data. It is defined as

$$\lambda(t) = p(t)/P(T > t),$$

and interpreted as the rate of dying at time t among the survivors. For example, if T follows an exponential distribution with mean θ, the hazard function of T is

$$\lambda(t) = 1/\theta.$$

This inverse relationship is sensible, since a short lifetime implies a large hazard. Because the hazard function is constant the exponential model may not be appropriate for living organisms, where the hazard is typically higher at both the beginning and the end of life. Models can be naturally put in hazard form, and the likelihood function can be computed based on the following relationships:

$$\lambda(t) = -\frac{d \log P(T > t)}{dt} \quad (11.1)$$

$$\log P(T > t) = -\int_0^t \lambda(u)du \quad (11.2)$$

$$\log p(t) = \log \lambda(t) - \int_0^t \lambda(u)du. \quad (11.3)$$

Given censored data $(y_1, \delta_1, x_1), \ldots, (y_n, \delta_n, x_n)$, where δ_i is the event indicator, and the underlying t_i has density $p_{\theta_i}(t_i)$, the log-likelihood function contribution of (y_i, δ_i, x_i) is

$$\begin{aligned} \log L_i &= \delta_i \log p_{\theta_i}(y_i) + (1-\delta_i) \log P_{\theta_i}(y_i) \\ &= \delta_i \log \lambda_i(y_i) - \int_0^{y_i} \lambda_i(u)du \quad (11.4) \end{aligned}$$

where the parameter θ_i is absorbed by the hazard function. Only uncensored observations contribute to the first term.

The most commonly used model in survival analysis is the proportional hazard model of the form

$$\lambda_i(t) = \lambda_0(t) e^{x_i' \beta}. \quad (11.5)$$

In survival regression or comparison studies the baseline hazard $\lambda_0(t)$ is a nuisance parameter; it must be specified for a full likelihood analysis of the problem. Here is a remarkable property of the model that avoids the need to specify $\lambda_0(t)$: if lifetimes T_1 and T_2 have proportional hazards

$$\lambda_i(t) = \lambda_0(t)\eta_i$$

for $i = 1, 2$, respectively, then

$$P(T_1 < T_2) = \eta_1/(\eta_1 + \eta_2)$$

regardless of the shape of the baseline hazard function (Exercise 11.28). We can interpret the result this way: if $\eta_1 > \eta_2$ then it is more likely that subject 1 will die first.

Such a probability can be used in a likelihood function based on knowing the event $[T_1 < T_2]$ alone, namely only the ranking information is used, not the actual values. Such a likelihood is then a marginal likelihood. If x_i is the covariate vector associated with T_i and we model

$$\eta_i = e^{x_i'\beta},$$

then the likelihood of β based on observing the event $[T_1 < T_2]$ is

$$L(\beta) = \frac{e^{x_1'\beta}}{e^{x_1'\beta} + e^{x_2'\beta}},$$

which is free of any nuisance parameter.

In general, given a sample $T_1, \ldots, T_n$ from a proportional hazard model

$$\lambda_i(t) = \lambda_0(t) e^{x_i'\beta},$$

the probability of a particular configuration $(i_1, \ldots, i_n)$ of $(1, \ldots, n)$ is

$$P(T_{i_1} < T_{i_2} < \cdots < T_{i_n}) = \prod_{j=1}^{n} e^{x_{i_j}'\beta} / (\sum_{k \in R_j} e^{x_k'\beta}), \qquad (11.6)$$

where R_j is the list of subjects where $T \geq T_{i_j}$, which is called the 'risk set' at time T_{i_j}. It is easier to see this in an example for $n = 3$: for $(i_1, i_2, i_3) = (2, 3, 1)$ we have

$$P(T_2 < T_3 < T_1) = \frac{e^{x_2'\beta}}{e^{x_1'\beta} + e^{x_2'\beta} + e^{x_3'\beta}} \times \frac{e^{x_3'\beta}}{e^{x_1'\beta} + e^{x_3'\beta}} \times \frac{e^{x_1'\beta}}{e^{x_1'\beta}}.$$

The likelihood of the regression parameter β computed from this formula is called the Cox partial likelihood (Cox 1972, 1975), the main tool of survival analysis. With this likelihood we are only using the ranking observed in the data. In the example with $n = 3$ above we only use the information that subject 2 died before subject 3, and subject 3 died before subject 1. If the baseline hazard can be any function, it seems reasonable that there is very little extra information beyond the ranking information in the data. In fact, it has been shown that, for a wide range of underlying hazard functions, the Cox partial likelihood loses little or no information (Efron 1977).

Extension to the censored data case is intuitively obvious by thinking of the risk set at each time of death: a censored value only contributes to the risk sets of the *prior* uncensored values, but it cannot be compared with later values. For example, if we have three data values 1, 5+ and 7, where 5+ is censored, we only know that 1 is less than 5+ and 7, but we cannot distinguish between 5+ and 7. In view of this, the same likelihood formula (11.6) holds.

Example 11.7: The following toy dataset will be used to illustrate the construction of the Cox likelihood for a two-group comparison.

i	x_i	y_i	δ_i
1	0	10	0
2	0	5	1
3	0	13	1
4	1	7	1
5	1	21	1
6	1	17	0
7	1	19	1

Assume a proportional hazard model

$$\lambda_i(t) = \lambda_0(t)e^{x_i\beta}.$$

We first sort the data according to the observed y_i.

i	x_i	y_i	δ_i
2	x_2 =0	5	1
4	x_4 =1	7	1
1	x_1 =0	10	0
3	x_3 =0	13	1
6	x_6 =1	17	0
7	x_7 =1	19	1
5	x_5 =1	21	1

The Cox partial likelihood of β is

$$\begin{aligned} L(\beta) &= \frac{e^{x_2\beta}}{\sum_{i=1}^{7} e^{x_i\beta}} \\ &\times \frac{e^{x_4\beta}}{e^{x_4\beta} + e^{x_1\beta} + e^{x_3\beta} + e^{x_6\beta} + e^{x_7\beta} + e^{x_5\beta}} \\ &\times \frac{e^{x_3\beta}}{e^{x_3\beta} + e^{x_6\beta} + e^{x_7\beta} + e^{x_5\beta}} \\ &\times \frac{e^{x_7\beta}}{e^{x_7\beta} + e^{x_5\beta}}. \end{aligned}$$

Note that only uncensored cases can appear in the numerator of the likelihood. The Cox partial likelihood is generally too complicated for direct analytical work; in practice most likelihood quantities such as MLEs and their standard errors are computed numerically. In this example, since x_i is either zero or one, it is possible to simplify the likelihood further. □

Example 11.8: Suppose we assume the proportional hazard model for the rat data in Example 11.5:

$$\lambda(t) = \lambda_0(t)e^{\beta_1 x_i},$$

where $x_i = 0$ or 1 for Group 1 or 2, respectively. We do not need an intercept term β_0 since it is absorbed in the baseline hazard $\lambda_0(t)$. The estimate of β_1 is

$$\widehat{\beta}_1 = -0.569 \text{ (se} = 0.347),$$

giving a Wald statistic of $z = -1.64$, comparable with the result based on the normal model. The minus sign means the hazard of Group 2 is $e^{-0.569} = 0.57$ times the hazard of Group 1; recall that the rats in Group 2 live longer.

The estimated hazard ratio, however, is much smaller than what we get using a constant hazard assumption (i.e. the exponential model); in Example 11.5 we obtain a ratio of $129.1/153.8 = e^{-0.175} = 0.84$. Since the shape of the hazard function in the Cox regression is free the result here is more plausible, although the ratio 0.57 is no longer interpretable as a ratio of mean lifetimes. The proportional hazard assumption itself can be checked; see, for example, Grambsch and Therneau (1994). □

11.8 Poisson point processes

Many random processes are events that happen at particular points in time (or space); for example, customer arrivals to a queue, times of equipment failures, times of epilepsy attacks in a person, etc. Poisson point or counting processes form a rich class of models for such processes. Let $N(t)$ be the number of events up to time t, dt a small time interval, and $o(dt)$ a quantity of much smaller magnitude than dt in the sense $o(dt)/dt \to 0$ as $dt \to 0$. The function $N(t)$ captures the point process and we say $N(t)$ is a Poisson point process with intensity $\lambda(t)$ if

$$N(t+dt) - N(t) = \begin{cases} 1 & \text{with probability } \lambda(t)dt \\ 0 & \text{with probability } 1 - \lambda(t)dt \\ > 1 & \text{with probability } o(dt) \end{cases}$$

and $N(t+dt) - N(t)$ is independent of $N(u)$ for $u < t$; the latter is called the independent increment property, which is a continuous version of the concept of independent trials.

It is fruitful to think of a Poisson process informally this way: $N(t+dt) - N(t)$, the number of events between t and $t+dt$, is Poisson with mean $\lambda(t)dt$. In a regular Poisson process, as defined above, there is a maximum of one event that can conceivably happen in an interval dt, so $N(t+dt) - N(t)$ is also approximately Bernoulli with probability $\lambda(t)dt$. Immediately identifying $N(t+dt) - N(t)$ as Poisson leads to simpler heuristics.

What we observe is determined stochastically by the intensity parameter $\lambda(t)$. Statistical questions can be expressed via $\lambda(t)$. It is intuitive that there is a close connection between this intensity function and the hazard function considered in survival analysis. The developments here and in the next sections are motivated by Whitehead (1983), Lawless (1987) and Lindsey (1995).

The first problem is, given a set of observations $t_1, \ldots t_n$, what is the likelihood of $\lambda(t)$? To answer this we first need two results from the theory of point processes. Both results are intuitively obvious; to get a formal proof see, for example, Diggle (1983). For the first result, since the number of events in each interval is Poisson, the sum of events on the interval $(0, T)$ is

$$\begin{aligned} N(T) &= \sum_{0<t<T} \{N(t+dt) - N(t)\} \\ &\sim \text{Poisson with mean } \sum_t \lambda(t)dt. \end{aligned}$$

So, in the limit we have:

Theorem 11.1 *$N(T)$ is Poisson with mean $\int_0^T \lambda(t)dt$.*

Furthermore, the way $t_1, \ldots, t_n$ are arranged will be determined by $\lambda(t)$, as shown by the following.

Theorem 11.2 *Given $N(T) = n$, the times $t_1, \ldots, t_n$ are distributed like the order statistics of an iid sample from a distribution with density proportional to $\lambda(t)$.*

To make it integrate to one, the exact density is given by

$$\frac{\lambda(t)}{\int_0^T \lambda(u)du}.$$

Suppose we model $\lambda(t) = \lambda(t, \theta)$, where θ is an unknown parameter. For example,

$$\lambda(t) = \alpha e^{\beta t}$$

with $\theta = (\alpha, \beta)$. For convenience, let

$$\Lambda(T) \equiv \int_0^T \lambda(t)dt$$

be the cumulative intensity. Then, given the observation times $0 < t_1, \ldots, t_n < T$, the likelihood of the parameters is

$$\begin{aligned} L(\theta) &= P\{N(T) = n) \times p\{t_1, \ldots, t_n | N(t) = n\} \\ &= e^{-\Lambda(T)} \frac{\Lambda(T)^n}{n!} \times n! \prod_{i=1}^n \frac{\lambda(t_i)}{\Lambda(T)} \\ &= e^{-\Lambda(T)} \prod_{i=1}^n \lambda(t_i). \end{aligned}$$

It is instructive to follow a heuristic derivation of the likelihood, since it applies more generally for point processes. First, partition the time axis into tiny intervals of length dt, such that only a single event can conceivably occur. On each interval let $y(t) \equiv N(t + dt) - N(t)$; then the time series $y(t)$ is an independent Poisson series with mean $\lambda(t)dt$. Observing $N(t)$ for $0 < t < T$ is equivalent to observing the series $y(t)$, where $y(t) = 1$ at $t_1, \ldots, t_n$, and zero otherwise; since we are thinking of dt as very small, the series $y(t)$ is mostly zero. For example, using dt as the time unit, events at times $t_1 = 3$ and $t_2 = 9$ during observation period $T = 10$ mean the series $y(t)$ is (0,0,1,0,0,0,0,0,1,0).

Given the observation times $0 < t_1, \ldots, t_n < T$, we obtain the likelihood

$$L(\theta) = \prod_t p(y_t)$$

$$
\begin{aligned}
&= \prod_t \exp\{-\lambda(t)dt\}\lambda(t)^{y(t)} \\
&\approx \exp\{-\sum_t \lambda(t)dt\} \prod_{i=1}^{n} \lambda(t_i) \\
&\approx \exp\{-\int_0^T \lambda(t)dt\} \prod_{i=1}^{n} \lambda(t_i),
\end{aligned}
$$

as we have just seen.

The last heuristic can be put into more technical notation. Let $dN(t) \equiv y(t) = N(t+dt) - N(t)$; then the log-likelihood can be written as

$$
\begin{aligned}
\log L(\theta) &= -\int_0^T \lambda(t)\ dt + \sum_{i=1}^{n} \log \lambda(t_i) \qquad (11.7)\\
&= -\int_0^T \lambda(t)\ dt + \int_0^T \log \lambda(t)\ dN(t),
\end{aligned}
$$

where for any function $h(t)$

$$
\begin{aligned}
\int_0^T h(t)\ dN(t) &\approx \sum_t h(t)\ dN(t) \\
&= \sum_{i=1}^{n} h(t_i).
\end{aligned}
$$

Note that, because of the way $dN(t)$ is defined, the intensity function $\lambda(t)$ can include values of the process $N(t)$, or any other data available prior to time t, without changing the likelihood. Andersen et al. (1993) provide a rigorous likelihood theory for counting processes.

The close connection between Poisson intensity modelling and hazard modelling of survival data in Section 11.7 is now clear: the likelihood (11.7) reduces to (11.4) if we limit ourselves to absorbing events (events that can occur only once, such as deaths, or events that end the observation period). It is also clear that the Poisson process models are more general than the survival models as they allow multiple end-points per subject.

Example 11.9: The following data (from Musa *et al.* 1987) are the times of 136 failures (in CPU seconds) of computer software during a period of $T = 25.4$ CPU hours. At each failure the cause is removed from the system. The questions are how fast can bugs be removed from the system, and how many bugs are still in the system. The histogram in Figure 11.5(a) shows that the number of failures decreases quickly over time. □

3	33	146	227	342	351	353	444	556	571	709
759	836	860	968	1056	1726	1846	1872	1986	2311	2366
2608	2676	3098	3278	3288	4434	5034	5049	5085	5089	5089
5097	5324	5389	5565	5623	6080	6380	6477	6740	7192	7447
7644	7837	7843	7922	8738	10089	10237	10258	10491	10625	10982

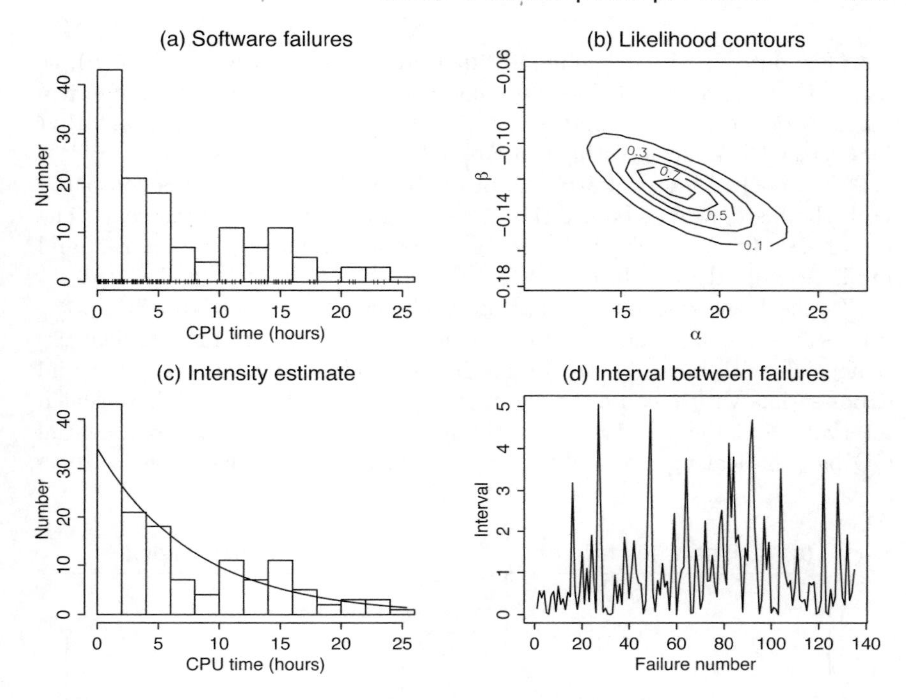

Figure 11.5: *(a) The vertical lines on the x-axis indicate the failure times; the histogram simply shows the number of failures in each 2-hour interval. (b) The likelihood of the parameters of an exponential decline model* $\lambda(t) = \alpha e^{\beta t}$. *(c) The fitted intensity function compared with the histogram. (d) The time series of scaled inter-failure times.*

```
11175 11411 11442 11811 12559 12559 12791 13121 13486 14708 15251
15261 15277 15806 16185 16229 16358 17168 17458 17758 18287 18568
18728 19556 20567 21012 21308 23063 24127 25910 26770 27753 28460
28493 29361 30085 32408 35338 36799 37642 37654 37915 39715 40580
42015 42045 42188 42296 42296 45406 46653 47596 48296 49171 49416
50145 52042 52489 52875 53321 53443 54433 55381 56463 56485 56560
57042 62551 62651 62661 63732 64103 64893 71043 74364 75409 76057
81542 82702 84566 88682
```

Assume a model $\lambda(t) = \alpha e^{\beta t}$, so

$$\begin{aligned} \Lambda(T) &= \int_0^T \alpha e^{\beta t} dt \\ &= \frac{\alpha}{\beta}(e^{\beta T} - 1) \end{aligned}$$

and, defining $\theta = (\alpha, \beta)$, we obtain

$$L(\theta) = \exp\left\{-\frac{\alpha}{\beta}(e^{\beta T} - 1)\right\} \prod_{i=1}^{136} (\alpha e^{\beta t_i}).$$

The contours of the likelihood function are given in Figure 11.5(b); as usual, they represent 10% to 90% confidence regions. Using a numerical optimization routine we find $\widehat{\alpha} = 17.79$ and $\widehat{\beta} = -0.13$, which means that for every CPU hour of testing the rate of failures is reduced by $(1 - e^{-0.13}) \times 100 = 12\%$. In Figure 11.5(c) the fit of the parametric model is compared with the histogram of failure times, showing a reasonable agreement. The parametric fit is useful for providing a simple summary parameter β, and prediction of future failures.

To check the Poisson assumption, we know theoretically that if a point process is Poisson, then the intervals between failures are independent exponentials with mean $1/\lambda(t)$, or the scaled intervals $\lambda(t_i)(t_i - t_{i-1})$ are iid exponentials with mean one. Figure 11.5(d) shows the plot of these scaled intervals, and Figure 11.6 shows the autocorrelation and the exponential QQ-plot, indicating reasonable agreement with Poisson behaviour.

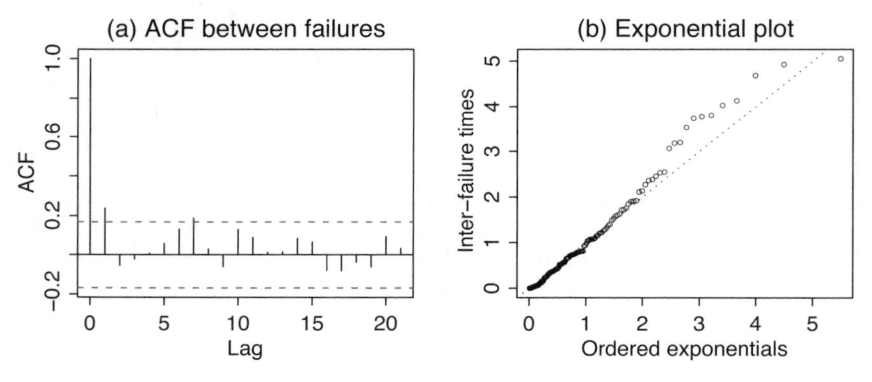

Figure 11.6: *Diagnostic check of the Poisson assumption: (a) autocorrelation between inter-event times and (b) exponential plot of inter-event times.*

11.9 Replicated Poisson processes

In biomedical applications we often deal with multiple processes, each generated by a subject under study. Since the notation can become cumbersome we will discuss the general methodology in the context of a specific example. Figure 11.7 shows a dataset from a study of treatment of epilepsy, where patients were randomized to either active or placebo groups. Because of staggered entry to the study, patients have different follow-up periods. The patients' families were asked to record the time of epileptic attacks during follow-up.

We will analyse this dataset using several methods of increasing complexity. Note the flexibility offered by the more complex methods in handling general intensity functions and covariates. Also, we will meet the Cox partial likelihood again.

Subject	x_i	T_i	n_i	Time of events										
1	active	12	3	2.6	3.3	7.2								
2	active	5	2	3.5	4.4									
3	active	7	4	1.5	1.6	2.2	6.1							
4	active	14	3	12.1	12.4	13.4								
5	active	10	5	0.7	2.6	3.9	6.9	7.8						
6	active	10	2	5.3	6.3									
7	active	12	1	10.2										
8	active	8	3	0.2	3.2	7.7								
9	active	11	3	0.1	2	3.2								
10	active	8	3	0.1	3.2	3.7								
11	placebo	11	4	2.3	7.9	8	8.8							
12	placebo	11	7	5.1	5.2	6.1	6.5	7.9	9.9	10.9				
13	placebo	8	6	0.5	0.8	1.9	2.7	5.4	7.2					
14	placebo	16	8	1.4	4.3	5	6	7.8	8.4	9.2	11.2			
15	placebo	11	11	0.3	0.3	1.9	1.9	2.7	3.1	3.9	5.3	7	8.8	10.1
16	placebo	7	8	1.2	2.6	3.5	4.7	5.3	5.7	5.9	6.1			
17	placebo	15	7	0.8	1.5	4.3	4.4	5.1	12.1	14				
18	placebo	9	7	0.1	0.1	1	3.6	5.4	6.3	8.7				
19	placebo	7	4	0.9	2.2	5.2	6.6							
20	placebo	4	2	2.2	3.2									
21	placebo	6	6	0.5	1.3	1.3	1.7	2.9	5.6					
22	placebo	4	1	1.4										

Figure 11.7: *The epilepsy data example. Patient i was followed for T_i weeks, and there were n_i events during the follow-up.*

Method 1

The first method will take into account the different follow-up periods among patients, but not use the times of attacks, and it cannot be generalized if we want to consider more covariates. Assume the event times within each patient follow a Poisson point process. Let λ_a and λ_p be the rate of attacks in the active and the placebo groups, i.e. assume that the rate or intensity is constant over time. Let

$$\begin{aligned} n_i &= \text{total number of attacks patient } i \\ y_a &= \sum_{i \in \text{active}} n_i \\ y_p &= \sum_{i \in \text{placebo}} n_i. \end{aligned}$$

Then

$$\begin{aligned} n_i &\sim \text{Poisson}(T_i \lambda) \\ y_a &\sim \text{Poisson}(\sum_{i \in \text{active}} T_i \lambda_a) \\ y_p &\sim \text{Poisson}(\sum_{i \in \text{placebo}} T_i \lambda_p), \end{aligned}$$

where λ in the first equation is either λ_a or λ_p. The parameter of interest is

$$\theta = \lambda_a/\lambda_p. \tag{11.8}$$

We can summarize the observed data in the following table:

	Active	Placebo
y_a or y_p	29	71
$\sum T_i$	97	109

Proceeding as in the aspirin data example in Section 4.7, conditional on $y_a + y_p$ the distribution of y_a is binomial with parameter $y_a + y_p = 100$ and probability

$$\frac{97\lambda_a}{97\lambda_a + 109\lambda_p} = \frac{97\theta}{97\theta + 109}.$$

So, the conditional likelihood is

$$L(\theta) = \left(\frac{97\theta}{97\theta + 109}\right)^{29} \left(1 - \frac{97\theta}{97\theta + 109}\right)^{71},$$

shown in Figure 11.8(a). We can verify that the MLE of θ is

$$\widehat{\theta} = (29/97)/(71/109) = 0.46.$$

The standard error is $\text{se}(\widehat{\theta}) = 0.10$, but the quadratic approximation is poor; the reader can verify that $\log\theta$ has a more regular likelihood. The likelihood of the null hypothesis H_0: $\theta = 1$ is tiny (approximately $e^{-6.9}$), leading to the conclusion that the active treatment has led to fewer attacks of epilepsy.

Method 2: Poisson regression

We now use the Poisson regression method, which could easily accommodate some covariates, but still does not use any information about event times. Let $x_i = 1$ if patient i belongs to the active group and zero otherwise. Using the same assumptions as in Method 1, the number of attacks n_i is Poisson with mean

$$\mu_i = T_i \exp(\beta_0 + \beta_1 x_i)$$

or

$$\log \mu_i = \log T_i + \beta_0 + \beta_1 x_i.$$

In generalized linear modelling $\log T_i$ is called an offset term. The likelihood of (β_0, β_1) can be computed as before, and the reader can verify the following summary table. Note that $e^{\widehat{\beta_1}} = e^{-0.78} = 0.46 = \widehat{\theta}$ as computed by Method 1.

Effect	Parameter	Estimate	se
Intercept	β_0	−0.43	0.11
Treatment	β_1	−0.78	0.21

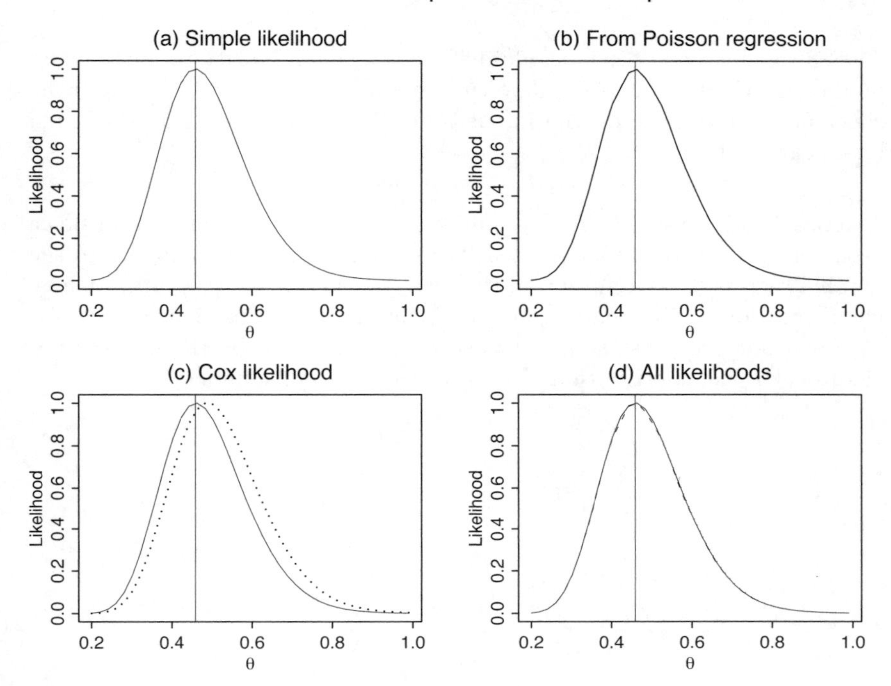

Figure 11.8: *Analysis of the epilepsy data using three methods: (a) Likelihood of θ using Method 1. (b) Profile likelihood of θ from Poisson regression. (c) Approximate (dotted) and true (solid) Cox partial likelihood. (d) All three likelihoods.*

To get a comparable likelihood function for the parameter $\theta = e^{\beta_1}$, we can compute the profile likelihood

$$L(\beta_1) = \max_{\beta_0} L(\beta_0, \beta_1),$$

and then evaluate $L(\theta) = L(\beta_1)$ at $\theta = e^{\beta_1}$. This likelihood function is given in Figure 11.8(b); it coincides with the likelihood given by Method 1.

Method 3

The previous methods make no use of the times of attacks, and they assume constant intensity for the Poisson processes for each subject. Neither can be generalized to overcome these limitations, so we will now consider a method that will address them at the expense of more complicated modelling.

We start by assuming the attacks for a patient follow a Poisson point process with intensity $\lambda_x(t)$, where x is the covariate vector. A useful general model for problems of this type is a proportional intensity model

$$\lambda_x(t) = \lambda_0(t, \alpha) g(x, \beta),$$

where $\lambda_0(t, \alpha)$ is the baseline intensity function with unknown parameter α. This is the analogue of the proportional hazard model in survival analysis.

The effect of covariate x is to modify the baseline intensity proportionally by a constant $g(x,\beta)$. The unknown regression parameter β expresses the effect of the covariate on the intensity level, for example using the usual log-linear model

$$\lambda_x(t) = \lambda_0(t,\alpha)e^{x'\beta}.$$

The baseline intensity $\lambda_0(t,\alpha)$ requires a parameter α, which is a nuisance parameter. The proportional intensity assumption is also analogous to the parallel regression assumption in the analysis of covariance; it may or may not be appropriate depending on the actual intensity functions.

From our previous theory, denoting $t_{i1},\ldots,t_{in_i}$ to be the event times of subject i, the contribution of this subject to the likelihood is

$$L_i(\alpha,\beta) = e^{-\Lambda_{x_i}(T_i)}\prod_{j=1}^{n_i}\lambda_{x_i}(t_{ij}),$$

where

$$\begin{aligned} \Lambda_{x_i}(T_i) &= \int_0^{T_i}\lambda_{x_i}(t)dt \\ &= g(x_i,\beta)\int_0^{T_i}\lambda_0(t,\alpha)dt \\ &= g(x_i,\beta)\Lambda_0(T_i,\alpha). \end{aligned}$$

So

$$\begin{aligned} L_i(\alpha,\beta) &= e^{-g(x_i,\beta)\Lambda_0(T_i,\alpha)}\{g(x_i,\beta)\Lambda_0(T_i,\alpha)\}^{n_i}\prod_{j=1}^{n_i}\frac{\lambda_0(t_{ij},\alpha)}{\Lambda_0(T_i,\alpha)} \\ &\equiv L_{1i}(\alpha,\beta)L_{2i}(\alpha), \end{aligned}$$

where

$$L_{1i}(\alpha,\beta) \equiv e^{-g(x_i,\beta)\Lambda_0(T_i,\alpha)}\{g(x_i,\beta)\Lambda_0(T_i,\alpha)\}^{n_i}.$$

The total likelihood from all, say m, individuals is

$$\begin{aligned} L(\alpha,\beta) &= \prod_{i=1}^{m} L_i(\alpha,\beta) \\ &= \prod_{i=1}^{m} L_{1i}(\alpha,\beta)\prod_i L_{2i}(\alpha) \\ &\equiv L_1(\alpha,\beta)L_2(\alpha). \end{aligned}$$

Hence, in proportional intensity models, the information about β is contained in the first term $L_1(\alpha,\beta)$; this is the likelihood based on the number of events from each individual

$$N_i \sim \text{Poisson}(\Lambda_0(T_i,\alpha)g(x_i,\beta)). \qquad (11.9)$$

If we assume constant intensity, this reduces to the first method.

Having to model the baseline intensity $\lambda_0(t, \alpha)$ is literally a nuisance, since it is not directly relevant to the question of treatment comparisons. The data for estimating $\lambda_0(t, \alpha)$ are provided by the set of event times t_{ij}'s. The data structure makes it difficult to specify an appropriate model for $\lambda_0(t, \alpha)$; for example, we cannot simply plot the histogram of the event times. Fitting a model for the current example is left as Exercise 11.33. Given such a model, the decomposition of the likelihood above suggests the following method of estimation. (Actual implementation of the procedure is left as Exercise 11.34.)

1. Estimate α from $L_2(\alpha)$, which is a conditional likelihood given n_i's. This is fully determined by the set of event times t_{ij}'s.
2. Use $\widehat{\alpha}$ to compute $\Lambda_0(T_i, \widehat{\alpha})$.
3. Estimate β in the Poisson regression (11.9) based on the data (n_i, x_i) with $\Lambda_0(T_i, \widehat{\alpha})$ as an offset term.

To get further simplification, in particular to remove the nuisance parameter α, let us assume for the moment that $T_i \equiv T$. We first need the result that if X_i, for $i = 1, \ldots, m$, are independent Poisson(λ_i), then the conditional distribution of $(X_1, \ldots, X_m)$ given $\sum X_i$ is multinomial with parameters $(\pi_1, \ldots, \pi_m)$, where $\pi_i = \lambda_i / \sum_{j=1}^m \lambda_j$. This is applied to

$$N_i \sim \text{Poisson}(\Lambda_0(T, \alpha) g(x_i, \beta)).$$

Letting $n = \sum_i n_i$, we now have

$$\begin{aligned} L_1(\alpha, \beta) &= P(N_1 = n_1, \ldots, N_m = n_m) \\ &= P(N_1 = n_1, \ldots, N_m = n_m | \sum N_i = n) P(\sum N_i = n) \\ &= \prod_{i=1}^m \left(\frac{g(x_i, \beta)}{\sum_{j=1}^m g(x_j, \beta)} \right)^{n_i} P(\sum N_i = n) \\ &\equiv L_{10}(\beta) L_{11}(\alpha, \beta), \end{aligned} \tag{11.10}$$

where

$$L_{10}(\beta) = \prod_{i=1}^m \left(\frac{g(x_i, \beta)}{\sum_{j=1}^m g(x_j, \beta)} \right)^{n_i}.$$

Finding the exact formula for $L_{11}(\alpha, \beta)$ is left as Exercise 11.35. Now $L_{10}(\beta)$ is only a function of the parameter of interest β. Intuitively, if the baseline intensity $\lambda_0(t, \alpha)$ can be of *any shape*, then there is little information about β in the total number of events $\sum n_i$. This reasoning is similar to conditioning on the sum in the comparison of two Poisson means.

If we use the common log-linear model

$$\lambda_x(t) = \lambda_0(t, \alpha) e^{x'\beta},$$

then

$$L_{10}(\beta) = \prod_{i=1}^{m} \left(\frac{e^{x'_i\beta}}{\sum_{j=1}^{m} e^{x'_j\beta}} \right)^{n_i},$$

exactly the Cox partial likelihood for this particular setup.

In general, when $T_i \neq T$, and assuming a Poisson process with proportional intensity, the Cox partial likelihood is defined as the following. Note that it allows the covariate to change over time. Let

$$\begin{aligned} t_{ij} &= \text{the } j\text{'th event of subject } i, \\ x_{ij} &= \text{the covariate value of subject } i \text{ at time } t_{ij}, \\ R_{ij} &= \text{the set of subjects still at risk at time } t_{ij}. \end{aligned}$$

Then

$$L_{10}(\beta) = \prod_{i=1}^{m} \prod_{j=1}^{n_i} \left(\frac{e^{x'_{ij}\beta}}{\sum_{k \in R_{ij}} e^{x'_{kj}\beta}} \right).$$

We can arrive at this likelihood by partitioning the time axis at T_i's and using, under the Poisson process, the independence of non-overlapping time intervals. The Cox partial likelihood looks more complicated, but it does correspond to the formula we derive before for the proportional hazard model. The significant contribution of the current approach is that it allows for multiple outcomes from each subject, i.e. the end-point does not have to be death, but it can be a recurrent event such as relapse. Furthermore, subjects can go in and out of the risk set depending whether they are being followed (events are being recorded) or not.

To apply this approach to the epilepsy data, consider first the approximate Cox partial likelihood, *pretending* that the follow-up period T_i's are the same for all subjects. Let the covariate $x_i = 1$ if i belongs to the active therapy group and $x_i = 0$ otherwise. Then

$$L_{10}(\beta) = \prod_{i=1}^{m} \left(\frac{e^{x'_i\beta}}{\sum_{j=1}^{m} e^{x'_j\beta}} \right)^{n_i},$$

where $e^{x'_i\beta} = \theta$ if $x_i = 1$ and $e^{x'_i\beta} = 1$ otherwise; then $\sum_{j=1}^{m} e^{x'_j\beta} = 10\theta + 12$, so

$$L_{10}(\beta) = \left(\frac{\theta}{10\theta + 12} \right)^{y_a} \left(\frac{1}{10\theta + 12} \right)^{y_p},$$

the same as the likelihood given by Method 1 had we used $T_i \equiv T$. Figure 11.8(c) shows the approximate and the true Cox partial likelihood for the dataset. Computation of the true Cox partial likelihood is left as an exercise. As shown in Figure 11.8(d), in this case the likelihoods from all methods virtually coincide.

11.10 Discrete time model for Poisson processes

In this section we describe in detail the close connection between censored survival data and the Poisson process, and the unifying technique of Poisson regression to handle all cases. To avoid unnecessary technicalities we consider discrete time models. While continuous time models are more elegant mathematically, in real studies we measure time in a discrete way, say hourly or daily. Hence a discrete Poisson process model is actually quite natural, and its statistical analysis reduces to the standard Poisson regression. Given a natural time unit dt we can model the outcomes $y(t)$ as a Poisson series with mean $\lambda(t)dt$. This model is valid not just as an approximation of the continuous model; in particular, and there is no need for special handling of ties, which are a problem under continuous time models. The stated results here generally extend to continuous time as a limit of discrete time as dt tends to zero.

The main advantage of the Poisson regression approach for survival-type data is the flexibility in specifying models. Moreover, the model elements generally have a clear interpretation. The price is that each outcome value, a single number y_i, generates a long array $y_i(t)$, so the overall size of the problem becomes large. For a single Poisson process the observed $y(t)$ is a Poisson time series; if dt is small enough then $y(t)$ takes 0-1 values, but this is not a requirement. The log-likelihood contribution from a single series is

$$\sum_t y(t)\log\lambda(t) - \sum_t \lambda(t)dt.$$

To see the generality of this setup, suppose we observe survival data (y_i, δ_i) as described in Section 11.5; δ_i is the event indicator, which is equal to one if y_i is a true event time. After partitioning time by interval unit dt we convert each observed y_i into a 0–1 series $y_i(t)$. For example, $(y_i = 5, \delta_i = 1)$ converts to $y_i(t) = (0,0,0,0,1)$, while $(y_i = 3, \delta_i = 0)$ converts to (0,0,0). Therefore, the log-likelihood contribution of (y_i, δ_i) is

$$\log L_i = \delta_i \log\lambda_i(y_i) - \sum_{t=1}^{y_i} \lambda_i(t)dt.$$

In the limit it can be written as

$$\log L_i = \delta_i \log\lambda_i(y_i) - \int_0^{y_i} \lambda_i(t)dt,$$

exactly the log-likelihood (11.4) we derive in Section 11.7 for general survival data. Any survival regression model that can be expressed in terms of a hazard function has a discrete version as a Poisson model.

Now for convenience we set the time unit $dt \equiv 1$, and let $y_i(t)$ be Poisson with mean $\lambda_i(t)$. To analyse the effect of covariate x_i on the intensity function $\lambda_i(t)$ we may consider, for example, the log-link

$$\log\lambda_i(t) = \alpha(t) + x_i'\beta,$$

which implies a proportional intensity model with baseline intensity

$$\lambda_0(t) = e^{\alpha(t)}.$$

The nuisance parameter $\alpha(t)$ needs further modelling.

Example 11.10: The exponential regression model for survival data (with a maximum of one event per subject) is equivalent to specifying a constant baseline intensity $\lambda_0(t) \equiv \lambda$, or

$$\log \lambda_i(t) = \alpha_0 + x_i'\beta,$$

with $\alpha(t) = \alpha_0 = \log \lambda$. The general extreme-value model, or the Weibull model, is equivalent to specifying a log-linear model

$$\log \lambda_i(t) = \alpha_0 + \alpha_1 \log t + x_i'\beta$$

or

$$\log \lambda_0(t) = \alpha(t) \equiv \alpha_0 + \alpha_1 \log t.$$

To get a specific comparison, recall the survival regression model in Section 11.6 for the underlying survival time t_i as

$$\log t_i = \beta_{0w} + x_i'\beta_w + \sigma W_i,$$

where W_i's are iid with standard extreme-value distribution. Then we have the following relationships (Exercise 11.37):

$$\begin{aligned} \alpha_0 &= -\beta_0/\sigma - \log \sigma \\ \alpha_1 &= 1/\sigma - 1 \\ \beta &= -\beta_w/\sigma. \end{aligned}$$

For multiple events, we can simply modify $\log t$ to $\log z_t$, where z_t is the time elapsed since the last event. Other models can be specified for the baseline intensity function. □

The Cox proportional hazard model is associated with nonparametric $\alpha(t)$. We obtain it by setting $\alpha(t)$ to be a categorical parameter, with one parameter value for each time t. This is useful for modelling, since we can then compare the parametric versus Cox models.

The Cox partial likelihood itself can be derived as a profile likelihood. Suppose the length of observation time is T_i and $y_i(t)$ is nonzero at event times $t_{i1}, \ldots, t_{in_i}$ and zero otherwise, and denote $y_{ij} \equiv y_i(t_{ij})$. If there is no tie, $y_{ij} = 1$ for all i and j. The notation will become very cumbersome here; see the epilepsy data example to get a concrete idea. The log-likelihood contribution of the i'th series is

$$\begin{aligned} \log L_i &= -\sum_{t=1}^{T_i} \lambda_i(t) + \sum_{j=1}^{n_i} y_{ij} \log \lambda_i(t_{ij}) \\ &= -\sum_{t=1}^{T_i} e^{\alpha(t) + x_{it}'\beta} + \sum_{j=1}^{n_i} y_{ij}\{\alpha(t_{ij}) + x_{ij}'\beta\}, \end{aligned}$$

where x_{it} is the value of the covariate x_i at time t and x_{ij} is the value at time t_{ij}. The overall log-likelihood is

$$\log L(\alpha, \beta) = \sum_i \log L_i.$$

To get the profile likelihood of β, we need to find the MLE of $\alpha(t)$ at each value β. To this end we analyse the event and nonevent times separately. At a nonevent time $t \neq t_{ij}$ the estimate $\widehat{\alpha}(t)$ satisfies

$$-\sum_i e^{\widehat{\alpha}(t)+x'_{it}\beta} = 0$$

or $e^{\widehat{\alpha}(t)} = 0$.

At an event time $t = t_{ij}$ we need to keep track of the multiplicity of $\alpha(t_{ij})$ in the first term of the log-likelihood. This can be seen graphically by drawing parallel lines to represent the observation intervals for the subjects. We define the risk set R_{ij} as the set of subjects still under observations at time t_{ij}. We can write the total log-likelihood as

$$\begin{aligned}
\log L(\alpha, \beta) &= -\sum_{i=1}^{n} \sum_{t \neq (t_{i1} \ldots t_{in_i})} e^{\alpha(t)+x'_{it}\beta} \\
&\quad - \sum_{i=1}^{n} \sum_{j=1}^{n_i} \sum_{k \in R_{ij}} e^{\alpha(t_{ij})+x'_{kj}\beta} \\
&\quad + \sum_{i=1}^{n} \sum_{j=1}^{n_i} y_{ij}\{\alpha(t_{ij}) + x'_{ij}\beta\},
\end{aligned}$$

so, in this case $\widehat{\alpha}(t_{ij})$ satisfies

$$\sum_{k \in R_{ij}} e^{\widehat{\alpha}(t_{ij})+x'_{kj}\beta} = y_{ij}$$

or

$$e^{\widehat{\alpha}(t_{ij})} = \frac{y_{ij}}{\sum_{k \in R_{ij}} e^{x'_{kj}\beta}}.$$

Substituting $\widehat{\alpha}(t)$ for all values of t we get the profile likelihood of β, which is proportional to

$$L(\beta) = \prod_{i=1}^{n} \prod_{j=1}^{n_i} \left(\frac{e^{x'_{ij}\beta}}{\sum_{k \in R_{ij}} e^{x'_{kj}\beta}} \right)^{y_{ij}}.$$

If there is no tie, $y_{ij} = 1$ for all i and j, and

$$L(\beta) = \prod_{i=1}^{n} \prod_{j=1}^{n_i} \left(\frac{e^{x'_{ij}\beta}}{\sum_{k \in R_{ij}} e^{x'_{kj}\beta}} \right),$$

exactly the Cox partial likelihood stated in the previous section. Cox's (1972) treatment of ties does not correspond to the result shown above; see Whitehead (1980) for a discussion.

From the derivation it is clear that we can estimate the baseline intensity or hazard function by

$$\widehat{\lambda}_0(t) = e^{\widehat{\alpha(t)}},$$

which takes nonzero values only at the event times. A sensible continuous estimate can be found by smoothing; see Chapter 18.

Example 11.11: For two-sample problems we can arrange the data to limit the problem size. This is because there are multiples of $e^{\alpha(t)+x_i\beta}$ according to $x_i = 0$ or 1. The multiplicity is simply the number of subjects at risk at each event time; this will enter as an offset term in the Poisson regression. Furthermore, to produce Cox regression results we should only consider the statistics at the event times; including time points where there is no event will produce numerical problems, since in this case $e^{\widehat{\alpha(t)}} = 0$.

For the epilepsy data example in the previous section suppose we choose the time unit $dt = 1$ week. Then the data can be summarized as in Table 11.2. The index g in the table now refers to week-by-treatment grouping, with a total of 29 groups with at least one event; R_g is the corresponding number at risk at the beginning of the week; y_g is the number of events during the week.

g	x_g	week$_g$	R_g	y_g	g	x_g	week$_g$	R_g	y_g
1	1	0	10	4	15	0	0	12	9
2	1	1	10	2	16	0	1	12	11
3	1	2	10	4	17	0	2	12	7
4	1	3	10	7	18	0	3	12	5
5	1	4	10	1	19	0	4	10	4
6	1	5	9	1	20	0	5	10	12
7	1	6	9	3	21	0	6	9	6
8	1	7	8	3	22	0	7	7	5
9	1	8	6	0	23	0	8	6	5
10	1	9	6	0	24	0	9	5	2
11	1	10	4	1	25	0	10	5	2
12	1	11	3	0	26	0	11	2	1
13	1	12	1	2	27	0	12	2	1
14	1	13	1	1	28	0	13	2	0
					29	0	14	2	1

Table 11.2: *Data setup for Poisson regression of the epilepsy example. The time intervals are of the form $[k, k+1)$; for example, an event at $t = 1.0$ is included in week-1, not week-0.*

Now assume y_g is Poisson with mean μ_g and use the log-link

$$\log \mu_g = \log R_g + \alpha(\text{week}_g) + \beta_1 x_g,$$

where $\log R_g$ is an offset, and β_1 is the treatment effect. The function $\alpha(\text{week}_g)$ is a generic function expressing the week effect. For example,

$$\alpha(\text{week}_g) \equiv \alpha_0$$

Effect	Parameter	Estimate	se
intercept	α_0	−0.247	0.284
treatment	β_1	−0.767	0.221
week-1	α_1	0.000	0.392
week-2	α_2	−0.167	0.410
week-3	α_3	−0.080	0.400
week-4	α_4	−0.828	0.526
week-5	α_5	0.160	0.392
week-6	α_6	−0.134	0.434
week-7	α_7	−0.045	0.450
week-8	α_8	−0.317	0.526
week-9	α_9	−1.112	0.760
week-10	α_{10}	−0.580	0.641
week-11	α_{11}	−0.975	1.038
week-12	α_{12}	0.444	0.641
week-13	α_{13}	−0.655	1.037
week-14	α_{14}	−0.446	1.040

Table 11.3: *Estimates from Poisson regression of the epilepsy data example.*

for constant intensity function, or

$$\alpha(\text{week}_g) \equiv \alpha_0 + \alpha_1 \times \log \text{week}_g$$

for log-linear time effect, etc. The most general function is setting 'week' to be a categorical variable, with a single parameter for each week; for convenience we set $\alpha_0 \equiv \alpha(0)$, and for $j > 0$ define $\alpha_j = \alpha(j) - \alpha(0)$ as the week-j effect relative to week-0. These options can be compared using the AIC.

For categorical week effects the parameter estimates are given in Table 11.3. The estimate of the treatment effect $\widehat{\beta}_1 = -0.767$ (se $= 0.221$) is similar to the estimates found by different methods in the previous section. Fitting various other models for $\alpha(\text{week}_g)$ is left as an exercise. □

11.11 Exercises

Exercise 11.1: Simulate AR(1) processes shown in Example 11.1, and verify the likelihoods given in Figure 11.1.

Exercise 11.2: In the previous exercise, compare the Fisher information of $\widehat{\phi}_1$ based on the full likelihood $L(\theta)$ and the conditional likelihood $L_2(\theta)$.

Exercise 11.3: For the AR(2) model

$$x_t = \theta_0 + \phi_1 x_{t-1} + \phi_2 x_{t-2} + e_t,$$

where e_t's are iid $N(0, \sigma^2)$. Derive the full and conditional likelihood of the parameters.

Exercise 11.4: Suppose we observe time series data $x_1, \ldots, x_n$, which we identify as an MA(1) series

$$x_t = \theta_0 + e_t - \theta_1 e_{t-1}.$$

Assuming e_t is $N(0, \sigma^2)$ derive the likelihood of the parameters. Is there a form of likelihood that is simpler to compute as in the AR models? Discuss a practical method to compute the likelihood. (Hint: to get the joint density of $x_1, \ldots, x_n$, first consider a transformation from $e_0, e_1, \ldots, e_n$ to $e_0, x_1, \ldots, x_n$, then evaluate the marginal density of $x_1, \ldots, x_n$.)

Exercise 11.5: Verify the standard errors given for the MLEs in Example 11.2.

Exercise 11.6: Use the likelihood approach to test a general hypothesis $H_0 : \theta_{01} = \theta_{11}$ to the data in Example 11.2. Compare the result using the standard error approximation. Interpret what the hypothesis means.

Exercise 11.7: A time series x_t is Markov of order 2, if the conditional distribution of x_t given its past depends only on the last two values. That is,

$$p(x_t|x_{t-1}, \ldots) = p(x_t|x_{t-1}, x_{t-2}).$$

(a) Describe the parameters of the simplest second-order Markov chain with 0-1 outcomes.

(b) Find the likelihood of the parameters given an observed series $x_1, \ldots, x_n$. Identify the 'easy' part of the likelihood.

(c) Estimate the parameters for the data in Example 11.2.

(d) Using the likelihood ratio test, check whether the first-order model is adequate.

Exercise 11.8: Investigate the association between the asthma episodes and the pollution level using the data given in Section 11.2. Compare the results of the two types of regression mentioned in the section.

Exercise 11.9: Verify the summary tables given in Section 11.3.

Exercise 11.10: Draw the profile likelihood of the odds-ratio parameters for the data shown by the 2×2 tables in Section 11.3. Compute the approximate 95% CI for the odds ratio. Compare the χ^2 tests with the Wald tests.

Exercise 11.11: For the data in Section 11.3, derive the likelihood of the parameters p and q for both groups given the summary table (there is a total of four parameters: p_0 and q_0 for the control group and p_1 and q_1 for the treatment group). Use the likelihood ratio test to test the hypothesis H_0: $q_0 = q_1$, and compare with the result given in the section.

Exercise 11.12: In Example 11.3 suppose the follow-up observation of patient 2 of the control group is missing at week 6, so we observe

w0	w2	w4	w6	w8
5	5	5	–	5

Derive the likelihood contribution of this patient.

Exercise 11.13: In Example 11.3, describe a regression model to take into account some possible baseline differences in the two groups such as in age or prior history of other related diseases. Note that the purpose of the analysis is still to compare the response to treatment.

Exercise 11.14: Describe ways to check whether the simplified transition probability matrix given in Section 11.3 is sensible for the data.

Exercise 11.15: The data in Figure 11.9 were collected by Dr. Rosemary Barry in a homeophatic clinic in Dublin from patients suffering from arthritis. Baseline information include age, sex (1= male), arthritis type (RA= rheumathoid arthritis, OA= ostheo-arthritis), and the number of years with the symptom. The pain score was assessed during a monthly followup and graded from 1 to 6 (high is worse) with -9 to indicate missing. All patients were under treatment, and only those with a baseline pain score greater than 3 and a minimum of six visits are reported here. Investigate the trend of pain score over time and the effect of the covariates.

No	Age	Sex	Type	Years	Pain scores											
1	41	0	RA	10	4	4	5	3	3	5						
2	34	0	RA	0	5	2	2	3	2	3	2					
3	53	1	RA	1	5	4	3	2	2	2	1	2	1			
4	38	0	RA	12	6	6	5	6	5	5	5	6				
5	51	0	RA	2	5	5	5	5	4	4						
6	70	0	RA	40	5	4	4	3	3	3	3	3				
7	74	0	RA	5	4	4	5	4	4	3	5	−9	6	5	4	3
8	56	0	RA	1	5	4	−9	5	5	5	4	4	3	3	3	−9
9	57	0	RA	33	5	5	6	5	6	6	−9	6	5	5	5	
10	65	1	RA	18	5	6	6	6	6	6						
11	61	0	RA	12	6	3	5	2	2	2	5	5				
12	64	0	RA	10	4	3	4	4	4	4	3	−9	6	6		
13	47	0	RA	10	5	3	4	4	3	3						
14	59	0	RA	1	6	5	5	6	4	4						
15	54	1	RA	2	6	5	3	4	4	6						
16	74	0	RA	14	4	4	4	4	−9	−9	−9					
17	57	0	RA	2	4	3	3	2	1	2	2	2				
18	86	0	RA	5	4	3	3	2	4	4	4	4	4	5	3	
19	69	0	RA	39	4	2	4	4	5	2	4	6	5			
20	45	0	RA	7	4	4	4	−9	4	4	4					
21	45	0	RA	20	5	5	4	4	5	4	3					
22	70	1	RA	6	6	6	6	6	6	6	6					
23	38	0	RA	0	4	5	4	2	2	3	3	2	2	4		
24	68	0	RA	16	6	−9	4	4	4	3						
25	18	1	RA	1	4	4	3	1	−9	−9	−9	1	1			
26	58	0	RA	1	5	4	3	4	3	3						
27	62	0	RA	1	4	3	5	4	3	4	2					
28	56	0	OA	6	4	5	6	3	5	4	3	3	5			
29	68	0	OA	10	5	5	4	3	2	2	2	2	2	1		
30	64	1	OA	5	5	4	3	5	−9	4						
31	49	0	OA	8	4	4	4	3	3	3	3					
32	66	0	OA	5	4	3	4	3	3	4						
33	70	0	OA	7	4	2	1	1	1	1	2	1				
34	61	0	OA	5	6	−9	3	3	3	5	−9					
35	41	0	OA	15	5	4	3	4	3	2	1	1	2			
36	57	0	OA	4	5	6	6	3	5	6	5	3	5			
37	49	0	OA	4	5	2	2	3	1	2	1	1	1	1	2	1
38	57	0	OA	14	4	4	4	3	5	3	5	3	4	4	3	2
39	78	0	OA	4	5	5	3	4	−9	5						
40	61	0	OA	20	4	4	3	3	3	3	2	3	3	2	2	
41	82	0	OA	40	5	−9	−9	3	3	3	4	5	6	6		
42	48	0	OA	1	5	3	1	1	1	1	2					
43	51	0	OA	2	5	3	3	3	2	3	3					
44	54	0	OA	2	4	4	4	3	2	2	2	2	2			
45	54	0	OA	1	5	5	3	6	6	6						
46	68	0	OA	15	6	5	5	6	3	2						
47	70	1	OA	15	4	4	4	5	5	4						
48	63	1	OA	1	4	4	4	3	2	2	2	1				
49	56	0	OA	4	6	5	3	3	3	3	4	4				
50	66	1	OA	5	4	4	3	3	2	2						
51	64	1	OA	2	5	2	1	1	1	1	1	−9				
52	53	1	OA	2	4	3	3	3	3	3	2	3				
53	58	0	OA	5	4	4	4	4	3	3	2	2	2	2	3	
54	65	1	OA	30	5	5	6	6	6	6	6	6				
55	74	0	OA	40	6	3	1	1	2	1	2	1	2			
56	60	0	OA	57	4	3	4	2	3	3	2	2	2	−9	−9	
57	88	0	OA	10	4	5	5	5	6	4	3	3	3	5		
58	66	0	OA	10	4	2	2	2	1	4	4	3	4			
59	71	0	OA	20	4	4	4	3	4	3						
60	66	0	OA	6	5	−9	4	5	4	3	4					

Figure 11.9: *Arthritis data*

Exercise 11.16: Show that the Ising model in Section 11.4 does imply the conditional probability in the logistic form in both the one-dimensional and two-dimensional models.

Exercise 11.17: Verify the maximum pseudo-likelihood estimates and the standard errors of the parameters (α, β) given in Example 11.4.

Exercise 11.18: To appreciate that the product of conditional probabilities is not really a probability specification, describe how you might simulate a realization of an Ising model given a certain parameter (α, β). (Being able to simulate data is important for Monte Carlo tests.)

Exercise 11.19: Bartlett (1971) reported the absence/presence of a certain plant in a 24×24 square region, reflecting some interaction or competition among *Carex arenaria* plant species.

```
0 1 1 1 0 1 1 1 1 1 0 0 0 1 0 0 1 0 0 1 1 0 1 0
0 1 0 0 1 1 1 0 0 1 0 0 0 0 0 0 0 1 1 1 0 0 0 1
1 1 1 1 0 1 1 0 0 0 0 0 0 0 1 1 1 1 0 0 0 0 0 0
0 0 1 1 1 1 1 0 1 0 1 0 0 0 0 0 1 1 0 0 0 0 0 0
0 1 1 0 1 1 1 1 1 0 0 0 0 1 1 1 0 1 1 0 0 0 0 0
0 1 0 0 0 1 0 1 0 1 1 1 1 1 0 0 0 0 1 1 0 0 0 0
0 1 0 1 1 0 1 0 1 0 0 0 1 0 0 1 1 0 0 1 0 0 0 0
0 0 0 0 0 0 1 1 0 1 0 0 0 0 0 0 1 0 0 0 0 0 0 0
0 0 0 0 0 1 0 1 1 0 0 0 0 0 0 0 1 0 0 0 1 0 0 0
0 0 0 0 0 0 0 1 0 1 0 0 0 0 0 0 1 0 0 0 1 0 0 0
0 0 0 0 0 1 0 0 0 1 1 0 0 0 0 1 1 0 0 0 0 0 0 0
0 0 0 0 1 0 0 0 0 0 1 1 0 0 0 0 1 0 0 0 0 0 0 0
0 0 0 1 0 0 0 0 0 0 0 0 1 0 0 1 0 0 0 0 0 0 1 0
0 0 1 0 0 0 1 1 0 1 0 1 1 1 1 1 1 1 0 0 0 0 1 1
0 1 0 0 1 1 0 0 0 0 0 0 0 1 1 0 0 1 0 1 0 1 0 1
0 0 0 0 0 0 0 0 0 0 0 1 0 0 1 1 0 0 0 1 1 0 0 1
1 0 0 0 0 0 0 1 0 0 1 0 1 0 1 0 0 0 0 0 0 0 0 0
0 0 0 0 0 0 1 0 0 0 0 0 0 1 1 1 0 0 1 1 0 1 0 1
0 1 0 0 0 0 1 0 0 0 0 0 0 1 1 0 0 1 1 1 1 1 0 1
1 0 0 0 0 0 0 0 0 0 0 0 0 1 0 0 0 0 1 0 0 1 0 1
0 0 0 1 0 1 0 0 0 0 0 0 1 0 0 0 0 0 1 0 0 0 0 1
1 0 0 0 0 1 1 0 1 1 0 0 1 0 0 0 0 0 0 0 0 0 0 0
0 0 1 0 0 0 0 0 0 0 0 1 1 0 1 0 0 0 1 0 0 0 0 1
0 0 0 0 1 1 0 0 1 0 0 0 0 0 0 0 0 0 1 0 0 0 0 0
```

(a) Perform and interpret the logistic regression analysis of the Ising model on the dataset.

(b) Test the randomness of the pattern by grouping the data, say into 3×3 or 4×4 squares, and test whether the number of plants on each square follows a Poisson model. What do you expect to see if there is clustering (positive dependence) or if there is negative dependence?

Exercise 11.20: Extend the pseudo-likelihood approach to analyse spatial count data, i.e. y_k is Poisson rather than Bernoulli as in the text. Simulate simple one- and two-dimensional Poisson data and apply the methodology. For a more advanced exercise, a real dataset is given in Table 6 of Breslow and Clayton (1993).

Exercise 11.21: Find the profile likelihood of $\theta = \theta_1/\theta_2$ for the rat data example using the exponential model; report the approximate 95% CI for θ. Compare the Wald statistic based on $\widehat{\theta}$ and $\log \widehat{\theta}$; which is more appropriate?

Exercise 11.22: Perform the group comparison using the normal model as described in Section 11.5. For simplicity, assume the two groups have common variance σ^2, and eliminate σ^2 by replacing it with its estimate from uncensored data (i.e. use the estimated likelihood). Find the profile likelihood of $\mu_1 - \mu_2$.

Exercise 11.23: Repeat the previous exercise, but with the unknown σ^2 estimated using the likelihood from all the data.

Exercise 11.24: Verify the regression analysis performed in Example 11.6.

Exercise 11.25: Derive theoretically the Fisher information for the regression parameters (β_0, β_1) under the exponential and the general extreme-value distributions in Section 11.6.

Exercise 11.26: Compute the Fisher information for the observed data, and verify the standard errors given in Example 11.6. Check the quadratic approximation of the profile likelihood of the slope parameter β_1.

Exercise 11.27: Find the profile likelihood of the scale parameter σ in Example 11.6, and verify the likelihood ratio statistic $W = 44.8$. Report also the Wald test of H_0: $\sigma = 1$, and discuss whether it is appropriate.

Exercise 11.28: If independent lifetimes T_1 and T_2 have proportional hazards, say $\lambda_i(t) = \lambda_0(t)\eta_i$ for $i = 1, 2$ respectively, then show that

$$P(T_1 < T_2) = \eta_1/(\eta_1 + \eta_2)$$

regardless of the shape of the baseline hazard function $\lambda_0(t)$. Generalize the result for $P(T_{i_1} < T_{i_2} < \cdots < T_{i_n})$.

Exercise 11.29: To repeat the exercise given Example 11.7, suppose we observe the following dataset:

i	x_i	y_i	δ_i
1	0	10	1
2	0	5	1
3	0	13	0
4	1	7	1
5	1	21	1
6	1	17	1
7	1	19	0

Assume a proportional hazard model

$$\lambda_i(t) = \lambda_0(t)e^{x_i\beta}.$$

(a) Assume the baseline hazard $\lambda_0(t) \equiv \lambda$, i.e. a constant hazard. Compute the profile likelihood of β.

(b) Compare the profile likelihood in part (a) with the Cox partial likelihood.

(c) Simplify the Cox partial likelihood as much as possible and try to interpret the terms.

(d) Compute the Fisher information of β from the Cox partial likelihood and compare with the information if there is no censoring (i.e. replace the two censored cases with $\delta_i = 1$).

Exercise 11.30: Verify the Cox regression analysis in Example 11.8. Compare the Cox partial likelihood with the profile likelihood found using the exponential model in Exercise 11.21. Check the quadratic approximation of the Cox partial likelihood.

Exercise 11.31: Verify the analysis of software failure data given in Example 11.9. Provide the standard error for $\widehat{\beta}$. Predict the number of bugs still to be found in the system.

Exercise 11.32: Different methods to analyse the epilepsy data in Section 11.9 lead to very similar likelihoods. Provide some explanation, and think of other situations where the three methods would give different results.

Exercise 11.33: For the data in Figure 11.7 find a reasonably simple parametric model for the baseline intensity $\lambda_0(t, \alpha)$ assuming a proportional hazard model.

Exercise 11.34: Assume an exponential decay model as the underlying parameter model for the epilepsy data in Section 11.9. Apply the iterative procedure suggested in Method 3 to estimate the regression parameter β. Compare the likelihood with those obtained by the other methods.

Exercise 11.35: Find the exact formula for $L_{11}(\alpha, \beta)$ in (11.10).

Exercise 11.36: Compare the three methods of analysis for the data in Figure 11.10. The structure is the same as in the previous dataset, but now there is also an additional covariate.

Subject	age_i	x_i	T_i	n_i	Time of events
1	16	active	14	1	1.6
2	22	active	14	3	2 3.2 4.7
3	20	active	1	0	
4	22	active	6	1	0.8
5	21	active	27	1	5
6	17	active	1	0	
7	18	active	1	0	
8	18	active	1	1	0
9	17	active	1	0	
10	22	active	15	2	1.1 5.9
11	21	placebo	13	5	3.9 3.9 4.3 5.5 11.8
12	21	placebo	4	1	0.9
13	22	placebo	6	6	0.5 0.5 0.8 1.2 1.2 2
14	17	placebo	18	1	9.2
15	18	placebo	13	3	2.2 2.5 4.7
16	21	placebo	8	3	0.3 1.4 4.5
17	20	placebo	4	2	3.1 3.4
18	21	placebo	14	0	
19	24	placebo	5	0	
20	24	placebo	17	0	
21	22	placebo	8	4	1.8 6.8 7.2 7.6
22	24	placebo	1	1	0.5

Figure 11.10: *Another sample dataset from an epilepsy clinical trial.*

Exercise 11.37: Prove the relationships stated in Example 11.10 between the coefficients of the Weibull and Poisson regression models.

Exercise 11.38: Verify the relationships stated in Example 11.10 using the rat data example given in Section 11.5.

Exercise 11.39: Verify the data setup in Table 11.2 for the Poisson regression of the epilepsy data. Analyse the data using a constant and log-linear week effect, and compare the different models using the AIC.

12

EM Algorithm

Finding maximum likelihood estimates usually requires a numerical method. Classical techniques such as the Newton–Raphson and Gauss–Newton algorithms have been the main tools for this purpose. The motivation for such techniques generally comes from calculus. A statistically motivated EM (Expectation–Maximization) algorithm appears naturally in problems where

- some parts of the data are missing, and analysis of the *incomplete data* is somewhat complicated or nonlinear;
- it is possible to 'fill in' the missing data, and analysis of the *complete data* is relatively simple.

The notion of 'missing data' does not have to mean data that are actually missing, but any incomplete information. The name 'EM' was coined by Dempster *et al.* (1977) in a seminal paper that showed many algorithms in use at the time were specific examples of a general scheme.

12.1 Motivation

Example 12.1: Consider a two-way table of y_{ij} for $i = 1, 2$ and $j = 1, 2, 3$ with one missing cell y_{23}:

10	15	17
22	23	–

Suppose we consider a linear model

$$y_{ij} = \mu + \alpha_i + \beta_j + e_{ij},$$

where $\sum_i \alpha_i = \sum_j \beta_j = 0$, and the e_{ij}'s are an iid sample from $N(0, \sigma^2)$. The MLEs of μ, α_i and β_j are the minimizer of the sum of squares

$$\sum_{ij} (y_{ij} - \mu - \alpha_i - \beta_j)^2$$

subject to the constraints, where the sum is over the available data $y = (y_{11}, y_{21}, y_{12}, y_{22}, y_{13})$. There is no closed form solution for the MLEs, but they can be found using any regression package by setting the design matrix X appropriately for the parameters μ, α_1, β_1 and β_2:

$$X = \begin{pmatrix} 1 & 1 & 1 & 0 \\ 1 & -1 & 1 & 0 \\ 1 & 1 & 0 & 1 \\ 1 & -1 & 0 & 1 \\ 1 & 1 & -1 & -1 \end{pmatrix}.$$

The solution $(X'X)^{-1}X'y$ is

$$\begin{aligned} \widehat{\mu} &= 19 \\ \widehat{\alpha}_1 &= -5 \\ \widehat{\beta}_1 = -3, & \quad \widehat{\beta}_2 = 0, \end{aligned}$$

and $\widehat{\alpha}_2 = -\widehat{\alpha}_1$, and $\widehat{\beta}_3 = -\widehat{\beta}_1 - \widehat{\beta}_2$. If required, the MLE of σ^2 is

$$\widehat{\sigma}^2 = \frac{1}{5}\sum_{ij}(y_{ij} - \widehat{\mu} - \widehat{\alpha}_i - \widehat{\beta}_j)^2,$$

where the summation is over the available data. In practice the unbiased estimate of σ^2 is usually preferred; in this case we simply replace the divisor $n = 5$ by $n - p = 5 - 4 = 1$.

Had there been no missing data there is a simple closed form solution:

$$\begin{aligned} \widehat{\mu} &= \overline{y} \\ \widehat{\alpha}_i &= \overline{y}_{i.} - \overline{y} \\ \widehat{\beta}_j &= \overline{y}_{.j} - \overline{y}. \end{aligned}$$

How can we use these simple 'complete data' formulae to help us estimate the parameters from the incomplete data?

One way to do that is first to 'fill in' the missing data y_{23}, for example by the average of the available data y, then compute the parameter estimates according to the complete data formulae. This constitutes one cycle of an iteration. The iteration continues by recomputing the missing data

$$\widehat{y}_{23} = \widehat{\mu} + \widehat{\alpha}_2 + \widehat{\beta}_3$$

and the parameter estimates until convergence. In this example, starting with $\widehat{y}_{23} = 17.4$ we obtain:

Iteration	$\widehat{\mu}$	$\widehat{\alpha}_1$	$\widehat{\beta}_1$	$\widehat{\beta}_2$
1	17.4	−3.4	−1.4	1.6
2	17.933	−3.933	−1.933	1.067
3	18.289	−4.289	−2.289	0.711
10	18.958	−4.958	−2.958	0.042
15	18.995	−4.995	−2.995	0.005
21	19.000	−5.000	−3.000	0.000

Thus the algorithm arrives at the solution without inverting the $X'X$ matrix. Given these estimates we can compute $\widehat{\sigma}^2$ similarly as before. □

12.2 General specification

The previous example has all the ingredients of an EM algorithm. The key notions are the incomplete data y and complete data x $(= (y, y_{23})$

above). In general $y = h(x)$ for some array-valued function $h(\cdot)$; this means that y is completely determined by x, but not vice versa. For example, if $x = (x_1, x_2, x_3)$, any of the following is a form of incomplete data:

$$\begin{aligned} y &= (x_1, x_2) \\ y &= (x_1 + x_2, x_2 + x_3) \\ y &= (x_1, x_2 + 2x_3). \end{aligned}$$

The key idea is that some information is lost by going from x to y; this 'loss of information' will be made specific later in terms of Fisher information.

In problems where EM is relevant, the available dataset will be denoted by y. The problem is to estimate θ from the likelihood based on y:

$$L(\theta; y) = p_\theta(y).$$

The dependence on y is made explicit, so we can distinguish it from

$$L(\theta; x) = p_\theta(x),$$

the likelihood based on x. The EM algorithm obtains the MLE $\widehat{\theta}$ by the following iteration: Start with an initial value θ^0, and

- *E-step:* compute the conditional expected value

$$Q(\theta) = Q(\theta|\theta^0) \equiv E\{\log L(\theta; x)|y, \theta^0\}$$

- *M-step: maximize* $Q(\theta)$ to give an updated value θ^1, then go to the E-step using the updated value, and iterate until convergence.

Example 12.2: The famous genetic example from Rao (1973, page 369) assumes that the phenotype data

$$y = (y_1, y_2, y_3, y_4) = (125, 18, 20, 34)$$

is distributed according to the multinomial distribution with probabilities

$$\left\{\frac{1}{2} + \frac{\theta}{4}, \frac{(1-\theta)}{4}, \frac{(1-\theta)}{4}, \frac{\theta}{4}\right\}.$$

The log-likelihood based on y is

$$\log L(\theta; y) = y_1 \log(2 + \theta) + (y_2 + y_3) \log(1 - \theta) + y_4 \log \theta, \qquad (12.1)$$

which does not yield a closed form estimate of θ.

Now we treat y as incomplete data from $x = (x_1, \ldots, x_5)$ with multinomial probabilities

$$\left\{\frac{1}{2}, \frac{\theta}{4}, \frac{(1-\theta)}{4}, \frac{(1-\theta)}{4}, \frac{\theta}{4}\right\}.$$

Here $y_1 = x_1 + x_2$, $y_2 = x_3$, $y_3 = x_4$ and $y_4 = x_5$. The log-likelihood based on x is

$$\log L(\theta; x) = (x_2 + x_5)\log\theta + (x_3 + x_4)\log(1-\theta), \tag{12.2}$$

which readily yields

$$\widehat{\theta} = \frac{x_2 + x_5}{x_2 + x_3 + x_4 + x_5},$$

so the 'complete data' x is simpler than y.

In this example, the E-step is to find

$$\begin{aligned} Q(\theta) &= E(x_2 + x_5|y, \theta^0)\log\theta + E(x_3 + x_4|y, \theta^0)\log(1-\theta) \\ &= \{E(x_2|y, \theta^0) + x_5\}\log\theta + (x_3 + x_4)\log(1-\theta), \end{aligned}$$

so we only need to compute

$$\widehat{x}_2 = E(x_2|y, \theta^0).$$

Since $x_1 + x_2 = y_1$, the conditional distribution of $x_2|y_1$ is binomial with parameters $y_1 = 125$ and probability

$$p^0 = \frac{\theta^0/4}{1/2 + \theta^0/4}$$

so

$$\widehat{x}_2 = y_1\frac{\theta^0/4}{1/2 + \theta^0/4}. \tag{12.3}$$

The M-step yields an update

$$\theta^1 = \frac{\widehat{x}_2 + x_5}{\widehat{x}_2 + x_3 + x_4 + x_5}. \tag{12.4}$$

The algorithm iterates between (12.3) and (12.4). From the last category of y we may obtain a starting value: $\theta^0/4 = 34/197$ or $\theta^0 = 0.69$. The first five iterates are 0.690, 0.635, 0.628, 0.627, 0.627, giving the MLE $\widehat{\theta} = 0.627$. □

There is a similarity between the Newton–Raphson and EM algorithms. With the Newton–Raphson algorithm, we obtain a quadratic approximation of the objective function $f(\theta)$ around an initial estimate θ^0:

$$q(\theta) = f(\theta^0) + f'(\theta^0)(\theta - \theta^0) + \frac{1}{2}f''(\theta^0)(\theta - \theta^0)^2,$$

and find the update θ^1 as the maximizer of $q(\theta)$. The algorithm converges quickly if $f(\theta)$ is well approximated by $q(\theta)$. With the EM algorithm, the objective function $\log L(\theta; y)$ is approximated by $Q(\theta)$.

For the genetic data example above, Figures 12.1(a)–(c) show $\log L(\theta; y)$ and $Q(\theta)$ starting with $\theta^0 = 0.2$. Figure 12.1(d) shows the climb of the likelihood surface by the iterates $\theta^0, \theta^1, \ldots$. Intuitively, a better approximation of $\log L(\theta; y)$ by $Q(\theta)$ implies faster convergence of the EM algorithm.

12.3 Exponential family model

In general, each step of the EM algorithm requires some analytical derivation, and there is no guarantee that a practical EM algorithm exists for

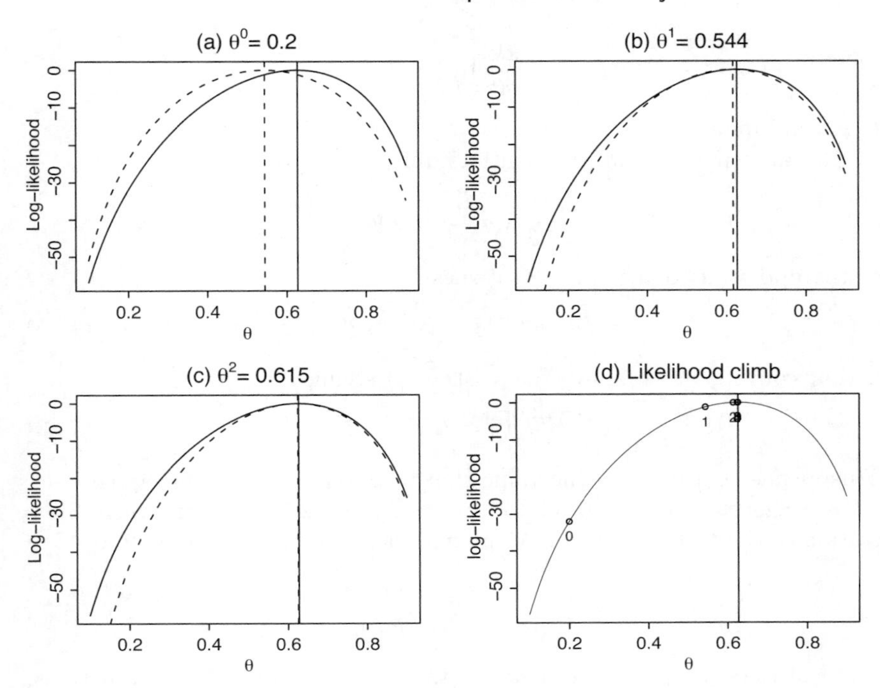

Figure 12.1: *(a)–(c) Log* $L(\theta; y)$ *(solid line) and* $Q(\theta)$ *(dashed line) at successive values of* θ^0, θ^1 *and* θ^2. *(d) The climb on* $\log L(\theta; y)$ *by the iterations.*

a particular incomplete data problem. The algorithm is particularly simple, and theoretically illuminating, if the complete data x is in the full exponential family:

$$\log L(\theta; x) = \theta' T - A(\theta),$$

where $T \equiv T(x)$ is a p-vector of sufficient statistics. (Having a full exponential family is not a necessary condition for the use of EM. For example, the algorithm works well in Example 12.2, even though the model is in a curved exponential family. However, the results in this section do not apply to that example.)

At the n'th iteration the E-step is to find

$$\begin{aligned} Q(\theta|\theta^n) &= E\{\log L(\theta; x)|y, \theta^n\} \\ &= \theta' E(T|y, \theta^n) - A(\theta), \end{aligned}$$

which reduces to finding ('filling in') the conditional expected value

$$\widehat{T} = E\{T|y, \theta^n\}.$$

For the M-step, taking the derivative of $Q(\theta|\theta^n)$ with respect to θ, and setting it to zero, we solve the equation

$$\frac{\partial}{\partial\theta}A(\theta) = \widehat{T}$$

to get an update θ^{n+1}.

Recall that for a full exponential family

$$\frac{\partial}{\partial\theta}A(\theta) = E(T|\theta),$$

so the updating equation for θ satisfies

$$E(T|\theta^{n+1}) = E(T|y,\theta^n), \tag{12.5}$$

and at convergence we have the MLE $\widehat{\theta}$ satisfying

$$E(T|\widehat{\theta}) = E(T|y,\widehat{\theta}). \tag{12.6}$$

This means that $\theta = \widehat{\theta}$ is the value that makes T and y uncorrelated.

As suggested by Navidi (1997), we can use (12.5) for a graphical illustration of the EM algorithm. Assume a scalar parameter and define

$$\begin{aligned} h_1(\theta) &\equiv E(T|y,\theta) \\ h_2(\theta) &\equiv E(T|\theta). \end{aligned}$$

The EM iteration progresses towards the intersection of the two functions; see Figure 12.2.

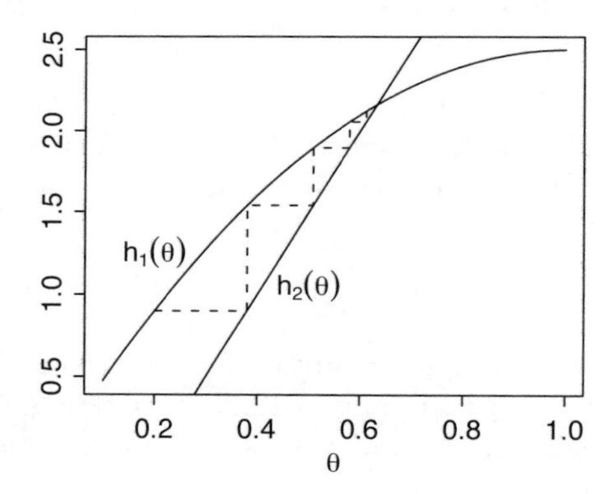

Figure 12.2: *A graphical illustration of the EM iterative process using $h_1(\theta)$ and $h_2(\theta)$.*

It is instructive to see in more detail the behaviour of these functions. In general, the conditional density of x given y is

$$p_\theta(x|y) = \frac{p_\theta(x)}{p_\theta(y)}$$

since y is completely determined by x. So, with obvious notations,

$$\log L(\theta; x|y) = \log L(\theta; x) - \log L(\theta; y) \tag{12.7}$$

$$= \theta T - A(\theta) - \log L(\theta; y),$$

which is also in the exponential family. Taking the derivative with respect to θ, and taking conditional expectation, we obtain

$$E(T|y,\theta) - A'(\theta) - S(\theta; y) = 0$$

so

$$h_1(\theta) = E(T|y,\theta) = A'(\theta) + S(\theta; y)$$

and

$$h_2(\theta) \equiv E(T|\theta) = A'(\theta).$$

The slopes of these functions are

$$\begin{aligned} h_1'(\theta) &= A''(\theta) - I(\theta; y) \\ h_2'(\theta) &= A''(\theta), \end{aligned}$$

where $I(\theta; y)$ is the Fisher information based on y. Since

$$\begin{aligned} \text{var}(T) &= A''(\theta) > 0 \\ \text{var}(T|y) &= A''(\theta) - I(\theta; y) > 0 \end{aligned}$$

both $h_1(\theta)$ and $h_2(\theta)$ are increasing functions of θ. Furthermore $I(\theta; y) > 0$ for θ near $\widehat{\theta}$, so $h_2(\theta)$ has a steeper slope.

Taking the conditional expectation of (12.7) given y yields

$$E\{\log L(\theta; x|y)|y, \theta^0\} = Q(\theta|\theta^0) - \log L(\theta; y). \tag{12.8}$$

Derivatives of $E\{\log L(\theta; x|y)|y, \theta^0\}$ behave like the expected score and Fisher information (Section 8.3); for example,

$$\begin{aligned} I(\theta; x|y) &\equiv -\partial^2 E\{\log L(\theta; x|y)|y, \theta^0\}/\partial\theta^2 \\ &= -E\{\partial^2 \log L(\theta; x|y)/\partial\theta^2|y, \theta^0\}. \end{aligned}$$

Defining

$$I(\theta; x) \equiv -\partial^2 Q(\theta|\theta^0)/\partial\theta^2,$$

we have from (12.8)

$$I(\theta; x|y) = I(\theta; x) - I(\theta; y)$$

or

$$I(\theta; y) = I(\theta; x) - I(\theta; x|y). \tag{12.9}$$

This intuitively means that the information in the incomplete data y is equal to the information in the complete data x minus the extra information in x which is not in y. This is a form of the so-called 'missing information principle' (Orchard and Woodbury 1972).

Near the solution $\widehat{\theta}$ we have

$$\begin{aligned} E(T|\theta) &\approx E(T|\widehat{\theta}) - I(\widehat{\theta}; x)(\theta - \widehat{\theta}) \\ E(T|y,\theta) &\approx E(T|y,\widehat{\theta}) - I(\widehat{\theta}; x|y)(\theta - \widehat{\theta}). \end{aligned}$$

Assuming $\theta^n \to \widehat{\theta}$ as $n \to \infty$, and in view of (12.5) and (12.6), we have

$$\frac{\theta^{n+1} - \widehat{\theta}}{\theta^n - \widehat{\theta}} \approx \frac{I(\widehat{\theta}; x|y)}{I(\widehat{\theta}; x)}.$$

Smaller $I(\widehat{\theta}; x|y)$, meaning less missing information in y relative to x, implies faster convergence. This is a more precise expression of the previous notion that the speed of convergence is determined by how close $Q(\theta|\theta^0)$ is to $\log L(\theta; y)$.

12.4 General properties

One of the most important properties of the EM algorithm is that its step always increases the likelihood:

$$L(\theta^{n+1}; y) \geq L(\theta^n; y). \tag{12.10}$$

This makes EM a numerically stable procedure as it climbs the likelihood surface; in contrast, no such guarantee exists for the Newton–Raphson algorithm. The likelihood-climbing property, however, does not guarantee convergence.

Another practical advantage of the EM algorithm is that it usually handles parameter constraints automatically. This is because each M-step produces an MLE-type estimate. For example, estimates of probabilities are naturally constrained to be between zero and one.

The main disadvantages of the EM algorithm compared with the competing Newton–Raphson algorithm are

- the convergence can be very slow. As discussed above, the speed of covergence is determined by the amount of missing information in y relative to x. It is sometimes possible to manipulate the complete data x to minimize the amount of missing information (Meng and van Dyk 1997), but there is no explicit general technique to achieve it. There are other proposals to accelerate the algorithm to make it closer to the Newton–Raphson algorithm (Lange 1995).
- there are no immediate standard errors for the estimates. If there is an explicit log-likelihood function $\log L(\theta; y)$, then one can easily find the observed Fisher information $I(\widehat{\theta}; y)$ numerically, and find the standard errors from the inverse of $I(\widehat{\theta}; y)$. (This assumes that the standard errors are meaningful quantities for inference about θ, or that $L(\theta; y)$ is regular.) In Section 12.9 we will show that it is possible to find the Fisher information $I(\widehat{\theta}; y)$ from the complete-data function $Q(\theta|\theta^0)$.

To prove (12.10), recall from (12.8) that

$$E\{\log L(\theta;x|y)|y,\theta^0\} = Q(\theta|\theta^0) - \log L(\theta;y). \tag{12.11}$$

Let $h(\theta|\theta^0) \equiv E\{\log L(\theta;x|y)|y,\theta^0\}$. From the information inequality (Theorem 9.5), for any two densities $f(x) \neq g(x)$ we have

$$E_g \log f(X) \leq E_g \log g(X).$$

Applying this to the conditional density of $x|y$,

$$h(\theta|\theta^0) \leq h(\theta^0|\theta^0),$$

and at the next iterate θ^1 we have

$$Q(\theta^1|\theta^0) - \log L(\theta^1;y) \leq Q(\theta^0|\theta^0) - \log L(\theta^0;y),$$

or

$$\log L(\theta^1;y) - \log L(\theta^0;y) \geq Q(\theta^1|\theta^0) - Q(\theta^0|\theta^0) \geq 0.$$

The right-hand side is positive by definition of θ^1 as the maximizer of $Q(\theta|\theta^0)$. In fact, the monotone likelihood-climbing property is satisfied as long as we choose the next iterate that satisfies $Q(\theta^1|\theta^0) \geq Q(\theta^0|\theta^0)$.

Starting at the solution $\widehat{\theta}$, we have

$$Q(\theta|\widehat{\theta}) = \log L(\theta;y) + h(\theta|\widehat{\theta}).$$

Since $\log L(\theta;y)$ and $h(\theta|\widehat{\theta})$ are both maximized at $\theta = \widehat{\theta}$, so is $Q(\theta|\widehat{\theta})$. This means $\widehat{\theta}$ is a fixed point of the EM algorithm. Unfortunately this argument does not guarantee the convergence of the EM algorithm. Wu (1983) shows that for curved exponential families the limit points (where the algorithm might converge to) are the stationary points of the likelihood surface, including all the local maxima and saddlepoints. In particular, this means the EM algorithm can get trapped in a local maximum. If the likelihood surface is unimodal and $Q(\theta|\theta^0)$ is continuous in both θ and θ^0, then the EM algorithm is convergent. In complex cases it is important to try several starting values, or to start with a sensible estimate.

12.5 Mixture models

Let $y = (y_1, \ldots, y_n)$ be an iid sample from a mixture model with density

$$p_\theta(u) = \sum_{j=1}^{J} \pi_j p_j(u|\theta_j) \tag{12.12}$$

where the π_j's are unknown mixing probabilities, such that $\sum_j \pi_j = 1$, $p_j(u|\theta_j)$'s are probability models, and θ_j's are unknown parameters. Hence

θ is the collection of π_j's and θ_j's. The log-likelihood based on the observed data y is

$$\log L(\theta; y) = \sum_i \log \left\{ \sum_{j=1}^{J} \pi_j p_j(y_i|\theta_j) \right\}.$$

Because of the constraints on π_j's, a simplistic application of the Newton–Raphson algorithm is prone to failure.

Mixture models are used routinely in parametric or model-based clustering methodology (McLachlan and Basford 1988). The main interest is to classify the observed data y into different clusters or subpopulations. The EM algorithm is natural here, since one of the by-products of the algorithm is the probability of belonging to one of the J populations.

Example 12.3: Table 12.1 gives the waiting time (in minutes) of $N = 299$ consecutive eruptions of the Old Faithful geyser in Yellowstone National Park. See Azzalini and Bowman (1990) for a detailed analysis of the data. The bimodal

80	71	57	80	75	77	60	86	77	56	81	50	89	54	90	73	60	83
65	82	84	54	85	58	79	57	88	68	76	78	74	85	75	65	76	58
91	50	87	48	93	54	86	53	78	52	83	60	87	49	80	60	92	43
89	60	84	69	74	71	108	50	77	57	80	61	82	48	81	73	62	79
54	80	73	81	62	81	71	79	81	74	59	81	66	87	53	80	50	87
51	82	58	81	49	92	50	88	62	93	56	89	51	79	58	82	52	88
52	78	69	75	77	53	80	55	87	53	85	61	93	54	76	80	81	59
86	78	71	77	76	94	75	50	83	82	72	77	75	65	79	72	78	77
79	75	78	64	80	49	88	54	85	51	96	50	80	78	81	72	75	78
87	69	55	83	49	82	57	84	57	84	73	78	57	79	57	90	62	87
78	52	98	48	78	79	65	84	50	83	60	80	50	88	50	84	74	76
65	89	49	88	51	78	85	65	75	77	69	92	68	87	61	81	55	93
53	84	70	73	93	50	87	77	74	72	82	74	80	49	91	53	86	49
79	89	87	76	59	80	89	45	93	72	71	54	79	74	65	78	57	87
72	84	47	84	57	87	68	86	75	73	53	82	93	77	54	96	48	89
63	84	76	62	83	50	85	78	78	81	78	76	74	81	66	84	48	93
47	87	51	78	54	87	52	85	58	88	79							

Table 12.1: *The waiting time between eruptions of the Old Faithful geyser in Yellowstone National Park.*

nature of the distribution, as shown in Figure 12.3, suggests a mixture of two processes. We model the data as coming from a normal mixture

$$\pi_1 N(\mu_1, \sigma_1^2) + \pi_2 N(\mu_2, \sigma_2^2).$$

Here $\pi_2 = 1 - \pi_1$, so $\theta = (\pi_1, \mu_1, \sigma_1, \mu_2, \sigma_2)$. The log-likelihood function is

$$\log L(\theta; y) = \sum_i \log\{\pi_1 \phi(y_i, \mu_1, \sigma_1^2) + (1 - \pi_1)\phi(y_i, \mu_2, \sigma_2^2)\},$$

where $\phi(y, \mu, \sigma^2)$ is the density of $N(\mu, \sigma^2)$. □

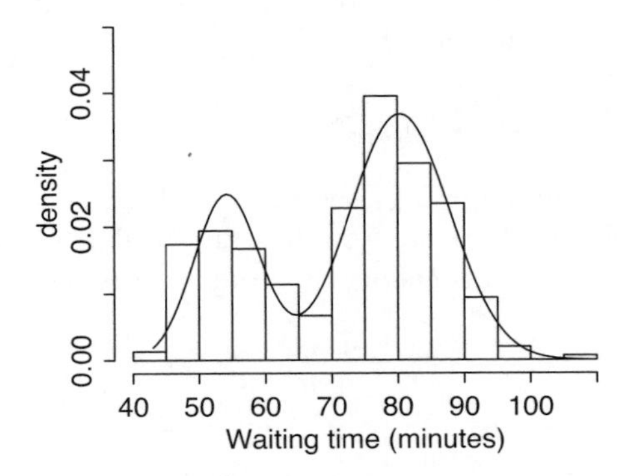

Figure 12.3: *The histogram of the geyser waiting time and the parametric density estimate (solid line) based on the mixture of two normals.*

One interpretation of the mixture model (12.12) is that y_i comes from one of the J populations, but we do not know which one. Had we observed the indicator z_i of where y_i is coming from, each θ_j could be estimated separately, and the estimation of θ would become trivial.

We define the 'complete data' $x = (x_1, \ldots, x_n)$, where $x_i = (y_i, z_i)$. The marginal probability of Z_i is $P(Z_i = j) = \pi_j$; conditional on $z_i = j$, assume y_i has density $p_j(u|\theta_j)$. Now let

$$\log L(\theta; x) = \sum_i \log L(\theta; x_i)$$

where the contribution of x_i to the log-likelihood is

$$\begin{aligned}\log L(\theta; x_i) &= \log p_{z_i}(y_i|\theta_{z_i}) + \log \pi_{z_i} \\ &= \sum_{j=1}^{J} \{I(z_i = j) \log p_j(y_i|\theta_j) + I(z_i = j) \log \pi_j\} \quad (12.13)\end{aligned}$$

and $I(z_i = j) = 1$ if $z_i = j$ and zero otherwise. So, with starting value θ^0, the E-step consists of finding the conditional probabilities

$$\begin{aligned}\widehat{p}_{ij} &= E\{I(Z_i = j)|y_i, \theta^0\} \\ &= P(Z_i = j|y_i, \theta^0) \\ &= \frac{\pi_j^0 p_j(y_i|\theta_j^0)}{p_{\theta^0}(y_i)} \\ &= \frac{\pi_j^0 p_j(y_i|\theta_j^0)}{\sum_k \pi_k^0 p_k(y_i|\theta_k^0)}.\end{aligned}$$

This is the estimated probability of y_i coming from population j; in clustering problems it is the quantity of interest. It is immediate that $\sum_j \widehat{p}_{ij} = 1$ for each i.

From (12.13) the M-step update of each θ_j is based on *separately* maximizing the weighted log-likelihood

$$\sum_i \widehat{p}_{ij} \log p_j(y_i|\theta_j).$$

While there is no guarantee of a closed-form update, this is a major simplication of the problem. Explicit formulae are available, for example, for the normal model. We can also show the update formula

$$\pi_j^1 = \frac{\sum_i \widehat{p}_{ij}}{n}$$

for the mixing probabilities.

Example 12.3: continued. For the Old Faithful data the weighted likelihood of the j'th parameter $\theta_j = (\mu_j, \sigma_j)$ is

$$-\frac{1}{2}\sum_i \widehat{p}_{ij}\left\{\log \sigma_j^2 + \frac{(y_i-\mu_j)^2}{\sigma_j^2}\right\},$$

which yields the following weighted averages as updates:

$$\mu_j^1 = \frac{\sum_i \widehat{p}_{ij} y_i}{\sum_i \widehat{p}_{ij}}$$

$$\sigma_j^{2(1)} = \frac{\sum_i \widehat{p}_{ij}(y_i-\mu_j^1)^2}{\sum_i \widehat{p}_{ij}}.$$

Starting with ($\pi_1^0 = 0.3, \mu_1^0 = 55, \sigma_1^0 = 4, \mu_2^0 = 80, \sigma_2^0 = 7$) we obtain the following iterations:

Iteration	π_1	μ_1	σ_1	μ_2	σ_2
1	0.306	54.092	4.813	80.339	7.494
2	0.306	54.136	4.891	80.317	7.542
3	0.306	54.154	4.913	80.323	7.541
5	0.307	54.175	4.930	80.338	7.528
10	0.307	54.195	4.946	80.355	7.513
15	0.308	54.201	4.951	80.359	7.509
25	0.308	54.203	4.952	80.360	7.508

Hence we obtain

$$(\widehat{\pi}_1 = 0.308, \widehat{\mu}_1 = 54.203, \widehat{\sigma}_1 = 4.952, \widehat{\mu}_2 = 80.360, \widehat{\sigma}_2 = 7.508),$$

giving the density estimate

$$\widehat{p}(u) = \widehat{\pi}_1 \phi(u, \widehat{\mu}_1, \widehat{\sigma}_1^2) + (1-\widehat{\pi}_1)\phi(u, \widehat{\mu}_2, \widehat{\sigma}_2^2).$$

Figure 12.3 compares this parametric density estimate with the histogram. □

12.6 Robust estimation

As described in Section 6.9 we can perform a robust regression analysis by assuming a heavy-tailed error distribution. The EM algorithm applied to this problem becomes an IWLS algorithm.

Suppose $y_1, \ldots, y_n$ are independent with locations $\mu_1, \ldots, \mu_n$ and a common scale parameter, such that

$$\mu_i = x_i'\beta,$$

or we write it as a regression model

$$y_i = x_i'\beta + e_i,$$

where the error e_i has a t_k-distribution with unknown scale σ and degrees of freedom k. Hence the total parameter is $\theta = (\beta, \sigma, k)$. From the t_k-density function, the contribution of y_i to the observed-data likelihood is

$$\begin{aligned}\log L(\theta; y_i) &= \log\Gamma(k/2 + 1/2) - \log\Gamma(k/2) - \frac{1}{2}\log k - \frac{1}{2}\log\sigma^2 \\ &\quad -\frac{k+1}{2}\log\left\{k + \frac{(y_i - \mu_i)^2}{\sigma^2}\right\}. \end{aligned} \tag{12.14}$$

The way to proceed with the EM algorithm may not be immediately obvious here, but recall that we may write

$$e_i = \sigma z_i / \sqrt{w_i},$$

where z_i is $N(0,1)$, and w_i is χ_k^2/k independent of z_i. So, if we knew w_i the regression problem would reduce to a normal-based regression problem.

Defining the 'complete data' as $x_i = (y_i, w_i)$, for $i = 1, \ldots, n$, the contribution of x_i to the complete data likelihood is

$$L(\theta; x_i) = p(y_i|w_i)p(w_i).$$

The conditional distribution $y_i|w_i$ is normal with mean μ_i and variance σ^2/w_i; the density of w_i is

$$p(w) = \frac{1}{2^{k/2}\Gamma(k/2)} w^{k/2-1} e^{-kw/2}.$$

Hence

$$\log L(\theta; x_i) = v(k) + \frac{k-1}{2}\log w_i - \frac{kw_i}{2} - \frac{1}{2}\log\sigma^2 - \frac{w_i(y_i - \mu_i)^2}{2\sigma^2},$$

where $v(k)$ is a function involving k only.

The E-step consists of finding $E(\log w_i|y_i, \theta^0)$ and $E(w_i|y_i, \theta^0)$. There is no closed form result for the former, thus raising a question regarding the practicality of the algorithm. Note, however, that $E(\log w_i|y_i, \theta^0)$ is only needed for updating k, while updating β and σ^2 only requires $E(w_i|y_i, \theta^0)$.

What we can do instead is to consider the estimation of β and σ^2 at each fixed k, such that we get a profile likelihood of k from $\log L(\theta; y)$ in (12.14). The MLE of k is then readily available from the profile likelihood.

Thus the EM algorithm can be performed at each k. The E-step reduces to finding

$$\widehat{w}_i \equiv E(w_i|y_i, \beta^0, \sigma^{2(0)}).$$

We can show that the conditional distribution of $w_i|y_i$ is $\chi^2_{k+1}/(k+d_i^2)$, where

$$d_i^2 = \frac{(y_i - \mu_i^0)^2}{\sigma^{2(0)}},$$

so

$$\widehat{w}_i = \frac{k+1}{k+d_i^2}.$$

For β and σ^2 only, the relevant term of the E-step function is

$$Q = -\frac{n}{2}\log\sigma^2 - \frac{1}{2\sigma^2}\sum_i \widehat{w}_i(y_i - \mu_i)^2.$$

This is the usual likelihood associated with weighted least squares, so the update of β is

$$\beta^1 = (X'WX)^{-1}X'Wy,$$

where the weight matrix W is a diagonal matrix of $\widehat{w}_i$, and the update of σ^2 is

$$\sigma^{2(1)} = \frac{1}{n}\sum_i \widehat{w}_i(y_i - \mu_i^1)^2,$$

thus completing the M-step.

For one-sample problems, where $y_1, \ldots, y_n$ have a common location μ, the update formula for μ is simply a weighted average

$$\mu^1 = \frac{\sum_i \widehat{w}_i y_i}{\sum_i \widehat{w}_i}.$$

The weight $\widehat{w}_i$ is small if y_i is far from μ^0, so outlying observations are downweighted. This is the source of the robustness in the estimation.

12.7 Estimating infection pattern

Some diseases such as AIDS or hepatitis are transmitted by intermittent exposures to infective agents, and a person once infected remains infected for life. Reilly and Lawlor (1999) describe an application of the EM algorithm in the identification of contaminated blood products that led to hepatitis C among rhesus-negative women.

Suppose, at a certain point in time, the exposure history and the infection status of a person are known, but the precise time of infection (in the person's past) is not known. For example, the exposure history of a person over a 10-year period might be

$$0\ 1\ 0\ 0\ 1\ 0\ 0\ 0\ 1\ 0,$$

meaning that this person was exposed at years 2, 5 and 9. If the final infection status (0 or 1) is only ascertained at year 11, a dataset for eight individuals may look like

```
0 1 0 0 1 0 0 0 1 0 1
0 0 0 0 0 1 0 0 0 0 1
0 0 0 1 1 0 1 0 0 0 1
0 0 0 1 0 1 0 1 1 0 0
0 1 0 0 0 0 0 1 0 0 0
1 1 0 0 0 0 0 0 0 0 0
0 0 0 0 0 0 0 0 1 1 0
0 0 0 0 1 0 0 1 0 0 0
```

where the last column indicates the infection status at year 11; to distinguish it from infections at various exposures, we will call it the *final status.* With the real data it is not necessary for the individuals to have the same length of exposure time, but, because of the definition of the exposure variable, we can simply augment the necessary zeros where there are no exposures. It is assumed, however, that every subject's final status is known at the time of analysis. What we want to estimate is the infection rate in each year.

For individual i, let y_i be the final status indicator, and z_{ij} be the exposure indicator at time $j = 1, \ldots, J$. We denote by p_j the infection rate at time j, and assume infections at different exposures are independent. A person can of course be infected more than once, but there is only one final status. Thus the full parameter set is $\theta = (p_1, \ldots, p_J)$, and these probabilities do not have to add up to one.

The contribution of y_i to the observed-data log-likelihood is

$$\log L(\theta; y_i) = y_i \log \pi_i + (1 - y_i) \log(1 - \pi_i)$$

where

$$\begin{aligned} \pi_i &= P(y_i = 1) \\ &= P(\text{subject } i \text{ is infected at least once during the exposures}) \\ &= 1 - \prod_k (1 - z_{ij} p_j). \end{aligned}$$

To use the EM algorithm, notice that the problem is trivial if we know whether or not subject i is infected at each time of exposure. Let (unobserved) x_{ij} be the infection indicator for person i at time j, and let $x_i = (x_{i1}, \ldots, x_{iJ})$. If $z_{ij} = 1$ then x_{ij} is Bernoulli with parameter p_j. Obviously $x_{ij} = 0$ if $z_{ij} = 0$, so overall x_{ij} is Bernoulli with parameter $z_{ij} p_j$. The observed data $y_i = \max(x_{i1}, \ldots, x_{iJ})$; so, we define $x = (x_1, \ldots, x_n)$ as the complete data. The contribution of x_i to the complete-data log-likelihood is

$$\begin{aligned} \log L(\theta; x_i) &= \sum_{j=1}^{J} \{x_{ij} \log(z_{ij} p_j) + (1 - x_{ij}) \log(1 - z_{ij} p_k)\} \\ &= \sum_{j=1}^{J} \{z_{ij} x_{ij} \log p_j + z_{ij}(1 - x_{ij}) \log(1 - p_k)\}. \end{aligned}$$

The E-step consists of taking the conditional expectation $E(x_{ij}|y_i, \theta^0)$. First, if $y_i = 0$ the subject cannot have been infected in the past, so $E(x_{ij}|y_i = 0, \theta^0) = 0$ for all j. So we only need to derive the case where $y_i = 1$:

$$\begin{aligned} E(x_{ij}|y_i = 1) &= P(x_{ij} = 1|y_i = 1) \\ &= \frac{P(x_{ij} = 1)}{P(y_i = 1)} \\ &= \frac{z_{ij}p_j}{1 - \prod_k(1 - z_{ik}p_k)}. \end{aligned}$$

So, for the E-step we compute

$$\widehat{x}_{ij} = \frac{z_{ij}p_j^0}{1 - \prod_k(1 - z_{ik}p_k^0)} \tag{12.15}$$

if $y_i = 1$, and $\widehat{x}_{ij} = 0$ if $y_i = 0$. The E-step function is

$$Q(\theta|\theta^0) = \sum_{i=1}^{n}\sum_{j=1}^{J}\{z_{ij}\widehat{x}_{ij}\log p_j + z_{ij}(1 - \widehat{x}_{ij})\log(1 - p_j)\}.$$

Taking the derivative with respect to p_j and setting it to zero, we obtain the M-step update

$$p_j^1 = \frac{\sum_{i=1}^n z_{ij}\widehat{x}_{ij}}{\sum_{i=1}^n z_{ij}}. \tag{12.16}$$

This formula has a sensible interpretation: the numerator represents the number of exposures at time j which resulted in infection, and the denominator is the total number of exposures at time j. The EM algorithm iterates between (12.15) and (12.16) until convergence.

12.8 Mixed model estimation*

Let an N-vector y be the outcome variate, X and Z be $N \times p$ and $N \times q$ design matrices for the fixed effects parameter β and random effects b. The standard linear mixed model specifies

$$y = X\beta + Zb + e \tag{12.17}$$

where e is $N(0, \Sigma)$, b is $N(0, D)$, and b and e are independent. The variance matrices Σ and D are parameterized by an unknown variance parameter θ. The full set of (fixed) parameters is (β, θ).

Estimation of those parameters would be simpler if b is observed, so the natural complete data is $x = (y, b)$. The likelihood of the fixed parameters based on $x = (y, b)$ is

$$L(\beta, \theta; x) = p(y|b)p(b).$$

From the mixed model specification, the conditional distribution of y given b is normal with mean

$$E(y|b) = X\beta + Zb$$

and variance Σ. The random parameter b is normal with mean zero and variance D, so we have

$$\begin{aligned}
\log L(\beta, \theta; x) &= -\frac{1}{2}\log|\Sigma| - \frac{1}{2}(y - X\beta - Zb)'\Sigma^{-1}(y - X\beta - Zb) \\
&\quad -\frac{1}{2}\log|D| - \frac{1}{2}b'D^{-1}b \\
&= -\frac{1}{2}\log|\Sigma| - \frac{1}{2}e'\Sigma^{-1}e - \frac{1}{2}\log|D| - \frac{1}{2}b'D^{-1}b, \quad (12.18)
\end{aligned}$$

using $e = y - X\beta - Zb$.

The E-step of the EM algorithm consists of taking the conditional expected values

$$\begin{aligned}
E(e'\Sigma^{-1}e|y, \beta^0, \theta^0) &= \text{trace } \Sigma^{-1}E(ee'|y, \beta^0, \theta^0) \\
E(b'D^{-1}b|y, \beta^0, \theta^0) &= \text{trace } D^{-1}E(bb'|y, \beta^0, \theta^0).
\end{aligned}$$

Now, for any random vector U with mean μ and variance V we have

$$E(UU') = \mu\mu' + V.$$

(The univariate version is familiar: $EU^2 = \mu^2 + V$.) From $e = y - X\beta - Zb$, and defining

$$\begin{aligned}
\widehat{b} &\equiv E(b|y, \beta^0, \theta^0) \\
V_b &\equiv \text{var}(b|y, \beta^0, \theta^0),
\end{aligned}$$

we have

$$\begin{aligned}
\widehat{e} &\equiv E(e|y, \beta^0, \theta^0) = y - X\beta - Z\widehat{b} \\
V_e &\equiv \text{var}(e|y, \beta^0, \theta^0) = ZV_bZ',
\end{aligned}$$

so

$$\begin{aligned}
E(ee'|y, \beta^0, \theta^0) &= \widehat{e}\widehat{e}' + ZV_bZ' \\
E(bb'|y, \beta^0, \theta^0) &= \widehat{b}\widehat{b}' + V_b.
\end{aligned}$$

Hence the E-step reduces to finding $\widehat{b}$ and V_b. Some knowledge of multivariate normal theory is needed here, but mostly it involves a careful rewriting of the univariate theory to take account of the size of the vectors or matrices. To start,

$$\text{cov}(y, b) = \text{cov}(X\beta + Zb + e, b) = Z\text{cov}(b, b) = ZD,$$

or $\text{cov}(b, y) = DZ'$. Suppressing the parameters, the conditional expected value $E(b|y)$ is the usual regression estimate

$$E(b|y) = Eb + \text{cov}(b, y)\{\text{var}(y)\}^{-1}(y - Ey)$$

$$\begin{aligned} &= DZ'(\Sigma + ZDZ')^{-1}(y - X\beta) \\ &= (Z'\Sigma^{-1}Z + D^{-1})^{-1}Z'\Sigma^{-1}(y - X\beta), \end{aligned} \tag{12.19}$$

where we have used $Eb = 0$, $Ey = X\beta$ and $\text{var}(y) = \Sigma + ZDZ'$. The last step uses the following matrix identity:

$$\begin{aligned} DZ'(\Sigma + ZDZ')^{-1} &= (Z'\Sigma^{-1}Z + D^{-1})^{-1}(Z'\Sigma^{-1}Z + D^{-1})DZ'(\Sigma + ZDZ')^{-1} \\ &= (Z'\Sigma^{-1}Z + D^{-1})^{-1}Z'\Sigma^{-1}(ZDZ' + \Sigma)(\Sigma + ZDZ')^{-1} \\ &= (Z'\Sigma^{-1}Z + D^{-1})^{-1}Z'\Sigma^{-1}. \end{aligned}$$

This identity will be used again to derive the variance fomula. In the E-step, all of the unknown parameters in (12.19) are set at the current value; that is,

$$\widehat{b} = (Z'\Sigma_0^{-1}Z + D_0^{-1})^{-1}Z'\Sigma_0^{-1}(y - X\beta^0), \tag{12.20}$$

with Σ_0 and D_0 indicating the current values of Σ and D.

From the normal regression theory we obtain the residual variance

$$\begin{aligned} \text{var}(b|y) &= \text{var}(b) - \text{cov}(b, y)\{\text{var}(y)\}^{-1}\text{cov}(y, b) \\ &= D - DZ'(\Sigma + ZDZ')^{-1}ZD \\ &= D - (Z'\Sigma^{-1}Z + D^{-1})^{-1}Z'\Sigma^{-1}ZD \\ &= \{I - (Z'\Sigma^{-1}Z + D^{-1})^{-1}Z'\Sigma^{-1}Z\}D \\ &= \{(Z'\Sigma^{-1}Z + D^{-1})^{-1}D^{-1}\}D \\ &= (Z'\Sigma^{-1}Z + D^{-1})^{-1}. \end{aligned}$$

Hence at the current values we have

$$V_b = (Z'\Sigma_0^{-1}Z + D_0^{-1})^{-1}. \tag{12.21}$$

Combining all the components, we obtain the E-step function

$$\begin{aligned} Q &\equiv Q(\beta, \theta|\beta^0, \theta^0) = E\log(\beta, \theta; x|y, \beta^0, \theta^0) \\ &= -\frac{1}{2}\log|\Sigma| - \frac{1}{2}\widehat{e}'\Sigma^{-1}\widehat{e} - \frac{1}{2}\log|D| - \frac{1}{2}\widehat{b}'D^{-1}\widehat{b} \\ &\quad -\frac{1}{2}\text{trace}\{(Z'\Sigma^{-1}Z + D^{-1})(Z'\Sigma_0^{-1}Z + D_0^{-1})^{-1}\} \end{aligned} \tag{12.22}$$

where β enters the function through $\widehat{e} = y - X\beta - Z\widehat{b}$.

The M-step of the EM algorithm involves maximizing Q with respect to β and θ. Taking the derivative of Q with respect to β

$$\frac{\partial Q}{\partial \beta} = X'\Sigma^{-1}(y - X\beta - Z\widehat{b})$$

and setting it to zero, we can first update β^0 by solving the weighted least-squares equation

$$(X'\Sigma^{-1}X)^{-1}\beta = X'\Sigma^{-1}(y - Z\widehat{b}). \tag{12.23}$$

Σ is a function of the unknown θ, which is to be solved/updated jointly with β, but we can simply set it at the current value Σ_0 to solve for β, and then use the updated β when we update the variance parameter θ.

There is no general updating formula for the variance parameters θ, except for specific covariance structures as given in the following example. Because of their similarity, see the numerical examples given in Section 17.5 for further examples of the EM algorithm.

Example 12.4: Suppose we have a normal mixed model with

$$\begin{aligned}\Sigma &= \sigma^2 A \\ D &= \sigma_b^2 R,\end{aligned}$$

where A and R are known matrices of rank N and q respectively, so $\theta = (\sigma^2, \sigma_b^2)$; in many applications A is an identity matrix. From the properties of determinant,

$$\begin{aligned}|\Sigma| &= \sigma^{2N}|A| \\ |D| &= \sigma_b^{2q}|R|,\end{aligned}$$

so, after dropping irrelevant constant terms,

$$\begin{aligned}Q &= -\frac{N}{2}\log\sigma^2 - \frac{1}{2\sigma^2}\widehat{e}'A^{-1}\widehat{e} - \frac{q}{2}\log\sigma_b^2 - \frac{1}{2\sigma_b^2}\widehat{b}'R^{-1}\widehat{b} \\ &\quad -\frac{1}{2}\text{trace}\{(\sigma^{-2}Z'A^{-1}Z + \sigma_b^{-2}R^{-1})(Z'\Sigma_0^{-1}Z + D_0^{-1})^{-1}\},\end{aligned}$$

where Σ_0 and D_0 are computed using the current values $\sigma^{2(0)}$ and $\sigma_b^{2(0)}$. The derivatives of Q are

$$\begin{aligned}\frac{\partial Q}{\partial\sigma^2} &= -\frac{N}{2\sigma^2} + \frac{1}{2\sigma^4}\widehat{e}'A^{-1}\widehat{e} \\ &\quad +\frac{1}{2\sigma^4}\text{trace}\{(Z'\Sigma_0^{-1}Z + D_0^{-1})^{-1}Z'A^{-1}Z\} \\ \frac{\partial Q}{\partial\sigma_b^2} &= -\frac{q}{2\sigma_b^2} + \frac{1}{2\sigma_b^4}\widehat{b}'R^{-1}\widehat{b} \\ &\quad +\frac{1}{2\sigma_b^4}\text{trace}\{(Z'\Sigma_0^{-1}Z + D_0^{-1})^{-1}R^{-1}\}.\end{aligned}$$

Setting these to zero, we obtain the M-step updating formulae:

$$\sigma^{2(1)} = \frac{1}{N}[\widehat{e}'A^{-1}\widehat{e} + \text{trace}\{(Z'\Sigma_0^{-1}Z + D_0^{-1})^{-1}Z'A^{-1}Z\}] \quad (12.24)$$

$$\sigma_b^{2(1)} = \frac{1}{q}[\widehat{b}'R^{-1}\widehat{b} + \text{trace}\{(Z'\Sigma_0^{-1}Z + D_0^{-1})^{-1}R^{-1}\}], \quad (12.25)$$

where, following the discussion after (12.23), we set $\widehat{e} = y - X\beta^1 - Z\widehat{b}$. Thus the complete EM algorithm starts with some estimates β^0 (e.g. ordinary least-squares estimate) and θ^0, and iterates through (12.20), (12.23), (12.24) and (12.25). □

12.9 Standard errors

One weakness of the EM algorithm is that it does not automatically provide standard errors for the estimates. If the observed data likelihood $L(\theta; y)$ is available, one can find the standard errors from the observed Fisher information $I(\widehat{\theta}; y)$ as soon as $\widehat{\theta}$ is obtained. $I(\widehat{\theta}; y)$ can be computed analytically or numerically; at this stage we are not trying to solve any equation, so this direct method is straightforward.

There have been many proposals for expressing $I(\theta; y)$ in terms of complete-data quantities. The hope is that we obtain simpler formulae, or quantities that can be computed as a by-product of the EM algorithm (e.g. Louis 1982; Meng and Rubin 1991; Oakes 1999). It turns out that it is possible to express the observed score and Fisher information in terms of derivatives of the E-step function $Q(\theta|\theta^0)$, although it does not necessarily mean that they are simpler than the derivatives of $\log L(\theta; y)$.

We will follow Oakes (1999), but for simplicity only consider a scalar parameter θ; the multiparameter case involves only a slight change of notation to deal with vectors and matrices properly. Recall from (12.11) that

$$\log L(\theta; y) = Q(\theta|\theta^0) - h(\theta|\theta^0). \tag{12.26}$$

Taking the derivatives of (12.26) with respect to θ we get

$$S(\theta; y) = \frac{\partial Q(\theta|\theta^0)}{\partial\theta} - \frac{\partial h(\theta|\theta^0)}{\partial\theta}, \tag{12.27}$$

and

$$\begin{aligned} I(\theta; y) &= -\frac{\partial^2 Q(\theta|\theta^0)}{\partial\theta^2} + \frac{\partial^2 h(\theta|\theta^0)}{\partial\theta^2} \\ &= I(\theta; x) - I(\theta; x|y), \end{aligned} \tag{12.28}$$

the missing information principle we found earlier in (12.9). We might try to use (12.28) to get $I(\theta; y)$.

Example 12.5: In the genetic data example (Example 12.2) the conditional distribution of x given y reduces to that of (x_1, x_2) given y_1. This is binomial with parameters y_1 and $\{2/(2+\theta), \theta/(2+\theta)\}$, so

$$\log L(\theta; x|y) = x_1 \log\left(\frac{2}{2+\theta}\right) + x_2 \log\left(\frac{\theta}{2+\theta}\right)$$

and

$$h(\theta|\theta^0) = \widehat{x}_1 \log\left(\frac{2}{2+\theta}\right) + \widehat{x}_2 \log\left(\frac{\theta}{2+\theta}\right), \tag{12.29}$$

where $\widehat{x}_2 = y_1\theta^0/(2+\theta^0)$, and $\widehat{x}_1 = y_1 - \widehat{x}_2$. Deriving $I(\theta; x|y)$ from (12.29) is not simpler than deriving $I(\theta; y)$ from $\log L(\theta; y)$ shown in (12.1). □

Now $\partial h(\theta|\theta^0)/\partial\theta = 0$ at $\theta = \theta^0$, since $h(\theta|\theta^0)$ is maximized at $\theta = \theta^0$. So, from (12.27),

$$S(\theta^0; y) = \left.\frac{\partial Q(\theta|\theta^0)}{\partial\theta}\right|_{\theta=\theta^0},$$

expressing the observed score in terms of the first derivative of Q. For independent data $y_1, \ldots, y_n$ this relationship holds at each y_i; hence

$$S(\theta^0; y_i) = \left.\frac{\partial Q_i(\theta|\theta^0)}{\partial \theta}\right|_{\theta=\theta^0},$$

with $Q_i = E\{\log L(\theta; x_i)|y_i, \theta^0\}$. We can use this result to provide an estimate of the observed Fisher information, since

$$I(\widehat{\theta}; y) \approx \sum_i S^2(\widehat{\theta}; y_i).$$

Now watch carefully as we are going to treat both θ and θ^0 as free variables. Taking the derivative of (12.27) with respect to θ^0 gives

$$0 = \frac{\partial^2 Q(\theta|\theta^0)}{\partial\theta\partial\theta^0} - \frac{\partial^2 h(\theta|\theta^0)}{\partial\theta\partial\theta^0},$$

but, from the definition of $h(\theta|\theta^0)$,

$$\begin{aligned}\frac{\partial^2 h(\theta|\theta^0)}{\partial\theta\partial\theta^0} &= \frac{\partial}{\partial\theta^0} E\left\{\frac{\partial}{\partial\theta}\log p_\theta(X|y)|y, \theta^0\right\} \\ &= \frac{\partial}{\partial\theta^0}\int\left\{\frac{\partial}{\partial\theta}\log p_\theta(x|y)\right\} p_{\theta^0}(x|y)dx \\ &= E\left\{\frac{\partial}{\partial\theta}\log p_\theta(X|y)\frac{\partial}{\partial\theta^0}\log p_{\theta^0}(X|y)|y, \theta^0\right\}.\end{aligned}$$

So,

$$\begin{aligned}\left.\frac{\partial^2 Q(\theta|\theta^0)}{\partial\theta\partial\theta^0}\right|_{\theta=\theta^0} &= E\left[\left\{\frac{\partial}{\partial\theta^0}\log p_{\theta^0}(X|y)\right\}^2 |y, \theta^0\right] \\ &= I(\theta^0; x|y). \qquad (12.30)\end{aligned}$$

From (12.28) and (12.30) we have a general formula

$$I(\theta^0; y) = -\left.\left\{\frac{\partial^2 Q(\theta|\theta^0)}{\partial\theta^2} + \frac{\partial^2 Q(\theta|\theta^0)}{\partial\theta\partial\theta^0}\right\}\right|_{\theta=\theta^0},$$

expressing the observed Fisher information in terms of derivatives of Q.

Example 12.6: In the full exponential family described in Section 12.3

$$Q(\theta|\theta^0) = \theta E(T|y, \theta^0) - A(\theta),$$

so

$$\frac{\partial^2 Q(\theta|\theta^0)}{\partial\theta^2} = -A''(\theta)$$

and

$$\frac{\partial^2 Q(\theta|\theta^0)}{\partial\theta\partial\theta^0} = \frac{\partial}{\partial\theta^0}E(T|y, \theta^0),$$

so

$$I(\theta; y) = A''(\theta) - \frac{\partial}{\partial\theta}E(T|y, \theta),$$

although we could have obtained this directly from the more general formula $I(\theta; y) = I(\theta; x) - I(\theta; x|y)$.

Alternatively, for independent data $y_1, \ldots, y_n$, we have

$$S(\theta^0; y_i) = \left. \frac{\partial Q_i(\theta|\theta^0)}{\partial \theta} \right|_{\theta=\theta^0} = E(T_i|y_i, \theta^0) - A_i'(\theta^0),$$

where $T_i \equiv T(x_i)$ and $A_i(\theta)$ is the A-function associated with x_i, so

$$\begin{aligned} I(\theta; y) &\approx \sum_i \{E(T_i|y_i, \theta) - A_i'(\theta)\}^2 \\ &= \sum_i \{E(T_i|y_i, \theta) - E(T_i|\theta)\}^2, \end{aligned}$$

to be evaluated at $\theta = \widehat{\theta}$. □

12.10 Exercises

Exercise 12.1: Verify the missing information principle

$$I(\theta; y) = I(\theta; x) - I(\theta; x|y)$$

in the genetic model in Example 12.2.

Exercise 12.2: Recording the difference in maximal solar radiation between two geographical regions over time produced the following (sorted) data:

```
-26.8  -3.5  -3.4  -1.2   0.4   1.3   2.3   2.7   3.0   3.2   3.2
  3.5   3.6   3.9   4.2   4.4   5.0   6.5   6.7   7.1   8.1  10.5
 10.7  24.0  32.8
```

Fit a t-distribution to the data, and estimate the location, scale and degrees of freedom using the EM algorithm.

Exercise 12.3: Perform the regression analysis of the stack-loss data in Section 6.9 using the EM algorithm developed in Section 12.6. Choose the degrees of parameter k from the data by finding the profile likelihood.

Exercise 12.4: For the genetic data example (Example 12.2), assuming $\widehat{\theta} = 0.627$ is available, compute the observed Fisher information directly using the observed-data likelihood and using the indirect formulae derived in Section 12.9.

Exercise 12.5: For the normal mixture example (Example 12.3), assuming $\widehat{\theta}$ is available, compute the observed Fisher information directly using the observed data likelihood and using the indirect formulae derived in Section 12.9.

Exercise 12.6: Suppose p-vectors $y_1, \ldots, y_n$ are an iid sample from the multivariate normal distribution with mean μ and variance matrix Σ. The first element y_{11} of y_1 is missing. Show the likelihood based on the available data, and discuss the direct estimation of μ and Σ from it. Derive the E- and M-steps of the EM algorithm to estimate μ and Σ.

Exercise 12.7: Refer to the problem of estimating infection pattern in Section 12.7. The following table (read by row) shows a larger dataset of exactly the same structure as the one described in that section. There are 100 subjects represented in the data, each one contributed a series of length eleven. Apply the EM algorithm for estimating the infection pattern. Compute also the standard errors of the estimates.

```
0 0 0 0 1 0 1 0 0 0 0   0 0 1 0 0 1 0 1 0 0 0
1 0 0 0 1 0 1 1 1 0 1   0 1 0 0 0 1 0 0 0 0 0
0 0 0 0 0 1 1 0 1 0 0   0 0 0 0 1 0 0 0 0 1 0
0 0 0 0 0 0 0 0 0 0 0   0 0 0 0 0 0 0 0 0 0 0
0 1 0 0 0 0 0 0 0 0 0   0 1 0 0 1 0 0 0 0 1 0
1 0 0 0 0 0 0 0 0 1 1   0 1 0 0 0 0 0 0 0 0 0
0 0 0 0 0 0 0 0 1 0 0   1 0 0 0 0 0 0 1 0 0 1
0 0 0 0 1 0 0 0 1 0 1   0 0 0 0 1 0 0 0 0 0 0
0 0 1 0 0 0 0 0 0 0 0   0 0 0 0 0 0 0 0 1 0 0
0 0 0 0 0 0 0 0 0 0 0   0 1 1 1 0 0 0 0 1 1 1
1 0 0 0 0 1 0 0 0 0 0   0 1 0 0 0 0 0 1 0 0 0
0 0 0 1 0 0 0 0 0 0 0   1 0 1 0 0 0 0 0 0 1 0
0 0 0 1 0 0 0 1 0 0 0   1 0 0 0 1 0 0 0 0 1 0
0 0 1 0 0 0 0 0 0 1 0   0 0 1 0 0 0 0 0 1 0 0
0 0 0 0 0 1 0 0 0 0 1   0 0 0 0 0 0 0 0 0 1 0
0 0 0 0 0 0 0 1 0 0 0   0 0 0 0 1 0 0 0 1 0 0
0 0 0 0 0 0 0 1 0 0 1   0 0 0 0 0 0 0 0 1 0 0
1 0 0 0 0 0 0 0 1 0 0   0 0 0 0 1 1 0 0 1 0 1
0 0 0 0 1 0 0 0 1 0 0   0 0 1 0 0 1 0 0 0 0 0
1 0 0 0 1 0 0 0 1 0 0   1 0 0 1 1 1 0 0 0 0 0
0 0 0 0 1 0 0 0 0 0 0   1 0 0 0 0 1 0 0 0 0 0
1 0 0 0 0 1 0 0 0 0 0   0 0 0 0 1 0 1 0 0 0 1
0 1 0 0 1 0 0 0 0 0 1   0 0 0 0 0 0 0 0 0 0 0
0 0 1 0 0 0 0 0 0 0 0   0 0 0 0 0 0 0 0 0 1 0
0 0 0 0 0 0 1 1 0 0 1   0 1 0 1 0 0 0 1 0 1 0
0 1 0 0 0 0 0 0 0 0 0   0 0 0 0 0 0 1 0 1 0 0
1 0 0 1 0 0 0 1 0 1 1   0 1 0 0 0 0 0 0 0 0 0
0 0 1 0 0 0 0 0 1 1 1   0 0 0 0 0 1 1 0 0 1 1
0 0 0 0 0 0 1 0 0 0 0   1 0 0 0 0 0 1 0 0 1
0 0 1 0 0 0 0 0 0 0 0   0 0 0 0 0 0 0 0 0 0 0
1 0 0 0 0 0 0 0 0 0 0   0 0 0 0 0 1 0 0 0 1 0
1 0 0 0 0 1 1 0 0 1 0   0 0 1 0 0 1 0 0 0 0 1
0 0 0 0 0 0 0 0 1 0 0   0 0 1 0 0 0 0 0 0 0 0
1 0 0 0 0 0 0 0 0 0 0   0 0 0 0 0 0 1 0 0 0 0
0 0 1 0 0 0 0 0 0 0 0   1 0 0 0 0 0 1 1 0 0 0
0 0 1 0 0 0 0 0 0 0 0   1 0 1 0 0 1 0 0 0 0 0
0 1 0 0 0 0 1 0 0 0 1   0 1 0 0 1 0 1 0 0 0 0
1 0 1 1 1 0 0 1 0 0 1   0 0 0 0 0 0 0 0 0 0 0
0 0 0 1 0 1 0 1 0 0 1   0 0 1 0 1 1 0 0 0 1 0
0 1 0 0 0 1 1 1 0 1 0   0 0 0 0 0 1 0 0 0 0 0
0 0 0 0 0 0 0 0 0 0 0   0 0 1 1 0 0 0 1 0 1 1
0 0 1 0 0 0 0 0 0 0 0   1 0 0 0 1 0 0 0 0 0 0
0 0 0 0 0 1 0 0 1 0 0   0 0 0 1 0 0 0 1 0 0 1
0 1 0 0 0 0 0 1 1 0 1   0 0 0 0 0 0 0 0 1 0 0
0 0 0 0 0 0 0 1 0 0 1   0 0 1 0 1 0 0 0 0 0 0
0 0 0 0 0 0 1 0 1 0 0   0 0 1 0 0 0 0 0 0 1 0
0 0 0 0 0 1 0 0 1 0 0   0 0 0 0 0 0 0 0 0 0 0
0 0 0 1 1 0 0 0 0 0 1   0 0 0 0 0 1 0 0 0 0 0
0 0 0 0 0 0 0 0 0 1 0   0 0 1 1 0 0 0 0 0 0 0
0 0 1 0 0 0 0 0 0 1 0   1 1 1 0 0 1 0 0 0 0 0
```

Exercise 12.8: In repeated measures experiments we typically model the vectors of observations $y_1, \ldots, y_n$ as

$$y_i | b_i \sim N(x_i \beta + z_i b_i, \sigma_e^2 I_{n_i}).$$

That is, each y_i is a vector of length n_i, measured from subject i; x_i and z_i are known design matrices; β is a vector of p *fixed* parameters; and b_i is a *random* effect associated with individual i. Assume that b_i's are iid $N(0, \sigma_b^2)$. We want to derive the EM algorithm to estimate the unknown parameters β, σ_e^2 and σ_b^2.

(a) What is the log-likelihood based on the observed data? (Hint: what is the marginal distribution of y_i?)

(b) Consider the collection (y_i, b_i) for $i = 1, \ldots, n$ as the 'complete data'. Write down the log-likelihood of the complete data. Hence show that the E-step of the EM algorithm involves computing $\widehat{b_i} \equiv E(b_i|y_i)$ and $\widehat{b_i^2} \equiv E(b_i^2|y_i)$.

(c) Given the current values of the unknown parameters, show that

$$\begin{aligned}
\widehat{b_i} &= \left(z_i'z_i + \frac{\sigma_e^2}{\sigma_b^2}\right)^{-1} z_i'(y_i - x_i\beta) \\
\widehat{b_i^2} &= (\widehat{b_i})^2 + \sigma_e^2 \left(z_i'z_i + \frac{\sigma_e^2}{\sigma_b^2}\right)^{-1}.
\end{aligned}$$

(d) Show the M-step updates for the unknown parameters β, σ_b^2 and σ_e^2.

13

Robustness of likelihood specification

In Section 9.4 we have shown that, under general conditions, the likelihood approach leads to the best possible inference. The notion of 'best', e.g. achieving the Cramér-Rao lower bound, is measured within a particular model. The likelihood approach requires a full specification of the probability structure, and it is difficult to quantify the uncertainty associated with model selection. The question we will address is how sensitive the result is to the correctness of an assumed model.

The important perspective we take in this chapter is to view an assumed model only as a working model, not as the true model that generates the data. Akaike (1974) considers this an important extension of the 'maximum likelihood principle'. A recent monograph that covers this topic is Burnham and Anderson (1998). There is a large Bayesian literature associated with model selection quite relevant to this chapter; see, for example Kass and Raftery (1995).

13.1 Analysis of Darwin's data

The following table shows the outcomes of a classic experiment by Darwin (1876), which was meant to show the virtue of cross-fertilization. The values are the heights of paired self- and cross-fertilized plants and the measurement of interest x is the height difference.

Cross	Self	x	Cross	Self	x
23.5	17.4	6.1	18.3	16.5	1.8
12.0	20.4	−8.4	21.6	18.0	3.6
21.0	20.0	1.0	23.3	16.3	7.0
22.0	20.0	2.0	21.0	18.0	3.0
19.1	18.4	0.7	22.1	12.8	9.3
21.5	18.6	2.9	23.0	15.5	7.5
22.1	18.6	3.5	12.0	18.0	−6.0
20.4	15.3	5.1			

Standard normal theory computation yields $\overline{x} = 2.61$ with standard error 1.22, and $t = 2.14$ with 14 degrees of freedom, which is borderline significant. Assuming $x_1, \ldots, x_n$ are iid $N(\mu, \sigma^2)$ the profile likelihood of μ is

$$\log L(\mu) = -\frac{n}{2}\log \hat{\sigma}_\mu^2, \tag{13.1}$$

with $\hat{\sigma}_\mu^2 = \frac{1}{n}\sum_i (x_i - \mu)^2$. But the data are not really normal: in particular note two outliers on the left-tail of the distribution shown in Figure 13.1(a). Dropping the outliers from the analysis would make the conclusion stronger, but it would raise objections.

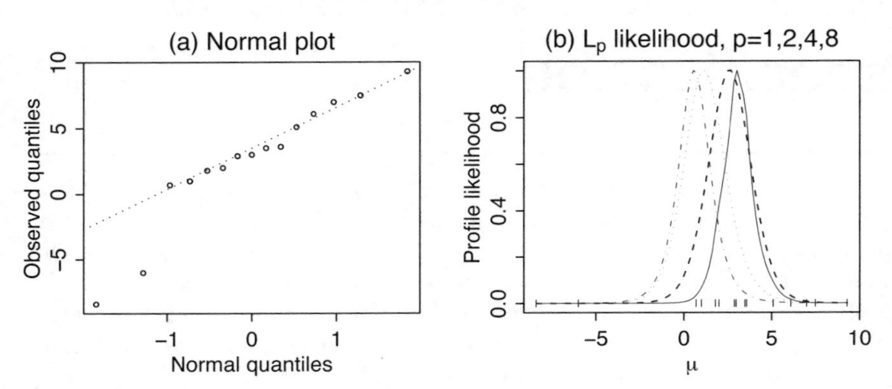

Figure 13.1: *(a) There are two outliers in Darwin's cross-fertilization data. (b) The profile likelihood of the location μ using the L_p-model for $p = 1, 2, 4$ and 8 (solid, dashed, dotted, dot–dashed respectively).*

To see how sensitive the likelihood is to the normality assumption, consider a generalized normal model called the power exponential or L_p-model with standard density

$$f_0(x) = \frac{1}{\Gamma(1+1/p)2^{1+1/p}}\exp\left\{-\frac{1}{2}|x|^p\right\}, \quad -\infty < x < \infty.$$

If $p = 2$ we have the standard normal model, while $p = 1$ gives the double-exponential model (long tail), and $p \to \infty$ gives the uniform model (short tail). The location-scale family generated by the L_p-model is

$$f(x) = \frac{1}{\Gamma(1+1/p)2^{1+1/p}}\frac{1}{\sigma}\exp\left\{-\frac{1}{2}\left|\frac{x-\mu}{\sigma}\right|^p\right\}.$$

To obtain the profile likelihood of μ and p, we first find the MLE of σ^p at fixed μ:

$$\hat{\sigma}_\mu^p = \frac{p}{2n}\sum |x_i - \mu|^p$$

and get

$$\log L(\mu, p) = -n\log\Gamma(1+1/p) - n(1+1/p)\log 2 - n\log\hat{\sigma}_\mu - n/p.$$

This is drawn in Figure 13.1(b) for $p = 1$, 2, 4 and 8. These likelihood functions are dramatically different, indicating a sensitivity to the choice of p.

Which p is best? If we are not interested in p itself, it is a nuisance parameter. It can be eliminated, for example, by taking the profile likelihood over σ^2 and p simultaneously. But even with this approach there is still a question whether the L_p-family is 'rich enough' for the observed data; for example, the Cauchy distribution is not in the L_p-family.

We will consider a simpler version of the profile likelihood by first estimating p from the data. The profile likelihood of p is found by maximizing $L(\mu, p)$ above over μ:

p	1	2	3	4
$-\log L(p)$	42.89	44.02	44.68	44.50

giving the MLE $\widehat{p} = 1$. So the best we can achieve within the L_p-model is by using $p = 1$, which is sensible since L_1 density has the heaviest tail.

Going back to the original problem, using $p = 1$, the profile likelihood of μ indicates that cross-fertilization is superior; the likelihood of H_0: $\mu = 0$ is 0.6%. This is stronger evidence than that under the normal assumption, where the likelihood of H_0 is 12%.

13.2 Distance between model and the 'truth'

By the 'truth' we mean the underlying distribution or probability structure that generates the data. We will discuss

- what we get if we assume a wrong model, and
- what we are actually doing when we maximize likelihood assuming a wrong model.

These are important since in real data analysis whatever we assume will always be wrong; for example, there is no such thing as exactly normal data. Generally, we will get biased estimates. Some *features* of the distribution, however, may still be consistently estimated. How wrong a model is can be measured in terms of the Kullback–Leibler distance.

Definition 13.1 *Suppose f is an assumed model density, and g is the true density. The Kullback–Leibler distance is*

$$\begin{aligned} D(g, f) &= E_g \log \frac{g(X)}{f(X)} \\ &= \int g(x) \log \frac{g(x)}{f(x)} dx, \end{aligned}$$

where the expected value assumes X is distributed with density $g(x)$.

We have shown in Theorem 9.5 that

$$D(g, f) \geq 0$$

with equality iff $g(x) = f(x)$. This intuitively means that, in the average, the log-likelihood is larger under the true model. The inequality implies we can use $D(g,f)$ as a distance measure, although it is not a symmetric distance: $D(g,f) \neq D(f,g)$. We can make it symmetric by defining

$$J(f,g) = D(f,g) + D(g,f),$$

but there is no need for that in our applications. The quantity

$$-E_g \log g(X) = -\int g(x) \log g(x) dx$$

is called the *entropy* of distribution g; it is used in communication theory as a fundamental measure of information or complexity. In this context $D(g,f)$ is called *cross-entropy*.

Example 13.1: For discrete distributions, say $g(x_i) = P(X = x_i)$,

$$D(g,f) = \sum_i g(x_i) \log \frac{g(x_i)}{f(x_i)}.$$

Given data the quantity $g(x_i)$ is associated with the observed frequencies, and $f(x_i)$ with the model-based frequencies, so the data version of $D(g,f)$ is associated with the likelihood ratio statistic

$$W = 2 \sum O \log \frac{O}{E}.$$

We have seen this in Section 9.12 as a goodness-of-fit statistic for the model $f(x_i)$, measuring a distance between data and model. □

Example 13.2: Suppose $g(x)$ is the density of $N(0,1)$, or $x \sim N(0,1)$, and $f(x)$ is the density of $N(\mu, \sigma^2)$. The Kullback–Leibler distance is

$$\begin{aligned} D(g,f) &= E_g \log g(X) - E_g \log f(X) \\ &= \frac{1}{2}\left(\log \sigma^2 + \frac{1+\mu^2}{\sigma^2} - 1\right). \end{aligned}$$

Figures 13.2(a)–(b) show $D(g,f)$ as a function of μ and σ^2. The distance between $N(0,1)$ and $N(\mu,1)$ gives a sense of size for $D \equiv D(g,f)$:

$$\mu = \sqrt{2D},$$

so for example $D = 2$ is equivalent to the distance between $N(0,1)$ and $N(2,1)$. □

Example 13.3: Suppose the true distribution is gamma(4,1), so

$$g(x) = \frac{1}{6} x^3 e^{-x}, \quad x > 0,$$

and we model the data as $N(\mu, \sigma^2)$. Let $X \sim$ gamma(4,1); then

$$D(g,f) = E_g \log g(X) - E_g \log f(X)$$

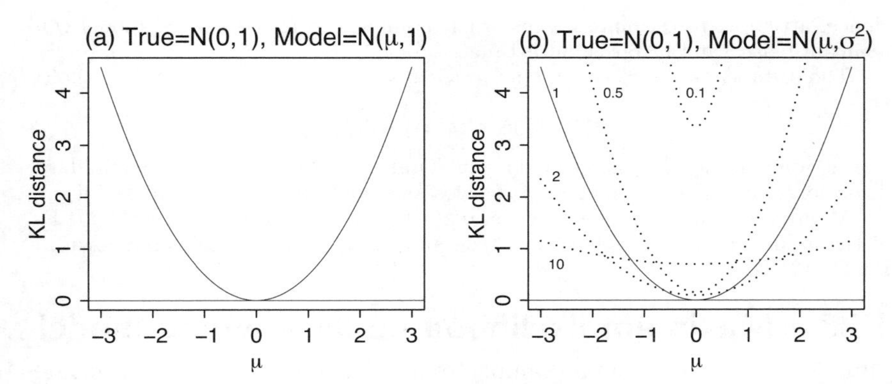

Figure 13.2: *(a) Kullback–Leibler distance between the true $N(0,1)$ and $N(\mu,1)$. (b) Kullback–Leibler distance between the true $N(0,1)$ and $N(\mu,\sigma^2)$, for $\sigma = 0.1, 0.5, 1, 2$ and 10.*

$$= \quad E(-\log 6 + 3\log X - X) - \frac{1}{2}E\left\{\log(2\pi\sigma^2) + \frac{(X-\mu)^2}{\sigma^2}\right\}.$$

General analytical evaluation of $D(g,f)$ can be complicated, but a Monte Carlo estimate can be computed easily: generate $x_i \sim$ gamma(4,1) and take the sample average of $\log g(x_i) - \log f(x_i)$.

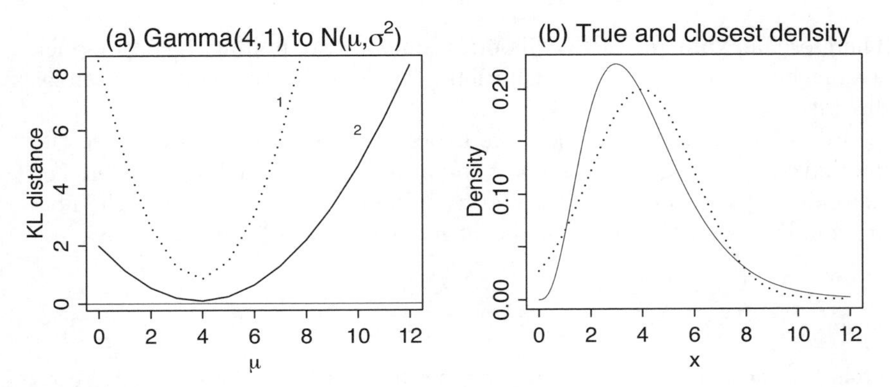

Figure 13.3: *(a) Kullback–Leibler distance between the true gamma(4,1) and $N(\mu,\sigma^2)$, for $\sigma = 1$ and 2. (b) The true density (solid) and the density of $N(4,4)$, the closest model.*

Figure 13.3(a) shows the Kullback–Leibler distance $D(g,f)$ as a function of μ and σ^2. The closest model $N(4,4)$ matches gamma(4,1) in terms of the mean and variance, but *not the higher-order moments*:

Moment	gamma(4,1)	$N(4,4)$
$E(X-\mu)^3$	7.3	0
$E(X-\mu)^4$	65.5	$48{=}3\sigma^4$

Hence, attempts to estimate higher-order moments using the normal model will result in biased and inconsistent estimates.

The distance of the closest model (estimated using the Monte Carlo method) is

$$D(G(4,1), N(4,4)) = 0.08.$$

This is comparable with the normal mean distance of $\sqrt{2 \times 0.08} = 0.4$ in standard deviation units. The parameters of the closest model can also be computed using the Monte Carlo method: generate gamma(4,1) data and then compute the MLEs of the assumed normal model. The true and the closest densities are shown in Figure 13.3(b). □

13.3 Maximum likelihood under a wrong model

Suppose $x_1, \ldots, x_n$ are an iid sample from an unknown $g(x)$ and our model is $f_\theta(x)$. Maximizing the likelihood is equivalent to

$$\text{maximizing } \frac{1}{n}\sum_i \log f_\theta(x_i),$$

which in large samples or in the average is equivalent to

$$\text{minimizing } \{-E_g \log f_\theta(X)\}.$$

Since $E_g \log g(X)$ is an (unknown) constant with respect to θ, it is also equivalent to

$$\text{minimizing } D(g,f) = E_g \log g(X) - E_g \log f_\theta(X).$$

Therefore, maximizing the likelihood is equivalent to finding the best model, the one closest to the true distribution in the sense of the Kullback–Leibler distance.

From the previous examples we expect the MLE $\widehat{\theta}$ to converge to the maximizer of $E_g \log f_\theta(X)$, i.e. to the parameter of the closest model. For example, suppose $x_1, \ldots, x_n$ are iid from gamma(4,1) and we use the normal model $N(\mu, \sigma^2)$. Here the closest model is $N(4,4)$, and we have

$$\begin{aligned} \widehat{\mu} &\to 4 \\ \widehat{\sigma}^2 &\to 4. \end{aligned}$$

Thus the mean and variance of the true distribution are consistently estimated. This is an example where a 'wrong' model would still yield consistent estimates of useful population parameters. Such estimates are said to be robust with respect to model mis-specification.

Using a wrong model, we will generally get biased or inconsistent estimates, but we might also lose efficiency.

Example 13.4: Suppose $x_1, \ldots, x_n$ are iid gamma$(\alpha, 1)$ with true $\alpha = 4$. If a gamma model is assumed, we have

$$\log L(\alpha) = (\alpha - 1)\sum_i \log x_i - \sum x_i - n \log \Gamma(\alpha)$$

$$S(\alpha) = \sum \log x_i - n\frac{\partial \log \Gamma(\alpha)}{\partial \alpha}$$
$$\mathcal{I}(\alpha) = n\frac{\partial^2 \log \Gamma(\alpha)}{\partial \alpha^2}.$$

In particular $\mathcal{I}(4) = 0.284n$, so the asymptotic variance of the MLE is $1/(0.284n) = 3.52/n$.

Under the normal assumption $\widehat{\mu} = \overline{x}$ is a consistent estimate of α, but

$$\text{var}(\overline{X}) = \text{var}(X_1)/n = 4/n > 3.52/n. \ \square$$

Example 13.5: The problem of bias may appear even for low-order moments. In this example we choose a seemingly reasonable model, but we get a biased estimate of the mean parameter. Suppose the true distribution is exponential with mean 1, and we use the log-normal model: $\log X$ is $N(\mu, \sigma^2)$. This may be considered reasonable since $X > 0$, and it is commonly used for survival data. The best model here is log-normal with parameters

$$\mu = -0.577$$
$$\sigma^2 = 1.644,$$

achieving $D(g, f) = 0.08$. (We can find the closest model using a Monte Carlo procedure: generate $x_1, \ldots, x_N$ from Exponential(1), then estimate μ and σ^2 by the mean and variance of $\log x_i$. Theoretically, $\log X$ has the extreme-value distribution.)

Now, suppose we have an iid sample $x_1, \ldots, x_n$, and we estimate $\theta \equiv EX$ by the estimated mean of the log-normal distribution:

$$\widehat{\theta} = e^{\widehat{\mu} + \widehat{\sigma^2}/2}.$$

Hence $\widehat{\theta}$ converges to

$$e^{\mu+\sigma^2/2} = e^{-0.577+1.644/2} = 1.278,$$

but using the true exponential model $EX = 1$. Furthermore, using the Delta method, assuming log-normality, we can show that (Exercise 13.1)

$$\text{var}(\widehat{\theta}) = \text{var}(e^{\widehat{\mu}+\widehat{\sigma^2}/2}) = 4.89/n,$$

while if we know the true distribution

$$\text{var}(\widehat{\theta}) = \text{var}(\overline{X}) = 1/n.$$

In this example $E \log X$ is consistently estimated using the wrong likelihood, but not EX since it is sensitive to the assumed model. $\square$

We have seen that different parameters (or features) of the distribution have different sensitivity to model mis-specification. It is desirable that at least the inference on the parameter of interest is robust/insensitive to wrong model assumptions.

13.4 Large-sample properties

Consistency property

Let $\widehat{\theta}$ be the MLE based on an assumed model $f_\theta(x)$. From Section 13.3, the parameter θ_0 being estimated by the maximum likelihood procedure is the maximizer of

$$\lambda(\theta) \equiv E_X \log f_\theta(X).$$

We expect $\widehat{\theta}$ to converge to θ_0. The proof in the scalar case is similar to the consistency proof we discuss in Section 9.3. If the likelihood has several maxima, then $\widehat{\theta}$ can be any one of the maximizers. (We will only cover the scalar case; more conditions are needed in the multiparameter case, but we will not cover it here.)

Theorem 13.1 *Suppose θ_0 is an isolated maximizer of $\lambda(\theta)$, and $f_\theta(x)$ is continuous in θ. Based on the iid sample $x_1, \ldots, x_n$, there exists a sequence of $\widehat{\theta}$ that converges in probability to θ_0.*

Proof: Let

$$\lambda_n(\theta) \equiv \frac{1}{n} \sum_i \log f_\theta(x_i).$$

For any $\epsilon > 0$, using the law of large numbers, we have

$$\lambda_n(\theta_0) \xrightarrow{p} \lambda(\theta_0),$$

but

$$\lambda_n(\theta_0 - \epsilon) \xrightarrow{p} \lambda(\theta_0 - \epsilon) < \lambda(\theta_0)$$

and

$$\lambda_n(\theta_0 + \epsilon) \xrightarrow{p} \lambda(\theta_0 + \epsilon) < \lambda(\theta_0).$$

This means, if the sample size is large enough, there is a large probability that there is a maximum in the interval $(\theta_0 - \epsilon, \theta_0 + \epsilon)$. □

Note that the limit point is θ_0; this may or may not be the true parameter of interest. Comments regarding the weakness of the theorem in Section 9.3 also apply. In particular, there is no guarantee that the global MLE is actually consistent for θ_0. However, if the MLE is unique for every n or as n tends to infinity, then it is consistent.

Distribution theory

Let $\widehat{\theta}$ be a consistent estimate of θ_0 assuming the model $f_\theta(x)$ as described above. Now allow θ to be a vector. We define

$$\mathcal{J} \equiv E \left(\frac{\partial \log f_\theta(X)}{\partial \theta} \right) \left(\frac{\partial \log f_\theta(X)}{\partial \theta'} \right) \bigg|_{\theta=\theta_0}$$

and

$$\mathcal{I} \equiv -E\frac{\partial^2 \log f_\theta(X)}{\partial\theta\partial\theta'}\bigg|_{\theta=\theta_0}.$$

The expected value is taken with respect to the true but unknown distribution. These two matrices are the same if $f_{\theta_0}(x)$ is the true model.

Theorem 13.2 *Based on an iid sample $x_1, \ldots, x_n$, and assuming regularity conditions stated in Section 9.4,*

$$\sqrt{n}(\widehat{\theta} - \theta_0) \xrightarrow{d} N(0, \mathcal{I}^{-1}\mathcal{J}\mathcal{I}^{-1}).$$

Proof: The log-likelihood of θ is

$$\log L(\theta) = \sum_i \log f_\theta(x_i),$$

so, expanding the score function around $\widehat{\theta}$ we obtain

$$\frac{\partial \log L(\theta)}{\partial\theta} = \frac{\partial \log L(\theta)}{\partial\theta}\bigg|_{\theta=\widehat{\theta}} + \frac{\partial^2 \log L(\theta^*)}{\partial\theta\partial\theta'}(\theta - \widehat{\theta}) \tag{13.2}$$

$$= \frac{\partial^2 \log L(\theta^*)}{\partial\theta\partial\theta'}(\theta - \widehat{\theta}), \tag{13.3}$$

where $|\theta^* - \theta| \leq |\theta - \widehat{\theta}|$. Let

$$y_i \equiv \frac{\partial \log f_\theta(x_i)}{\partial\theta},$$

so the left-hand side of (13.2) is a sum of an iid variate y_i, with mean

$$\begin{aligned} EY_i &= E\frac{\partial \log f_\theta(X_i)}{\partial\theta} \\ &= \frac{\partial E \log f_\theta(X)}{\partial\theta} = \lambda'(\theta). \end{aligned}$$

At $\theta = \theta_0$ we have $EY_i = 0$ and variance

$$\text{var}(Y_i) = \mathcal{J} = E\left(\frac{\partial \log f_\theta(X)}{\partial\theta}\right)\left(\frac{\partial \log f_\theta(X)}{\partial\theta'}\right)\bigg|_{\theta=\theta_0}.$$

By the central limit theorem, at $\theta = \theta_0$ we have

$$\frac{1}{\sqrt{n}}\sum_i Y_i \xrightarrow{d} N(0, \mathcal{J}).$$

Since $\widehat{\theta} \xrightarrow{p} \theta_0$ we have

$$\frac{1}{n}\frac{\partial^2 \log L(\theta^*)}{\partial\theta\partial\theta'} = \frac{1}{n}\sum_i \frac{\partial^2 \log f_{\theta^*}(x_i)}{\partial\theta\partial\theta'}$$

$$\stackrel{p}{\to} \quad E\frac{\partial^2 \log f_\theta(X)}{\partial\theta\partial\theta'}\bigg|_{\theta=\theta_0} = -\mathcal{I}.$$

From the Taylor series expansion (13.3)

$$\frac{1}{\sqrt{n}}\frac{\partial \log L(\theta_0)}{\partial\theta} = \frac{1}{n}\frac{\partial^2 \log L(\theta^*)}{\partial\theta\partial\theta'} \times \sqrt{n}(\theta_0 - \widehat{\theta}),$$

so, using Slutsky's theorem,

$$\sqrt{n}(\widehat{\theta} - \theta_0) \stackrel{d}{\to} N(0, \mathcal{I}^{-1}\mathcal{J}\mathcal{I}^{-1}).$$

The formula $\mathcal{I}^{-1}\mathcal{J}\mathcal{I}^{-1}$ is called a robust variance formula, since it is a correct variance regardless whether of the assumed model $f_\theta(x)$ is correct or not. Assuming the model is correct we get $\mathcal{J} = \mathcal{I}$, and the formula reduces to the usual inverse Fisher information $\mathcal{I}^{-1}$, usually called the 'naive' variance formula. □

As a corollary, which we will use later,

$$\begin{aligned} E\{n(\widehat{\theta} - \theta_0)'\mathcal{I}(\widehat{\theta} - \theta_0)\} &= \text{trace}[\mathcal{I}E\{n(\widehat{\theta} - \theta_0)(\widehat{\theta} - \theta_0)'\}] \\ &\stackrel{d}{\to} \text{trace}(\mathcal{I}\mathcal{I}^{-1}\mathcal{J}\mathcal{I}^{-1}) \\ &= \text{trace}(\mathcal{J}\mathcal{I}^{-1}). \end{aligned} \tag{13.4}$$

Example 13.6: Let the count data $x_1, \ldots, x_n$ be an iid sample from some distribution, and assume a Poisson model with mean θ. We have

$$\log f_\theta(x_i) = -\theta + x_i \log \theta,$$

so

$$\lambda(\theta) = E\{\log f_\theta(X)\} = -\theta + E(X)\log\theta$$

and $\theta_0 = EX$ maximizes $\lambda(\theta)$. It is also clear that $\widehat{\theta} = \overline{x}$ is a consistent estimate of the population mean EX. The first and second derivatives of the log-likelihood are

$$\begin{aligned} \frac{\partial}{\partial\theta}\log f_\theta(x_i) \equiv y_i &= -1 + x_i/\theta, \\ \frac{\partial^2}{\partial\theta^2}\log f_\theta(x_i) &= -x_i/\theta^2 \end{aligned}$$

and

$$\begin{aligned} \mathcal{J} &= \text{var}(Y_i) = \text{var}(X)/\theta_0^2 \\ \mathcal{I} &= EX/\theta_0^2 = 1/\theta_0. \end{aligned}$$

The robust variance formula gives

$$\mathcal{I}^{-1}\mathcal{J}\mathcal{I}^{-1} = \text{var}(X)$$

as the asymptotic variance of $\sqrt{n}(\widehat{\theta}-\theta_0)$, which is correct regardless of what the underlying distribution is. In contrast, the naive formula using the inverse of the Fisher information gives

$$\mathcal{I}^{-1} = \theta_0$$

as the asymptotic variance of $\sqrt{n}(\widehat{\theta}-\theta_0)$; this is generally incorrect, except if the variance happens to match the mean. □

13.5 Comparing working models with the AIC

Specifying a correct model is important for a consistent inference and efficiency. Unfortunately, in practice, it is not always easy to know if we have a correct model. What is easier to do is to compare models, and decide which is the best (most supported by the data) among them. This is especially simple if the models are 'nested', so model selection is equivalent to parameter testing. For example,

$$\begin{aligned}\text{Model 1:} &\quad \mu = \beta_0 + \beta_1 x_1\\ \text{Model 2:} &\quad \mu = \beta_0 + \beta_1 x_1 + \beta_2 x_2.\end{aligned}$$

But what if the models are not nested, for example

Model 1: the error is normal

Model 2: the error is Cauchy?

Considering both as 'working' models as opposed to true models, in principle we should choose the one closest to the true distribution; that is, we should minimize the Kullback–Leibler distance to the true distribution. We show earlier that minimizing the Kullback–Leibler distance is equivalent to maximizing $E_g \log f(X)$. So based on a sample we should choose a model that maximizes the assumed log-likelihood

$$\sum_i \log f(x_i)$$

across possible models f.

Example 13.7: Suppose we observe an iid sample

```
-5.2 -1.9 -1.0 -0.7 -0.3 0.0 0.4 0.5 2.3 3.3
```

and consider these two models

$$\begin{aligned}\text{Model 1:} &\quad x_i \sim N(0,1)\\ \text{Model 2:} &\quad x_i \sim \text{Cauchy}(0,1).\end{aligned}$$

The observed log-likelihoods are

$$\begin{aligned}\text{Model 1:} \sum_i \log f(x_i) &= -5\log(2\pi) - \frac{1}{2}\sum_i x_i^2\\ &= -33.6\end{aligned}$$

$$\text{Model 2: } \sum_i \log f(x_i) = -10\log\pi - \sum_i \log(x_i^2+1)$$
$$= -22.2,$$

so we should choose the Cauchy model. This is sensible because of the outlying observation (−5.2). The procedure appears equivalent to the classical likelihood ratio procedure, but it is based on a different reasoning. Here we consider both models on equal footing as 'working' models, with the true model being unknown.

It is not clear how to get a probability-based inference (i.e. P-value) for this model comparison. The natural hypotheses are H_0: normal or Cauchy models are equally good for the data, versus H_1: one model is better than the other. The null hypothesis does not specify a distribution, so we cannot compute a P-value for the observed likelihood ratio. From Section 2.6 the Fisherian compromise is to base inference on the pure likelihood comparison. □

To understand why we cannot apply the usual probability-based theory, suppose we have a model $f_\theta(x)$ that is rich enough to contain the true distribution $g(x) = f_{\theta_0}(x)$. In model comparison, we compare two parameter values θ_1 versus θ_2, neither of which is the true parameter. We do not have a theory on how $L(\theta_2)/L(\theta_1)$ behaves. (This, however, is not a realistic analogy for model selection. If $f_\theta(x)$ is available, then the best parameter estimate is the MLE $\widehat{\theta}$, not θ_1 or θ_2, so it is possible to know whether both θ_1 and θ_2 have low likelihood relative to $\widehat{\theta}$, and we have a better alternative if the data do not support both θ_1 and θ_2. There is no such luxury in model selection, we may not be able to know if our set of pre-specifed models will contain the true model.)

Choosing between a normal versus a Cauchy model is more difficult, since it is unclear how to parameterize the problem so that both models are represented by two parameter values. While the two models are contained in a t_k-family, the choice of such a family is arbitrary, and there is no guarantee that the chosen family contains the true distribution. There are infinitely many families that can contain two specific models $f_1(x)$ and $f_2(x)$: we can construct a mixture model

$$f_\lambda(x) = \lambda f_1(x) + (1-\lambda) f_2(x),$$

for $0 \le \lambda \le 1$, or a different mixture

$$f_\lambda(x) = c(\lambda) f_1^\lambda(x) f_2^{1-\lambda}(x).$$

We can also add arbitrary models and define other mixtures.

The most difficult problem, yet common enough in practice, is comparing non-nested models with the nuisance parameters involved. The rich family of distributions available for modelling has exaggerated the problem. For example, to model positive outcome data we might consider:

Model 1: GLM with normal family and identity link function with possible heteroscedastic errors.

Model 2: GLM with gamma family and log-link function.

In this case the nuisance parameters are the usual parameters in the GLM.

We can use the general principle of profiling over the nuisance parameters. Specifically, suppose we want to compare K models $f_k(\cdot, \theta_k)$ for $k = 1, \ldots, K$, given data $x_1, \ldots, x_n$, where θ_k is the parameter associated with model k. The parameter dimension is allowed to vary between models, and the interpretation of the parameter can be model dependent. We then

- find the best θ_k in each model via the standard maximum likelihood, i.e. by choosing
$$\widehat{\theta}_k = \text{argmax}_\theta \sum_i \log f_k(x_i, \theta_k).$$
This is the profiling step.
- choose the model k that maximizes the log-likelihood
$$\sum_i \log f_k(x_i, \widehat{\theta}_k).$$
As a crucial difference with the comparison of nested models, here *all the constants in the density function must be kept.* This can be problematic when we use likelihood values reported by a standard statistical software, since it is not always clear if they are based on the full definition of the density, including all the constant terms.

Example 13.8: Recall Darwin's plant data in Section 13.1. The distribution is characterized by two outliers on the left-tail. Consider the following models:

$$\text{Model 1 = normal:} \quad f(x) = \frac{1}{\sqrt{2\pi\sigma^2}} \exp\left\{-\frac{1}{2\sigma^2}(x-\mu)^2\right\}$$
$$\text{Model 2 = double exponential:} \quad f(x) = \frac{1}{2\sigma} \exp\left\{-\frac{|x-\mu|}{\sigma}\right\}$$
$$\text{Model 3 = Cauchy:} \quad f(x) = \frac{1}{\pi\sigma\{1+(x-\mu)^2/\sigma^2\}}.$$

We know that the normal model is not plausible. All models have two parameters; we can apply the standard maximum likelihood procedure on each model and obtain the following table (Exercise 13.3):

Model	$\widehat{\mu}$	$\widehat{\sigma}$	$\sum_i \log f_k(x_i, \widehat{\theta}_k)$
Normal	2.61	4.55	−44.02
D.exp.	3.00	3.21	−42.89
Cauchy	3.13	1.95	−44.10

The double-exponential model is found to be the best among the three. The Cauchy is almost of equal distance to the true distribution as the normal model. □

The AIC

Suppose $x_1, \ldots, x_n$ are an iid sample from an unknown distribution g, and $\widehat{\theta}_k$ is the MLE given a model f_k. From previous considerations, we want a model that maximizes $E\{\log f_k(Z, \theta_k)\}$, where $Z \sim g$. Since θ_k is unknown, we will settle with one that maximizes

$$Q_k = E\{\log f_k(Z, \widehat{\theta}_k)\},$$

where the expectation is taken over Z and $\widehat{\theta}_k$, and Z is assumed independent of $\widehat{\theta}_k$. We cannot naively compare the maximized log-likelihood

$$\log L(\widehat{\theta}_k) = \sum_i \log f_k(x_i, \widehat{\theta}_k)$$

since it is a biased quantity: the same data are used to compute $\widehat{\theta}_k$ by maximizing the likelihood. In particular, if two models are nested, the one with more parameters will always yield a larger $\log L(\widehat{\theta}_k)$, even though it is not necessarily a better model. This means we should not compare the maximized likelihood if the number of parameters varies between models.

There are two better options. One is to use the jackknife or leave-one-out cross-validation method (Stone 1974, 1977); see Section 5.2. Let $\widehat{\theta}_{ki}$ be the MLE of θ_k after removing x_i from the data. Then a less biased estimate of Q_k is

$$c_k = \frac{1}{n} \sum_i \log f_k(x_i, \widehat{\theta}_{ki}).$$

The other is to correct the bias analytically. This is achieved by the AIC formula

$$\text{AIC}(k) = -2 \sum_i \log f_k(x_i, \widehat{\theta}_k) + 2p,$$

where p is the number of parameters. Under regularity conditions, we show in the next section that AIC(k) is an approximately unbiased estimate of $-2nQ_k$, so we select a model that minimizes AIC(k). In Section 13.6 we will also show that the AIC is in fact equivalent to the cross-validation method.

We can interpret the first term in AIC(k) as a measure of data fit and the second as a penalty. If we are comparing models with the same number of parameters then we only need to compare the profile likelihood; for example, see the analysis of Darwin's data above.

As in Example 13.7, in most applications of the AIC there is usually no available P-value. Inference is taken from the perspective that we are comparing several working models of equal footing, not necessarily containing the true model. The advantage of this view is that we do not have to limit the AIC to comparison of nested models, which makes the AIC a general purpose model selection tool. However, as discussed in Section 3.5, a pure AIC-based comparison is potentially weaker than probability-based inference.

Example 13.9: Consider a multiple regression model

$$y_i \sim N(x_i'\beta, \sigma^2)$$

for $i = 1, \ldots, n$, where $\beta \in R^p$. The MLEs are

$$\begin{aligned}\widehat{\beta} &= (X'X)^{-1}X'Y \\ \widehat{\sigma}_p^2 &= \frac{1}{n}\sum_i (y_i - x_i'\widehat{\beta})^2,\end{aligned}$$

so the maximized log-likelihood is

$$\log L(\widehat{\beta}, \widehat{\sigma}^2) = -\frac{n}{2}\log(2\pi\widehat{\sigma}^2) - \frac{n}{2}$$

and

$$\text{AIC}(p) = n\log(2\pi\widehat{\sigma}_p^2) + n + 2p,$$

which is usually simplified as

$$\text{AIC}(p) = n\log(\widehat{\sigma}_p^2) + 2p.$$

As we increase p (add the number of predictors) we reduce the estimated error variance $\widehat{\sigma}_p^2$, but we add a penalty. □

Using a prediction-based computation specific to regression analysis (Exercise 13.4), we have the so-called C_p criterion, given by

$$C_p = n\widehat{\sigma}_p^2 + 2p\sigma_0^2, \tag{13.5}$$

where σ_0^2 is the true error variance. In practice σ_0^2 is usually estimated from the largest or the most unbiased model. Note the differences with the AIC:

- the AIC does not require knowledge of σ_0^2, and
- C_p only applies to regression model selection.

13.6 Deriving the AIC

In Section 3.5 we discussed Lindsey's derivation of the AIC from the principle that inferences in different dimensions should be compatible. The main advantage is that no asymptotic theory is needed. An alternative derivation here shows that minimizing the AIC corresponds to minimizing the Kullback–Leibler distance.

The purpose of the AIC is to compare different models $f_k(x, \theta_k)$ given data $x_1, \ldots, x_n$ from the unknown true model $g(x)$. For convenience we will assume the data are an iid sample $x_1, \ldots, x_n$; the AIC formula itself is true more generally such as for regression problems. We want to find an estimate of

$$Q_k = E\{\log f_k(Z, \widehat{\theta}_k)\},$$

where the expectation is taken over Z and $x_1, \ldots, x_n$, and Z is assumed independent of $x_1, \ldots, x_n$.

The log-likelihood of θ_k assuming a working model $f_k(x, \theta_k)$ is

$$\log L(\theta_k) = \sum_i \log f_k(x_i, \theta_k).$$

Let θ_{k0} be the solution of $\lambda'(\theta_k) = 0$, where $\lambda(\theta_k) = E\{\log f_k(Z, \theta_k)\}$. From Theorem 13.1, θ_{k0} is the parameter being estimated by $\widehat{\theta}$. Let

$$\mathcal{J}_k \equiv E\left(\frac{\partial \log f_k(Z, \theta_k)}{\partial \theta_k}\right)\left(\frac{\partial \log f_k(Z, \theta_k)}{\partial \theta'_k}\right)\Bigg|_{\theta_k=\theta_{k0}}$$

and

$$\mathcal{I}_k = -E\frac{\partial^2 \log f_k(Z, \theta_k)}{\partial \theta_k \partial \theta'_k}\Bigg|_{\theta_k=\theta_{k0}}.$$

Recall from (13.4)

$$E\{n(\widehat{\theta}_k - \theta_{k0})'\mathcal{I}_k(\widehat{\theta}_k - \theta_{k0})\} \approx \text{trace}(\mathcal{J}_k\mathcal{I}_k^{-1}).$$

Sample space formulae

A second-order Taylor expansion of the log-likelihood around $\widehat{\theta}_k$s yields

$$\begin{aligned}\log L(\theta_k) &= \log L(\widehat{\theta}_k) + \frac{\partial \log L(\widehat{\theta}_k)}{\partial \theta_k}(\theta_k - \widehat{\theta}_k) \\ &\quad + \frac{1}{2}(\theta_k - \widehat{\theta}_k)'\frac{\partial^2 \log L(\theta_k^*)}{\partial \theta_k \partial \theta'_k}(\theta_k - \widehat{\theta}_k) \\ &\approx \log L(\widehat{\theta}_k) - \frac{1}{2}n(\theta_k - \widehat{\theta}_k)'\mathcal{I}_k(\theta_k - \widehat{\theta}_k),\end{aligned}$$

so, at $\theta_k = \theta_{k0}$,

$$\log L(\theta_{k0}) \approx \log L(\widehat{\theta}_k) - \frac{1}{2}\text{trace}(\mathcal{J}_k\mathcal{I}_k^{-1}).$$

Hence,

$$E\{\log L(\theta_{k0})\} \approx E\{\log L(\widehat{\theta}_k)\} - \frac{1}{2}\text{trace}(\mathcal{J}_k\mathcal{I}_k^{-1}),$$

or, in terms of $\lambda(\theta_k)$, we have

$$n\lambda(\theta_{k0}) \approx E\{\log L(\widehat{\theta}_k)\} - \frac{1}{2}\text{trace}(\mathcal{J}_k\mathcal{I}_k^{-1}). \tag{13.6}$$

Model space formulae

Expanding $\lambda(\theta_k)$ around θ_{k0}

$$\begin{aligned}\lambda(\theta_k) &= \lambda(\theta_{k0}) + \frac{\partial \lambda(\theta_{k0})}{\partial \theta_k}(\theta_k - \theta_{k0}) \\ &\quad + \frac{1}{2}(\theta_k - \theta_{k0})'\frac{\partial^2 \lambda(\theta_k^*)}{\partial \theta_k \partial \theta'_k}(\theta_k - \theta_{k0})\end{aligned}$$

$$\approx \quad \lambda(\theta_{k0}) - \frac{1}{2}(\theta_k - \theta_{k0})'\mathcal{I}_k(\theta_k - \theta_{k0}).$$

Now, at $\theta_k = \widehat{\theta}_k$, we have

$$Q_k = E\lambda(\widehat{\theta}_k) \approx \lambda(\theta_{k0}) - \frac{1}{2n}\text{trace}(\mathcal{J}_k\mathcal{I}_k^{-1}). \tag{13.7}$$

So, combining (13.6) and (13.7), we obtain

$$nQ_k \approx E\{\log L(\widehat{\theta}_k)\} - \text{trace}(\mathcal{J}_k\mathcal{I}_k^{-1}),$$

and we establish that

$$\log L(\widehat{\theta}_k) - \text{trace}(\mathcal{J}_k\mathcal{I}_k^{-1})$$

is approximately unbiased for nQ_k.

The AIC formula is based on assuming $\mathcal{J}_k \approx \mathcal{I}_k$, so $\text{trace}(\mathcal{J}_k\mathcal{I}_k^{-1}) \approx p$, the number of parameters of the k'th model, and

$$\text{AIC}(k) = -2\log L(\widehat{\theta}_k) + 2p.$$

If the trace approximation is appropriate we can view the AIC as an unbiased estimator of $-2nQ_k$. This is the case, for example, if the assumed model is rich enough to contain the true model, or the best model $f_k(x, \theta_{k0})$ is close enough to the true model. In general, however, the true value of the trace can be far from the number of parameters, so the AIC may not be a good estimator of $-2nQ_k$ (although, from the likelihood point of view, this does not invalidate the AIC as a model selection criterion).

Example 13.10: In Example 13.6 we use the Poisson model to model count data. Assuming that the Poisson model is the k'th model, we can get the trace exactly from

$$\begin{aligned} \mathcal{J}_k &= \text{var}(X)/\theta_{k0}^2 \\ \mathcal{I}_k &= E(X)/\theta_{k0}^2, \end{aligned}$$

so

$$\text{trace}(\mathcal{J}_k\mathcal{I}_k^{-1}) = \text{var}(X)/E(X).$$

Hence, the trace depends on the underlying distribution, and it can be far from the number of parameters. □

Connection with the cross-validation score

Stone (1974, 1977) proposed the cross-validation method for model selection and proved its equivalence with the AIC method. Let $\widehat{\theta}_{ki}$ be the MLE of θ_k after removing x_i from the data. An approximately unbiased estimate of Q_k is

$$c_k = \frac{1}{n}\sum_i \log f_k(x_i, \widehat{\theta}_{ki}).$$

This is obvious, since the expected value of each summand is

$$E\{\log f_k(X_i, \widehat{\theta}_{ki})\} \approx E\{\log f_k(Z, \widehat{\theta}_k)\}.$$

A direct proof is instructive, and it provides an alternative proof of the AIC. By definition $\widehat{\theta}_{ki}$ is the maximizer of $\log L(\theta_k) - \log L_i(\theta_k)$, where

$$\log L_i(\theta_k) = \log f_k(x_i, \theta_k)$$

is the individual contribution to the log-likelihood. Expanding $\log f_k(x_i, \widehat{\theta}_{ki})$ around $\widehat{\theta}_k$, we get

$$\begin{aligned} nc_k &= \sum_i \log f_k(x_i, \widehat{\theta}_k) + \sum_i \frac{\partial \log L_i(\theta_k^*)}{\partial \theta_k'}(\widehat{\theta}_{ki} - \widehat{\theta}_k) && (13.8) \\ &= \log L(\widehat{\theta}_k) + \sum_i \frac{\partial \log L_i(\theta_k^*)}{\partial \theta_k'}(\widehat{\theta}_{ki} - \widehat{\theta}_k), && (13.9) \end{aligned}$$

where $|\theta_k^* - \widehat{\theta}_k| \leq |\widehat{\theta}_{ki} - \widehat{\theta}_k|$. To express $(\widehat{\theta}_{ki} - \widehat{\theta}_k)$ in terms of other quantities, we expand the score function around $\widehat{\theta}_k$

$$\frac{\partial \log L(\widehat{\theta}_{ki})}{\partial \theta_k} = \frac{\partial^2 \log L(\widehat{\theta}_k)}{\partial \theta_k \partial \theta_k'}(\widehat{\theta}_{ki} - \widehat{\theta}_k) \quad (13.10)$$

and, from the definition of $\widehat{\theta}_{ki}$,

$$\frac{\partial \log L(\widehat{\theta}_{ki})}{\partial \theta_k} = \frac{\partial \log L_i(\widehat{\theta}_{ki})}{\partial \theta_k}. \quad (13.11)$$

Recognizing

$$\frac{\partial^2 \log L(\widehat{\theta}_k)}{\partial \theta_k \partial \theta_k'} \approx -n\mathcal{I}_k,$$

from (13.10) and (13.11) we get

$$(\widehat{\theta}_{ki} - \widehat{\theta}_k) \approx -\frac{1}{n}\mathcal{I}_k^{-1}\frac{\partial \log L_i(\widehat{\theta}_{ki})}{\partial \theta_k}.$$

From (13.9) we now have

$$\begin{aligned} nc_k &= \log L(\widehat{\theta}) - \frac{1}{n}\sum_i \frac{\partial \log L_i(\theta_k^*)}{\partial \theta_k'}\mathcal{I}_k^{-1}\frac{\partial \log L_i(\widehat{\theta}_{ki})}{\partial \theta_k} \\ &\approx \log L(\widehat{\theta}) - \text{trace}(\mathcal{J}_k\mathcal{I}_k^{-1}), \end{aligned}$$

since both $\widehat{\theta}_{ki}$ and θ_k^* converge to θ_{k0}. If we make the same approximation $\text{trace}(\mathcal{J}_k\mathcal{I}_k^{-1}) \approx p$, we obtain

$$\text{AIC}(k) \approx -2nc_k$$

and c_k itself is an approximately unbiased estimate for Q_k.

13.7 Exercises

Exercise 13.1: In Example 13.5 show that the variance of $\widehat{\theta}$ assuming the log-normal model is $4.89/n$.

Exercise 13.2: Suppose the true model is gamma(4,1), and we model the data by $N(\mu, \sigma^2)$. Find the bias and variance in the MLE of the following parameters; compare also the robust and naive variance formulae.

(a) The mean of the distribution.

(b) First quartile $\theta = F^{-1}(0.25)$ of the population.

(c) Median $\theta = F^{-1}(0.5)$.

Exercise 13.3: Verify the likelihood comparison in Example 13.8.

Exercise 13.4: Suppose y_i has mean μ_i and variance σ_0^2, where $\mu_i = x_i'\beta$ and β is of length p. The predicted value of μ_i is $\widehat{\mu}_i = x_i\widehat{\beta}$. Define the total prediction error as

$$R = \sum_i (\widehat{\mu}_i - \mu_i)^2.$$

Show that, up to a constant term, the C_p criterion (13.5) is an unbiased estimate of R.

Exercise 13.5: For Darwin's data in Section 13.1 consider a contaminated normal model

$$(1-\pi)N(\mu, \sigma_1^2) + \pi N(\mu, \sigma_2^2)$$

for the height difference, where we expect σ_2^2 to be much larger than σ_1^2; the probability π reflects the amount of contamination. The total parameter is $\theta = (\pi, \mu, \sigma_1^2, \sigma_2^2)$.

(a) Write down the likelihood function based on the observed data.

(b) Find the MLEs of the parameters and their standard errors.

(c) Find the profile likelihood of μ.

(d) Using the AIC, compare the contaminated normal model with the models in Example 13.8.

Exercise 13.6: This is a simulation study where y_i, for $i = 1, \ldots, n$, is generated according to

$$y_i = e^{-x_i} + e_i,$$

and x_i is equispaced between -1 and 2, and e_i's are iid $N(0, \sigma^2)$.

(a) Simulate the data using $n = 40$ and $\sigma^2 = 0.25$, and fit the polynomial model

$$Ey_i = \sum_{k=0}^{p} \beta_k x_i^k.$$

Use the AIC to choose an appropriate order p.

(b) For one simulated dataset, fit the model using $p = 1, 2, \ldots, 8$. Plot the resulting regression curves on eight separate panels. Repeat the process 50 times, starting each time with a new dataset and finishing with plotting the curves on the panels. Compare the bias and variance characteristics of the regression curves at different p. What value of p is chosen most often? What is the bias and variance characteristics of the regression curves associated with the optimal p?

14

Estimating equations and quasi-likelihood

There are many practical problems for which a complete probability mechanism is too complicated to specify, hence precluding a full likelihood approach. This is typically the case with non-Gaussian time series, including repeated measures or longitudinal studies, and image analysis problems. From our discussion in Chapter 13 we can specify convenient working models that do not necessarily match the underlying model. It makes sense, however, to require that the model and methodology achieves a robust inference for the parameters of interest.

The term 'robust' means many things in statistics. Typically we qualify what the robustness is against: for example, the median is robust against outliers; the sample mean is *not* robust against outliers, but inference for the mean might be robust against model mis-specification. It is important to keep in mind what we are protected against when we say something is robust.

'Working models' can be specified at two levels:

- likelihood level: we treat the likelihood only as an objective function for estimation or model comparison,
- score equation level: we specify an equation and solve it to produce an estimate. We call the approach an estimating equation approach or M-estimation.

The estimating equation approach allows a weaker and more general specification, since it does not have to be associated with a proper likelihood. Rather than trying to specify the whole probability structure of the observations, in this approach we can focus on the parameter of interest. The advantage of a likelihood-level specification is that, as described in Section 13.5, it can be used in the AIC for model comparison.

The method of moments is an example of the estimating equation approach. The modern theory was first explored by Durbin (1960) for a time series problem, and more generally by Godambe (1960). Given data $y_1, \ldots, y_n$, the estimating equation approach specifies that estimate $\widehat{\theta}$ is the solution of

$$\sum_i \psi(y_i, \theta) = 0,$$

where the estimating function $\psi(y_i, \theta)$ is a known function. The parameter of interest determines the choice of $\psi(y_i, \theta)$. For its closeness (in idea) with the MLE, $\widehat{\theta}$ is called an 'M-estimate'.

One of the most important estimating equations is associated with GLM: given observation y_i we assume

$$\begin{aligned} Ey_i &= \mu_i(\beta) \\ \text{var}(y_i) &= \phi v_i(\beta) \end{aligned}$$

for known functions $\mu_i(\cdot)$ and $v_i(\cdot)$ of an unknown regression parameter β. The unknown parameter ϕ is a dispersion parameter. The estimate of β is the solution of

$$\sum_{i=1}^{n} \frac{\partial \mu_i}{\partial \beta} v_i^{-1}(y_i - \mu_i) = 0. \tag{14.1}$$

The equation applies immediately to multivariate outcomes, in which case it is called a generalized estimating equation (GEE). The objective function associated with (14.1) is called the quasi-likelihood, with two notable features:

- In contrast with a full likelihood, we are not specifying any probability structure, but only the mean and variance functions. We may call it a *semi-parametric* approach, where parameters other than those of interest are left as free as possible. By only specifying the mean and the variance we are letting the shape of the distribution remain totally free. This is a useful strategy for dealing with non-Gaussian multivariate data.
- With limited modelling, the range of possible inference is also limited. In particular, the approach is geared towards producing a point estimate of β. Testing and CI usually rely on the asymptotic normality of the estimate, making it equivalent to Wald-type inference. Typically there is no exact inference, but more accurate inference may be available via bootstrapping or empirical likelihood (Section 5.6 or Chapter 15). Model comparison is limited to nested ones, since there is no available AIC.

Another important class of quasi-likelihood estimation is the so-called Gaussian estimation in time series analysis. For a time series $y_1, \ldots, y_n$, an exact likelihood computation is usually too complicated except for the Gaussian case. Hence, given a model, for example an ARMA model (Section 11.1), we simply use the Gaussian likelihood as a working model, even for non-Gaussian data. Again, only the mean and (co)variance functions are used to derive the likelihood.

Finally, the so-called 'robust estimation' forms another major branch of the estimating equation approach. The main concern is the development of procedures, or modification of classical procedures, which are resistant to outlier or unusual observations. We will discuss this in Section 14.5.

14.1 Examples

One-sample problems

Estimating a population mean is the simplest nontrivial statistical problem. Given an iid sample $y_1, \ldots, y_n$, assume that

$$\begin{aligned} E(y_i) &= \mu \\ \mathrm{var}(y_i) &= \sigma^2. \end{aligned}$$

From (14.1) $\widehat{\mu}$ is the solution of

$$\sum_{i=1}^{n} 1 \cdot \sigma^{-2} \cdot (y_i - \mu) = 0,$$

which yields $\widehat{\mu} = \overline{y}$.

This example shows clearly the advantages and disadvantages of the estimating equation compared with a full likelihood approach:

- The estimate is consistent for a very wide class of underlying distributions, namely any distribution with mean μ. In fact, the sample does not even have to be independent. Compare this with the biased likelihood-based estimate in Example 13.5 if we assume a wrong model.
- We have to base inference on asymptotic considerations, since there is no small sample inference. One might use the bootstrap (Section 5.6) or empirical likelihood (Chapter 15) for a reliable inference.
- There is a potential loss of efficiency compared with a full likelihood inference under some distribution assumption; see Example 13.5.
- There is no standard prescription how to estimate the variance parameter σ^2. Other principles may be needed, for example using the method-of-moments estimate

  $$\widehat{\sigma}^2 = \frac{1}{n} \sum_i (x_i - \overline{x})^2,$$

  still without making any distributional assumption. The estimating equation can be extended to include the dispersion parameter.

Linear models

Given an independent sample (y_i, x_i) for $i = 1, \ldots, n$, let

$$\begin{aligned} E(y_i) &= x_i'\beta \equiv \mu_i(\beta) \\ \mathrm{var}(y_i) &= \sigma_i^2 \equiv v_i(\beta). \end{aligned}$$

The estimating equation for β is

$$\sum_i x_i \sigma_i^{-2} (y_i - x_i'\beta) = 0,$$

giving us the weighted least-squares estimate

$$\widehat{\beta} = (\sum_i x_i x_i'/\sigma_i^2)^{-1} \sum_i x_i y_i/\sigma_i^2$$

$$= (X'V^{-1}X)^{-1}X'V^{-1}Y,$$

by defining the $n \times p$ design matrix X as $X = [x_1 \dots x_n]'$, the variance matrix $V = \text{diag}[\sigma_i^2]$, and outcome vector $Y = (y_1, \dots, y_n)'$.

Poisson regression

For independent count data y_i with predictor vector x_i, suppose we assume that

$$\begin{aligned} E(y_i) &= \mu_i = e^{x_i'\beta} \\ \text{var}(y_i) &= \mu_i. \end{aligned}$$

The estimating equation (14.1) for β is

$$\sum_i e^{x_i'\beta} x_i e^{-x_i'\beta}(y_i - \mu_i) = 0$$

or

$$\sum_i x_i(y_i - e^{x_i'\beta}) = 0,$$

requiring a numerical solution (Section 14.2). The estimating equation here is exactly the score equation under the Poisson model.

There are two ways to interpret this. Firstly, the estimate based on Poisson likelihood is robust with respect to the distribution assumption up to the correct specification of the mean and variance functions. Secondly, the estimating equation method is efficient (i.e. producing an estimate that is equal to the best estimate, which is the MLE), if the true distribution is Poisson. This is a specific instance of the robustness and efficiency of the quasi-likelihood method.

General quasi-likelihood models

With the general quasi-likelihood approach, for a certain outcome y_i and predictor x_i, we specify using a known function $f(\cdot)$ and $v(\cdot)$

$$E(y_i) = \mu_i = f(x_i'\beta)$$

or

$$h(\mu_i) = x_i'\beta,$$

where $h(\mu_i)$ is the link function, and

$$\text{var}(y_i) = \phi v(\mu_i).$$

We can generate any standard GLM using either the estimating equation or full likelihood approach, but, as described earlier, the underlying modelling philosophies are different. This is important to keep in mind, though in practice the distinction tends to be blurred. One crucial distinction is that the likelihood ratio statistic, such as deviance, is not available for the estimating equation approach.

Likelihood-level specification

It is useful, whenever possible, to think of the previous estimating equation approach in likelihood terms, since from Section 13.5 we can still compare models (via the AIC) using an assumed likelihood. The likelihood connection is provided by the general exponential models described in Section 6.5; some of the details are repeated here for completeness.

Consider a model where the i'th contribution to the log-likelihood is assumed to be

$$\log L_i = \{y_i\theta_i - A(\theta_i)\}/\phi + c(y_i, \phi) \tag{14.2}$$

with known function $A(\cdot)$. The function $c(y_i, \phi)$ is implicit in the definition of $A(\cdot)$ since the density must integrate to one. In standard quasi-likelihood estimation $c(y_i, \phi)$ does not have to be specified; explicit formulae are available for some models, otherwise we can use the extended quasi-likelihood below.

The score function and the Fisher information are

$$\begin{aligned} S_i &= \frac{\partial}{\partial\theta_i}\log L_i = \{y_i - A'(\theta_i)\}/\phi \\ I_i &= -\frac{\partial^2}{\partial\theta_i^2}\log L_i = A''(\theta_i)/\phi. \end{aligned}$$

If $A(\theta_i)$ is chosen such that

$$E(y_i) = A'(\theta_i) \equiv \mu_i$$

and

$$\text{var}(y_i) = \phi A''(\theta_i) \equiv \phi v_i(\mu_i),$$

the likelihood satisfies the regular properties: $ES_i = 0$ and $\text{var}(S_i) = E(I_i)$. A regression model with link function $h(\cdot)$ is

$$h(\mu_i) = x_i'\beta.$$

The scale of $h(\cdot)$ is called the linear predictor scale, and the choice $\theta_i = h(\mu_i)$ is called the canonical-link function.

We now show that the score equation $S(\beta) = 0$ is equal to the estimating equation (14.1):

$$\begin{aligned} S(\beta) &= \frac{\partial \log L}{\partial\beta} \\ &= \sum_i \frac{\partial \log L}{\partial\theta_i} \times \frac{\partial\theta_i}{\partial\beta} \\ &= \phi^{-1}\sum_i \frac{\partial\theta_i}{\partial\beta}(y_i - \mu_i). \end{aligned}$$

Since $\mu_i = A'(\theta_i)$

$$\frac{\partial\mu_i}{\partial\beta} = A''(\theta_i)\frac{\partial\theta_i}{\partial\beta}$$

$$= v_i \frac{\partial \theta_i}{\partial \beta}.$$

So, we obtain

$$S(\beta) = \phi^{-1} \sum_i \frac{\partial \mu_i}{\partial \beta} v_i^{-1} (y_i - \mu_i),$$

and we are done. This means that the exponential family likelihood (14.2) is robust, meaning consistent for a much wider class of distributions than specified by the likelihood itself, as long as the mean and the variance models are correctly specified.

Extended quasi-likelihood

Assuming a quasi-likelihood model, the log-likelihood contribution from a single observation y_i is

$$\log L(\theta_i, \phi) = \{y_i \theta_i - A(\theta_i)\}/\phi + c(y_i, \phi).$$

The function $c(y_i, \phi)$ may not be available explicitly, so direct estimation of ϕ is not possible. In Sections 4.9 and 6.6 we describe an approximation

$$\log L(\theta_i, \phi) \approx -\frac{1}{2} \log\{2\pi\phi v(y_i)\} - \frac{1}{2\phi} D(y_i, \mu_i), \tag{14.3}$$

using the deviance

$$D(y_i, \mu_i) = 2 \log \frac{L(y_i, \phi = 1; y_i)}{L(\mu_i, \phi = 1; y_i)},$$

where $L(\mu_i, \phi = 1; y_i)$ is the likelihood of μ_i based on a single observation y_i, assuming $\phi = 1$. In this setting the name extended quasi-likelihood is apt (Nelder and Pregibon 1987).

14.2 Computing $\widehat{\beta}$ in nonlinear cases

The standard IWLS algorithm to solve the estimating equation can be viewed as one of the following:

- Gauss–Newton algorithm
- Weighted least-squares
- Newton–Raphson algorithm, which coincides with the first two provided the link function is canonical. This was derived in Section 6.7.

Despite having the same formulae, these approaches are based on different motivations. The statistical content of the last two methods immediately suggests the variance of the estimate.

Gauss–Newton algorithm

This is a general algorithm for solving nonlinear equations. We solve

$$\sum_i \frac{\partial \mu_i}{\partial \beta} v_i^{-1}(y_i - \mu_i) = 0$$

by first linearizing μ_i around an initial estimate β^0 and evaluating v_i at the initial estimate. Let $h(\mu_i) = x_i'\beta$ be the linear predictor scale. Then

$$\frac{\partial \mu_i}{\partial \beta} = \frac{\partial \mu_i}{\partial h}\frac{\partial h}{\partial \beta} = \frac{\partial \mu_i}{\partial h} x_i$$

so

$$\begin{aligned} \mu_i &\approx \mu_i^0 + \frac{\partial \mu_i}{\partial \beta}(\beta - \beta^0) \\ &= \mu_i^0 + \frac{\partial \mu_i}{\partial h} x_i'(\beta - \beta^0) \end{aligned}$$

and

$$y_i - \mu_i = y_i - \mu_i^0 - \frac{\partial \mu_i}{\partial h} x_i'(\beta - \beta^0).$$

Putting these into the estimating equation, we obtain

$$\sum_i \frac{\partial \mu_i}{\partial h} v_i^{-1} x_i \{y_i - \mu_i^0 - \frac{\partial \mu_i}{\partial h} x_i'(\beta - \beta^0)\} = 0$$

which we solve for β as the next iterate. Thus

$$\beta^1 = \beta^0 + A^{-1}b,$$

where

$$\begin{aligned} A &= \sum_i \left(\frac{\partial \mu_i}{\partial h}\right)^2 v_i^{-1} x_i x_i' \\ b &= \sum_i \frac{\partial \mu_i}{\partial h} v_i^{-1} x_i (y_i - \mu_i^0). \end{aligned}$$

A better way of expressing the update formula is

$$\beta^1 = (X'\Sigma^{-1}X)^{-1}X'\Sigma^{-1}Y,$$

where X is the design matrix of predictor variables, Σ is a diagonal matrix with elements

$$\Sigma_{ii} = \left(\frac{\partial h}{\partial \mu_i}\right)^2 \phi v_i$$

and Y is a 'working vector' of

$$Y_i = x_i'\beta^0 + \frac{\partial h}{\partial \mu_i}(y_i - \mu_i^0).$$

So the Gauss–Newton iteration is the same as the IWLS algorithm in Section 6.7.

Weighted least-squares

The most direct way to derive the IWLS algorithm is as follows. Consider a semi-linear model

$$y_i = f(x_i'\beta) + e_i$$

and denote $\mu_i = f(x_i'\beta)$ or the link function $h(\mu_i) = x_i'\beta$. For the variance specification, let

$$\text{var}(y_i) = \text{var}(e_i) = \phi v_i.$$

To solve for β, first linearize μ_i around the initial estimate β^0

$$\mu_i = \mu_i^0 + \frac{\partial \mu_i}{\partial h} x_i'(\beta - \beta^0),$$

from which we get

$$y_i = \mu_i^0 + \frac{\partial \mu_i}{\partial h}(x_i'\beta - x_i'\beta^0) + e_i$$

and, upon rearrangement,

$$x_i'\beta^0 + \frac{\partial h}{\partial \mu_i}(y_i - \mu_i^0) = x_i'\beta + \frac{\partial h}{\partial \mu_i} e_i.$$

By defining

$$e_i^* = \frac{\partial h}{\partial \mu_i} e_i$$

and

$$Y_i \equiv x_i'\beta^0 + \frac{\partial h}{\partial \mu_i}(y_i - \mu_i^0)$$

we have a linear model

$$Y_i = x_i'\beta + e_i^*,$$

where

$$\text{var}(e_i^*) = \left(\frac{\partial h}{\partial \mu_i}\right)^2 \phi v_i = \Sigma_{ii}.$$

The value e_i^* is called the *working residual* and $\text{var}(e_i^*)$ the *working variance*, which are quantities associated with the linear predictor scale. The updating formula is given by the weighted least-squares

$$\beta^1 = (X'\Sigma^{-1}X)^{-1}X'\Sigma^{-1}Y,$$

so the algorithm also coincides with the IWLS.

14.3 Asymptotic distribution

In the pure estimating equation approach, the inference for the parameter of interest typically relies on the asymptotic results. We can distinguish two approaches:

- Assuming the variance specification is correct. This is the so-called 'naive' approach.
- Making no assumption about the correctness of the variance specification. This is useful for situations where even specifying the variance is difficult (e.g. longitudinal studies, which we will discuss later). If the variance specification is near correct then these two approaches will be close.

First note the likelihood view of the estimating equation:

$$S(\beta) = \phi^{-1} \sum_i \frac{\partial \mu_i}{\partial \beta} v_i^{-1} (y_i - \mu_i)$$

implies $E_\beta S(\beta) = 0$ if $E(y_i) = \mu_i$ regardless of the variance specification. In view of Theorem 13.1, consistency of $\widehat{\beta}$ depends only on the correctness of this assumption.

Assuming correct variance

If $\text{var}(y_i) = \phi v_i$ then

$$\begin{aligned} \text{var}\{S(\beta)\} &= \phi^{-1} \sum_i \frac{\partial \mu_i}{\partial \beta} v_i^{-1} v_i v_i^{-1} \frac{\partial \mu_i}{\partial \beta'} \\ &= \phi^{-1} \sum_i \frac{\partial \mu_i}{\partial \beta} v_i^{-1} \frac{\partial \mu_i}{\partial \beta'}. \end{aligned}$$

Using a similar derivation to that given for the Newton–Raphson algorithm in Section 6.7, the expected Fisher information is

$$\begin{aligned} E\left\{-\frac{\partial S(\beta)}{\partial \beta}\right\} &= \phi^{-1} \sum_i \frac{\partial \mu_i}{\partial \beta} v_i^{-1} \frac{\partial \mu_i}{\partial \beta'}, \\ &= X'\Sigma^{-1}X, \end{aligned}$$

using X and Σ defined previously, so the usual likelihood theory that

$$\text{var}\{S(\beta)\} = \mathcal{I}(\beta)$$

holds. This is called the naive variance formula. We can expect the standard distribution theory for $\widehat{\beta}$ to hold, i.e. approximately,

$$\widehat{\beta} \sim N(\beta, (X'\Sigma^{-1}X)^{-1}).$$

Example 14.1: (Poisson regression) Suppose we specify the standard Poisson model (dispersion parameter $\phi = 1$) with a log-link function for our outcome y_i:

$$Ey_i = \mu_i = e^{x_i'\beta}$$

$$\text{var}(y_i) = v_i = \mu_i.$$

Then the working variance is

$$\begin{aligned}\Sigma_{ii} &= \left(\frac{\partial h}{\partial \mu_i}\right)^2 v_i \\ &= (1/\mu_i)^2 \mu_i = 1/\mu_i,\end{aligned}$$

so approximately

$$\widehat{\beta} \sim N(\beta, \sum_i \mu_i x_i x_i'),$$

where observations with large means get a large weight. □

Not assuming the variance specification is correct

It is clearer to view the problem from the least-squares approach. Using the true β as the starting value we can derive the working variate and the working variance as

$$\begin{aligned}Y_i &= x_i'\beta + \frac{\partial h}{\partial \mu_i}(y_i - \mu_i) \\ &= x_i'\beta + e_i^* \\ \Sigma_{ii} &= \left(\frac{\partial h}{\partial \mu_i}\right)^2 \phi v_i,\end{aligned}$$

and, setting $\Sigma = \text{diag}[\Sigma_{ii}]$, the regression estimate is

$$\widehat{\beta} = (X'\Sigma^{-1}X)^{-1}X'\Sigma^{-1}Y.$$

But now $\text{var}(Y) = \Sigma_Y \neq \Sigma$. Assuming regularity conditions, we expect that $\widehat{\beta}$ is approximately normal with mean β and variance

$$\text{var}(\widehat{\beta}) = (X'\Sigma^{-1}X)^{-1}X'\Sigma^{-1}\Sigma_Y\Sigma^{-1}X(X'\Sigma^{-1}X)^{-1}.$$

This formula does not simplify any further; it is called a robust variance formula, to indicate that it is correct even if the assumed variance of y_i is not correct. It is a special case of the robust variance formula derived in Section 13.4.

In practice we can estimate Σ_Y by a diagonal matrix $\text{diag}[(e_i^*)^2]$, so the middle term of the variance formula can be estimated by

$$\text{est}(X'\Sigma^{-1}\Sigma_Y\Sigma^{-1}X) = \sum_i \frac{(e_i^*)^2}{\Sigma_{ii}^2} x_i x_i'.$$

This formula automatically takes care of overdispersion, so it is preferable to the naive formula.

14.4 Generalized estimating equation

The estimating equation we have developed so far is useful for univariate outcomes. The need to consider multivariate outcomes arises in repeated measures or longitudinal studies, where a study subject contributes a cluster or a series of measurements. Classical multivariate statistics has been mostly limited to either nonparametric descriptive techniques such as principal component analysis, or normal-based parametric models; furthermore, there is usually a rigid requirement of equal-length data vectors.

New applications, especially in medical statistics, have generated questions and datasets that cannot be addressed using classical multivariate statistics. Extension to nonnormal models, such as the exponential family, has been slow because of the inherent difficulty to model dependence structure in a natural way. Full likelihood modelling of repeated measures data can be approached using the generalized linear mixed models in Section 17.8. Other classes of models are described for example in Lindsey (2000).

For a less-than-full likelihood approach, Zeger and Liang (1986) proposed the so-called generalized estimating equation (GEE) technique. They recognized that the previous estimating equation

$$\sum_i \frac{\partial \mu_i}{\partial \beta} V_i^{-1}(y_i - \mu_i) = 0 \tag{14.4}$$

is immediately applicable for multivariate outcome data, where y_i and μ_i are now vector valued, and V_i is a variance matrix. Most of the previous theories for the univariate outcome data apply, but obviously there is more complexity in model specification and computation of estimates. It is clearest to introduce the idea and notations through an example.

Example 14.2: This example is from Zeger and Liang (1986), where the reader can also find the actual results of the analysis. A sample of N pairs of mother and baby were followed over a period to investigate the effect of the mother's stress level on her baby's health status. The outcome data y_i for the i'th baby is a time series of length n_i of the health status at day $t_1, \ldots, t_{n_i}$. The covariate vector for the i'th pair at time t is x_{it}:

1. mother's stress level at day $t-1$
2. mother's stress level at day $t-2$
3. mother's stress level at day $t-3$
4. household size
5. mother's race
6. mother's employment status
7. marital status.

In the study x_{it} was recorded over time $t_1, \ldots, t_{n_i}$, and put into an $n_i \times p$ matrix x_i; the number of predictors p equals seven in this example. The vector size n_i and the times of follow-up do not have to be common for different subjects, nor do the times of follow-up need to be equally spaced. □

The estimating equation approach, as before, involves a specification of the mean and variance function. To simplify our presentation we assume that the experimental units are independent; this means we need to be concerned only with the dependence within the units, and the overall variance matrix is blocked diagonal. If the units are not independent, the problem is not conceptually more difficult: dependent units simply form a larger cluster, so the current technique applies to the larger clusters, but the specification and practical computations do become more complicated.

Assume that the mean vector $Ey_i = \mu_i$ satisfies

$$h(\mu_i) = x_i'\beta,$$

where β is a $p\times 1$ vector of regression parameters and the link function $h(\mu_i)$ applies element by element on the vector μ_i. For example, if the outcome y_{it} is binary, then a natural model is

$$\text{logit}\ \mu_i = x_i'\beta.$$

Most of the usual considerations when modelling univariate outcomes apply, and the parameter estimates from the GEE are consistent as long as the mean function is correctly specified.

To complete the model specification, we need to define an $n_i \times n_i$ covariance matrix of y_i, which in general will be a function of μ_i. It is doubtful that we would be able to guess what the 'correct' form should be. Specifying a complete distribution of y_i would be even harder. This is where the concept of specifying a 'wrong' but convenient variance function is useful. The reward for specifying a correct variance is efficiency, and the price of specifying a wrong variance is a loss of efficiency, so convenience must be balanced against efficiency.

Specifying the variance of y_{it} is a univariate problem we have seen before, i.e.

$$\text{var}(y_{it}) = \phi v(\mu_i)$$

for some known function $v(\cdot)$; for example, if y_{it} is binary

$$v(\mu_i) = \mu_i(1 - \mu_i).$$

So the covariance matrix specification is reduced to that of a correlation structure.

In general, let us denote the 'working' correlation matrix of y_i by

$$\text{cor}(y_i) = R_i(\alpha),$$

where α is some unknown parameter. The choice of $R_i(\alpha)$ is dictated by the application or by the nature of the repeated measures. For example

1. Uncorrelated/identity structure

$$R_i(\alpha) = 1_{n_i}$$

where 1_{n_i} is an $n_i \times n_i$ identity matrix. This choice is simplest and least efficient; it is equivalent to ignoring the repeated measures structure. It is useful to generate a starting value for an iterative procedure, where we try to identify the correct correlation structure from the residuals.

2. Exchangeable structure

$$R_i(\alpha) = \begin{pmatrix} 1 & \alpha & \dots & \alpha \\ \alpha & 1 & \dots & \alpha \\ \vdots & \vdots & \dots & \vdots \\ \alpha & \alpha & \cdots & 1 \end{pmatrix}.$$

   This is also known as the random effects structure, or split-plot structure, or compound symmetry, or spherical symmetry. This assumption is sensible if the repeated measures are taken on similar subunits, such as family members.

3. Stationary first-order autoregressive structure

$$\mathrm{cor}(y_{it}, y_{it'}) = \alpha^{|t-t'|}$$

   for $|\alpha| < 1$, which is sensible for time series data.

4. Nonparametric structure

$$\mathrm{cor}(y_{it}, y_{it'}) = R_{tt'}$$

   for any t and t'. This choice is reasonable if there are enough individuals to provide a stable estimate.

If A is the diagonal matrix of the variances, then

$$\mathrm{var}(y_i) = \phi V_i = \phi A^{1/2} R_i(\alpha) A^{1/2}.$$

Given the mean and variance specification, the GEE estimate $\widehat{\beta}$ is the solution of the multivariate version of the estimating equation:

$$\sum_i \frac{\partial \mu_i}{\partial \beta} V_i^{-1}(y_i - \mu_i) = 0.$$

Both the computation of $\widehat{\beta}$ given the variance structure parameters α (e.g. the Gauss–Newton algorithm) and the theory given in the previous section extend naturally here.

Computing $\widehat{\beta}$ given ϕ and α

We can derive the Gauss–Newton algorithm in a similar way as for the univariate case; we only need to interpret the previous formulae appropriately in terms of vectors and matrices. Starting with β^0, first compute the $n_i \times 1$ working vector

$$Y_i = x_i' \beta^0 + \frac{\partial h}{\partial \mu_i}(y_i - \mu_i^0),$$

where the product with the derivative vector $\partial h / \partial \mu_i$ is computed element by element. The variance of Y_i is

$$\Sigma_{ii} = \phi \frac{\partial h}{\partial \mu_i} V_i \frac{\partial h}{\partial \mu_i'}.$$

By stacking $Y_1, \ldots, Y_n$ into a column vector Y, and similarly $x_1, \ldots, x_n$ into X, and putting Σ_{ii} into a blocked diagonal Σ, the update formula is simply

$$\beta^1 = (X'\Sigma^{-1}X)^{-1}X'\Sigma^{-1}Y.$$

This formula is only nice to look at; in practice we must exploit the block diagonal structure of Σ to compute Σ^{-1}.

Computing ϕ and α

Given $\widehat{\beta}$ we can estimate α and ϕ using the method of moments. For example, we can define the normalized residual

$$r_{it} = \frac{y_{it} - \widehat{\mu}_{it}}{\sqrt{\widehat{v}_i}},$$

so $\text{var}(r_{it}) = \phi$ and

$$\widehat{\phi} = \frac{\sum_i \sum_t r_{it}^2}{\sum_i n_i - p},$$

where p is the number of regression parameters. As an example of estimating α, suppose we assume an exchangeable correlation structure; then

$$\text{cor}(y_{it}, y_{it'}) \approx \text{cor}(r_{it}, r_{it'}) \approx \phi^{-1} E r_{it} r_{it'},$$

so we can estimate α by

$$\widehat{\alpha} = \widehat{\phi}^{-1} \frac{\sum_i \sum_{t \neq t'} r_{it} r_{it'}}{\sum_i n_i(n_i - 1) - p}.$$

If we assume a nonparametric structure we can use

$$R_{tt'} = \frac{1}{\widehat{\phi}(N-p)} \sum_i r_{it} r_{it'}.$$

14.5 Robust estimation

It is well known that the sample mean is sensitive to outliers. Classical robust estimation theory provides a way of looking at the sensitivity of the sample mean. If we use the estimating function $\psi(y, \theta) = y - \theta$, then the solution of the estimating equation

$$\sum_i \psi(y_i, \theta) = 0$$

is $\widehat{\theta} = \overline{y}$. For location parameters $\psi(y, \theta)$ is a function of $(y - \theta)$, i.e. $\psi(y, \theta) \equiv \psi(y - \theta)$. We will limit the discussions in this section to location parameter problems.

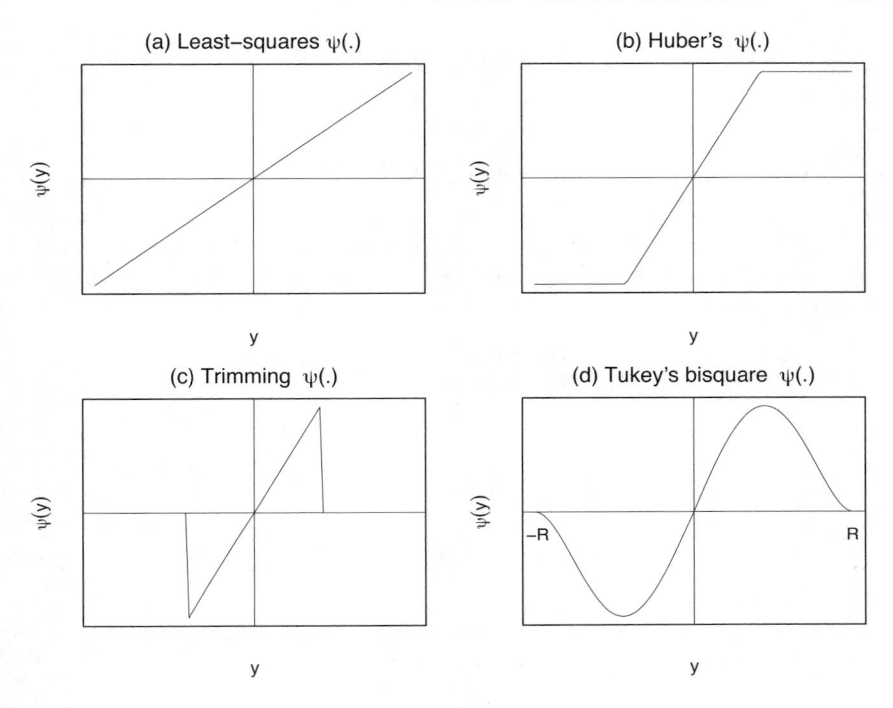

Figure 14.1: *Some examples of* $\psi(\cdot)$.

As a function of y we can think of $\psi(y-\theta)$ as a residual. To make $\sum(y_i-\theta)=0$, outliers will have a large influence. The idea of robust estimation is simple: reduce the influence of the outlying values by capping the function $\psi(\cdot)$.

Figure 14.1(a) shows $\psi(\cdot)$ for the least-squares method, bestowing outlying values with unlimited influence. In contrast, Figure 14.1(b) shows the famous Huber ψ function. Using this function, values far from the centre of the data contribute the same error as $\theta \pm k$. To solve $\sum_i \psi(y_i, \theta) = 0$ we can use the following iterative procedure: start with θ^0, then compute

$$y_i^* = \begin{cases} y_i & |y_i - \theta^0| < k \\ \theta^0 + k & y_i > \theta^0 + k \\ \theta^0 - k & y_i > \theta^0 - k \end{cases}$$

and update $\theta^1 = \sum_i y_i^*/n$. $\widehat{\theta}$ in this case is called the 'Winsorized' mean of y. The so-called tuning parameter k determines the properties of the procedure. As $k \to \infty$ then $\widehat{\theta} \to \overline{y}$ and as $k \to 0$ then $\widehat{\theta}$ goes to the median of y. The choice of $k = 1.345\sigma$, where σ is the dispersion parameter, corresponds to 95% efficiency if the true distribution is normal. (By the efficiency we mean $\text{var}(\overline{y})/\text{var}(\widehat{\theta}) \approx 0.95$, where we know $\overline{y}$ is the best estimate in the normal case.)

Example 14.3: For Darwin's data in Section 13.1 we can verify that

$$\begin{aligned}
\overline{x} &= 2.61 \\
\text{median} &= 3.0 \\
\widehat{\theta}_{k=2} &= 3.13 \\
\widehat{\theta}_{k=3.2} &= 3.29. \;\square
\end{aligned}$$

To estimate k from the data, first view the estimating equation as a score equation. The associated likelihood is that of a normal centre plus an exponential tail, thus

$$\frac{\partial \log L_i}{\partial \theta} = \psi(y_i, \theta)$$

so

$$\begin{aligned}
\log L_i &= \int \psi(y_i, \theta) d\theta \\
&= \begin{cases} -\frac{1}{2}(y_i - \theta)^2 & |y_i - \theta| \leq k \\ -\frac{1}{2}k(2|y_i - \theta| - k) & |y_i - \theta| > k \end{cases}
\end{aligned}$$

and the parameter k can be estimated by maximum likelihood.

A more dramatic modification is given by

$$\psi(y, \theta) = \begin{cases} |y - \theta| & |y - \theta| < k \\ 0 & \text{otherwise} \end{cases}$$

shown in Figure 14.1(c). Intuitively this forces the outlying observations to have no influence, which can be achieved by dropping them. There is no likelihood associated with this ψ function. Solving

$$\sum_{i=1}^{n} \psi(y_i, \theta) = 0$$

can be done iteratively: start with θ^0, then update

$$\theta^1 = \frac{1}{m} \sum_i y_i$$

for i such that $|y_i - \theta^0| < k$.

With this choice, the function $\sum \psi(y_i, \theta)$ is numerically complicated, since it is not monotone, so it can have many solutions. For symmetric distributions, the solution $\widehat{\theta}$ is approximately the same as the so-called 'trimmed' means, which is easier to compute: for example, to get a 5% trimmed mean we first drop the top and bottom 5% of the values before computing the average. (Incidentally this rule is adopted in some sport competitions, where a performance is evaluated by several judges, then the lowest and the highest scores are dropped, and the rest added as the final score.)

Other variations of the $\psi(\cdot)$ function are the so-called 'soft' trimming functions. An example is the Tukey's bisquare function

$$\psi(y,\theta) = (y-\theta)\{1-(y-\theta)^2/R^2\}_+^2$$

shown in Figure 14.1(d). The choice $R = 4.685\sigma$ gives 95% efficiency under the normal distribution.

Median as a robust estimate

Consider the extreme version of Huber ψ function: $\psi(y,\theta) = \text{sign}(y-\theta)$, that is

$$\psi(y,\theta) = \begin{cases} 1 & y > \theta \\ 0 & y = \theta \\ -1 & y < \theta. \end{cases}$$

Intuitively all points are given equal-sized residuals: plus one or minus one. So the solution of $\sum_i \psi(y_i,\theta) = 0$ is given by the sample median.

Quantile estimation

Now consider for a fixed and known p, where $0 < p < 1$,

$$\psi(y,\theta) = \begin{cases} p/(1-p) & y > \theta \\ 0 & y = \theta \\ -1 & y < \theta. \end{cases}$$

The solution $\widehat{\theta}_p$ of $\sum_i \psi(y_i,\theta) = 0$ is the sample $100p$ percentile, which is given by the k'th order statistic $y_{(k)}$ with $k = \lfloor 100p \rfloor$.

Robust regression model and computation using IWLS

The robust approach can be extended to the regression setup to analyse a predictor–outcome relationship. Suppose we have a model

$$y_i = x_i'\beta + e_i.$$

The estimate $\widehat{\beta}$ is called a robust regression estimate or an M-estimate if it solves

$$\sum_i x_i \psi(y_i - x_i'\beta) = 0,$$

for some choice of function $\psi(\cdot)$. In principle, we can also take into account the variance of the outcome and nonlinear link functions by considering the robust version of the estimating equation:

$$\begin{aligned} Ey_i &= \mu_i = f(x_i'\beta) \\ \text{dispersion}(y_i) &= \phi v_i \end{aligned}$$

where the 'dispersion' does not have to be a variance. Then $\widehat{\beta}$ is the solution of

$$\sum_i \frac{\partial \mu_i}{\partial \beta} v_i^{-1} \psi(y_i - \mu_i) = 0. \tag{14.5}$$

Certain choices of $\psi(\cdot)$ will protect $\widehat{\beta}$ against outliers in the outcome variable.

Equation (14.5) provides an alternative interpretation of robust methods, and indicates that *we can use the Gauss–Newton or IWLS algorithm for routine computations.* Defining

$$w_i \equiv w(y_i, \mu_i) = \frac{\psi(y_i - \mu_i)}{y_i - \mu_i}$$

and

$$V_i^{-1} \equiv w_i v_i^{-1}.$$

we can write (14.5) as

$$\sum_i \frac{\partial \mu_i}{\partial \beta} V_i^{-1}(y_i - \mu_i) = 0,$$

the standard estimating equation.

So $\psi(\cdot)$ simply modifies the weight for $(y_i - \mu_i)$ depending on how large it is. The Gauss–Newton algorithm applies immediately with a modified weight formula. For example, for Huber ψ function the weight is

$$w(y, \mu) = \begin{cases} 1, & |y - \mu| \leq k \\ k/|y - \mu|, & |y - \mu| > k, \end{cases}$$

so observations with large residuals get reduced weights.

Likelihood-based robust estimation

We now compare the $\psi(\cdot)$ functions implied by the likelihood models. The previous discussion facilitates the robust interpretation of the models, and the likelihood set-up points to an immediate inferential method and adaptive estimation of the tuning parameter. In the likelihood approach, a model, hence $\psi(\cdot)$, is suggested by the actual data distribution. Furthermore, a model is subject to criticism or selection via the AIC.

For simplicity, we set the scale parameter to one in the following examples. Assuming $y \sim N(\theta, 1)$, we have

$$\begin{aligned} \log L(\theta) &= -\frac{1}{2}(y - \theta)^2 \\ \psi(y, \theta) &= (y - \theta). \end{aligned}$$

The function is shown in Figure 14.2(a), the same as the least-squares $\psi(\cdot)$, illustrating the lack of robustness against outliers.

For the Cauchy model $y \sim \text{Cauchy}(\theta, 1)$, we have

$$\begin{aligned} \log L(\theta) &= -\log\{1 + (y - \theta)^2\} \\ \psi(y, \theta) &= \frac{2(y - \theta)}{1 + (y - \theta)^2}, \end{aligned}$$

showing a soft trimming action on large residuals, comparable with Tukey's bisquare. The discussion following (14.5) also shows how to use the Gauss–Newton or IWLS algorithm to compute Cauchy-based regression models.

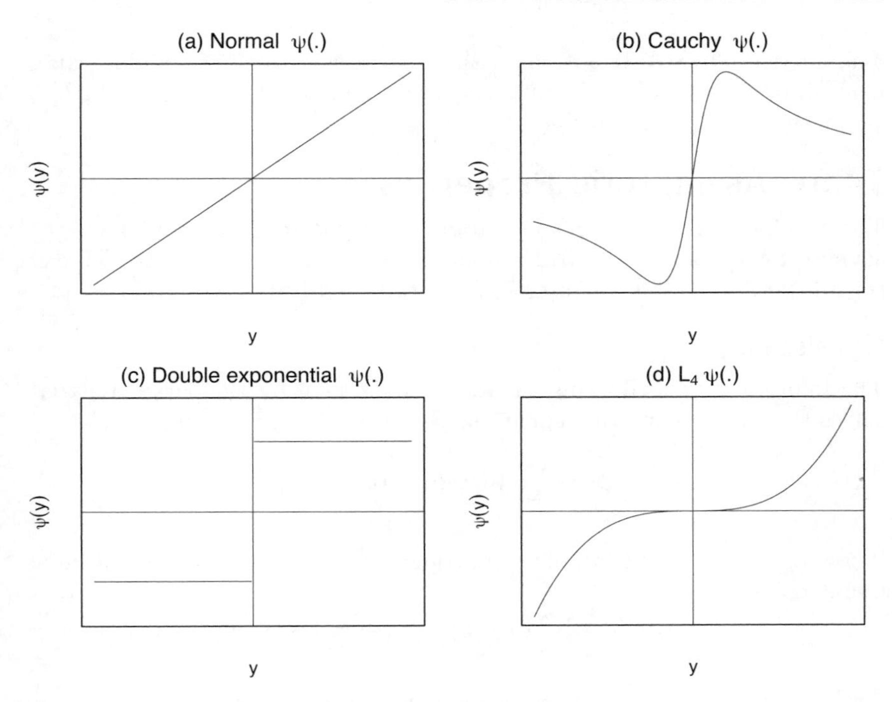

Figure 14.2: *The $\psi(\cdot)$ functions associated with some common models.*

As described in Section 6.9, the Cauchy model can be generalized to a family of t-distribution with k degrees of freedom. This would include both the normal ($k \to \infty$) and the Cauchy ($k = 1$) models. All models with $k > 2$ have heavier tails than the normal model. The degrees of freedom parameter k is a tuning parameter to control the robust modification, and it can be estimated from data using a maximum likelihood procedure; see Lange *et al.* (1989) for further discussion and many applications.

A double-exponential model, $y \sim \text{DE}(\theta)$, implies

$$\begin{aligned} \log L(\theta) &= -|y - \theta| \\ \psi(y, \theta) &= \text{sign}(y - \theta). \end{aligned}$$

Figure 14.2(c) shows that this model is associated with the median estimation. Estimation using this model is called L_1 estimation, and it can be extended to the L_p model, with

$$\begin{aligned} \log L(\theta) &= -\frac{1}{2}|y - \theta|^p \\ \psi(y, \theta) &= \begin{cases} \frac{p}{2}|y - \theta|^{p-1} & \theta < y \\ -\frac{p}{2}|y - \theta|^{p-1} & \theta > y. \end{cases} \end{aligned}$$

This model allows distributions with either longer tail (e.g. double exponential) or shorter tails (e.g. uniform) than the normal tail. Figure 14.2(d) shows the $\psi(\cdot)$ function for $p = 4$.

14.6 Asymptotic Properties

The results in this section are parallel to those in Section 13.4 on the asymptotic properties of MLEs under a wrong model. The reader can consult Serfling (1980, Chapter 7) for details and further results.

Consistency

The solution of an estimating equation is consistent for the parameter that solves the 'true' estimating equation. Recall that $\widehat{\theta}$ is the solution of

$$\sum_i \psi(y_i, \theta) = 0.$$

If $y_1, \ldots, y_n$ are an iid sample from some population, by the law of large numbers

$$\lambda_n(\theta) \equiv \frac{1}{n} \sum_i \psi(y_i, \theta) \to \lambda(\theta) \equiv E\psi(Y, \theta).$$

Let θ_0 be the (unique) solution of the true estimating equation

$$\lambda(\theta) = 0.$$

Then we expect $\widehat{\theta} \to \theta_0$.

For example, let $\psi(y, \theta) = \text{sign}(y - \theta)$ so $\widehat{\theta}$ is the sample median. Then

$$\begin{aligned} \lambda(\theta) &= \int \psi(y, \theta) f(y) dy \\ &= -\int_{y<\theta} f(y)dy + \int_{y>\theta} f(y)dy \\ &= 1 - 2F(\theta). \end{aligned}$$

Solving $\lambda(\theta) = 0$ yields θ_0 to be the true median. The sample median is consistent for the true median in view of the following basic result.

Theorem 14.1 *If $\psi(y, \theta)$ is monotone in θ, then $\widehat{\theta}$ is consistent for θ_0.*

If $\psi(\cdot)$ is not monotone we get a weaker result, since there could be multiple solutions. The proof of the following theorem is left as an exercise. It is useful to recall the proof of the consistency of the MLE given in Section 9.3.

Theorem 14.2 *If $\psi(y, \theta)$ is continuous and bounded in θ then* there exists *a consistent solution.*

Asymptotic normality

The following result is the most important distribution theory for M-estimation.

Theorem 14.3 *Let $y_1, \ldots, y_n$ be an iid sample from a distribution F and $\widehat{\theta}$ is the solution of $\sum_i \psi(y_i, \theta) = 0$. Assume that*

A1. $\psi(y, \theta)$ is monotone in θ.

A2. $\lambda(\theta)$ is differentiable at θ_0 and $\lambda'(\theta_0) \neq 0$.

A3. $E\psi^2(Y, \theta)$ is finite and continuous around θ_0.

If $\widehat{\theta}$ is the solution of $\sum_i \psi(y_i, \theta) = 0$ then

$$\sqrt{n}(\widehat{\theta} - \theta_0) \to N(0, \sigma^2)$$

with

$$\sigma^2 = \frac{E\psi^2(Y, \theta_0)}{\{\lambda'(\theta_0)\}^2}.$$

Assumption A1 guarantees the uniqueness of the solution of $\lambda(\theta) = 0$ and consistency of $\widehat{\theta}$; A2 and A3 are needed to guarantee σ^2 is finite.

Example 14.4: We have seen that the sample quantile is an M-estimate for the estimating function

$$\psi(y, \theta) = \begin{cases} p/(1-p) & y > \theta \\ 0 & y = \theta \\ -1 & y < \theta, \end{cases}$$

where p is a known value between zero and one. To get an asymptotic distribution theory for $\widehat{\theta}$ we need the following quantities:

1. $\lambda(\theta) = E\psi(Y, \theta)$, where $Y \sim F$.
2. $\lambda'(\theta_0)$, where $\lambda(\theta_0) = 0$.
3. $E\psi^2(Y, \theta_0)$.

Straightforward computations give

$$\begin{aligned} \lambda(\theta) &= \int \psi(y, \theta) f(y) dy \\ &= -\int_{y<\theta} f(y) dy + \frac{p}{1-p} \int_{y>\theta} f(y) dy \\ &= -F(\theta) + \frac{p}{1-p}(1 - F(\theta)) \\ &= \frac{p - F(\theta)}{1-p}. \end{aligned}$$

The solution of $\lambda(\theta) = 0$ is $\theta_0 = F^{-1}(p)$, and we also have

$$\lambda'(\theta_0) = -\frac{f(\theta_0)}{1-p}.$$

Furthermore,

$$\begin{aligned} E\psi^2(Y,\theta_0) &= \int \psi^2(y,\theta_0)f(y)dy \\ &= \int_{y<\theta_0} f(y)dy + \left(\frac{p}{1-p}\right)^2 \int_{y>\theta_0} f(y)dy \\ &= F(\theta_0) + \left(\frac{p}{1-p}\right)^2 (1-F(\theta_0)) \\ &= \frac{p}{1-p}. \end{aligned}$$

So we get an important distribution theory for the p'th sample quantile:

$$\sqrt{n}(\widehat{\theta}-\theta_0) \to N\left(0, \frac{p(1-p)}{f(\theta_0)^2}\right). \tag{14.6}$$

In particular for median we have

$$\sqrt{n}(\widehat{\theta}-\theta_0) \to N\left(0, \frac{1}{4f(\theta_0)^2}\right). \tag{14.7}$$

The distribution depends on the value of the density at the true quantile; this is a 'local' property as opposed to a 'global' one such as the variance that affects the distribution of the sample mean. Because of its locality, the distribution of the sample median is not affected by the tail of the distribution.

Example 14.5: For Darwin's data ($n = 15$) we found the sample median $\widehat{\theta} = 3.0$. The standard error is

$$\text{se}(\widehat{\theta}) = \frac{1}{\sqrt{4\widehat{f}(3)^2 n}} = \frac{1}{2\widehat{f}(3)\sqrt{n}}.$$

To get $\widehat{f}(3)$, suppose $n(3,b)$ is the number of sample values that fall between $3-b$ and $3+b$. It is clear $n(3,b)$ is binomial with parameters $n = 15$ and probability

$$p = P(3-b < X < 3+b) \approx 2bf(3),$$

so we can estimate $f(3)$ by $\widehat{f}(3) = n(3,b)/(2nb)$. For a range of b we can verify

b	$\widehat{f}(3)$	se
0.2	0.33	$1.5/\sqrt{n}$=0.39
0.5	0.13	$3.8/\sqrt{n}$=0.98
1.0	0.13	$3.8/\sqrt{n}$=0.98
1.5	0.13	$3.8/\sqrt{n}$=0.98
2.0	0.10	$5.0/\sqrt{n}$=1.29

For a wide range of value of b between 0.5 and 1.5 the estimate $\widehat{f}(3)$ is quite stable, so it is reasonable to conclude that the standard error of $\widehat{\theta}$ is 0.98.

For comparison, we have shown before that the double exponential model is best for this dataset. Here we have the density function

$$f(y) = \frac{1}{2\sigma}e^{-|y-\theta|/\sigma}.$$

Given the data we get the MLEs

$$\begin{aligned}\widehat{\theta} &= 3.0 \\ \widehat{\sigma} &= \frac{1}{n}\sum_i |y_i - \widehat{\theta}| = 3.2\end{aligned}$$

so the parametric estimate $\widehat{f}(\widehat{\theta}) = 1/(2\widehat{\sigma})$ and the standard error of the sample median is

$$\begin{aligned}\text{se}(\widehat{\theta}) &= 1/\sqrt{4\widehat{f}(\widehat{\theta})^2 n} \\ &= \widehat{\sigma}/\sqrt{n} = 3.2/\sqrt{n} = 0.83,\end{aligned}$$

slightly smaller than the nonparametric estimate, which could be expected due to the pointedness of the density function at the median. □

Sample version of the variance formula

The variance formula in the asymptotic distribution can be easily estimated as follows. Given $y_1, \ldots, y_n$ let

$$\begin{aligned}J &= \widehat{E}\psi^2(Y, \theta_0) \\ &= \frac{1}{n}\sum_i \psi^2(y_i, \widehat{\theta}),\end{aligned}$$

where in the vector case we would simply use $\psi(y_i, \widehat{\theta})\psi'(y_i, \widehat{\theta})$ in the summation, and

$$\begin{aligned}I &= -\widehat{E}\frac{\partial}{\partial\theta}\psi(Y, \theta_0) \\ &= -\frac{1}{n}\sum_i \frac{\partial}{\partial\theta}\psi(y_i, \widehat{\theta}).\end{aligned}$$

Then, approximately,

$$\sqrt{n}(\widehat{\theta} - \theta_0) \sim N(0, I^{-1}JI^{-1}),$$

which is also a sample version of Theorem 13.2. The sample variance $\widehat{\sigma}^2 \equiv I^{-1}JI^{-1}$ is the robust variance formula, also known as the 'sandwich' estimator, with I^{-1} as the bread and J the filling. In a likelihood setting I is the usual observed Fisher information. In general, however, there is no Fisher information interpretation of these quantities.

15

Empirical likelihood

Given $x_1, \ldots, x_n$ from $N(\theta, \sigma^2)$ where σ^2 is unknown, we can obtain an appropriate likelihood for θ by profiling over σ^2. What if the normal assumption is in doubt, and we do not want to use any specific parametric model? Is there a way of treating the whole shape of the distribution as a nuisance parameter, and still get a sensible likelihood for the mean? Yes, the *empirical* or *nonparametric* likelihood is the answer. This is a significant result: as 'shape' is an infinite-dimensional nuisance parameter, there is no guarantee that it can be 'profiled out'.

Without making any distributional assumption about the shape of the population we know that asymptotically

$$\frac{\sqrt{n}}{s}(\overline{x} - \theta) \rightarrow N(0, 1).$$

This means we can get an approximate likelihood for the mean based on the asymptotic normal distribution:

$$\log L(\theta) = -\frac{n}{2s^2}(\overline{x} - \theta)^2.$$

It is a 'nonparametric' likelihood in the sense that, up to the normal approximation, it is appropriate for a very large class of distributions specified only by the mean. The empirical likelihood achieves the same property, but it is based on better approximations of the distribution of the sample mean. In general we can evaluate an empirical likelihood using

- the profile likelihood
- the bootstrap-based likelihood
- the exponential family model.

We will discuss these in detail in the following sections.

15.1 Profile likelihood

The empirical likelihood (Owen 1988) is originally defined as the profile likelihood

$$L(\theta) = \sup_{F_\theta} \prod_{i=1}^{n} p_i(\theta) \tag{15.1}$$

where the supremum is taken over all possible distributions F_θ on $x_1, \ldots, x_n$, such that the functional $t(F_\theta) = \theta$. The distribution F_θ is characterized by a set of probabilities $\{p_i(\theta)\}$ on $x_1, \ldots, x_n$.

The term 'functional' $t(F)$ simply means a particular feature of a distribution; for example, the mean of F is a functional

$$t(F) = \int x dF(x).$$

(Note: the symbol '$dF(x)$' is shorthand for '$f(x)dx$' in the continuous case or the probability mass '$p(x)$' in the discrete case.) Other functionals are, for example, the variance, skewness and kurtosis for a univariate distribution, or the correlation coefficient for a bivariate distribution.

To motivate the term 'empirical likelihood', recall the empirical distribution function (EDF). Given $x_1, \ldots, x_n$ (assuming no tie) from an unknown distribution F, the EDF F_n is the nonparametric MLE of F as it maximizes the likelihood

$$\begin{aligned} L(F) &= \prod_{i=1}^{n} P(X_i = x_i) \\ &= \prod_{i=1}^{n} p_i, \end{aligned}$$

subject to the conditions $p_i \geq 0$, for all i, and $\sum_i p_i = 1$.

F_n is a discrete distribution that assigns probability $p_i = 1/n$ to each data point. To show that it does maximize the likelihood, we can use the Lagrange multiplier technique to maximize

$$Q = \sum_i \log p_i + \lambda(\sum_i p_i - 1).$$

Taking the derivative

$$\frac{\partial Q}{\partial p_i} = \frac{1}{p_i} + \lambda$$

setting it to zero, we get

$$\lambda p_i = -1,$$

so $\lambda \sum_i p_i = -n$ or $\lambda = -n$. This immediately yields the solution $p_i = 1/n$ for each i.

Empirical likelihood of the mean

Given an iid sample $x_1, \ldots, x_n$ from F, the empirical likelihood of the mean θ is the profile likelihood

$$L(\theta) = \sup \prod_{i=1}^{n} p_i(\theta)$$

where the supremum is taken over all discrete distributions on $x_1, \ldots, x_n$ with mean θ, satisfying

$$\text{(i)} \qquad P(X = x_i) = p_i(\theta) \geq 0$$

$$\text{(ii)} \qquad \sum_i p_i(\theta) = 1$$

$$\text{(iii)} \qquad \sum_i x_i p_i(\theta) = \theta.$$

At each fixed value of θ, finding $L(\theta)$ is a constrained optimization problem. For simplicity we will suppress the dependence of p_i on θ. Let

$$Q = \sum_i \log p_i + \psi(\sum_i p_i - 1) + \lambda(\sum_i x_i p_i - \theta),$$

so

$$\frac{\partial Q}{\partial p_i} = \frac{1}{p_i} + \psi + \lambda x_i.$$

Setting $\partial Q/\partial p_i = 0$ we get

$$p_i = \frac{-1}{\psi + \lambda x_i} \tag{15.2}$$

and also, after multiplying by p_i,

$$1 + \psi p_i + \lambda x_i p_i = 0.$$

Summing the last equation over i, we obtain

$$n + \psi \sum_i p_i + \lambda \sum_i x_i p_i = 0,$$

and, using $\sum_i p_i = 1$ and $\sum_i x_i p_i = \theta$,

$$n + \psi + \lambda\theta = 0$$

so

$$\psi = -n - \lambda\theta.$$

Substituting this in (15.2) yields the solution

$$p_i = \frac{1}{n - \lambda(x_i - \theta)},$$

where λ must satisfy

$$\sum_i x_i p_i = \theta,$$

or

$$\sum_i \frac{x_i}{n - \lambda(x_i - \theta)} = \theta,$$

or, again using $\sum p_i = 1$,

$$\sum_i \frac{x_i - \theta}{n - \lambda(x_i - \theta)} = \sum_i (x_i - \theta) p_i = 0. \tag{15.3}$$

Equation (15.3) gives a defining property of θ, and indicates how to extend the empirical likelihood to estimating equation problems.

Now let us define, for fixed θ,

$$g(\lambda) \equiv \sum_i \frac{x_i - \theta}{n - \lambda(x_i - \theta)}.$$

So, we have

$$g'(\lambda) = \sum_i \frac{(x_i - \theta)^2}{\{n - \lambda(x_i - \theta)\}^2} > 0,$$

implying $g(\lambda)$ is strictly increasing. The constraint $p_i \geq 0$ for all i implies that λ must be in the interval

$$\frac{n}{x_{(1)} - \theta} < \lambda < \frac{n}{x_{(n)} - \theta},$$

where $x_{(1)}$ and $x_{(n)}$ are the minimum and maximum of x_i's, and it is assumed that $x_{(1)} < \theta < x_{(n)}$. It can be seen that $g(\lambda)$ ranges from $-\infty$ to ∞ as λ varies in the allowed interval, so the equation $g(\lambda) = 0$ will always produce a unique solution.

The empirical likelihood $L(\theta)$ is maximized at $\theta = \overline{x}$, implying the sample mean is the nonparametric MLE of the population mean. At $\theta = \overline{x}$ we get $\lambda = 0$ and $p_i = 1/n$ (giving the EDF). It makes sense to compute the likelihood on an interval around $\overline{x}$. Furthermore, for the purpose of frequentist calibration, Owen (1988) showed that the usual asymptotic distribution of the likelihood ratio statistic holds. That is,

$$2 \log \frac{L(\widehat{\theta})}{L(\theta)} \rightarrow \chi_1^2.$$

This means that the empirical likelihood can be interpreted and calibrated like the ordinary likelihood.

Example 15.1: Suppose we observe $n = 5$ data points

2.3 2.4 4.5 5.4 10.1

and we are interested in the empirical likelihood for the mean. The following table shows the various quantities at several values of θ.

θ	λ			p_i			$\log L(\theta) = \sum_i \log p_i$
3	−3.85	0.43	0.37	0.09	0.07	0.03	−10.33
4	−0.85	0.28	0.27	0.18	0.16	0.10	−8.40
4.94	0	0.20	0.20	0.20	0.20	0.20	−8.05
6	0.52	0.14	0.15	0.17	0.19	0.35	−8.34
7	0.97	0.10	0.11	0.13	0.15	0.50	−9.08

At $\theta = \overline{x} = 4.94$ we have $\lambda = 0$ and $p_i = 0.2$. We can see that a large value of θ would yield a distribution with more weight on the right, and vice versa for a small value of θ. Figure 15.1 shows the plot of the likelihood. Its asymmetry is due to the skewness of the data; from the likelihood we can get a naturally asymmetric CI around $\overline{x}$. □

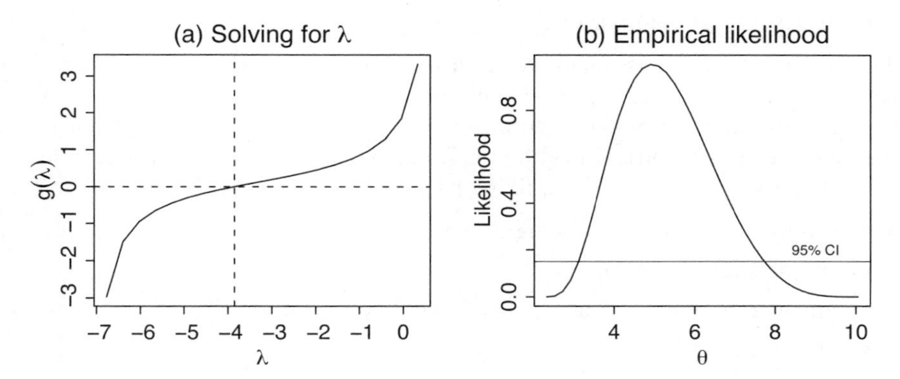

Figure 15.1: *(a) At* $\theta = 3$ *the equation* $g(\lambda) = 0$ *yields the solution* $\lambda = -3.85$. *(b) Empirical likelihood of the mean* θ.

15.2 Double-bootstrap likelihood

In some sense an empirical likelihood is a likelihood based on an estimate rather than the whole data. Suppose we have a dataset x, an estimate $t(x)$ for a scalar parameter θ, and a nuisance parameter η such that

$$p_{\theta,\eta}(x) = p_\theta(t)p_{\theta,\eta}(x|t).$$

The empirical likelihood is very close to the likelihood based on $p_\theta(t)$, so it is a form of marginal likelihood. This leads to the idea of double bootstrapping to generate the empirical likelihood (Davison *et al.* 1992).

The bootstrap methodology (see Sections 5.2, 5.3 and 5.6) was invented to provide a (simple) computational-based inference in complex statistical problems. It is an exchange between analysis and computer power as it replaces thinking with computing (which makes economic sense since a thinker is more expensive than a computer). The bootstrap method *is* simple if all we want is a standard error, or a CI based on the percentile method. However, in its development, more complicated schemes are necessary to get a better inference than what is merely provided by the standard error or the percentile method (Efron 1987). The ambitious goal is to improve on the Wald-type CI, $\widehat{\theta} \pm 1.96\text{se}(\theta)$, purely by computational means.

It is instructive to imagine a 3D plot of a bivariate function

$$h(\theta, t) \equiv p_\theta(t).$$

A slice at $\theta = \widehat{\theta}$ gives a density function $p_{\widehat{\theta}}(t) = h(\widehat{\theta}, t)$; a slice at the observed value $t = t_{obs}$ provides the likelihood $L(\theta) = h(\theta, t_{obs})$. The standard bootstrap procedure is:

- resample from the data x, yielding a new dataset x^*
- compute $t^* \equiv t(x^*)$

- repeat a large number of times, so we have a collection of t^*'s

The collection of t^*'s is a sample from $p_{\hat{\theta}}(t)$, so from such a collection we can estimate $p_{\hat{\theta}}(t)$, but not the likelihood $L(\theta)$.

The following double-bootstrap procedure generates a likelihood that is very close to the empirical likelihood. Imagine that each bootstrap sample x^* yields a new $\theta^* \equiv t^*$. Then, fixing x^*,

- resample from x^*, producing a new dataset x^{**}
- compute $t^{**} \equiv t(x^{**})$
- repeat a number of times, so we have a collection of t^{**}'s
- estimate $p_{\theta^*}(t)$ from such a collection.

The above procedure is performed at each first-stage sample x^*. In principle, by piecing together $p_{\theta^*}(t)$ for different values of θ^* we can build the function $h(\theta, t)$, and evaluate the likelihood $L(\theta) = h(\theta, t_{obs})$.

The double-bootstrap method is computationally intensive. If we use $B_1 = 1000$ bootstrap replications for the first stage and another $B_2 = 1000$ bootstrap replications for the second stage, in total we require 1,000,000 replications. Two layers of smoothing are required: one to estimate $p_{\theta^*}(t)$ at $t = t_{obs}$ for each θ^*, and the other to smooth $L(\theta^*)$. The first smoothing operation must always be performed, but the unsmoothed $L(\theta^*)$ might be displayed to show the shape of the likelihood. The usual advantage of the bootstrap is that we can consider more complicated statistics or sampling models than achievable by analytical methods.

Example 15.2: Consider $n = 24$ intervals in hours between repairs and failures of an aircraft air-conditioning equipment, given in Example T of Cox and Snell (1981):

```
50  44 102  72  22  39    3  15  197  188   79   88
46   5   5  36  22 139  210  97   30   23   13   14
```

and assume that they are an iid sample from some distribution. Figure 15.2 shows the unsmoothed and smoothed bootstrap likelihood of the mean; in this example we use $B_1 = 200$ and $B_2 = 40$. The standard empirical likelihood tracks the bootstrap likelihood closely. Indeed, for a large class of M-estimators, Davison *et al.* (1992) show that the bootstrap and empirical likelihoods agree up to terms of order $O(n^{-0.5})$. Therefore, we can think of the bootstrap likelihood as a numerical way to produce an empirical likelihood. □

Example 15.3: The following table shows the average scholastic aptitude tests (SAT) and the grade point average (GPA) from 15 law schools as reported in Table 4 of Efron (1987). The data are plotted in Figure 15.3(a). We are interested in the correlation coefficient ρ between the two scores; the observed correlation is 0.78.

Inference for the correlation coefficient is a benchmark problem of the bootstrap methodology; it is 'the lightning rod' in the sense that failures in the methodology will hit this problem first. The standard empirical likelihood of ρ is given in Owen (1990). Construction of the bootstrap likelihood proceeds the same way as before. Figure 15.3(b) shows the unsmoothed and smoothed log-likelihood of ρ. Since estimates of correlation are biased (in this case upwards), the likelihood is not maximized at the observed correlation. □

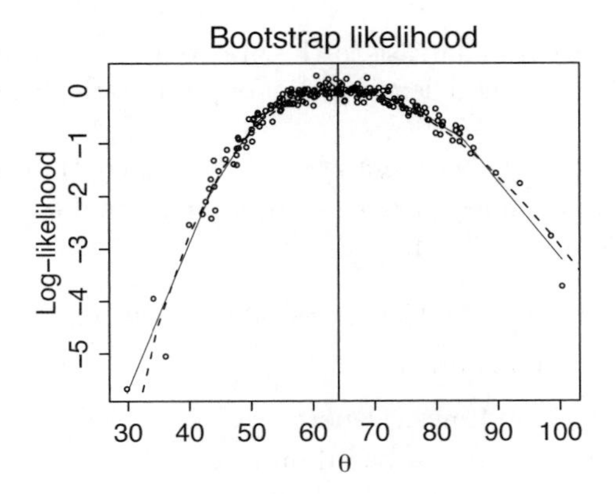

Figure 15.2: *The scattered points are the unsmoothed bootstrap log-likelihood, the solid curve is the smoothed bootstrap log-likelihood, and the dashed curve is the empirical likelihood.*

SAT	GPA	SAT	GPA
576	3.39	651	3.36
635	3.30	605	3.13
558	2.81	653	3.12
578	3.03	575	2.74
666	3.44	545	2.76
580	3.07	572	2.88
555	3.00	594	2.96
661	3.43		

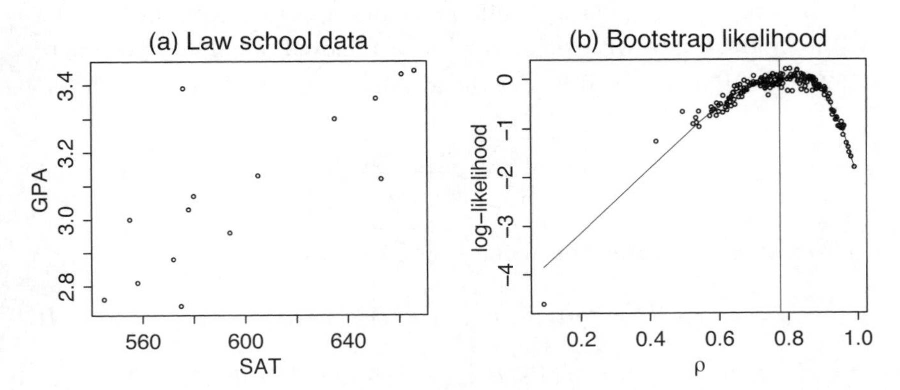

Figure 15.3: *(a) The scatter plot of SAT versus GPA scores. The sample correlation is $\widehat{\rho} = 0.78$. (b) The unsmoothed and smoothed log-likelihood of the correlation coefficient.*

15.3 BC$_a$ bootstrap likelihood

The BC$_a$ (bias corrected accelerated) method for setting bootstrap CIs was introduced by Efron (1987) to overcome the weaknesses of the percentile method (Section 5.6). In contrast to the previous method of double bootstrapping, the BC_a method leads to a nonparametric likelihood from a single bootstrap,

Let $x \equiv (x_1, \ldots, x_n)$ be a random sample from a continuous distribution F, $\theta \equiv t(F)$ be a real-valued parameter of interest, and $\widehat{\theta} = t(x)$ and $\widehat{F}$ be the estimates of θ and F. Then

(a) generate bootstrap data x^* by resampling from $\widehat{F}$.

(b) Compute $t(x^*)$ and call it t^*.

(c) Repeat (a) and (b) B times. Denote by G the true distribution function of $t(x^*)$ and $\widehat{G}$ its estimate based on $t_1^*, \ldots, t_B^*$.

(d) Compute the bias correction term z_0 by

$$z_0 = \Phi^{-1}\{\widehat{G}(\widehat{\theta})\}$$

where $\Phi(z)$ is the standard normal distribution function.

(e) Compute the term a, the rate of change of the standard deviation of the normalized parameter (see below). For the mean parameter estimation we have

$$a = \frac{\sum_i (x_i - \overline{x})^3}{6\{\sum_i (x_i - \overline{x})^2\}^{3/2}},$$

which is a measure of skewness of the data. For a general parameter, a is computed using a jackknife procedure (Section 5.2). Let x_{-i} be the original data with the i'th item x_i removed, $\widehat{\theta}_{(i)}$ be the estimate of θ based on x_{-i}, and $\widehat{\theta}_{(\cdot)}$ be the average of $\widehat{\theta}_{(i)}$'s. Then

$$a = \frac{\sum_i (\widehat{\theta}_{(\cdot)} - \widehat{\theta}_{(i)})^3}{6\{\sum_i (\widehat{\theta}_{(\cdot)} - \widehat{\theta}_{(i)})^2\}^{3/2}}.$$

Given these quantities the $100(1-\alpha)\%$ BC_a CI of θ is

$$\widehat{G}^{-1}\{\Phi(z[\alpha])\} < \theta < \widehat{G}^{-1}\{\Phi(z[1-\alpha])\}, \tag{15.4}$$

where $z[\alpha] \equiv z_0 + (z_0 + z^\alpha)/\{1 - a(z_0 + z^\alpha)\}$ and $z^\alpha \equiv \Phi^{-1}(\alpha)$.

For $B = \infty$ the interval (15.4) is exact if there exists a normalizing transform $\phi = h(\theta)$ such that

$$\frac{\widehat{\phi} - \phi}{\sigma_\phi} \sim N(-z_0, 1), \tag{15.5}$$

where $\sigma_\phi = 1 + a\phi$. Generally it is more accurate than the Wald interval, since (15.5) involves a better approximation than the standard CLT.

To construct a BC_a likelihood, first construct the likelihood function of ϕ based on model (15.5), then use the inverse-transform to compute the likelihood of θ. Specifically, if $\{\phi, L(\phi)\}$ is the graph of ϕ-likelihood, where

$$\log L(\phi) = -\log \sigma_\phi - \frac{(\widehat{\phi} - \phi + z_0\sigma_\phi)^2}{2\sigma_\phi^2},$$

then $\{h^{-1}(\phi), L(\phi)\}$ is the graph of θ-likelihood. The BC_a likelihood is defined as

$$L_B(\theta) \equiv L[\Phi^{-1}\{\widehat{G}(\theta)\}], \tag{15.6}$$

where the observed $\widehat{\phi} = \Phi^{-1}\{\widehat{G}(\widehat{\theta})\}$ is used in computing $L(\phi)$. This employs

$$\phi = h(\theta) = \Phi^{-1}\{\widehat{G}(\theta)\}$$

as the normalizing transform, which can be justified as follows: if $\widehat{\theta}$ has distribution $G(\cdot)$, then $G(\widehat{\theta})$ is standard uniform and $\Phi^{-1}\{G(\widehat{\theta})\}$ is standard normal, so $\Phi^{-1}\{G(\cdot)\}$ is an exact normalizing transform for $\widehat{\theta}$.

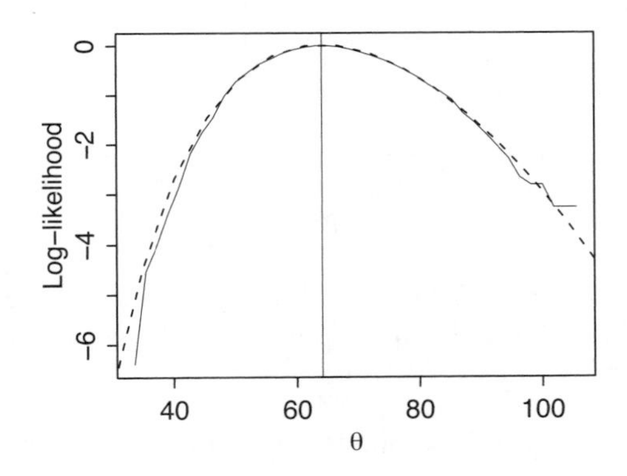

Figure 15.4: *BC_a likelihood (solid line) and the standard empirical likelihood (dashed line) of the mean of the aircraft failure data.*

For the aircraft failure data in Example 15.2, Figure 15.4 shows the BC_a and the empirical likelihoods for the mean parameter θ. In this example the number of bootstrap replications is $B = 1000$. In Pawitan (2000) it is shown that, for a general class of M-estimators, the BC_a likelihood is very close to the empirical likelihood, so it can be considered as a numerical way to produce the latter.

Even in small samples, the bootstrap distribution G may be regarded as continuous, so generally no smoothing is needed in computing $L_B(\theta)$. In practice we only need to ensure the bootstrap replication B is large enough for accurate CI computations; Efron (1987) indicated that $B = 1000$ is adequate. Theoretically, the validity of BC_a likelihood is inherited from the

BC$_a$ method; Efron (1987) proved the validity of the BC$_a$ method for any regular one-parameter problem, including those with nuisance parameters.

The bootstrap likelihood can also be used to approximate a *parametric* likelihood, especially a profile likelihood for a scalar parameter. The computation is the same as before, but instead of using the empirical distribution, we generate bootstrap data according to a parametric model. This is convenient in multiparameter problems, since a parametric profile likelihood is sometimes difficult to derive, or tedious to compute. For example, to compute the profile likelihood of the correlation coefficient ρ in a bivariate normal model, the full likelihood is a function of $(\mu_x, \sigma_x^2, \mu_y, \sigma_y^2, \rho)$. To get a profile likelihood for ρ we need to maximize over the 4D subspace of $(\mu_x, \sigma_x^2, \mu_y, \sigma_y^2)$ at each fixed value of ρ.

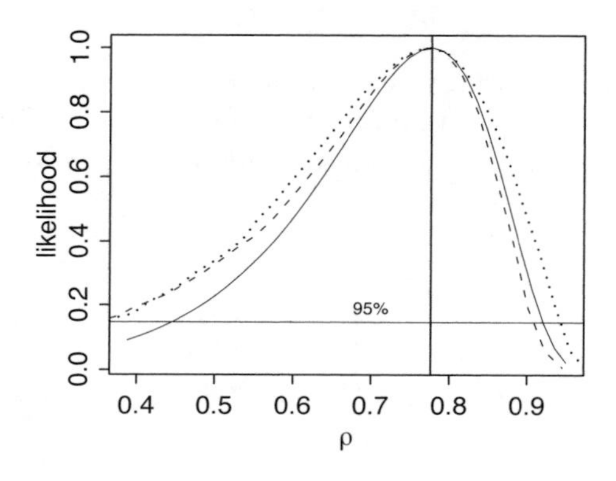

Figure 15.5: *Correlation of SAT and GPA: the parametric profile (solid), parametric bootstrap (dashed), nonparametric BC$_a$ bootstrap (dotted) likelihoods.*

For the law school data in Example 15.3, Figure 15.5 shows the exact parametric profile, parametric bootstrap, and nonparametric BC$_a$ likelihoods of the correlation coefficient. Both bootstrap-based likelihoods are based on 4000 bootstrap samples. For the parametric likelihoods the data are assumed to be bivariate normal. Compared with the double-bootstrap likelihood shown in Figure 15.3, the nonparametric BC$_a$ likelihood is closer to the parametric ones.

15.4 Exponential family model

The previous derivation of the empirical likelihood of the mean shows that the EDF F_n is embedded in the one-parameter family of distributions of the form

$$p_i = \frac{1}{n - \lambda(x_i - \theta)}, \tag{15.7}$$

where λ is a function of θ through

$$\sum_i x_i p_i = \theta.$$

Efron (1982) suggested a seemingly more natural exponential formula

$$p_i \equiv \frac{e^{\lambda x_i}}{\sum_j e^{\lambda x_j}} \tag{15.8}$$

where λ is a function of the mean θ also through

$$\sum_i x_i p_i = \theta.$$

The observed EDF is a member of both families (at $\lambda = 0$).

The log-likelihood of the mean θ based on model (15.8) is

$$\begin{aligned} \log L(\theta) &= \sum_i \log p_i \\ &= \sum_i \{\lambda x_i - \log \sum_j e^{\lambda x_j}\} \\ &= n\lambda\overline{x} - n \log \sum_j e^{\lambda x_j}, \end{aligned}$$

where θ enters through λ.

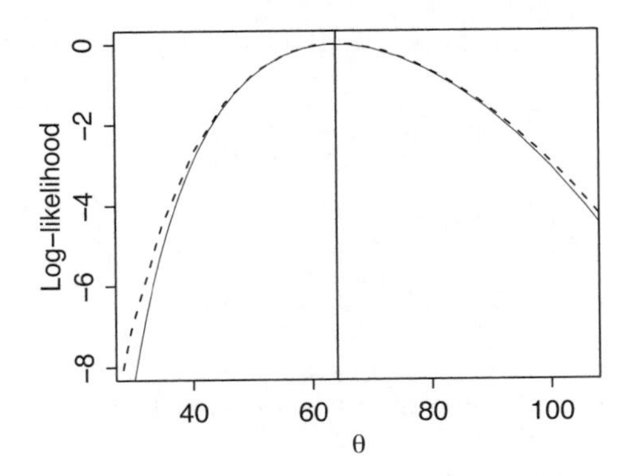

Figure 15.6: *Exponential family log-likelihood (solid) and the standard empirical log-likelihood (dashed) for the mean of the aircraft failure data in Example 15.2.*

The two models are in fact quite close and would yield similar likelihoods. For the aircraft failure data in Example 15.2 the two likelihoods are shown in Figure 15.6. Davison *et al.* (1992) and Monti and Ronchetti (1993) gave a comparison between the two likelihoods in terms on their asymptotic expansion.

15.5 General cases: M-estimation

The empirical likelihood can be extended to the estimating equation method described in Chapter 14. Suppose an estimate t is the solution of an estimating equation

$$\sum_i \psi(x_i, t) = 0.$$

To cover the multiparameter problems, $\psi(x_i, t)$ is assumed to be an array-valued function. The parameter θ being estimated is the solution of the theoretical estimating equation

$$E_F \psi(X, \theta) = 0.$$

As in the case of the population mean, the empirical likelihood of θ based on data $x_1, \ldots, x_n$ is the profile likelihood

$$L(\theta) = \sup \prod_i p_i$$

where the supremum is taken over all discrete distributions on $x_1, \ldots, x_n$ satisfying the requirements: $p_i \geq 0$ for all i, $\sum_i p_i = 1$ and

$$\sum_i \psi(x_i, \theta) p_i = 0,$$

a generalization of equation (15.3). Using a similar Lagrange multiplier technique as before we can show that the solution is of the form

$$p_i = \frac{1}{n - \lambda' \psi(x_i, \theta)}$$

where λ is a vector that satisfies

$$\sum_i \frac{\psi(x_i, \theta)}{n - \lambda' \psi(x_i, \theta)} = 0.$$

So, at each θ, we need to solve the last equation to find λ, then compute p_i and the likelihood.

In the multiparameter case, the empirical profile likelihood for individual parameters can be obtained by further optimization of the joint empirical likelihood. Alternatively, one can compute the overall optimization directly using some numerical procedure. Let $\theta = (\theta_1, \theta_2)$, where $\theta_2 \in R^q$ is the nuisance parameter; the empirical likelihood of θ_1 is

$$L(\theta_1) = \sup \prod p_i$$

where the supremum is taken over $\{p_1, \ldots, p_n\}$ and $\theta_2 \in R^q$, satisfying $p_i \geq 0$, $\sum_i p_i = 1$ and $\sum_i \psi(x_i, \theta) p_i = 0$. A number of constrained optimization programs are available, for example, the FORTRAN package NPSOL from Gill *et al.* (1986).

The empirical likelihood can be extended to linear and generalized linear models (Owen 1991; Kolaczyk 1994). Similar formulae apply using a proper definition of the estimating function $\psi(\cdot)$. Suppose we observe $(y_1, x_1), \ldots, (y_n, x_n)$, where $y_1, \ldots, y_n$ are independent outcomes and $x_1, \ldots, x_n$ are known predictors. Let $Ey_i = \mu_i$, and we have a link function

$$h(\mu_i) = x_i'\beta$$

and $\text{var}(y_i) = \phi v_i$. Now define

$$\psi_i = \psi_i(y_i, \beta) = \frac{\partial \mu_i}{\partial \beta} \frac{(y_i - \mu_i)}{\phi v_i}.$$

The estimate of β is the solution of the estimating equation

$$\sum_i \psi_i(y_i, \beta) = 0.$$

For example, for a Poisson regression model, suppose $\log \mu_i = x_i'\beta$ and $v_i = \mu_i$ $(\phi = 1)$; then

$$\psi_i = x_i(y_i - e^{x_i'\beta}).$$

The empirical likelihood of β is the profile likelihood

$$L(\beta) = \sup \prod_i p_i$$

where the supremum is taken over all discrete distributions satisfying $p_i \geq 0$, $\sum_i p_i = 1$ and

$$\sum_i \psi_i(y_i, \beta) p_i = 0.$$

If the dispersion parameter ϕ is unknown, it can be replaced by its estimate. The empirical likelihood for individual regression parameters can be computed using the same numerical technique described above, by enlarging the parameter space for optimization.

Inference based on the empirical likelihood is valid as long as the mean specification is correct, while the variance specification may be incorrect (Kolaczyk 1994). In the latter case the curvature of the empirical likelihood automatically leads to the robust variance estimate (Section 14.3).

Exponential family model

The exponential family model in Section 15.4 can also be generalized to derive a nonparametric likelihood associated with an estimating equation. For an iid sample $x_1, \ldots, x_n$ assume a probability model

$$P(X = x_i) = p_i = \frac{e^{\lambda' \psi(x_i, t)}}{\sum_j e^{\lambda' \psi(x_j, t)}}$$

where for fixed θ the vector λ satisfies $\sum_i \psi(x_i, \theta)p_i = 0$, or

$$\sum_i \psi(x_i, \theta)e^{\lambda'\psi(x_i,\theta)} = 0, \tag{15.9}$$

and t is the solution of an estimating equation

$$\sum_i \psi(x_i, t) = 0.$$

Implicitly we have assumed a model indexed by the parameter θ. The log-likelihood of θ is

$$\begin{aligned}
\log L(\theta) &= \sum_i \log p_i \\
&= \sum_i \lambda\psi(x_i, t) - n \log \sum_j e^{\lambda'\psi(x_j,t)} \\
&= -n \log \sum_j e^{\lambda'\psi(x_j,t)},
\end{aligned}$$

using $\sum_i \psi(x_i, t) = 0$. The quantity

$$K(\lambda) \equiv \log \left(\frac{1}{n} \sum_j e^{\lambda'\psi(x_j,t)} \right)$$

is known as the empirical cumulant generating function. So, up to an additive constant, we have an interesting relationship

$$\log L(\theta) = -nK(\lambda),$$

where λ is a function of θ through (15.9). For example, if $\psi(x, t) = x - t$ then

$$\log L(\theta) = n\lambda\overline{x} - n \log \sum_j e^{\lambda x_j}$$

as we have seen in Section 15.4. The close relationship between the exponential family likelihood and the standard empirical likelihood is explored by Monti and Ronchetti (1993).

15.6 Parametric versus empirical likelihood

Recall that the empirical likelihood is close to a likelihood based on an estimate rather than the whole data. If the estimate is sufficient for a particular parametric family, then the empirical likelihood will be close to the full parametric likelihood, otherwise there can be a large discrepancy. A detailed theoretical comparison between the empirical and the parametric likelihoods is given by DiCiccio *et al.* (1989).

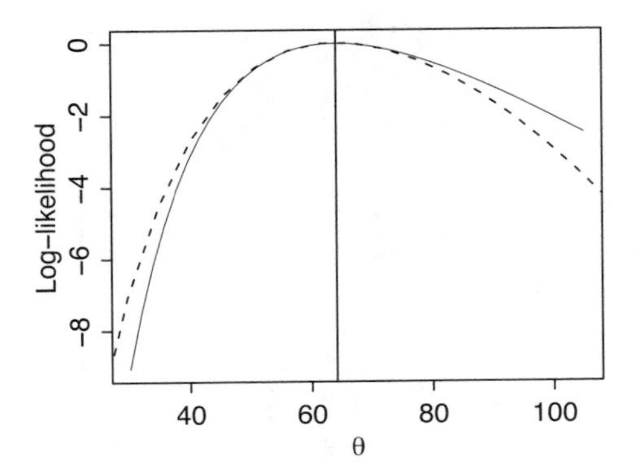

Figure 15.7: *Parametric log-likelihood (solid) based on the exponential distribution and the standard empirical log-likelihood (dashed) for the mean of the aircraft failure data in Example 15.2.*

Example 15.4: For the aircraft failure data (Example 15.2), Figure 15.7 compares the empirical likelihood versus a full parametric likelihood. To compute the latter, we assume the data are iid exponential with mean θ, so the log-likelihood is

$$\log L(\theta) = -n \log \theta - \sum_{i=1}^{n} x_i/\theta.$$

The two likelihoods are maximized in the same location, and there is a reasonable agreement around the MLE. DiCiccio *et al.* (1989) show that this is generally true for exponential family models, which is expected because of the sufficiency of estimates under such models. □

Example 15.5: We refer again to Darwin's self- and cross-fertilized plants in Section 13.1. The measurement of interest x is the height difference:

```
-8.4 -6.0  0.7  1.0  1.8  2.0  2.9  3.0
 3.5  3.6  5.1  6.1  7.0  7.5  9.3
```

Assume that the data are an iid sample from the double-exponential distribution with density

$$p(x) = \frac{1}{2\sigma} e^{-|x-\theta|/\sigma}.$$

The parametric MLE of θ is the sample median $m = 3$. For comparison, the sample mean is $\overline{x} = 2.61$. At fixed θ the MLE of σ is $\sum_i |x_i - \theta|/n$, so the profile log-likelihood for θ is

$$\log L(\theta) = -n \log \sum_i |x_i - \theta|.$$

Figure 15.8 compares the parametric likelihood with the empirical likelihood based on the mean. The mean of the distribution is θ, but the sample mean is not a sufficient estimate; this is the main source of the discrepancy, since the empirical likelihood is centred at the sample mean. Even the locations of the maxima differ. If the data are a sample from the double-exponential distribution,

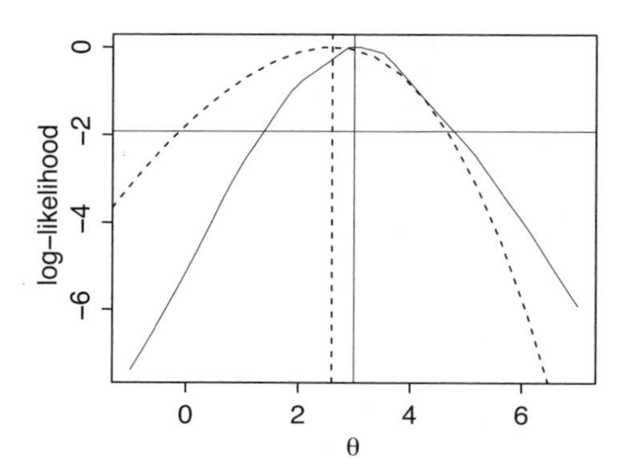

Figure 15.8: *Parametric log-likelihood (solid) based on the double-exponential distribution and the standard empirical log-likelihood (dashed) based on the sample mean.*

then the empirical likelihood based on the mean is not an efficient quantity for inference. Computing the empirical likelihood based on the sample median is left as an exercise. □

The last example shows that some care is needed in choosing an appropriate estimate; this requires a look at the observed data, since only an appropriate parametric model can suggest a proper estimate. Since there is an empirical likelihood associated with an estimating equation, justifying an empirical likelihood is the same as justifying an estimating equation. Using empirical likelihood it is possible to provide some inference for the model-free population parameters, but with a potential loss of efficiency if the observed data actually allow a better parametric procedure.

15.7 Exercises

Exercise 15.1: For the law school data in Example 15.3 verify the BC_a likelihood for the correlation coefficient. Compare this with the parametric profile likelihood and the parametric BC_a likelihood.

Exercise 15.2: For the law school data in Example 15.3 construct a bivariate empirical likelihood of the mean parameters. Compare this with the normal-based profile likelihood.

Exercise 15.3: For Darwin's data on self- and cross-fertilized plants, construct the empirical likelihood based on the sample median. (Hint: use an appropriate $\psi(\cdot)$ function.) Compare it with the parametric likelihood based on the double-exponential model. Compare it also with the BC_a likelihood based on the sample median.

16

Likelihood of random parameters

16.1 The need to extend the likelihood

The simplest random effects model is of the form

$$y_{ij} = \mu_i + e_{ij},$$

where μ_i is the random group i effect and e_{ij} is the error term. In classical linear models it is common to assume that μ_i's are an iid sample from $N(\mu, \sigma_\mu^2)$ and e_{ij}'s are an iid sample from $N(0, \sigma^2)$. Classical analysis of random effects models concentrates on the estimation of the variance component parameters σ^2 and σ_μ^2, but in many recent applications of statistics the main interest is estimating the random parameters μ_i's. These applications are characterized by the large number of parameters.

The classical likelihood approach does not go very far. In that approach we must take the view that for the data at hand the realized μ_i's are fixed parameters, so in effect we will be analysing a fixed effects model. This ignores the extra information that μ_i's are a sample from a certain distribution. It is now well established that treating the random parameters properly as random leads to better estimates than treating them as fixed parameters; this is demonstrated especially by the examples in Chapter 18. A proper likelihood treatment of random parameters necessitates an extended definition of the likelihood function.

There could be some questions regarding what is meant by 'the random parameters are a sample from a certain distribution'. In practice we can view the statement simply as a *model* for dealing with a large number of parameters, hence no actual sampling needs to have taken place. What we are modelling is that a collection of (parameter) values is distributed according to some shape, and we use the probability model to represent it. No such modelling is needed when we are dealing with only a few parameters.

To strip the random effects model to the bare minimum, assume that the only information we have is that μ is taken at random from $N(0, 1)$. Here is what we get if we view the likelihood simply as evidence about a fixed parameter; i.e. after μ is generated it is considered a fixed unknown value. Since there is no data, strictly we cannot compute a likelihood, but having no data is equivalent to collecting irrelevant data that do not involve μ, so the 'fixed-parameter-μ' (pure) likelihood is

$$L(\mu) = \text{constant},$$

i.e. we are in a state of ignorance. But actually *we are not*, since we know μ is generated according to $N(0,1)$; in this case this valid information is contextual, not evidence from data. Recall that the pure likelihood represents the current state of knowledge if we know absolutely nothing about the parameter prior to the experiment.

In this example it seems obvious and sensible that the information about μ is captured by the standard normal density

$$L(\mu) = \frac{1}{\sqrt{2\pi}} e^{-\mu^2/2}.$$

The extended definition of the likelihood should capture this uncertainty. One may interpret the likelihood as before, which is to give us a rational degree of belief where μ is likely to fall, but since it is also a standard density there is an objective interpretation in terms of frequencies. This 'no-data' likelihood of μ may be thought of as a prior density in a Bayesian setting (except here we do have an objective meaning to μ being a random outcome from $N(0,1)$, while in a Bayesian argument such a physical model for μ is not necessary). In this example we do not need to call the density function a likelihood, but the need does arise in prediction problems (Section 16.2), mixed effects models (Chapter 17) and smoothing problems (Chapter 18).

We have defined the likelihood to deal with uncertainty. Uncertainty is simply a lack of certainty: we can be uncertain about a binomial probability θ, about what will happen in the stock market tomorrow, or about what the result of a coin toss will be. We will treat the uncertainty about a fixed parameter differently from the uncertainty about a random outcome. Traditionally, the former is expressed by the likelihood and the latter by a density function. But modern applications suggest that the likelihood framework should be able to deal with both uncertainties. We will make this clear by examples. To summarize, we will distinguish two kinds of uncertainty:

(i) uncertainty about a fixed parameter, which is due to incomplete information in the data. The uncertainty is expressed by the likelihood function, which captures the *evidence* in the data about the model parameter. Within this context the likelihood is *not a density function.* In particular, the rule regarding transformation of parameters is governed by the invariance principle (Section 2.8). The traditional definition of likelihood (Section 2.1) and most of the classical likelihood theory falls in this category.

(ii) uncertainty about a random parameter. Here the likelihood should not just express the evidence about the parameter, but may include other contextual information; the actual definition is given in Section 16.3. For such a parameter the likelihood function can be interpreted as a density, so in general the mathematical rules regarding a probability density function apply; for example, it should integrate or sum to one. To emphasize the difference we might call the likelihood here an *extended likelihood*, while the likelihood for a fixed parameter is a *pure likelihood*. But such a terminology seems unnecessary since it is usually clear from the context whether we are dealing with fixed or random parameters, so except for emphasis we will simply use the term 'likelihood' to cover both.

16.2 Statistical prediction

Binomial prediction problem

The prototype of statistical problems in Section 1.1 involves a question of inference for fixed parameters. In 'The fundamental problems of practical statistics' Pearson (1920) wrote: 'An "event" has occurred p times out of $p+q=n$ trials, where we have no *a priori* knowledge of the frequency of the event in the total population of occurrences. What is the probability of its occurring r times in a further $r+s=m$ trials?' This is an extension of Bayes' problem in his 1763 *Essay*.

Though this is a prediction problem its nature is similar to our previous prototype, which is how to extract information from the data to be able to say something about an unobserved quantity. How we use the information eventually, whether for prediction or estimation, for description or for decision, is somewhat secondary.

It is interesting to see how this problem leads to a Bayesian point of view. To state Pearson's problem in current terminology: given θ, X and Y independent binomial(n,θ) and binomial(m,θ) respectively, what is the conditional distribution of Y given $X=x$? If we view θ as fixed, then Y is binomial(m,θ), which is useless since θ is unknown.

The appearance that x carries no information is rather surprising, since one should expect X to tell us about θ, so it should be informative about Y. One way to salvage this line of reasoning is to think of θ as random, with some probability density $f(\theta)$ on (0,1), the prior density of θ. The conditional distribution of Y given $X=x$ can be computed using the Bayes theorem

$$\begin{aligned} P(Y=y|X=x) &= \frac{P(X=x,Y=y)}{P(X=x)} \\ &= \frac{\int_\theta P(X=x,Y=y|\theta)f(\theta)d\theta}{\int_\theta P(X=x|\theta)f(\theta)d\theta}. \end{aligned}$$

The choice of $f(\theta)$ obviously matters. For example, one might use the uniform prior $f(\theta)=1$, giving

$$\begin{aligned} P(Y=y|X=x) &= \frac{\binom{n}{x}\binom{m}{y}\int_0^1 \theta^{x+y}(1-\theta)^{n+m-x-y}d\theta}{\binom{n}{x}\int_0^1 \theta^{x}(1-\theta)^{n-x}d\theta} \\ &= \binom{m}{y}\frac{B(x+y+1,n+m-x-y+1)}{B(x+1,n-x+1)} \\ &= \frac{m!(x+y)!(n+m-x-y)!(n+1)!}{x!(n-x)!y!(m-y)!(n+m+1)!}, \end{aligned}$$

where $B(a,b)$ is the beta function. For $n=10$, $x=8$ and $m=8$ the conditional probability $P(Y=y|X=x)$ is plotted in Figure 16.1.

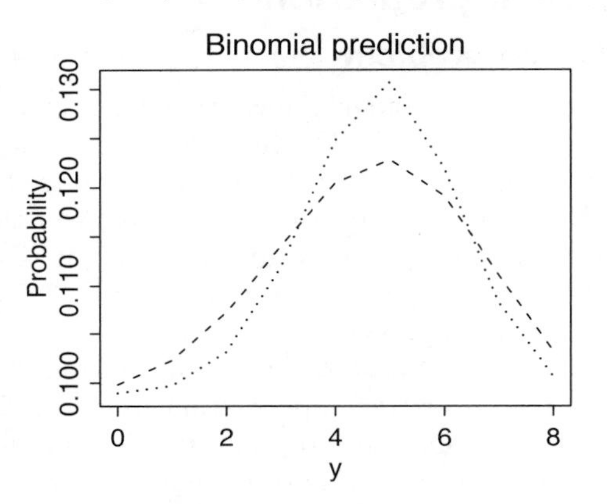

Figure 16.1: *The Bayesian(dashed line) and the plug-in solutions (dotted line) for the distribution of* Y*, which is binomial*$(8,\widehat{\theta}=8/10)$.

Here is a classical/pure likelihood solution. Imagine that the 'future y' is actually already known somewhere by somebody, but we do not know it. Now treat it as an *unknown fixed parameter.* The conditional distribution of X given y (and θ) is free of y, so on observing x the pure likelihood of y is

$$L(y) = \text{constant},$$

or there is no information about y in the data x. This is consistent with the property of likelihood as evidence about a fixed parameter. Here it is unfruitful since the problem is still unsolved, even though we 'feel' knowing x should tell us something about θ and hence the future y. The Bayesian solution seems a lot more natural.

An *ad hoc* solution is simply to specify that Y is binomial$(m,x/n)$, which is a short way of saying we want to estimate $P_\theta(Y=y)$ by $P_{\widehat{\theta}}(Y=y)$, using $\widehat{\theta}=x/n$. We may call the solution a *plug-in solution,* a typical frequentist solution. The weakness of this approach is obvious: it is not easy to account for the uncertainty of $\widehat{\theta}$ in the prediction, which can be substantial when n is small. In Figure 16.1 it appears that the plug-in solution is too optimistic (having too little variability) compared with the Bayesian solution.

To appreciate the last point consider instead a normal prediction problem. Suppose we observe $X_1, \ldots, X_n$ iid from $N(\mu, \sigma^2)$, where μ is not known but σ^2 is, and denote the sample average by $\overline{X}$. Then

$$X_{n+1} - \overline{X} \sim N\{0, \sigma^2(1 + 1/n)\},$$

from which we can get a correct $100(1-\alpha)\%$ prediction interval for X_{n+1} as

$$\overline{X} \pm z_{\alpha/2}\sigma\sqrt{1 + 1/n},$$

where $z_{\alpha/2}$ is the appropriate value from the normal table. By comparison, the plug-in method would prescribe

$$\overline{X} \pm z_{\alpha/2}\sigma,$$

which would be far from the correct solution if n is small.

16.3 Defining extended likelihood

The exact prediction solution is obvious in the normal case, but the method of deriving it does not follow from a general principle as in the Bayesian solution. The purpose of a 'general principle' is to make our thinking straightforward and transparent in complex problems. Thus the likelihood approach to solve the prediction problem is to treat the future y as an unknown random parameter, and *to treat such a parameter differently from a fixed parameter.*

Definition 16.1 *Assuming a statistical model $p_\theta(x, y)$, where $p_\theta(x, y)$ is a joint density function of observed data x and unobserved y given a fixed parameter θ, the likelihood function for θ and y is*

$$L(\theta, y) \equiv p_\theta(x, y).$$

This extended definition of likelihood agrees with Butler (1987), while recently Bjørnstad (1996) provided the theoretical justification that such a likelihood carries all of the information on the unknown parameters (θ, y). We can interpret it as an augmented likelihood, since we may write

$$L(\theta, y) = p_\theta(x|y)p_\theta(y),$$

where $p_\theta(x|y)$ is the pure likelihood term, and $p_\theta(y)$ is the contextual information that y is random. In mixed effects modelling the extended likelihood has been called *h-likelihood* (for hierarchical likelihood) by Lee and Nelder (1996), while in smoothing literature it is known as *penalized likelihood* (e.g. Green and Silverman 1993). The main advantage of this likelihood is that it is general, and it is well known that it leads to better estimates of the random parameters. In the current example we shall show that it can take into account the uncertainty of $\widehat{\theta}$ in the prediction of y.

There is also a close connection with Bayesian modelling, though the connection is only mathematical rather than philosophical. If θ is known, the likelihood of y alone based on x is

$$L(y) = p(x, y) = p(y|x)p(x) = \text{constant} \times p(y|x)$$

which is proportional to the posterior density of y given x. In this context the density $p(y)$ might be called *the prior likelihood* of y, and $L(y)$ is *the posterior likelihood* of y given x. The method is philosophically non-Bayesian since, to be applicable, the random variable y needs to have an observable distribution.

Example 16.1: Let us return to Pearson's binomial prediction problem in current terminology. Given θ, the observed x and the unobserved y are independent binomial with parameters (n, θ) and (m, θ); hence

$$p_\theta(x, y) = \binom{n}{x}\binom{m}{y}\theta^{x+y}(1-\theta)^{n+m-x-y}.$$

The likelihood for the unknown θ and y is

$$L(\theta, y) \equiv \binom{m}{y}\theta^{x+y}(1-\theta)^{n+m-x-y}.$$

Actual prediction of y with its prediction interval from $L(\theta, y)$ requires a method of 'removing' θ. □

Inference on individual parameters

In real prediction problems where θ is unknown, even for scalar y we have to deal with a multiparameter likelihood. Recall that there is always some complexity when dealing with such a problem (Chapter 10). The likelihood principle does not tell us how to get a separate likelihood for y while treating the parameter θ as a nuisance parameter. In some simple prediction problems Hinkley (1979) and Butler (1986) described specialized methods to get a predictive likelihood of y that is free of θ.

We will take the pragmatic approach and describe a general method that gives reasonable results in most cases. The key idea is that fixed and random parameters should be treated differently. Thus the general likelihood approach to remove a nuisance parameter is to compute a profile likelihood by

(i) *taking a maximum over a fixed parameter*, with possible modifications as described in Chapter 10;

(ii) *integrating out a random parameter.*

These conventions have a dramatic impact on the inference for individual parameters, so for consistency we have to state explicitly that

(iii) *inference for individual parameters should, whenever possible, be based on the profile likelihood.*

Only the second convention is new. The third is implicit when we introduce the notion of profile likelihood.

In view of the second convention the profile likelihood for θ is exactly the same as the pure likelihood based on x alone. That is,

$$L(\theta) = \int p_\theta(x, y)dy = p_\theta(x)$$

(or use summation for discrete y). To compute the profile likelihood for y, we maximize $L(\theta, y)$ over θ at each value of y:

$$L(y) = \max_\theta L(\theta, y).$$

Since y is a random variable the profile likelihood $L(y)$ can be interpreted objectively like a probability density function; for example, it can be normalized to sum to one, and a prediction statement or interval can be constructed by allocating appropriate probabilities. We will refer to profile likelihood $L(y)$ as the *predictive likelihood* of y. From the definition we can also see that, if x and y are independent, as the information on θ becomes more precise the (properly normalized) predictive likelihood $L(y)$ converges to $p_\theta(y)$.

If the predictive likelihood $L(y)$ is too difficult to compute we may consider the estimated likelihood $L(\widehat{\theta}, y)$ for inference on y, where $\widehat{\theta}$ is an estimate of θ; this is the case, for example, if y is high dimensional as in random effects models (Chapter 17).

Because of the second convention regarding treatment of random parameters, fixed-parameter ideas such as the invariance property of the MLE no longer apply. For example:

(i) If $\widehat{y}$ is the MLE of y, then $g(\widehat{y})$ is *not* necessarily the MLE of $g(y)$; see Exercises 16.3 and 16.4. This is not a weakness, it is a consequence of dealing with random quantities; a similar phenomenon occurs with expected values where, in general, $Eg(X) \neq g(EX)$.

(ii) The MLE of θ based on the joint likelihood $L(\theta, y)$ does not have to be the same as the MLE based on the profile likelihood $L(\theta)$; see Exercise 16.5. According to our third rule, inference on θ should be based on $L(\theta)$, not the joint likelihood $L(\theta, y)$.

Binomial prediction example

Returning to Pearson's prediction problem, to get the profile likelihood of y, we first find that at each fixed y the maximizer of $L(\theta, y)$ is

$$\widehat{\theta}(y) = (x + y)/(n + m),$$

which, as we expect intuitively, is the total number of successes divided by the total number of trials. Substituting $\widehat{\theta}(y)$ for θ in $L(\theta, y)$ yields

$$\begin{aligned} L(y) &= L(\widehat{\theta}(y), y) \\ &= \binom{m}{y} \frac{(x+y)^{x+y}(n+m-x-y)^{n+m-x-y}}{(n+m)^{n+m}}. \end{aligned}$$

For $n = 10$, $x = 6$ and $m = 8$, this solution is compared with the Bayesian and the plug-in solutions, where we can see that the profile likelihood does take into account the uncertainty in x; see Figure 16.2. In the previous normal prediction example the method produces the correct solution (Exercise 16.1).

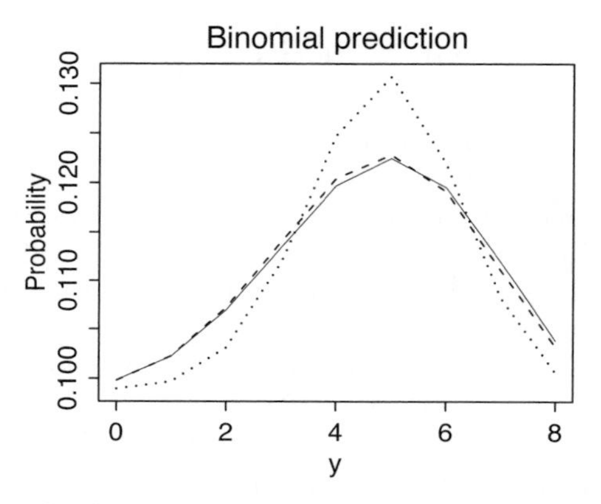

Figure 16.2: *Likelihood-based prediction (solid line) for future $Y \sim$ binomial$(8, \theta)$ with unknown θ, based on observing $n = 10$ and $x = 8$; the probabilities sum to one. Also shown are the Bayesian (dashed line) and the plug-in (dotted line) solutions for the distribution of Y.*

Example 16.2: In a certain region during the past five years, 3, 2, 5, 0 and 4 earthquakes (of at least a certain magnitude) were observed each year. We are interested in predicting y, the number of earthquakes next year.

Assuming a Poisson model, the likelihood of θ and y based on observing the data x is

$$L(\theta, y) = e^{-(n+1)\theta} \frac{\theta^{y+\sum x_i}}{y!}.$$

For each y the MLE of θ is $\widehat{\theta}(y) = (y + \sum x_i)/(n+1)$, so the profile likelihood for y is

$$L(y) = e^{-(n+1)\widehat{\theta}(y)} \frac{\widehat{\theta}(y)^{y+\sum x_i}}{y!}.$$

Figure 16.3 shows this profile likelihood, which is normalized to sum to one. Prediction intervals can be constructed from this likelihood; for example, there is a 4.1% probability that there will be more than five earthquakes next year. □

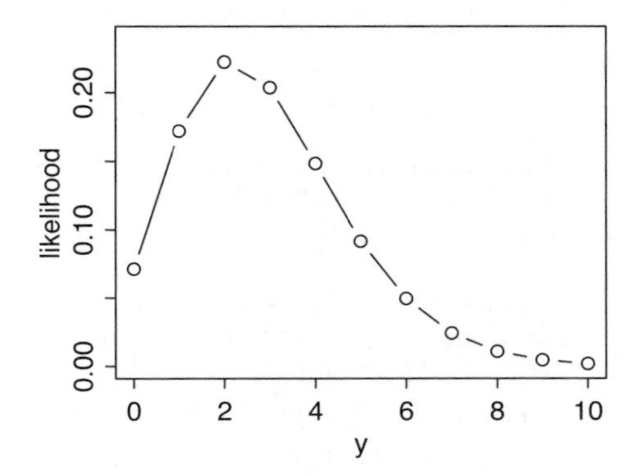

Figure 16.3: *Predictive likelihood of the number of earthquakes.*

16.4 Exercises

Exercise 16.1: Suppose we observe $x_1, \ldots, x_n$ iid from $N(\mu, \sigma^2)$, where μ is not known but σ^2 is. Denote the sample average by $\overline{x}$. Find the predictive likelihood of x_{n+1} and show that the $100(1-\alpha)\%$ prediction interval for x_{n+1} is

$$\overline{x} \pm z_{\alpha/2}\sigma\sqrt{1+1/n},$$

where $z_{\alpha/2}$ is the appropriate value from the normal table. Derive the prediction interval for the case where σ^2 is unknown, and identify it as a t-interval.

Exercise 16.2: Hinkley's (1979) predictive likelihood solution of the binomial prediction problem is

$$L(y) = \text{constant} \times \frac{\binom{n}{x}\binom{m}{y}}{\binom{m+n}{x+y}}.$$

In the specific example of $n = 10$, $x = 6$ and $m = 8$, show that this is very close to the profile likelihood solution.

Exercise 16.3: Suppose y is a random parameter to be predicted using the likelihood method. Explain theoretically and give a simple example that if $\widehat{y}$ is the MLE of y, then $g(\widehat{y})$ is *not* necessarily the MLE of $g(y)$.

Exercise 16.4: Let $x_1, \ldots, x_n$ be an iid sample from the exponential distribution with mean θ, having density

$$p_\theta(x) = \theta^{-1}e^{-x/\theta}.$$

Suppose y_1 and y_2 are iid future values from the same distribution.

(a) Find the profile likelihood of y_1 and y_2, jointly and separately. Predict y_1 and y_2 from the joint profile and from the individual profiles.

(b) Find the profile likelihood of $t = y_1 + y_2$ and the predicted value of t.

(c) Discuss here how the invariance property of the MLE does not apply.

(d) Let t be as in part (b) and $\sum_i x_i = s$. Hinkley's (1979) predictive likelihood of t is

$$L(t) = \text{constant} \times \frac{s^{n-1}t}{(s+t)^{n+1}}.$$

(e) For the following dataset compare Hinkley's likelihood with the profile likelihood in part (b) and the plug-in method:

```
1.9  2.3  5.1  2.0 12.9  0.2  6.2  6.0  3.7  2.2
```

Exercise 16.5: Let a scalar-valued y be a sample from the exponential distribution with mean θ. Conditional on y, x is assumed exponential with mean y, so x and y are not independent. Based on observing x:

(a) Find the MLE of θ and y from the joint likelihood $L(\theta, y)$.

(b) Find the profile likelihood and the MLE of θ and y separately. Comment on the difference between (a) and (b). Which is more appropriate as the MLE?

Exercise 16.6: Let $x_1, \ldots, x_n, x_{n+1}$ be an iid sample from the uniform distribution on $(0, \theta)$. Derive the prediction of x_{n+1} based on $x_1, \ldots, x_n$.

Exercise 16.7: Suppose (x_i, y_i) for $i = 1, \ldots, n$ are a bivariate sample, where, given x_i, the outcome y_i is normal with mean

$$E(y_i|x_i) = \mu_i = \beta_0 + \beta_1 x_i$$

and variance σ^2. Derive the predictive likelihood of y_{n+1} at x_{n+1} and find the MLE of y_{n+1}.

Exercise 16.8: The same setup as in the previous exercise, except the outcome y_i given x_i is now Poisson with mean μ_i, where

$$\log \mu_i = \beta_0 + \beta_1 x_i.$$

17

Random and mixed effects models

Regression models are the work horse of statistics, but while rich classes of models have been introduced to deal with various data types, in most cases the underlying mean structure is still the linear model

$$h(\mu) = X\beta,$$

where β is a fixed parameter of dimension p, and p is typically quite small. Such a model is called a fixed effects model.

The need to consider random effects models arises when the number of unknown parameters is large, while the number of observations per parameter is limited. For example, we follow 100 subjects over time, where from each subject we obtain two to four measurements and, to account for natural variability, we want to fit a linear growth model for each subject. Let y_{ij} be the j'th measurement of subject i, made at time t_{ij}, and consider the model

$$y_{ij} = b_{0i} + b_{1i}t_{ij} + e_{ij},$$

where b_{0i} and b_{1i} are the intercept and slope parameters for the i'th subject. In effect these parameters are random variables, and the model is called a *random effects model.* Since the random parameters have means which will be treated as fixed parameters, we may also call the model a *mixed effects model.* In this example we have 200 unknown parameters, but only a maximum of 400 observations. We can, of course, try to estimate the parameters using data from each subject separately, but if the error variance is large the estimates will be poor. Theoretically and empirically it has been shown that we get better estimates by treating the parameters as random variables.

The mixed model framework allows for the most complex model structure in statistics. It is not surprising that their theory and applications, especially in nonnormal cases, were the last to develop. Even as this is being written there are still issues, such as variance component estimation for nonnormal models, which are not yet resolved. We will start with an extension of the likelihood concept that captures the information about a random parameter, and then proceed with the relatively 'clean' area of normal mixed models. We close the chapter with the generalized linear mixed models (GLMM).

17.1 Simple random effects models

Theoretical analyses of random effects models very quickly become cumbersome, where often it is not possible to get any closed formula. This is fine in practice, where we can rely on statistical packages to produce numerically the results we want, but it can be a problem for those who want or need to see everything worked out in detail. We first describe the simplest nontrivial random effects model, where it is possible to arrive at explicit formulae. Familiarity with matrix algebra is assumed throughout the rest of the chapter.

Data example

Table 9.1 shows the repeated measurements of estrone level from five post-menopausal women. Several questions may be asked from the data: (i) Is there significant variation between women relative to within-woman variation? (ii) What is the reliability of the measurements? (iii) What is each woman's mean estrone concentration? The first two questions were addressed in Section 9.10.

Let $y_{ij} = 10 \times \log_{10} x_{ij}$ where x_{ij} is the raw estrone measurement. Consider a random effects model

$$y_{ij} = \mu + b_i + e_{ij}$$

where μ is a fixed overall mean parameter,

$$\begin{aligned} b_i &= \text{person effect, for } i = 1, \ldots, q = 5 \\ e_{ij} &= \text{residual effect, for } j = 1, \ldots, n = 16. \end{aligned}$$

Assume that b_i's are iid $N(0, \sigma_b^2)$, e_{ij}'s are iid $N(0, \sigma^2)$ and they are independent.

Likelihood function of σ^2 and σ_b^2

In Section 9.10 we derived the profile likelihood

$$\log L(\sigma^2, \sigma_b^2) = -\frac{q}{2}\{(n-1)\log\sigma^2 + \log(\sigma^2 + n\sigma_b^2)\} - \frac{1}{2}\left(\frac{\text{SSE}}{\sigma^2} + \frac{\text{SSB}}{\sigma^2 + n\sigma_b^2}\right),$$

where the (corrected) total, person and error sum-of-squares are

$$\begin{aligned} \text{SST} &= \sum_{ij}(y_{ij} - \overline{y})^2 \\ \text{SSB} &= \sum_i \{\sum_j (y_{ij} - \overline{y})\}^2/n \\ \text{SSE} &= \text{SST} - \text{SSB}, \end{aligned}$$

and $\overline{y}$ is the overall mean of the data.

Estimating the random effects

Classical analysis of random effects models concentrates on the estimation of the variance components σ^2 and σ_b^2, but recent interests are on the estimation of the random effects parameters. (Note: we will use both terms 'estimation' and 'prediction' for estimation of the random effects.)

For the one-way random effects model above, the full parameter space is

$$(\theta, b) \equiv (\mu, \sigma^2, \sigma_b^2, b_1, \ldots, b_5),$$

where the fixed parameter is $\theta \equiv (\mu, \sigma^2, \sigma_b^2)$. By Definition 16.1, the likelihood based on dataset y is

$$L(\theta, b) = p_\theta(y, b) = p_\theta(y|b)p_\theta(b).$$

Given b the outcomes y_{ij}'s are independent with mean

$$\mu_i = \mu + b_i$$

and variance σ^2, while b_i's are iid with mean zero and variance σ_b^2. Hence

$$\begin{aligned} \log L(\theta, b) &= -\frac{qn}{2}\log\sigma^2 - \frac{1}{2\sigma^2}\sum_{i=1}^{q}\sum_{j=1}^{n}(y_{ij} - \mu - b_i)^2 \\ & \quad -\frac{q}{2}\log\sigma_b^2 - \frac{1}{2\sigma_b^2}\sum_{i=1}^{q} b_i^2. \end{aligned} \tag{17.1}$$

Deriving the joint likelihood $L(\theta, b)$ is simpler than deriving $L(\theta)$. Estimates of b_i can be computed by directly maximizing $\log L$ with respect to (θ, b); this will be described below.

If we assume a fixed effects model, i.e. b_i's are fixed parameters, the log-likelihood of the unknown parameters is

$$\log L = -\frac{qn}{2}\log\sigma^2 - \frac{1}{2\sigma^2}\sum_{i=1}^{q}\sum_{j=1}^{n}(y_{ij} - \mu - b_i)^2,$$

which involves only the first two terms of (17.1). Using the constraint $\sum b_i = 0$ we can verify that the MLE of b_i is

$$\widehat{b}_i = \overline{y}_i - \overline{y}, \tag{17.2}$$

where $\overline{y}_i$ is the average of $y_{i1}, \ldots, y_{in}$, the MLE of μ is $\overline{y}$, and the MLE of μ_i is $\overline{y}_i$ regardless of the constraint on b_i's (Exercise 17.1). The constraint is not needed in the random effects model, but there we do need to specify $Eb_i = 0$.

Estimating b_i

Assume for the moment that θ is known. The derivative of the log-likelihood (17.1) at a fixed value of θ is

$$\frac{\partial \log L}{\partial b_i} = \frac{1}{\sigma^2}\sum_{j=1}^{n}(y_{ij} - \mu - b_i) - \frac{b_i}{\sigma_b^2}$$

and on setting it to zero we get

$$\left(\frac{n}{\sigma^2} + \frac{1}{\sigma_b^2}\right)\widehat{b}_i = \frac{1}{\sigma^2}\sum_{j=1}^{n}(y_{ij} - \mu),$$

or

$$\widehat{b}_i = \left(\frac{n}{\sigma^2} + \frac{1}{\sigma_b^2}\right)^{-1}\frac{n}{\sigma^2}(\overline{y}_i - \mu). \tag{17.3}$$

We can show that $\widehat{b}_i$ is also the conditional mean of the distribution of b_i given y (Exercise 17.2). There is a Bayesian interpretation for this estimate: if b_i is a fixed parameter, and we use a prior density $p_\theta(b)$, then $\widehat{b}_i$ is called a Bayesian estimate of b_i.

If the fixed parameters are known, the Fisher information for b_i is

$$\begin{aligned} I(b_i) &= -\frac{\partial^2 \log L}{\partial b_i^2} \\ &= \frac{n}{\sigma^2} + \frac{1}{\sigma_b^2} \end{aligned}$$

compared with n/σ^2 if b_i is assumed fixed. Consequently the standard error of $\widehat{b}_i$ under the random effects model is smaller than the standard error under the fixed effects model.

In practice the unknown θ is replaced by its estimate. We may use the MLE, but if the MLE is too difficult to compute other estimates may be used. Thus

$$\widehat{b}_i = \left(\frac{n}{\widehat{\sigma}^2} + \frac{1}{\widehat{\sigma}_b^2}\right)^{-1}\frac{n}{\widehat{\sigma}^2}(\overline{y}_i - \overline{y}). \tag{17.4}$$

Comparing this with (17.2) it is clear that the effect of the random effects assumption is to 'shrink' $\widehat{b}_i$ towards its zero mean; that is why the estimate is also called a shrinkage estimate. The estimate of μ_i is

$$\begin{aligned} \widehat{\mu}_i &= \overline{y} + \widehat{b}_i \\ &= \overline{y} + \left(\frac{n}{\widehat{\sigma}^2} + \frac{1}{\widehat{\sigma}_b^2}\right)^{-1}\frac{n}{\widehat{\sigma}^2}(\overline{y}_i - \overline{y}) \\ &= \left(\frac{n}{\widehat{\sigma}^2} + \frac{1}{\widehat{\sigma}_b^2}\right)^{-1}\left(\frac{n}{\widehat{\sigma}^2}\overline{y}_i + \frac{1}{\widehat{\sigma}_b^2}\widehat{\mu}\right) \end{aligned}$$

$$= \alpha\overline{y}_i + (1-\alpha)\overline{y}$$

where

$$\alpha = \left(\frac{n}{\widehat{\sigma}^2} + \frac{1}{\widehat{\sigma}_b^2}\right)^{-1} \frac{n}{\widehat{\sigma}^2}.$$

If n/σ^2 is large relative to $1/\sigma_b^2$ (i.e. there is a lot of information in the data about μ_i), then α is close to one and the estimated mean is close to the sample average. The estimate is called an empirical Bayes estimate, as it can be thought of as implementing a Bayes estimation procedure on the mean parameter μ_i, with a normal prior that has mean μ and variance σ_b^2. It is 'empirical' since the parameter of the prior is estimated from the data.

Application

To apply these results to the estrone data we will use from Section 9.10:

$$\begin{aligned} \widehat{\mu} &= \overline{y} = 14.175 \\ \widehat{\sigma}^2 &= 0.325 \\ \widehat{\sigma}_b^2 &= 1.395 \end{aligned}$$

and $n = 16$, so the shrinkage parameter is

$$\alpha = 0.986.$$

The sample means $\overline{y}_i$'s are

```
13.545   14.447   15.635   12.233   15.015
```

and we have the following shrinkage estimates of the individual means $\widehat{\mu}_i$'s

```
13.554   14.443   15.614   12.261   15.003
```

They are very close to the sample means in this example since α is close to one.

17.2 Normal linear mixed models

As in the standard fixed effects models, classical analyses of linear mixed models are almost totally based on the normal model. We will study the normal models in detail, since the results are theoretically 'clean' and well understood, and the formulae extend to the nonnormal cases with little change in notation.

Let y be an N-vector of outcome data, and X and Z be $N \times p$ and $N \times q$ design matrices for the fixed effects parameter β and random effects b. The standard linear model specifies

$$y = X\beta + Zb + e \tag{17.5}$$

where e is $N(0, \Sigma)$, b is $N(0, D)$, and b and e are independent. The variance matrices Σ and D are parameterized by an unknown variance component parameter θ.

Here is an equivalent formulation that allows an immediate extension to nonnormal models: conditional on an unobserved random effects b the outcome vector y is normal with mean

$$E(y|b) = X\beta + Zb$$

and variance Σ. The vector of random effects b is assumed normal with mean zero and variance D. The advantage of the formulation is that we do not specify an explicit error term e, which is not natural for nonnormal models.

Example 17.1: The one-way random effects model

$$y_{ij} = \mu + b_i + e_{ij},$$

for $i = 1, \ldots, q$ and $j = 1, \ldots, n$, can be written in the general form (17.5) with total data size $N = qn$ and

$$\begin{aligned} X &= \mathbf{1}_N \\ \beta &= \mu \\ Z &= \text{a matrix of } z_{ij} \text{ explained below} \\ b &= (b_1, \ldots, b_q)' \\ \Sigma &= \sigma^2 I_N \\ D &= \sigma_b^2 I_q. \end{aligned}$$

$\mathbf{1}_N$ is a column vector of N ones. The element z_{ij} of the matrix Z is equal to one if y_{ij} comes from the i'th group, and zero otherwise. The variance parameter is $\theta = (\sigma^2, \sigma_a^2)$. □

Estimation of the fixed parameters

From (17.5) the marginal distribution of y is normal with mean $X\beta$ and variance

$$V = \Sigma + ZDZ',$$

so the log-likelihood of the fixed parameters (β, θ) is

$$\log L(\beta, \theta) = -\frac{1}{2} \log |V| - \frac{1}{2}(y - X\beta)' V^{-1} (y - X\beta), \tag{17.6}$$

where the parameter θ enters through the marginal variance V.

For fixed θ, taking the derivative of the log-likelihood with respect to β, we find the estimate of β as the solution of

$$(X'V^{-1}X)\beta = X'V^{-1}y, \tag{17.7}$$

the well-known generalized or weighted least-squares formula. The profile likelihood of the variance parameter θ is

$$\log L(\theta) = -\frac{1}{2} \log |V| - \frac{1}{2}(y - X\widehat{\beta})' V^{-1} (y - X\widehat{\beta}), \tag{17.8}$$

where $\widehat{\beta}$ is computed in (17.7). The observed Fisher information of β is

$$I(\widehat{\beta}) = X'V^{-1}X,$$

from which we can find the standard error for $\widehat{\beta}$.

In Example 10.12 we derive a modified profile likelihood

$$\log L_m(\theta) = -\frac{1}{2}\log|V| - \frac{1}{2}\log|X'V^{-1}X| - \frac{1}{2}(y - X\widehat{\beta})'V^{-1}(y - X\widehat{\beta})$$

that takes into account the estimation of β. This matches the so-called restricted maximum likelihood (REML) adjustment; see Patterson and Thompson (1971), or Harville (1974).

An iterative procedure is needed if $\widehat{\beta}$ depends on θ. In the one-way random effects model $\widehat{\beta}$ does not depend on the variance components, but the analytical derivation of the (profile) likelihood of the variance component is quite complicated. In practice, we will usually not try to get any closed form for (17.8). Inference on θ is typically based on the MLE and its standard error, although we have seen that the standard error of a variance component estimate can be misleading. Numerical methods to compute the MLEs are described in Sections 17.4 and 17.5.

Estimation of the random effects

The log-likelihood of all the parameters is based on the joint density of (y, b); thus

$$L(\beta, \theta, b) = p(y|b)p(b).$$

From the mixed model specification, the conditional distribution of y given b is normal with mean

$$E(y|b) = X\beta + Zb$$

and variance Σ. The random effects b is normal with mean zero and variance D, so

$$\begin{aligned} \log L(\beta, \theta, b) &= -\frac{1}{2}\log|\Sigma| - \frac{1}{2}(y - X\beta - Zb)'\Sigma^{-1}(y - X\beta - Zb) \\ &\quad -\frac{1}{2}\log|D| - \frac{1}{2}b'D^{-1}b. \end{aligned} \tag{17.9}$$

Given the fixed parameters (β, θ), we take the derivative of the log-likelihood with respect to b:

$$\frac{\partial \log L}{\partial b} = Z'\Sigma^{-1}(y - X\beta - Zb) - D^{-1}b.$$

Setting this to zero, the estimate $\widehat{b}$ is the solution of

$$(Z'\Sigma^{-1}Z + D^{-1})b = Z'\Sigma^{-1}(y - X\beta). \tag{17.10}$$

This estimate is also known as the best linear unbiased predictor (BLUP). In practice we replace the unknown fixed parameters by their estimates as described above.

The second derivative matrix of the log-likelihood with respect to b is

$$\frac{\partial^2 \log L}{\partial b \partial b'} = -Z'\Sigma^{-1}Z - D^{-1},$$

so the observed Fisher information is equal to

$$I(\widehat{b}) = (Z'\Sigma^{-1}Z + D^{-1}). \tag{17.11}$$

Assuming the fixed effects are known, the standard errors of $\widehat{b}$ (also interpreted as the prediction error for a random parameter) can be computed as the square root of the diagonal elements of $I(\widehat{b})^{-1}$. Prediction intervals for b are usually computed element by element using

$$\widehat{b}_i \pm z_{\alpha/2}\ \mathrm{se}(\widehat{b}_i)$$

where $z_{\alpha/2}$ is an appropriate value from the normal table, and $\mathrm{se}(\widehat{b}_i)$ is the standard error of $\widehat{b}_i$.

Note that $I(\widehat{b})$ is not a function of b; this implies that the log-likelihood of b alone, assuming β and θ are known, is quadratic around $\widehat{b}$. Since b is random we can interpret the likelihood as a density function, so, thinking of $p(b)$ as the 'prior' density of b, the 'posterior' distribution of b is normal with mean $\widehat{b}$ and variance $(Z'\Sigma^{-1}Z + D^{-1})^{-1}$. This is the empirical Bayes intepretation of the formulae.

17.3 Estimating genetic value from family data*

One of the largest and most successful applications of the mixed models is in animal breeding, where the main objective is to identify animals of high genetic merit. New applications in genetic epidemiology may involve similar data structure, but to be specific we will use the language of animal breeding. The genetic merit of an animal is modelled as a random parameter, and the identification process involves estimation of this parameter. Some analytical complexity is introduced by the different degrees of relationship between the animals. A typical application of animal breeding, for example to identify dairy cows that have high fertility, involves thousands of animals with hundreds of thousands of records. We will show a very simple example that conceptually captures this application.

Let y be the combined outcome vector of a certain performance measure from all the animals; some animals may have repeated measures. Conditional on the genetic merit b, suppose y is normal with mean

$$E(y|b) = X\beta + Zb,$$

with appropriate design matrices X and Z, and variance $\Sigma = \sigma^2 I_N$. The matrix X might capture the effects of farm, age, sex, or other predictors. Z is typically a matrix of zeros and ones. Setting up these matrices is the same as in a standard regression analysis. The random effects parameter b is assumed normal with zero mean and variance $\sigma_g^2 R$, where R is a relationship matrix induced by the familial relationship between animals. The length of b is typically the same as the number of animals.

Relationships between animals are represented in a simple dataset called the pedigree dataset that identifies the parents of each animal. For example, suppose we have seven animals labelled 1, 2, ..., 7 with the following pedigree table:

Animal	Sire	Dam
1	.	.
2	.	.
3	1	2
4	1	.
5	4	3
6	.	.
7	5	6

where '·' means unknown. Animals 1, 2 and 6 have unknown parents; 3 is the offspring of 1 and 2; 4 has a known sire 1, but no known dam; 5 has known parents 3 and 4; 7 has known parents 5 and 6.

Let r_{ij}'s be the elements of R; the value of r_{ij} as a function of the pedigree data is defined by

1. *For diagonal elements:*

$$r_{ii} = 1 + \frac{1}{2} r_{\text{sire,dam}}$$

where 'sire' and 'dam' are those of the i'th animal. The second term on the right-hand side takes account of inbreeding; if there is no inbreeding then $r_{ii} = 1$, and R is a correlation matrix.

2. *For off-diagonal elements* r_{ij}, where j is of an older generation compared with i and

- both parents of i are known

$$r_{ij} = \frac{1}{2}(r_{j,\text{sire}} + r_{j,\text{dam}})$$

- only one parent of i is known

$$r_{ij} = \frac{1}{2} r_{j,\text{sire}}$$

or

$$r_{ij} = \frac{1}{2} r_{j,\text{dam}}$$

- both parents of i are unknown

$$r_{ij} = 0.$$

By definition $r_{ij} = r_{ji}$, so R is a symmetric matrix.

For the sample pedigree data we have

$$\begin{aligned} r_{11} &= 1 \\ r_{12} &= 0 \\ r_{13} = r_{31} &= \frac{1}{2}(r_{11} + r_{12}) = \frac{1}{2} \\ r_{14} = r_{41} &= \frac{1}{2} r_{11} = \frac{1}{2} \\ r_{15} = r_{51} &= \frac{1}{2}(r_{14} + r_{13}) = \frac{1}{2} \\ r_{16} &= 0 \\ r_{17} = r_{71} &= \frac{1}{2}(r_{15} + r_{16}) = \frac{1}{4} \end{aligned}$$

etc. We can verify that R is given by

$$R = \begin{pmatrix} 1 & 0 & 0.5 & 0.5 & 0.5 & 0 & 0.25 \\ & 1 & 0.5 & 0 & 0.25 & 0 & 0.125 \\ & & 1 & 0.25 & 0.625 & 0 & 0.3125 \\ & & & 1 & 0.625 & 0 & 0.3125 \\ & & & & 1.125 & 0 & 0.5 \\ & & & & & 1 & 0.5 \\ & & & & & & 1 \end{pmatrix}.$$

The lower half of R is computed by symmetry.

Given σ^2 and σ_g^2, the estimates are computed as the solution of the mixed model equations:

$$\begin{aligned} (X'V^{-1}X)\beta &= X'V^{-1}y \\ (Z'Z + \lambda R^{-1})b &= Z'(y - X\beta) \end{aligned}$$

where $V = \sigma^2 I_N + \sigma_g^2 ZRZ'$, and $\lambda = \sigma^2/\sigma_g^2$. In practice the length of b can be of the order of several thousands to tens of thousands, so clever numerical techniques are required (e.g. Henderson 1976).

17.4 Joint estimation of β and b

It is an interesting coincidence that the estimates of β in (17.7) and of b in (17.10) are the joint maximizer of $L(\beta, \theta, b)$ at fixed θ. This provides a useful heuristic when we deal with nonnormal cases. Specifically, the derivative of $\log L(\beta, \theta, b)$ with respect to β is

$$\frac{\partial \log L}{\partial \beta} = X'\Sigma^{-1}(y - X\beta - Zb).$$

Combining this with the derivative with respect to b and setting them to zero, we have

$$\begin{pmatrix} X'\Sigma^{-1}X & X'\Sigma^{-1}Z \\ Z'\Sigma^{-1}X & Z'\Sigma^{-1}Z + D^{-1} \end{pmatrix} \begin{pmatrix} \beta \\ b \end{pmatrix} = \begin{pmatrix} X'\Sigma^{-1}y \\ Z'\Sigma^{-1}y \end{pmatrix}. \tag{17.12}$$

The estimates we get from this simultaneous equation are exactly those we get from (17.7) and (17.10) (Exercise 17.3). The equation suggests the possibility of estimating β and b without computing the marginal variance V or its inverse.

This is given by the so-called Jacobi or Gauss–Seidel method in linear algebra, also known as the iterative backfitting algorithm in statistics. In the algorithm β and b are computed in turn as follows.

1. Start with an estimate of β, for example the ordinary least-squares estimate
$$\widehat{\beta} = (X'X)^{-1}X'y,$$
then iterate between 2 and 3 below until convergence.
2. Compute a corrected outcome
$$y^c = y - X\widehat{\beta}$$
and estimate b from a random effects model
$$y^c = Zb + e,$$
based on the same formula as before:
$$(Z'\Sigma^{-1}Z + D^{-1})b = Z'\Sigma^{-1}y^c.$$
3. Recompute a corrected outcome
$$y^c = y - Zb$$
and estimate β from a fixed effects model
$$y^c = X\beta + e,$$
which updates $\widehat{\beta}$ from the solution of
$$(X'\Sigma^{-1}X)\beta = X'\Sigma^{-1}y^c.$$

17.5 Computing the variance component via $\widehat{\beta}$ and $\widehat{b}$

The marginal likelihood formula (17.8) for the variance component parameter θ is not desirable due to the terms involving V or V^{-1}. We now show an alternative formula which may be easier to compute. First we can show the following series of identities:

$$V^{-1} = \Sigma^{-1} - \Sigma^{-1}Z(Z'\Sigma^{-1}Z + D^{-1})^{-1}Z'\Sigma^{-1}$$

$$\begin{aligned}
\widehat{b} &= DZ'V^{-1}(y - X\widehat{\beta}) \\
V^{-1}(y - X\widehat{\beta}) &= \Sigma^{-1}(y - X\widehat{\beta} - Z\widehat{b}) \qquad (17.13)\\
(y - X\widehat{\beta})'V^{-1}(y - X\widehat{\beta}) &= (y - X\widehat{\beta} - Z\widehat{b})'\Sigma^{-1}(y - X\widehat{\beta} - Z\widehat{b}) + \widehat{b}'D^{-1}\widehat{b} \\
|V| &= |\Sigma|\,|D|\,|Z'\Sigma^{-1}Z + D^{-1}|.
\end{aligned}$$

The first identity can be verified by multiplying V and V^{-1}. The next three identities are left as exercises, while the last identity on the determinant is based on the following partitioned matrix result (Rao 1973, page 32):

$$\begin{aligned}
\begin{vmatrix} \Sigma & Z \\ Z' & -D^{-1} \end{vmatrix} &= |-D^{-1}|\,|\Sigma + ZDZ'| \\
&= |\Sigma|\,|-Z'\Sigma^{-1}Z - D^{-1}|.
\end{aligned}$$

Hence we can rewrite (17.8) as

$$\begin{aligned}
\log L(\theta) &= -\frac{1}{2}\log|\Sigma| - \frac{1}{2}(y - X\widehat{\beta} - Z\widehat{b})'\Sigma^{-1}(y - X\widehat{\beta} - Z\widehat{b}) \\
&\quad -\frac{1}{2}\log|D| - \frac{1}{2}\widehat{b}'D^{-1}\widehat{b} - \frac{1}{2}\log|Z'\Sigma^{-1}Z + D^{-1}| \qquad (17.14)\\
&= \log L(\widehat{\beta}, \theta, \widehat{b}) - \frac{1}{2}\log|Z'\Sigma^{-1}Z + D^{-1}| \qquad (17.15)
\end{aligned}$$

where θ enters the function through Σ, D, $\widehat{\beta}$ and $\widehat{b}$. The formulae for $\widehat{\beta}$ and $\widehat{b}$ as functions of θ are given by (17.7) and (17.10). In view of the joint likelihood (17.9) it is important to note that (17.15) is a modified profile likelihood, with the extra term

$$-\frac{1}{2}\log|Z'\Sigma^{-1}Z + D^{-1}|,$$

where the matrix $Z'\Sigma^{-1}Z + D^{-1}$ is the Fisher information of b from (17.9).

Formula (17.14) *is* simpler than (17.8), since the matrices involved are simpler than V or V^{-1}. The quantity $(y - X\widehat{\beta} - Z\widehat{b})$ is the residual of the model given θ. In applications where $\widehat{b}$ can be computed reasonably fast, i.e. the matrix $Z'\Sigma^{-1}Z + D^{-1}$ can be inverted easily, the determinant is a usual by-product of the computation.

Iterative procedure

In practice it is convenient to use a derivative free optimization routine to maximize (17.14), where in the process we also get the estimates $\widehat{\beta}$ and $\widehat{b}$. Note that *computationally* we can view the whole estimation of β, b and θ as maximizing an objective function

$$\begin{aligned}
Q &= -\frac{1}{2}\log|\Sigma| - \frac{1}{2}(y - X\beta - Zb)'\Sigma^{-1}(y - X\beta - Zb) \\
&\quad -\frac{1}{2}\log|D| - \frac{1}{2}b'D^{-1}b - \frac{1}{2}\log|Z'\Sigma^{-1}Z + D^{-1}|.
\end{aligned}$$

(Q is not a log-likelihood, only a device to justify the algorithm below.) The score equations for β and b yield the usual formulae for $\widehat{\beta}$ and $\widehat{b}$ at fixed θ. Hence, we can apply an iterative algorithm as follows. Start with an estimate of the variance parameter θ, then:

1. Compute $\widehat{\beta}$ and $\widehat{b}$ using (17.7) and (17.10), or using the iterative backfitting algorithm given in Section 17.4.
2. Fixing β and b at the values $\widehat{\beta}$ and $\widehat{b}$, maximize Q to get an update of θ.
3. Iterate between 1 and 2 until convergence.

Step 2 in the algorithm is left open in general, although an explicit update of the variance component is possible for specific variance structures. In most mixed model applications it is common to assume that the variance matrices are of the form

$$\begin{aligned}\Sigma &= \sigma^2 A\\ D &= \sigma_b^2 R,\end{aligned}$$

where A and R are known matrices of rank N and q, respectively; the only unknown variance parameters are σ^2 and σ_b^2. In many applications A is simply an identity matrix. From the properties of the determinant,

$$\begin{aligned}|\Sigma| &= \sigma^{2N}|A|\\ |D| &= \sigma_b^{2q}|R|,\end{aligned}$$

so, after dropping irrelevant constant terms,

$$\begin{aligned}Q &= -\frac{N}{2}\log\sigma^2 - \frac{1}{2\sigma^2}e'A^{-1}e\\ &\quad -\frac{q}{2}\log\sigma_b^2 - \frac{1}{2\sigma_b^2}b'R^{-1}b - \frac{1}{2}\log|\sigma^{-2}Z'A^{-1}Z + \sigma_b^{-2}R^{-1}|,\end{aligned}$$

where $e = y - X\beta - Zb$ is the error vector. The derivatives of Q are

$$\begin{aligned}\frac{\partial Q}{\partial \sigma^2} &= -\frac{N}{2\sigma^2} + \frac{1}{2\sigma^4}e'A^{-1}e\\ &\quad +\frac{1}{2\sigma^4}\text{trace}\{(\sigma^{-2}Z'A^{-1}Z + \sigma_b^{-2}R^{-1})^{-1}Z'A^{-1}Z\}\\ \frac{\partial Q}{\partial \sigma_b^2} &= -\frac{q}{2\sigma_b^2} + \frac{1}{2\sigma_b^4}b'R^{-1}b\\ &\quad +\frac{1}{2\sigma_b^4}\text{trace}\{(\sigma^{-2}Z'A^{-1}Z + \sigma_b^{-2}R^{-1})^{-1}R^{-1}\}.\end{aligned}$$

Setting these to zero, and isolating σ^2 and σ_b^2 for a simple iteration, we obtain updating equations for Step 2 of the algorithm:

$$\sigma^2 = \frac{1}{N}[e'A^{-1}e + \text{trace}\{(\sigma^{-2}Z'A^{-1}Z + \sigma_b^{-2}R^{-1})^{-1}Z'A^{-1}Z\}]$$

$$\sigma_b^2 = \frac{1}{q}[b'R^{-1}b + \text{trace}\{(\sigma^{-2}Z'A^{-1}Z + \sigma_b^{-2}R^{-1})^{-1}R^{-1}\}],$$

where all unknown parameters on the right-hand side are evaluated at the last available values during the iteration, and only a single iteration is needed at Step 2. Alternatively, we may use

$$\sigma^2 = \frac{e'A^{-1}e}{N - \text{df}},$$

where

$$\text{df} = \text{trace}\{(Z'A^{-1}Z + \lambda R^{-1})^{-1}Z'A^{-1}Z\}$$

and $\lambda = \sigma^2/\sigma_b^2$. The quantity 'df' can be thought of as the model degrees of freedom. In practice the traces are the most difficult components to compute, and may require some approximations. The algorithm is the same as the EM algorithm described in Section 12.8, though the objective function Q is different.

Computing the variance of $\widehat{\beta}$

The above formulation also suggests a practical method to compute the variance of the fixed effects estimate

$$\text{var}(\widehat{\beta}) = (X'V^{-1}X)^{-1}$$

without computing V^{-1} explicitly. Using (17.13) we have

$$\begin{aligned} V^{-1}X &= \Sigma^{-1}\{X - Z(Z'\Sigma^{-1}Z + D^{-1})^{-1}Z'\Sigma^{-1}X\} \\ &= \Sigma^{-1}(X - Zb_x) \end{aligned}$$

where the term

$$b_x \equiv (Z'\Sigma^{-1}Z + D^{-1})^{-1}Z'\Sigma^{-1}X$$

is the estimate obtained from the same random effects model

$$X = Zb_x + e$$

using X as the outcome variable. This formula is convenient in applications where b_x can be computed easily. Since X is a matrix of p columns, we can simply compute b_x column by column according to the columns of X.

17.6 Examples

One-way random effects model

Consider the simplest random effects model in Section 17.1:

$$y_{ij} = \mu + b_i + e_{ij}$$

where μ is a fixed overall mean, b_i's are iid $N(0, \sigma_b^2)$ and e_{ij}'s are iid $N(0, \sigma_e^2)$. The subject index $i = 1, \ldots, q$, the measurement index $j = 1, \ldots, n$ and the size of the outcome vector y is $N = qn$.

The iterative procedure proceeds as follows. Evaluate the individual group mean $\overline{y}_i$ and variance $\widehat{\sigma}_i^2$, then use the variance of $\overline{y}_i$'s as the initial estimate of σ_b^2, and the average of $\widehat{\sigma}_i^2$'s as the initial estimate of σ_e^2. Start with $b_i \equiv 0$, then:

1. Compute $\lambda = \sigma_e^2/\sigma_b^2$, and

$$\begin{aligned} \mu &= \frac{1}{N}\sum_{ij}(y_{ij} - b_i) \\ b_i &= \frac{n(\overline{y}_i - \mu)}{n+\lambda}. \end{aligned}$$

2. Update

$$\sigma_e^2 = \frac{e'e}{N - \text{df}}$$

where e is the vector of $e_{ij} = y_{ij} - \mu - b_i$ and

$$\text{df} = \frac{qn}{n+\lambda},$$

and

$$\sigma_b^2 = \frac{1}{q}\sum_{i=1}^{q} b_i^2 + \left(\frac{n}{\sigma_e^2} + \frac{1}{\sigma_b^2}\right)^{-1}.$$

3. Iterate 1 and 2 until convergence.

Applying this to the data in Section 17.1, we start with $\sigma_b^2 = 1.7698$ and $\sigma_e^2 = 0.3254$ and at convergence we obtain

$$\begin{aligned} \widehat{\mu} &= 14.1751 \\ \widehat{b} &= (-0.6211,\ 0.2683,\ 1.4389,\ -1.914,\ 0.8279) \\ \widehat{\sigma}_b^2 &= 1.3955 \\ \widehat{\sigma}_e^2 &= 0.3254. \end{aligned}$$

The standard errors of $\widehat{b}_i$ are $\widehat{\sigma}_e/\sqrt{n+\lambda}$ compared with $\widehat{\sigma}_e/\sqrt{n}$ for the fixed effects estimate. In the estimation procedure we do not get a profile likelihood function of σ_b^2; if such a likelihood is required we might use the simpler version (17.15) rather than (17.8).

Factorial experiment

Johnson and Wichern (1992) reported an experiment in anaesthesiology, where each of 19 dogs was put under four different treatments. The treatments were all combinations from two factors with two levels each: CO_2 pressure (high or low) and hydrogen (presence or absence). Treatment 1 is high CO_2 pressure without hydrogen, treatment 2 low CO_2 pressure without hydrogen, treatment 3 high CO_2 pressure with hydrogen and treatment

	Treatment			
Dog	1	2	3	4
1	426	609	556	600
2	253	236	392	395
3	359	433	349	357
4	432	431	522	600
5	405	426	513	513
6	324	438	507	539
7	310	312	410	456
8	326	326	350	504
9	375	447	547	548
10	286	286	403	422
11	349	382	473	497
12	429	410	488	547
13	348	377	447	514
14	412	473	472	446
15	347	326	455	468
16	434	458	637	524
17	364	367	432	469
18	420	395	508	531
19	397	556	645	625

Table 17.1: *The sleeping dog experimental data from Johnson and Wichern (1992). Each measurement is the mean period between heartbeats (msec).*

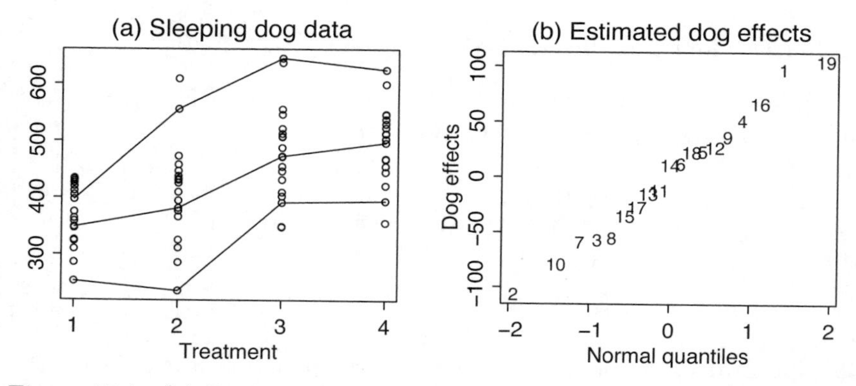

Figure 17.1: *(a) Repeated measures data from 19 dogs under four different treatments. The three lines are those from three dogs with the minimum, median and maximum averages of the four measurements. (b) Estimates of random dog effects.*

4 low CO_2 pressure with hydrogen. The data are given in Table 17.1 and plotted in Figure 17.1(a).

A standard analysis of variance model for this experiment is

$$y_{ij} = \mu + b_i + (H + C + HC)_j + e_{ij} \tag{17.16}$$

where b_i is the effect of dog i, $(H + C + HC)_j$ is the effect of treatment combination j, which is decomposed into the main effects of hydrogen and CO_2 pressure and their interaction, and e_{ij} is the error term. Assuming the dog effects are random, we can write (17.16) as a mixed model

$$y = X\beta + Zb + e$$

with appropriate matrices X and Z, such that β is of length 4: $\beta_1 = \mu$ is the constant term, β_2 the hydrogen effect (presence versus absence), β_3 the CO_2 effect (high versus low) and β_4 the interaction between hydrogen and CO_2.

Specifically, we first set y by stacking the columns of data; the first column of X is 1^{76}, i.e. a column of ones repeated 76 times; the second column $(-0.5^{38}, 0.5^{38})$; the third column $(0.5^{19}, -0.5^{19}, 0.5^{19}, -0.5^{19})$; and the fourth column is the product of the second and the third columns. Z is a 76×19 matrix obtained by stacking four identity matrices I_{19}.

The random effects b is assumed normal with mean zero and variance $\sigma_b^2 I_{q=19}$; the error e is normal with mean zero and variance $\sigma_e^2 I_{N=76}$, independent of b. We obtain the starting values by fitting fixed dog effects in the model:

$$\begin{aligned} \beta &= (438.75,\ 104.6579,\ -30.0263,\ 12.7895) \\ \sigma_b^2 &= 4241.5903 \\ \sigma_e^2 &= 1331.1274, \end{aligned}$$

where σ_b^2 is the variance of the fixed dog effects. At convergence of the iterative procedure we have

$$\begin{aligned} \widehat{\beta} &= (438.75,\ 104.6579,\ -30.0263,\ 12.7895) \\ \text{se}(\widehat{\beta}) &= (4.3482,\ 8.6965,\ 8.6965,\ 17.3930) \\ \widehat{\sigma}_b^2 &= 3580.5174 \\ \widehat{\sigma}_e^2 &= 1751.4804. \end{aligned}$$

The estimate $\widehat{\beta}_4 = 12.7895$ (se $= 17.3930$) indicates that the presence of hydrogen and CO_2 pressure show no interaction. Because the design matrix X is orthogonal we may interpret the main effects directly in this case. Both the presence of hydrogen ($\widehat{\beta}_2 = 104.6579$, se $= 8.6965$) and high CO_2 pressure ($\widehat{\beta}_3 = -30.0263$, se $= 8.6965$) have significant effects on the outcome. Figure 17.1(b) shows the random effects estimates of the 19 dogs, presented in a normal probability plot; the plot indicates which dogs are typical, and which are extreme.

Comparisons

If we assume the dog effects are fixed we obtain the following results:

$$\widehat{\beta}_{2,3,4} = (104.6579,\ -30.0263,\ 12.7895)$$

$$\text{se}(\widehat{\beta}_{2,3,4}) = (9.8643,\ 9.8643,\ 19.7286),$$

i.e. the same estimates but slightly larger standard errors. In practice, of course, there is very little to choose between the two approaches if the interest is only on the treatment comparisons.

Another approach is to compute separate contrasts representing the main and interaction effects. For example, to test for interaction, we compute the contrast

$$(y_{i1} + y_{i4}) - (y_{i2} + y_{i3}),$$

while to test for the hydrogen effect, we compute

$$(y_{i3} + y_{i4})/2 - (y_{i1} + y_{i2})/2.$$

Thus to test for interaction, which should be done first, each dog contributes a single measurement; from 19 dogs we get

```
-139   20  -66   79  -21  -82   44  154  -71   19
   -9   78   38  -87   34 -137   34   48 -179
```

yielding an average of -12.7895 (se $= 19.94$). The estimates for the main effects are also the same as before: 104.6579, -30.0263, but the standard errors are 11.14 and 8.27, respectively. This approach is the most 'nonparametric', i.e. it makes the fewest assumptions. The random effects model assumes, for example, that the error terms e_{ij}'s are iid, while no such assumption is made in the simple approach here.

17.7 Extension to several random effects

Extension of the computational procedure for more than one random factor is as follows; for convenience consider the case of two random factors. Let $b = (b_1, b_2)$ be the vector of random effects parameters. Suppose, conditional on b, the outcome y has mean

$$E(y|b) = X\beta + Z_1 b_1 + Z_2 b_2$$

and variance Σ. X, Z_1 and Z_2 are appropriate design matrices; β is the usual fixed effects parameter. Assume b_1 and b_2 are independent normal with mean zero and variance D_1 and D_2, respectively. The variance matrices Σ, D_1 and D_2 are functions of the variance parameter θ. The independence between b_1 and b_2 is a simplifying assumption that may not be correct; if they are in fact dependent then it is probably easier to consider them as a single random factor.

By defining a larger design matrix

$$Z = [Z_1 \ \ Z_2],$$

we can rewrite the two-component model as a one-component model

$$E(y|b) = X\beta + Zb,$$

so, conceptually, there is nothing new. The several-component model may look more natural, however, and it would help the model specification (e.g.

specifying matrices D_1 and D_2) and estimation. Logically the problem is likely to be more complicated since now we are dealing with more variables or factors, so the possibility of incorrect model specification is greater. The several-component model is also computationally more difficult.

Given θ, the estimates of β, b_1 and b_2 are the solution of

$$\begin{pmatrix} X'\Sigma^{-1}X & X'\Sigma^{-1}Z_1 & X'\Sigma^{-1}Z_2 \\ Z_1'\Sigma^{-1}X & Z_1'\Sigma^{-1}Z_1 + D_1^{-1} & Z_1'\Sigma^{-1}Z_2 \\ Z_2'\Sigma^{-1}X & Z_2'\Sigma^{-1}Z_1 & Z_2'\Sigma^{-1}Z_2 + D_2^{-1} \end{pmatrix} \begin{pmatrix} \beta \\ b_1 \\ b_2 \end{pmatrix} = \begin{pmatrix} X'\Sigma^{-1}y \\ Z_1'\Sigma^{-1}y \\ Z_2'\Sigma^{-1}y \end{pmatrix}. \tag{17.17}$$

The iterative backfitting algorithm for (17.17) evaluates each parameter in turn, while fixing the other two parameters. For example, start with an initial estimate of β and b_1, then

1. Compute a corrected data vector
$$y^c = y - X\beta - Z_1 b_1$$
and solve
$$(Z_2\Sigma^{-1}Z_2 + D_2^{-1})b_2 = Z_2'\Sigma^{-1}y^c \tag{17.18}$$
to get an updated estimate of b_2.
2. Compute
$$y^c = y - X\beta - Z_2 b_2$$
and update b_1 by solving
$$(Z_1\Sigma^{-1}Z_1 + D_1^{-1})b_1 = Z_1'\Sigma^{-1}y^c. \tag{17.19}$$
3. Compute
$$y^c = y - Z_1 b_1 - Z_2 b_2$$
and update β from
$$(X'\Sigma^{-1}X)\beta = X'\Sigma^{-1}y^c. \tag{17.20}$$
4. Repeat 1 to 3 until convergence.

Estimating variance components

The marginal distribution of y is normal with mean $X\beta$ and variance

$$V = \Sigma + Z_1 D_1 Z_1' + Z_2 D_2 Z_2'.$$

The profile log-likelihood of the variance parameter θ is

$$\log L(\theta) = -\frac{1}{2}\log|V| - \frac{1}{2}(y - X\widehat{\beta})'V^{-1}(y - X\widehat{\beta}), \tag{17.21}$$

where $\widehat{\beta}$ is computed above. This is similar to the previous formula (17.8) in the case of single random effects, though now V is more complicated.

Iterative procedure

An iterative algorithm to compute θ, β and b jointly can be derived in an important special case where Z_1 and Z_2 are orthogonal in the sense that $Z_1'\Sigma^{-1}Z_2 = 0$. Define the overall objective function

$$\begin{aligned} Q &= -\frac{1}{2}\log|\Sigma| - \frac{1}{2}e'\Sigma^{-1}e \\ &\quad -\frac{1}{2}\log|D_1| - \frac{1}{2}b_1'D_1^{-1}b_1 - \frac{1}{2}\log|Z_1'\Sigma^{-1}Z_1 + D_1^{-1}| \\ &\quad -\frac{1}{2}\log|D_2| - \frac{1}{2}b_2'D_2^{-1}b_2 - \frac{1}{2}\log|Z_2'\Sigma^{-1}Z_2 + D_2^{-1}|, \end{aligned}$$

where $e = y - X\beta - Z_1b_1 - Z_2b_2$. This can be derived from Q in Section 17.5, using the assumption that b_1 and b_2 are independent, and Z_1 and Z_2 are orthogonal. Now assume further that

$$\begin{aligned} \Sigma &= \sigma^2 A \\ D_1 &= \sigma_1^2 R_1 \\ D_2 &= \sigma_2^2 R_2, \end{aligned}$$

where A, R_1 and R_2 are known matrices of rank N, q_1 and q_2.

We take the derivatives of Q with respect to all the parameters, then isolate the parameters one at a time to arrive at the following algorithm. Start with some initial estimates of the variance component parameters σ^2, σ_1^2 and σ_2^2, and the mean parameters β and b_1, then:

1. Update b_2, b_1 and β using (17.18), (17.19) and (17.20).
2. Update the variance component parameters using

$$\begin{aligned} \sigma^2 &= \frac{1}{N}[e'A^{-1}e + \text{trace}\{(\sigma^{-2}Z_1'A^{-1}Z_1 + \sigma_1^{-2}R_1^{-1})^{-1}Z_1'A^{-1}Z_1\} \\ &\quad + \text{trace}\{(\sigma^{-2}Z_2'A^{-1}Z_2 + \sigma_2^{-2}R_2^{-1})^{-1}Z_2'A^{-1}Z_2\}] \\ \sigma_1^2 &= \frac{1}{q_1}[b_1'R_1^{-1}b_1 + \text{trace}\{(\sigma^{-2}Z_1'A^{-1}Z_1 + \sigma_1^{-2}R_1^{-1})^{-1}R_1^{-1}\}], \\ \sigma_2^2 &= \frac{1}{q_2}[b_2'R_2^{-1}b_2 + \text{trace}\{(\sigma^{-2}Z_2'A^{-1}Z_2 + \sigma_2^{-2}R_2^{-1})^{-1}R_2^{-1}\}], \end{aligned}$$

where all unknown terms on the right-hand side are evaluated at the last available value.

2′. Alternatively we may use

$$\sigma^2 = \frac{e'A^{-1}e}{N - \text{df}_1 - \text{df}_2}$$

where

$$\text{df}_i = \text{trace}\{(Z_i'A^{-1}Z_i + \lambda_i R_i^{-1})^{-1}Z_i'A^{-1}Z_i\}$$

and $\lambda_i = \sigma^2/\sigma_i^2$. The quantity df_i may be thought of as degrees of freedom.

3. Iterate 1 and 2 until convergence.

Example: growth curve analysis

Table 17.2 shows the height of the ramus (jaw) bones of 20 boys, measured at 8, 8.5, 9 and 9.5 years of age (Elston and Grizzle 1962). The data are plotted in Figure 17.2(a); the three lines shown are those for the three boys with minimum, median and maximum average measurements (i.e. take the average of four measurements from each boy, so we have 20 average values, then find those with the minimum, median and maximum averages).

Boy	Age 8	Age 8.5	Age 9	Age 9.5
1	47.8	48.8	49.0	49.7
2	46.4	47.3	47.7	48.4
3	46.3	46.8	47.8	48.5
4	45.1	45.3	46.1	47.2
5	47.6	48.5	48.9	49.3
6	52.5	53.2	53.3	53.7
7	51.2	53.0	54.3	54.5
8	49.8	50.0	50.3	52.7
9	48.1	50.8	52.3	54.4
10	45.0	47.0	47.3	48.3
11	51.2	51.4	51.6	51.9
12	48.5	49.2	53.0	55.5
13	52.1	52.8	53.7	55.0
14	48.2	48.9	49.3	49.8
15	49.6	50.4	51.2	51.8
16	50.7	51.7	52.7	53.3
17	47.2	47.7	48.4	49.5
18	53.3	54.6	55.1	55.3
19	46.2	47.5	48.1	48.4
20	46.3	47.6	51.3	51.8

Table 17.2: *Repeated measures of the height of the ramus (jaw) bones of 20 boys (Elston and Grizzle, 1962).*

The parallel lines indicate a strong subject effect. While there is a definite growth over time, the variability in growth rate is also of interest. Consider a linear model

$$y_{ij} = \beta_0 + \beta_1 t_{ij} + b_{0i} + b_{1i} t_{ij} + e_{ij}, \tag{17.22}$$

where y_{ij} is the ramus height of the i'th boy at follow-up time Age_{ij}. To reduce any correlation between the intercept and slope we centre the age variable to

$$t_{ij} = \text{Age}_{ij} - \text{ave(Age)}.$$

In this model $\beta_0 + b_{0i}$ and $\beta_1 + b_{1i}$ are the subject-specific intercept and slope of the i'th boy. Variability in these quantities is evident in Figure 17.2(b), which is based on separate estimation from each boy.

Now let y be the vector of y_{ij}, b_0 the vector of b_{0i}, b_1 the vector of b_{1i}, and $\beta = (\beta_0, \beta_1)$. The vector y is of length $N = 80$, and b_0 and b_1 are of length $q = 20$. Model (17.22) may be written in matrix form as

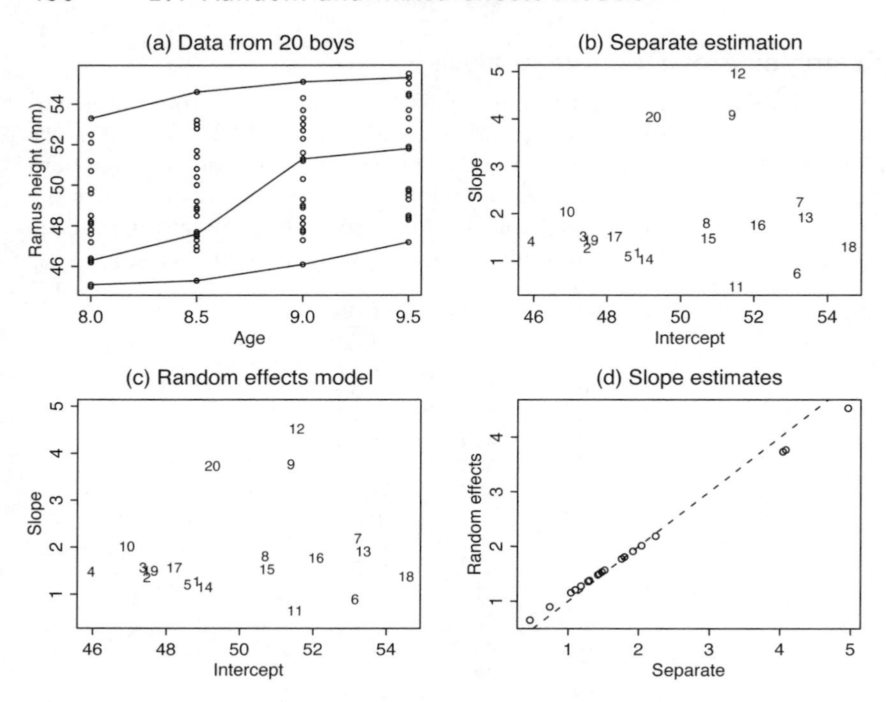

Figure 17.2: *(a) Repeated measures data of the height of ramus bones of 20 boys. (b) Intercepts and slopes estimated from each boy separately. (c) Intercepts and slopes estimated from the mixed effects model. (d) The slope estimates using the two different methods.*

$$y = X\beta + Z_0 b_0 + Z_1 b_1 + e$$

with appropriate design matrices X, Z_0 and Z_1. From our definition of t_{ij} above, Z_0 and Z_1 are orthogonal. Assume the random effects b_0 and b_1 are independent normal with mean zero and variance $\sigma_0^2 I_q$ and $\sigma_1^2 I_q$, and the independent error term e is normal with mean zero and variance $\sigma_e^2 I_N$.

The individual estimates of the intercepts and slopes provide sensible starting values for most of the parameters:

$$\begin{aligned}
\widehat{\beta}_0 &= \text{ave(intercepts)} = 50.075 \\
\widehat{\beta}_1 &= \text{ave(slopes)} = 1.866 \\
\widehat{\sigma}_0^2 &= \text{var(intercepts)} = 6.26 \\
\widehat{\sigma}_1^2 &= \text{var(slopes)} = 1.35.
\end{aligned}$$

The starting value for the residual variance σ_e^2 is computed from the average of the individual residual variances, giving $\widehat{\sigma}_e^2 = 0.13$.

At convergence of the iterative procedure we obtain the final estimates

$$\widehat{\beta}_0 = 50.075 \text{ (se} = 0.55)$$

$$\begin{aligned}
\widehat{\beta}_1 &= 1.866 \text{ (se} = 0.24) \\
\widehat{\sigma}_0^2 &= 5.90 \\
\widehat{\sigma}_1^2 &= 0.96 \\
\widehat{\sigma}_e^2 &= 0.20.
\end{aligned}$$

The standard errors for $\widehat{\beta}_0$ and $\widehat{\beta}_1$ are computed from $(X'V^{-1}X)^{-1}$, where

$$V = \widehat{\sigma}_0^2 Z_0 Z_0' + \widehat{\sigma}_1^2 Z_1 Z_1' + \widehat{\sigma}_e^2 I_N,$$

but the term $V^{-1}X$ is computed using the method given in Section 17.5.

Figure 17.2(c) shows the plot of $\widehat{\beta}_0 + \widehat{b}_{i0}$ versus $\widehat{\beta}_1 + \widehat{b}_{i1}$, which is comparable to Figure 17.2(b). Figure 17.2(d) shows $\widehat{\beta}_1 + \widehat{b}_{i1}$ from the separate estimation and from the mixed model. The lower-left and upper-right end of the plot show the shrinkage effect.

Comparisons

Let us first compare the previous results with a naive computation that ignores the dependence of the repeated measures. We simply fit a standard regression model

$$y = X\beta + e,$$

where e is assumed $\sigma_e^2 I_N$. Here we obtain the ordinary least-squares (OLS) estimate

$$\begin{aligned}
\widehat{\beta}_0 &= 50.075 \text{ (se} = 0.29) \\
\widehat{\beta}_1 &= 1.866 \text{ (se} = 0.51).
\end{aligned}$$

The estimates are the same, but the standard error of $\widehat{\beta}_1$ is twice that under the mixed effects model. The mixed model takes the dependence into account, and in this case it yields a more efficient estimate.

We can account for the dependence by computing a two-stage least-squares estimate as follows. In the first stage, compute the OLS estimate and the residuals $e = y - X\widehat{\beta}$. Let e_{ij} be the j'th error term of the i'th subject, then estimate the 4×4 covariance matrix C of $(y_{i1}, \ldots, y_{i4})$ with the element

$$\widehat{C}_{kl} = \frac{1}{n-1} \sum_{i=1}^{n=20} e_{ik} e_{il}.$$

From the data we obtain

$$\widehat{C} = \begin{pmatrix} 6.3300 & 6.1891 & 5.7770 & 5.5482 \\ 6.1891 & 6.4493 & 6.1534 & 5.9234 \\ 5.7770 & 6.1534 & 6.9180 & 6.9463 \\ 5.5482 & 5.9234 & 6.9463 & 7.4647 \end{pmatrix}.$$

Now create a large 80×80 block diagonal matrix V with $\widehat{C}$ on the diagonal. The second-stage estimate is

$$\widehat{\beta} = (X'V^{-1}X)^{-1}X'V^{-1}y$$

with variance matrix $(X'V^{-1}X)^{-1}$. Since C does not have any particular structure we cannot simplify V^{-1}. Here we obtain

$$\begin{aligned}\widehat{\beta}_0 &= 50.0496 \text{ (se} = 0.56) \\ \widehat{\beta}_1 &= 1.8616 \text{ (se} = 0.21),\end{aligned}$$

comparable with those obtained using the mixed effects model.

Using the mixed effects model, the covariance structure is implied by the model

$$y_{ij} = \beta_0 + \beta_1 t_{ij} + b_{0i} + b_{1i}t_{ij} + e_{ij}.$$

Writing $t_i = (t_{i1}, \ldots, t_{i4})'$, we have the implied covariance matrix

$$C_m = \sigma_0^2 J_4 + \sigma_1^2 t_i t_i' + \sigma_e^2 I_4,$$

where J_4 is a matrix of ones. Using the estimated parameter values:

$$\widehat{C}_m = \begin{pmatrix} 6.6388 & 6.0829 & 5.7225 & 5.3621 \\ 6.0829 & 6.1583 & 5.8426 & 5.7225 \\ 5.7225 & 5.8426 & 6.1583 & 6.0829 \\ 5.3621 & 5.7225 & 6.0829 & 6.6388 \end{pmatrix}.$$

The closeness of $\widehat{C}$ and $\widehat{C}_m$ indicates the adequacy of the mixed effects model. The main advantage of the mixed effects model is the availability of the individual intercepts and slopes. Also, there are applications where the subject-level formulation is more natural.

17.8 Generalized linear mixed models

The generalized linear mixed models (GLMM) extend the classical linear mixed models to nonnormal outcome data.

Example 17.2: The following table shows the number of seeds that germinated on 10 experimental plates; $n = 20$ seeds were planted on each plate.

Plate	1	2	3	4	5	6	7	8	9	10
Germination	6	3	10	11	16	5	9	9	4	10

If the number that germinated y_i is binomial(n, p), then the estimated mean is $n\widehat{p} = 8.3$ and variance $n\widehat{p}(1 - \widehat{p}) = 4.9$, but the observed variance is 15.1, much larger than the binomial variance.

To account for plate-to-plate variability, it may make sense to assume that y_i is binomial(n, p_i), but p_i is random; we might model

$$\text{logit } p_i = \beta_0 + b_i$$

where β_0 is a fixed parameter and b_i random. If the distribution of b_i is normal, we have a binomial–normal model. If b_i has a logit–beta distribution (or $e^{b_i}(1 + e^{b_i})^{-1}$ has a beta distribution), the outcome y_i follows a beta–binomial model. □

Example 17.3: Suppose the number of children y_i for family i is Poisson with mean μ_i, but we expect families to vary in their mean number of children (determined by factors such as the length of the marriage, education level of the parents, etc.). With random μ_i, we may want to model

$$\log \mu_i = \beta_0 + b_i$$

where β_0 is fixed, but b_i is a random parameter. If we assume b_i is normal the overall model for y_i is known as the Poisson–normal model. If b_i has a log–gamma distribution (or e^{b_i} has a gamma distribution), we have a Poisson–gamma model. In Section 4.5 we showed that the marginal distribution of the Poisson–gamma y_i is the negative binomial, so it is possible to get an explicit marginal likelihood of the fixed effects parameter. □

In Sections 4.2, 4.5 and 6.5 we have actually discussed the binomial and Poisson models with overdispersion. The new feature in GLMM is that we try to put more structure on the overdispersion pattern. If the random effects parameters are of interest, the GLMM specification is essential.

The simple mean model in the previous examples can be extended to a general regression of the form

$$h(\mu_i) = x_i'\beta + b_i,$$

where $h(\cdot)$ is a link function. In the examples, the germination data may come from an experiment with several treatment variables, or we may want to associate family size with socio-demographic variables. If the interest is on the fixed parameter β, the role of a random parameter b_i is to cover for some unmeasured variables, so that, conditional on b_i, the outcome y_i has a recognized distribution with mean μ_i, and we get a better inference for β.

General specification

The GLM part of GLMM specifies that conditional on b the outcome y_i is independently distributed according to an exponential family model (Section 6.5) of the form

$$\log p(y_i|b) = \frac{y_i\psi_i - A(\psi_i)}{\phi} + c(y_i, \phi).$$

Let $\mu_i \equiv E(y_i|b)$ and assume that

$$h(\mu_i) = x_i'\beta + z_i'b.$$

Writing μ as the vector of μ_i's and thinking of $h(\cdot)$ as a function that applies element by element, we can write the model more concisely as

$$h(\mu) = X\beta + Zb.$$

As described in Section 6.5 this model specification is equivalent to assuming particular mean and variance functions. That is, conditional on the random effects b, the outcome y_i is independently distributed with mean μ_i and variance $\phi v_i(\mu_i)$.

We complete the model specification by making an assumption on the random effects b. In general b does not have to be normally distributed, though such a distribution usually leads to simple computational formulae. Let θ include the dispersion parameter ϕ above and any other parameters needed to specify b.

The marginal variance of y_i is

$$\begin{aligned} \text{var}(y_i) &= E\{\text{var}(y_i|b)\} + \text{var}\{E(y_i|b)\} \\ &= \phi E\{v_i(\mu_i)\} + \text{var}(\mu_i). \end{aligned}$$

Overdispersion might come from both terms on the right-hand side; the first term accounts for the contribution of the dispersion family, and the second term is the contribution of the random effects.

Example 17.4: Suppose, conditional on (scalar) b, the (scalar) outcome y has mean μ and variance $\phi\mu$, where

$$\log \mu = \beta_0 + b,$$

and e^b is gamma with mean one and variance ν. The marginal mean of y is $Ey = e^{\beta_0}$, and the marginal variance of y is

$$\begin{aligned} \text{var}(y) &= \phi e^{\beta_0} + \nu e^{2\beta_0} \\ &= \phi Ey + \nu (Ey)^2. \end{aligned}$$

If $\phi = 1$ we have the Poisson–gamma or negative binomial model; if $\nu = 0$ we have the dispersion-family Poisson model. Hence the GLMM can fit potentially richer overdispersion patterns. In recent applications of GLMM, except in normal models, the parameter ϕ is usually set to one. Lee and Nelder (2000) give some applications that use the full model.

Example 17.5: Classical animal breeding programmes are usually limited to continuous outcomes (such as meat or milk yields) which are assumed normal, but there are increasingly many nonnormal characteristics that require GLMM techniques, for example calving difficulty in cows, ability to produce twins in sheep, disease resistance, fertility rate, performance of race horses, etc. The model in each case is usually of the form

$$h(\mu) = X\beta + Zb,$$

where the parameter of interest is the random parameter b, the vector genetic merit of the animals under observation. The fixed parameter β accounts for predictors such as age of animal, farm characteristics, etc. Applications in animal breeding typically involve thousands of related animals (see Section 17.3), demanding highly nontrivial computations. □

17.9 Exact likelihood in GLMM

While the marginal likelihood in the normal–normal GLMM is relatively simple (Section 17.2), the general case is notoriously difficult. The approximate computations will be described in the next section, and here we illustrate the computation of the exact marginal likelihood for the binomial–normal model. Suppose, conditional on b_i, the observed y_i is binomial with mean $\mu_i = n_i p_i$, where

$$\text{logit } p_i = x_i'\beta + b_i,$$

and b_i is a random sample from $N(0, \sigma^2)$.

The marginal probability of y_i is

$$\begin{aligned} P(y_i = k) &= E\{P(y_i = k|b_i)\} \\ &= \int \binom{n_i}{k} p^k (1-p)^{n_i - k} \phi(b/\sigma) db/\sigma \\ &\equiv p(n_i, k, \beta, \sigma^2), \end{aligned}$$

where logit $p = x_i'\beta + b$ and $\phi(u)$ is the standard normal density. Given the observed y_i's, the marginal log-likelihood of the fixed parameters is

$$\log L(\beta, \sigma^2) = \sum_i \log p(n_i, y_i, \beta, \sigma^2).$$

If $n_i \equiv n$, the likelihood reduces to

$$\log L(\beta, \sigma^2) = \sum_k m_k \log p(n, k, \beta, \sigma^2),$$

where m_k is the number of y_i's equal to k.

This illustrates the computational problem of exact likelihood inference in GLMM: in general there is no closed form solution for integrating out the random effects b. In practice it has to be computed numerically, for example using the Gaussian quadrature technique. This is feasible if the random effects are independent, so we only need to evaluate single integrals. If the random effects are correlated, the integrals become multidimensional, and the exact approach is no longer tractable.

Example 17.6: We return to the germination data analysed in Example 6.4. The data were from a 2×2 factorial experiment of seed variety and type of root extract. For comparison Table 17.3 shows again the summary result using the ordinary logistic model

$$\text{logit } p_i = x_i'\beta.$$

In particular the result shows a significant interaction term.

The deviance of this fixed effects model is 33.28 with 21 − 4 = 17 degrees of freedom; for binomial data this can be interpreted as a goodness-of-fit statistic, indicating overdispersion (P-value=0.01). Now consider a mixed model that accounts for the plate variability:

$$\text{logit } p_i = x_i'\beta + b_i,$$

where the plate effects b_i's are iid $N(0, \sigma_b^2)$. The results of the numerical optimization of the exact binomial–normal log-likelihood are shown in Table 17.3. The Gaussian quadrature method is used to compute the integral. The parameter estimates from the two models are comparable, but, as expected from an

	Ordinary		Mixed model	
Effects	Estimate	se	Estimate	se
Constant	−0.56	0.13	−0.55	0.17
Seed variety	0.15	0.22	0.10	0.28
Root extract	1.32	0.18	1.34	0.24
Interaction	−0.78	0.31	−0.81	0.38
Plate σ_b			0.24	0.11

Table 17.3: *Summary analysis of germination data using the ordinary logistic regression and the exact binomial–normal mixed model.*

overdispersed model, all of the standard errors from the mixed model are slightly larger.

The MLE of σ_b^2 is $\widehat{\sigma}_b^2 = 0.056$ (se $= 0.052$), but the standard error is not meaningful since the likelihood is not regular. Figure 17.3 shows the exact profile log-likelihood of σ_b^2, σ_b and $\log \sigma_b$ with their quadratic approximations. The best is for σ_b, except near zero. The exact likelihood interval for σ_b at 15% cutoff covers zero; the likelihood of H: $\sigma_b = 0$ is around 31%, indicating that there is no evidence against the ordinary logistic model. The Wald statistic based on $\widehat{\sigma}_b$ is $z = 0.24/0.11 = 2.18$.

The exact likelihood result is in conflict with the earlier goodness-of-fit test that shows evidence against the ordinary logistic regression. (An exact version of the χ^2 goodness-of-fit test is also significant; see Exercise 17.4.) In this case, because it is slightly more conservative, we may want to keep the binomial–normal model, although both models lead to similar conclusions. □

17.10 Approximate likelihood in GLMM

The basis of the likelihood approximation in GLMM is the extended likelihood that includes the random effects, plus the heuristics provided by the normal mixed models. Given the standard setup of GLMM in the last section, the joint likelihood of the parameters is

$$\log L(\beta, \theta, b) = \log p(y|b) + \log p(b)$$

where $p(y|b)$ is the likelihood based on the conditional distibution of y given b, and $p(b)$ is the likelihood based on the assumed distribution of the random effects. Lee and Nelder (1996) refer to this as the hierarchical likelihood or h-likelihood.

Example 17.7: For the binomial–normal model, conditional on b, let y_i be binomial(n_i, p_i) where

$$\text{logit } p_i = x_i'\beta + z_i'b.$$

Assume that the random effects b_i's are iid $N(0, \sigma^2)$; it is obvious how to modify the likelihood if b is multivariate normal. Given the data $y_1, \ldots, y_n$ we obtain

$$\log L(\beta, \theta, b) = \sum \{y_i \log p_i + (n_i - y_i) \log(1 - p_i)\} - \frac{1}{2\sigma^2} \sum_i b_i^2,$$

where p_i is understood to be a function of β and b, and $\theta \equiv \sigma^2$. For a fixed value of σ^2 this likelihood is trivial to compute or optimize. □

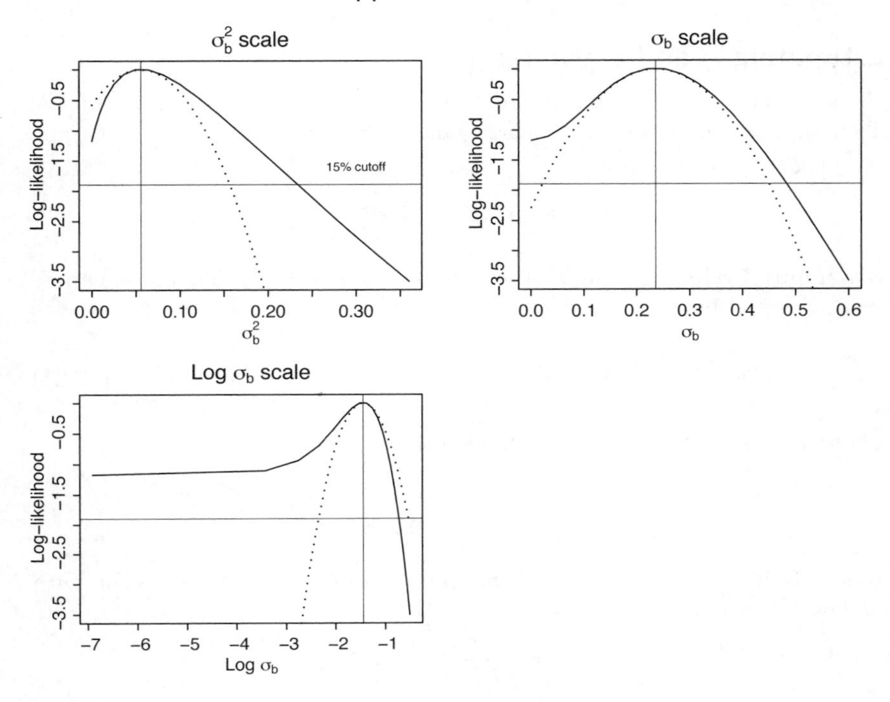

Figure 17.3: *Exact profile log-likelihoods (solid lines) of* σ_b^2, σ_b *and* $\log \sigma_b$, *and their quadratic approximations.*

Example 17.8: Suppose conditional on b the outcome y_i is Poisson with mean μ_i, where

$$\log \mu_i = x_i'\beta + b_i$$

and $u_i = e^{b_i}$ is iid gamma with mean 1 and shape parameter α. Given $y_1, \ldots, y_n$ we have

$$\log L(\beta, \theta, b) = \sum_i (-\mu_i + y_i \log \mu_i) + \sum_i \{\alpha b_i - \alpha e^{b_i} + \alpha \log \alpha - \log \Gamma(\alpha)\},$$

where $\theta \equiv \alpha$. □

The joint likelihood $L(\beta, \theta, b)$ is generally much easier to evaluate and optimize than the marginal likelihood $L(\beta, \theta)$. For the normal model in Section 17.4 it is shown that, at fixed value of θ, the estimate $\widehat{\beta}$ from $L(\beta, \theta, b)$ is the same as that from the marginal likelihood. The Poisson–gamma model above is another case where the joint likelihood estimate of β is the same as the marginal likelihood estimate. Breslow and Clayton (1993) present applications that support the practical viability of such an approach. Lee and Nelder (1996) show that under fairly general conditions the two estimates are asymptotically close.

Estimating β and b given θ

Given a fixed value of θ we can use a quadratic approximation of the likelihood to derive an IWLS algorithm; see Section 6.7. Let the outcome vector y have mean μ and

$$h(\mu) = X\beta + Zb.$$

Given initial values β^0 and b^0, the exponential family log-likelihood can be approximated by

$$-\frac{1}{2}\log|\Sigma| - \frac{1}{2}(Y - X\beta - Zb)'\Sigma^{-1}(Y - X\beta - Zb), \tag{17.23}$$

where Y is a working vector with elements

$$Y_i = x_i'\beta^0 + z_i'b^0 + \frac{\partial h}{\partial \mu_i}(y_i - \mu_i^0),$$

and Σ is a diagonal matrix of the variance of the working vector with elements

$$\Sigma_{ii} = \left(\frac{\partial h}{\partial \mu_i}\right)^2 \phi v_i(\mu_i^0),$$

where $\phi v_i(\mu_i^0)$ is the conditional variance of y_i given b. The derivative $\partial h/\partial \mu_i$ is also evaluated at the current values of β and b.

If the random effects parameter b is assumed normal with mean zero and variance $D = D(\theta)$, we have a familiar normal-based formula:

$$\begin{aligned}\log L(\beta,\theta,b) \quad &\approx \quad -\frac{1}{2}\log|\Sigma| - \frac{1}{2}(Y - X\beta - Zb)'\Sigma^{-1}(Y - X\beta - Zb) \\ &\quad -\frac{1}{2}\log|D| - \frac{1}{2}b'D^{-1}b. \end{aligned} \tag{17.24}$$

This yields the usual mixed model equations to update β and b:

$$\begin{pmatrix} X'\Sigma^{-1}X & X'\Sigma^{-1}Z \\ Z'\Sigma^{-1}X & Z'\Sigma^{-1}Z + D^{-1} \end{pmatrix} \begin{pmatrix} \beta \\ b \end{pmatrix} = \begin{pmatrix} X'\Sigma^{-1}Y \\ Z'\Sigma^{-1}Y \end{pmatrix}. \tag{17.25}$$

An iterative backfitting algorithm (Section 17.4) might be used to find the solution. The iteration continues by recomputing Y and Σ. Hence the computation of estimates in GLMM involves repeated applications of normal-based formulae.

Nonnormal random effects*

The estimation of nonnormal random effects is best derived on a case-by-case basis, as different models may have different optimal algorithms. However, as an illustration of the additional work needed, we show here a useful general technique.

If b is not normally distributed an extra step is needed first to approximate its log-likelihood by a quadratic form. Let b_i's be an iid sample from $p(\cdot)$, and let $\ell(b_i) = \log p(b_i)$. Using initial value b_i^0

$$\log p(b_i) \approx \log p(b_i^c) + \frac{1}{2}\ell''(b_i^0)(b_i - b_i^c)^2,$$

where

$$b_i^c = b_i^0 - \frac{\ell'(b_i^0)}{\ell''(b_i^0)}.$$

Now let $D^{-1} = \text{diag}[-\ell''(b_i^0)]$, the Fisher information matrix of b based on $p(\cdot)$, and let $\ell'(b^0)$ be the vector of $\ell'(b_i^0)$, so in vector notation

$$b^c = b^0 + D\ell'(b^0).$$

We can now write

$$\log p(b) = \log p(b^c) - \frac{1}{2}(b - b^c)'D^{-1}(b - b^c).$$

In the normal case D is the covariance matrix, and $b^c = 0$.

After combining this with the quadratic approximation of $\log p(y|b)$, we take the derivatives with respect to β and b and find the updating equation

$$\begin{pmatrix} X'\Sigma^{-1}X & X'\Sigma^{-1}Z \\ Z'\Sigma^{-1}X & Z'\Sigma^{-1}Z + D^{-1} \end{pmatrix} \begin{pmatrix} \beta \\ b \end{pmatrix} = \begin{pmatrix} X'\Sigma^{-1}Y \\ Z'\Sigma^{-1}Y + D^{-1}b^c \end{pmatrix}. \quad (17.26)$$

This is similar to (17.25), except for the term $D^{-1}b^c$.

Example 17.9: If e^{b_i} is iid gamma with mean 1 and shape parameter α,

$$\ell(b_i) = \alpha b_i - \alpha e^{b_i} + \alpha \log \alpha - \log \Gamma(\alpha)$$

and we have

$$\begin{aligned} \ell'(b_i) &= \alpha - \alpha e^{b_i} \\ \ell''(b_i) &= -\alpha e^{b_i} \end{aligned}$$

so, using the starting value b^0, we get $D^{-1} = \text{diag}[\alpha e^{b^0}]$ and

$$\begin{aligned} b^c &= b^0 + \frac{\alpha - \alpha e^{b^0}}{\alpha e^{b^0}} \\ &= b^0 + e^{-b^0} - 1. \ \square \end{aligned}$$

Example 17.10: Suppose b_i is iid Cauchy with location 0 and scale σ_b, so

$$\ell(b_i) = -\log\left(1 + \frac{b_i^2}{\sigma_b^2}\right),$$

and we have

$$\ell'(b_i) = -\frac{2b_i}{b_i^2 + \sigma_b^2}$$

$$\ell''(b_i) = -\frac{2(b_i^2 - \sigma_b^2)}{(b_i^2 + \sigma_b^2)^2},$$

and $D^{-1} = \text{diag}[-\ell''(b^0)]$ and

$$\begin{aligned} b^c &= b^0 - b^0 \frac{(b^0)^2 + \sigma_b^2}{(b^0)^2 - \sigma_b^2} \\ &= -\frac{2\sigma_b^2 b^0}{(b^0)^2 - \sigma_b^2}. \end{aligned}$$

In this Cauchy example we can derive another computational scheme. First note that

$$\ell'(b) = -D^{-1}b$$

where now $D^{-1} \equiv \text{diag}[2/(b^2+\sigma_b^2)]$. Combining this with the derivative of (17.23) with respect to b, we can update b using

$$(Z'\Sigma^{-1}Z + D^{-1})b = Z'\Sigma^{-1}(Y - X\beta),$$

where D is computed at the starting value b^0. This is simpler and probably more stable than the general method above. □

Estimating θ

Even in the normal case, the estimate of the variance parameter cannot be computed from the joint likelihood $L(\beta, b, \theta)$. Instead, we have shown in Section 17.5 that the profile likelihood of θ is equivalent to a modified profile from the joint likelihood:

$$\begin{aligned} \log L(\theta) &= \log L(\widehat{\beta}, \theta) \\ &= \log L(\widehat{\beta}, \theta, \widehat{b}) - \frac{1}{2}\log|Z'\Sigma^{-1}Z + D^{-1}|, \end{aligned}$$

where θ enters the function through Σ, D, $\widehat{\beta}$ and $\widehat{b}$.

The approximate methods of variance component estimation in GLMM have not yet settled to an agreed form, but a strong candidate is to estimate θ by maximizing a modified profile likelihood exactly as in the normal case (Lee and Nelder 1996):

$$\log L(\theta) = \log L(\widehat{\beta}, \theta, \widehat{b}) - \frac{1}{2}\log|Z'\Sigma^{-1}Z + D^{-1}|, \tag{17.27}$$

where $\widehat{\beta}$ and $\widehat{b}$ are computed at fixed θ as described before. There is one crucial difference: in the normal case Σ is typically a simple function of a variance component, but in GLMM Σ is also a function of the unknown mean μ, and hence of β and b. Since μ is unknown it will be convenient to compute Σ using $\widehat{\beta}$ and $\widehat{b}$.

A heuristic justification in terms of Laplace's integral approximation (Section 10.6) is given by Breslow and Clayton (1993). It has been shown in many examples that this method does provide solutions which are close to the exact marginal likelihood estimates, provided the variance component θ is not too large. The method tends to underestimate θ, and the problem can be severe for large θ.

Joint estimation of β, θ and b

As in the normal case, the approximate MLEs of β, θ and b are the joint maximizers of

$$Q(\beta,\theta,b) = \log L(\beta,\theta,b) - \frac{1}{2}\log|Z'\Sigma^{-1}Z + D^{-1}|.$$

To derive an iterative estimation procedure the first term is approximated by a quadratic form (17.24). However, in contrast with the normal mixed models, because of the dependence of Σ on β and b, we cannot immediately justify this iterative algorithm:

1. Compute $\widehat{\beta}$ and $\widehat{b}$ given θ by solving (17.25).
2. Fixing β and b at the values $\widehat{\beta}$ and $\widehat{b}$, update θ by maximizing Q.
3. Iterate between 1 and 2 until convergence.

The algorithm is appropriate, for example, if Σ is a slowly varying function of μ. This means we can ignore the derivative of the second term of Q with respect to β and b, so the first step is justified. This is not the only way to estimate θ; any derivative-free method can be used to optimize $\log L(\theta)$ in (17.27) directly.

In the important special case where $\phi = 1$, and

- Σ is assumed to be a slowly varying function of the mean μ, and
- b is normal with mean zero and variance $\sigma_b^2 R$, where R is a known matrix of rank q, so $\theta \equiv \sigma_b^2$,

we can derive an update formula for Step 2 of the algorithm as in Section 17.5:

$$\sigma_b^2 = \frac{1}{q}[b'R^{-1}b + \text{trace}\{(Z'\Sigma^{-1}Z + \sigma_b^{-2}R^{-1})^{-1}R^{-1}\}],$$

where all unknown quantities on the right-hand side are evaluated at the last available values.

Computation in the general case $\phi \neq 1$ can also follow the algorithm presented in Section 17.5 if we are willing to make the quadratic approximation (17.23) around the estimates $\widehat{\beta}$ and $\widehat{b}$ as a profile log-likelihood for ϕ. Alternatively we can use the extended quasi-likelihood approximation (4.5) for an explicit profile likelihood of ϕ.

Approximate inference

The simplest inference on β and b can be provided by computing the standard errors from the Fisher information matrix. By analogy with the normal model, given variance parameter θ, the Fisher information for β is

$$I(\widehat{\beta}) = X'V^{-1}X$$

where $V = \Sigma + ZDZ'$, and the Fisher information for b is

$$I(\widehat{b}) = (Z'\Sigma^{-1}Z + D^{-1}).$$

In practice these are evaluated at the estimated value of θ. These formulae imply (i) the inference on $\widehat{\beta}$ accounts for the fact that b, but not θ, is unknown; (ii) inference on $\widehat{b}$ does not account for the fact that both β and θ are estimated.

Standard errors of $\widehat{\theta}$ can be computed by taking derivatives of $\log L(\theta)$ in (17.27), but bear in mind that the likelihood function may not be regular.

Example 17.11: We apply the methodology to Crowder's data in Example 17.6. Recall that y_i is binomial(n_i, p_i) with

$$\text{logit } p_i = x_i'\beta + b_i,$$

so the design matrix Z is an identity matrix I_{21}, and b is normal with mean zero and variance $\sigma_b^2 I_{21}$.

We can use the ordinary GLM estimates as the starting value for β, and set $b = 0$ and $\sigma_b^2 = 0.01$. Then:

1. Update b by solving

$$(Z'\Sigma^{-1}Z + \sigma_b^{-2}I_q)b = Z'\Sigma^{-1}(Y - X\beta).$$

This reduces to

$$b_i = \frac{\Sigma_{ii}^{-1}(Y_i - x_i'\beta)}{\Sigma_{ii}^{-1} + \sigma_b^{-2}},$$

where Y is the working vector with elements

$$Y_i = x_i'\beta + b_i + \frac{y_i - n_i p_i}{n_i p_i(1 - p_i)}$$

and p_i is computed from

$$\text{logit } p_i = x_i'\beta + b_i.$$

The matrix Σ is diagonal with elements

$$\Sigma_{ii} = \frac{1}{n_i p_i(1 - p_i)}.$$

2. Update β by solving

$$(X'\Sigma^{-1}X)\beta = X'\Sigma^{-1}(Y - b)$$

after recomputing Y and Σ.

3. Update σ_b^2 using

$$\sigma_b^2 = \frac{1}{21}[b'b + \text{trace}\{(\Sigma^{-1} + \sigma_b^{-2}I_{21})^{-1}\}].$$

4. Iterate 1, 2 and 3 until convergence.

The results at convergence are shown in Table 17.4 under the 'Approximate' columns. For comparison the ordinary logistic regression and the 'Exact'

Effects	Ordinary Estimate	Ordinary se	Mixed models: Exact Estimate	Exact se	Mixed models: Approximate Estimate	Approximate se
Constant	−0.56	0.13	−0.55	0.17	−0.54	0.17
Seed variety	0.15	0.22	0.10	0.28	0.10	0.27
Root extract	1.32	0.18	1.34	0.24	1.33	0.23
Interaction	−0.78	0.31	−0.81	0.38	−0.80	0.38
Plate σ_b			0.24	0.11	0.23	0.11

Table 17.4: *Summary table for logistic regression analysis of germination data using ordinary GLM, the exact binomial–normal and approximate likelihood methods.*

binomial–normal mixed model results from Table 17.3 are also shown. In this example the approximate method yields results very close to the exact results.

To show the quality of the likelihood approximation, Figure 17.4 compares the approximate profile likelihood of σ_b computed according to (17.27) and its exact version computed in the previous section. □

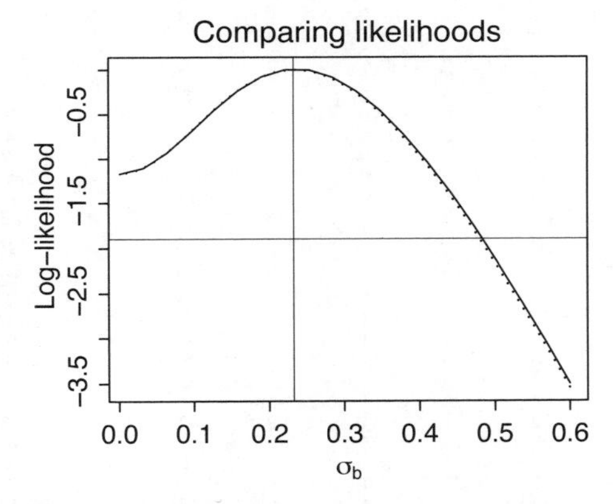

Figure 17.4: *Exact profile log-likelihood (solid) of σ_b and its approximation using (17.27).*

17.11 Exercises

Exercise 17.1: As stated after equation (17.2), show that the MLE of μ_i is $\overline{y}_i$ regardless of any constraint on the b_i's.

Exercise 17.2: Show that $\widehat{b_i}$ in (17.3) is the conditional mean of the distribution of b_i given y.

Exercise 17.3: Show that the solution of the mixed model equation (17.12) is exactly the same as (17.7) and (17.10).

Exercise 17.4: Compute the exact distribution of the χ^2 goodness-of-fit test for the germination data in Example 17.6. (Hint: think of the success–failure data from each group as a $k \times 2$ table. If the binomial model is correct, the distribution of the table entries conditional on the observed margins is hypergeometric; recall Fisher's exact test for the 2×2 table. Hence, we can generate realizations according to the hypergeometric model, compute a χ^2 statistic for each realization, and then repeat a large number of times.)

Exercise 17.5: Figure 17.5 shows the data from a clinical trial of an epileptic drug progabide (Thall and Vail, 1990). Patients were randomized to the active (Tr=1) or placebo (Tr=0) drug in addition to a standard treatment. Baseline data at entry include age and the number of seizures in the previous 8 weeks. The follow-up data y_1 to y_4 are the number of seizures during the two-week period prior to each of four visits.

ID	Age	Base	Tr	y_1	y_2	y_3	y_4
1	31	11	0	5	3	3	3
2	30	11	0	3	5	3	3
3	25	6	0	2	4	0	5
4	36	8	0	4	4	1	4
5	22	66	0	7	18	9	21
6	29	27	0	5	2	8	7
7	31	12	0	6	4	0	2
8	42	52	0	40	20	23	12
9	37	23	0	5	6	6	5
10	28	10	0	14	13	6	0
11	36	52	0	26	12	6	22
12	24	33	0	12	6	8	4
13	23	18	0	4	4	6	2
14	36	42	0	7	9	12	14
15	26	87	0	16	24	10	9
16	26	50	0	11	0	0	5
17	28	18	0	0	0	3	3
18	31	111	0	37	29	28	29
19	32	18	0	3	5	2	5
20	21	20	0	3	0	6	7
21	29	12	0	3	4	3	4
22	21	9	0	3	4	3	4
23	32	17	0	2	3	3	5
24	25	28	0	8	12	2	8
25	30	55	0	18	24	76	25
26	40	9	0	2	1	2	1
27	19	10	0	3	1	4	2
28	22	47	0	13	15	13	12
29	18	76	1	11	14	9	8
30	32	38	1	8	7	9	4
31	20	19	1	0	4	3	0
32	30	10	1	3	6	1	3
33	18	19	1	2	6	7	4
34	24	24	1	4	3	1	3
35	30	31	1	22	17	19	16
36	35	14	1	5	4	7	4
37	27	11	1	2	4	0	4
38	20	67	1	3	7	7	7
39	22	41	1	4	18	2	5
40	28	7	1	2	1	1	0
41	23	22	1	0	2	4	0
42	40	13	1	5	4	0	3
43	33	46	1	11	14	25	15
44	21	36	1	10	5	3	8
45	35	38	1	19	7	6	7
46	25	7	1	1	1	2	3
47	26	36	1	6	10	8	8
48	25	11	1	2	1	0	0
49	22	151	1	102	65	72	63
50	32	22	1	4	3	2	4
51	25	41	1	8	6	5	7
52	35	32	1	1	3	1	5
53	21	56	1	18	11	28	13
54	41	24	1	6	3	4	0
55	32	16	1	3	5	4	3
56	26	22	1	1	23	19	8
57	21	25	1	2	3	0	1
58	36	13	1	0	0	0	0
59	37	12	1	1	4	3	2

Figure 17.5: *Epilepsy data from Thall and Vail (1990)*

(a) Investigate the benefit of the treatment by analysing the number of seizures on the last visit only (y_4). Is the benefit apparent on the earlier visits?

(b) Use the Poisson regression to incorporate the baseline covariates in your analysis of the last visit. Interpret the deviance of the model.

(c) Analyse all visits together by including a fixed effect of time in the model. Consider a linear and a categorical model for time.

(d) Show graphically that there is a significant subject effect.

(e) Suppose μ_{it} is the mean number of seizures for subject i on visit t. Fit a GLMM where conditional on the subject effects, the outcome y_{it} is Poisson with mean μ_{it} satisfying

$$\log \mu_{it} = x_i'\beta + b_{io} + b_{i1}t,$$

where x_i is a vector of fixed covariates that include the treatment assignment, age, baseline counts and linear time effect; b_{io} and b_{i1} are the individual random intercepts and slopes. Assume that b_{io} and b_{i1} are independent normal with mean zero and variance σ_0^2 and σ_1^2. Derive the iterative algorithm to compute the estimates.

18

Nonparametric smoothing

Nonparametric smoothing or nonparametric function estimation grew enormously in the 1980s. The word 'nonparametric' has nothing to do with the classical rank-based methods, such as the Wilcoxon rank test, but it is understood as follows. The simplest relationship between an outcome y and a predictor x is given by a linear model

$$E(y) = \beta_0 + \beta_1 x.$$

This linear model is a parametric equation with two parameters β_0 and β_1. A nonparametric model would simply specify $E(y)$ as some function of x

$$E(y) = f(x).$$

The class of all possible $f(\cdot)$ is 'nonparametric' or infinite dimensional.

The literature on nonparametric smoothing is vast, and we cannot hope to do justice to it in a single chapter. We will focus on a general methodology that fits well with the likelihood-based mixed effects modelling. The approach is practical, treating functions with discrete rather than continuous index. This means it suffices to deal with the usual vectors and matrices, rather than function spaces and operators.

18.1 Motivation

Example 18.1: Figure 18.1 shows the scatter plot of SO_2 level and industrial activity; the latter is measured as the number of manufacturing enterprises employing 20 or more workers. In Section 6.8 we have shown that it is sensible to log-transform the data. Since our first instinct is that more industry leads to more pollution, when faced with this dataset, we might only consider a linear model (dotted line). A nonparametric regression estimate (solid line) suggests a quadratic model, shown in Section 6.8 to be well supported by the data. The nonparametric or quadratic fits are harder to interpret in this case, but in this empirical modelling there could be other confounding factors not accounted for by the variables. The idea is that we should let the data tell their story rather than impose our prejudice; with this attitude a nonparametric smoothing technique is an invaluable tool for exploratory data analysis. □

Ad hoc methods

Our general problem is as follows: given bivariate data $(x_1, y_1), \ldots, (x_N, y_N)$ we assume that conditional on x_i the outcome y_i is normal with mean

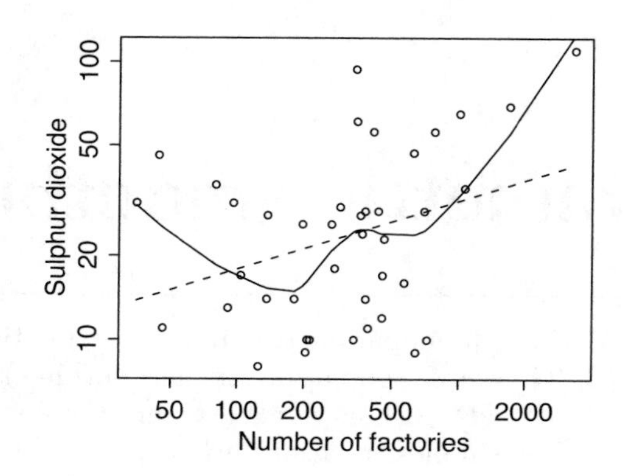

Figure 18.1: *Relationship between SO_2 level and industrial activity in 41 US cities. Shown are the nonparametric smooth (solid) and the linear regression (dased) estimates.*

$$E(y_i) = f(x_i)$$

and variance σ^2. We want to estimate the function $f(x)$ from the data.

The key idea of nonparametric smoothing is *local averaging*: $f(x)$ at a fixed value of x is the mean of y_i at x, so if there are many y_i's observed at x, then the estimate of $f(x)$ is the average of those y_i's. More often than not we have to compromise: the estimate of $f(x)$ is the average of y_i's for x_i's 'near' x. This can be implemented by partitioning the data, finding the nearest neighbours or kernel smoothing.

Partitioning

We can partition the range of the predictor x into n small intervals or bins, so that within an interval $f(x_i)$ is approximately constant, and y_i's are approximately iid with mean $f(x_i)$. We can then estimate $f(x_i)$ by the sample average of y_i's in the corresponding interval. The estimated function can be drawn as the polygon connecting the sample means from each interval.

As an example, Table 18.1 partitions the SO_2 data into 20 equispaced intervals (in log x). Note that some intervals are empty, but that does not affect the method. Figure 18.2 shows the nonparametric smoothing of the SO_2 level against the industry using different numbers of bins. The amount of smoothing is determined by the interval size, which has the following trade-off: if the interval is too large then the estimate might smooth out important patterns in $f(x)$, and the estimate is biased; but if it is too small the noise variance exaggerates the local variation and obscures the real patterns. The purpose of smoothing is to achieve a balance between

No.	x	Bin	Mid-x	y	No.	x	Bin	Mid-x	y
1	35	1	39	31	22	361	11	383	28
2	44	2	49	46	23	368	11	383	24
3	46	2	49	11	24	379	11	383	29
4	80	4	78	36	25	381	11	383	14
5	91	5	98	13	26	391	11	383	11
6	96	5	98	31	27	412	11	383	56
7	104	5	98	17	28	434	12	482	29
8	125	6	123	8	29	453	12	482	12
9	136	6	123	14	30	454	12	482	17
10	137	6	123	28	31	462	12	482	23
11	181	8	193	14	32	569	13	605	16
12	197	8	193	26	33	625	13	605	47
13	204	8	193	9	34	641	13	605	9
14	207	8	193	10	35	699	14	760	29
15	213	8	193	10	36	721	14	760	10
16	266	9	243	26	37	775	14	760	56
17	275	10	305	18	38	1007	15	954	65
18	291	10	305	30	39	1064	15	954	35
19	337	10	305	10	40	1692	18	1891	69
20	343	11	383	94	41	3344	20	2984	110
21	347	11	383	61					

Table 18.1: *Partitioning the SO_2 level (= y) data into 20 intervals/bins of the predictor variable x = industrial activities. 'Mid-x' is the midpoint (in log scale) of the interval. Note: throughout this section SO_2 is analysed in log scale.*

local bias and variance.

Nearest neighbours

The *nearest-neighbour method* simply prescribes, for any x,

$$\widehat{f}(x) = \frac{1}{k} \sum_{i \in n_k(x)} y_i,$$

where $n_k(x)$ is the neighbourhood of x that includes only the k values of x_i's nearest to x. Hence $\widehat{f}(x)$ is a simple average of y_i's for k nearest neighbours of x; larger values of k effect more smoothing. For example, using $k = 7$, at $x = 125$ we obtain the following nearest neighbours of x with the corresponding y:

x	125	136	137	104	96	91	181
y	8	14	28	17	31	13	14

giving an average log y of 2.79. The set of nearest neighbours needs to be computed at every value of x, making this method computationally more demanding than the partition method. For the plots in the top row of Figure 18.3 $\widehat{f}(x)$ is computed at the observed x_i's, but this is not necessary as it can be computed at a smaller subset of values. Note that the

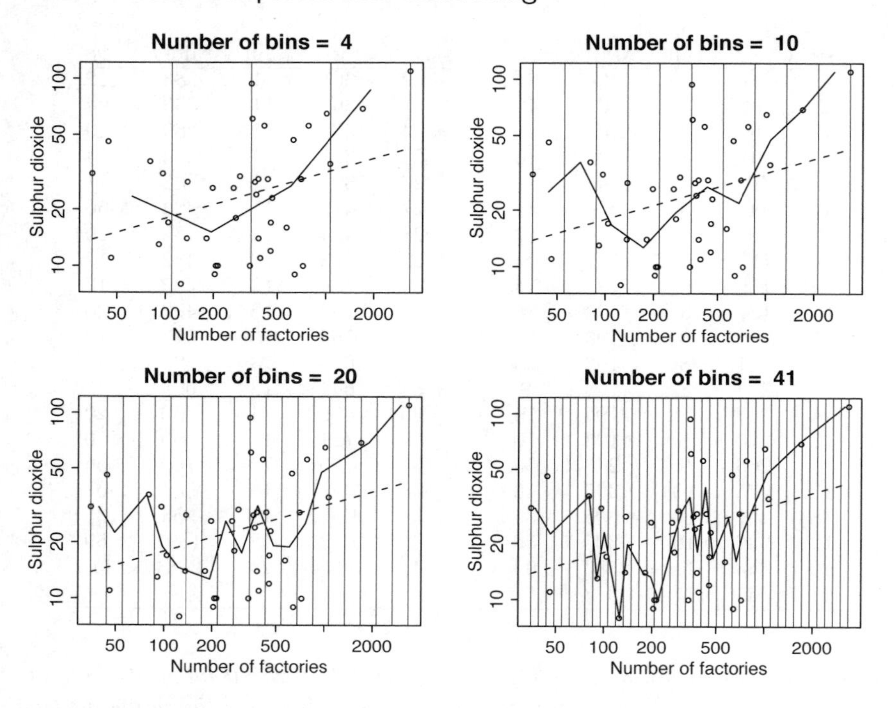

Figure 18.2: *Nonparametric smoothing of the SO_2 level against the industrial activity using simple partitioning of the predictor variable. The dashed line in each plot is the linear regression line.*

estimate at the boundaries is biased, especially as we increase the number of neighbours.

Kernel method

Using *the kernel method* one computes a weighted average

$$\widehat{f}(x) = \frac{\sum_{i=1}^{n} k(x_i - x) y_i}{\sum_{i=1}^{n} k(x_i - x)}$$

where the kernel function $k(x)$ is typically a symmetric density function. If we use the normal density the method is called Gaussian smoothing. The amount of smoothing is determined by the scale or width of the kernel; in Gaussian smoothing it is controlled by the standard deviation. The bottom row of Figure 18.3 shows the Gaussian smoothing of the SO_2 data using a standard deviation of 0.2 and 0.05 (note: x is also in log scale).

With the last two methods $\widehat{f}(x)$ can be computed at any x, while with the first method the choice of a partition or interval size determines the values of x for which $\widehat{f}(x)$ is available. This is a weakness of the first method, since if we want $f(x)$ for a lot of x values we have to make the intervals small, which in turn makes the estimation error large. The general

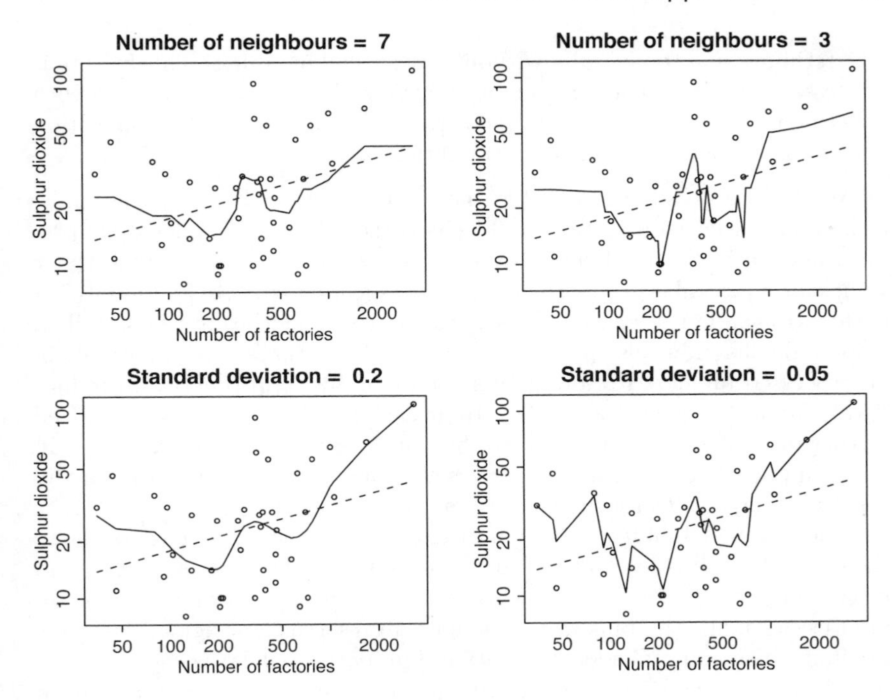

Figure 18.3: *Nonparametric smoothing of the SO_2 level against the industrial activity using the nearest-neighbour method (top row) and the kernel method (bottom row). The kernel method uses the normal density with the stated standard deviation.*

method in the next section overcomes this weakness. For each method the choice of the smoothing parameter is an important issue. The next section shows that the problem is simply a variance component estimation problem.

18.2 Linear mixed models approach

We will now put the nonparametric function estimation within the linear mixed model framework with likelihood-based methodology. Compared to the nearest-neighbour or the kernel method, the likelihood-based method is easier to extend to deal with

- non-Gaussian outcome data, such as Poisson or binomial data;
- different types of functions, such as functions with jump discontinuities or partly parametric models;
- the so-called 'inverse problems' (e.g. O'Sullivan 1986): the observed data y satisfies a linear model $Ey = X\beta$, where β is a smooth function and X is ill-conditioned.
- higher-dimensional smoothing problems, including image analysis and disease mapping. The mixed model approach deals with the boundary

estimation automatically without any special handling. This feature is essential when we are dealing with higher-dimensional smoothing with irregular boundaries as in a geographical map, where the application of the kernel method is not straightforward.

While it is possible to develop the theory where x_i's are not assumed to be equispaced (see Section 18.9), the presentation is simpler if we pre-bin the data in terms of the x values prior to analysis. So assume the x values form a regular grid; each x_i can be associated with several y-data, or perhaps none. This is exactly the same as partitioning the data as described before. Rather than just presenting the simple averages, the partition will be processed further. Pre-binning is commonly done in practice to reduce data volume, especially as the y-data at each x value may be summarized further into a few statistics such as the sample size, mean and variance. In many applications, such as time series or image analysis, the data usually come in a regular grid format.

The effect of binning is determined by the bin size: if it is too large then we introduce bias and lose some resolution of the original data, and in the limit as the bin size goes to zero we resolve the original data. In practice we make the bins small enough to preserve the original data (i.e. minimize bias and make local variation dominate), but large enough to be practical since there is a computational price for setting too many bins. We will not develop any theory to say how small is 'small enough', since in practice it is easy to recognize a large local variation, and if we are in doubt we can simply set it smaller. As a guideline, the degrees of freedom of the estimate (described in Section 18.5) should be much smaller than the number of bins.

So, after pre-binning, our problem is as follows. Given the observations (x_i, y_{ij}) for $i = 1, \ldots, n$ and $j = 1, \ldots, n_i$, where x_i's are equispaced, we assume that y_{ij}'s are normal with mean

$$E(y_{ij}) = f(x_i) = f_i$$

and variance σ^2. We want to estimate $f = (f_1, \ldots, f_n)$. Note that some n_i's may be zero. Smoothness or other properties of $f(x)$ will be imposed via some stochastic structure on f_i; this is discussed in the next section.

To put this in the linear mixed model framework, first stack the data y_{ij}'s into a column vector y. Conditional on b, the outcome y is normal with mean

$$E(y|b) = X\beta + Zb$$

and variance $\Sigma = \sigma^2 I_N$, where $N = \sum_i n_i$. The mixed model framework covers the inverse problems (O'Sullivan 1986) by defining Z properly. For our current problem we have

$$f_i = \beta + b_i,$$

and, for identifiability, assume that $E(b_i) = 0$. Here X is simply a column of ones of length N, and Z is an $N \times n$ design matrix of zeros and ones;

the row of Z associated with original data (x_i, y_{ij}) has value one at the i'th location and zero otherwise. The random effects b is simply the mean-corrected version of f. The actual estimate $\widehat{b}$ depends on the smoothness assumption of the function $f(x)$.

It is instructive to see what we get if we simply assume that b is a fixed effect parameter. The data structure is that of a one-way model

$$y_{ij} = \beta + b_i + e_{ij},$$

where, for identifiability, we typically assume $\sum_i n_i b_i = 0$. In this setup the regularity of the grid points $x_1, \ldots, x_n$ is not needed; in fact, we have to drop the x_i values where $n_i = 0$, since for those values $f(x)$ is not estimable. For simplicity, we just relabel the points for which $n_i > 0$ to $x_1, \ldots, x_n$, so we can use the same notation as before. The estimate of b_i is

$$\widehat{b}_i = \overline{y}_i - \overline{y}$$

where $\overline{y} = \sum_i y_{ij}/N$ is the grand mean and $\overline{y}_i$ is simply the average of the data of the i'th bin, and the estimate of f_i (regardless of the constraint) is

$$\widehat{f}_i = \overline{y}_i. \tag{18.1}$$

The variance of this estimate is σ^2/n_i. If n_i is small, which is likely if the bin size is small, the statistical noise in this simple formula would be large, obscuring the underlying patterns in the function $f(x)$. The purpose of smoothing is to reduce such noise and to reveal the patterns of $f(x)$.

18.3 Imposing smoothness using random effects model

The assumption about the random effects b depends on the nature of the function. If $f(x)$ is smooth, then the smoothness can be expressed by assuming that the differences

$$\Delta b_j = b_j - b_{j-1} \tag{18.2}$$

or the second differences

$$\Delta^2 b_j = b_j - 2b_{j-1} + b_{j-2} \tag{18.3}$$

are iid normal with mean zero and variance σ_b^2. In general we can define differencing of order d as $\Delta^d b_j$, and smoothness can be imposed by assuming that it is iid with some distribution.

For example, assuming $d = 1$, we have

$$b_j = b_{j-1} + e_j,$$

where e_j's are an iid sequence; this means b is a first-order random walk on the grid. Figure 18.4 shows some simulated normal random walks of

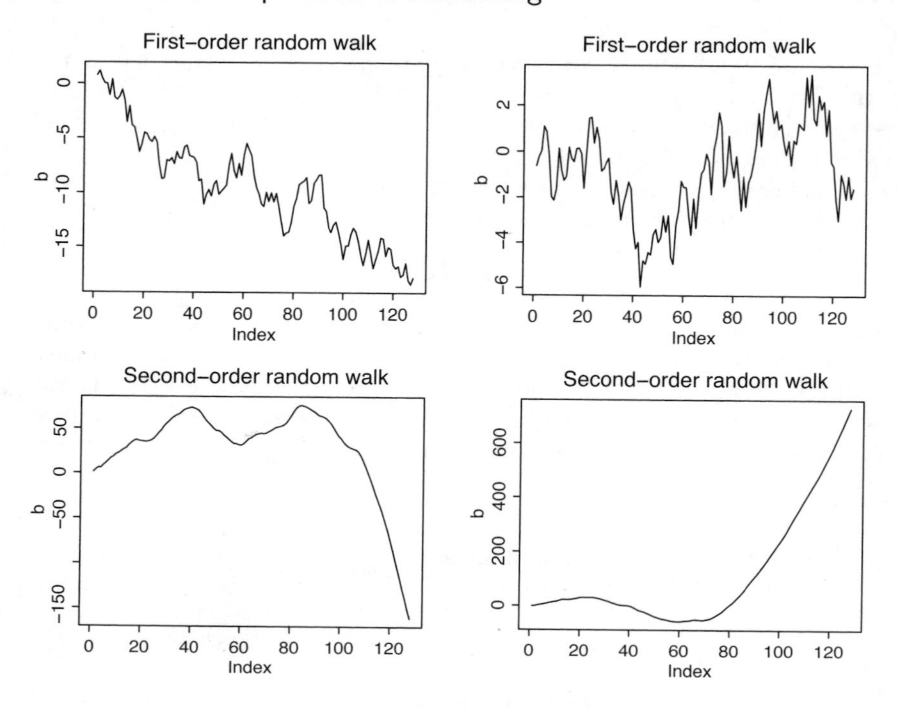

Figure 18.4: *Top row: simulated random walks of the form* $b_j = b_{j-1} + e_j$, *for* $j = 1, \ldots, 128$, *where* $b_1 \equiv 0$ *and* e_j*'s are iid* $N(0,1)$. *Bottom row:* $b_j = 2b_{j-1} - b_{j-2} + e_j$, *for* $j = 1, \ldots, 128$, *where* $b_1 = b_2 \equiv 0$ *and* e_j*'s are the same as before.*

order 1 and 2; it is clear that the trajectory of random walks of order 2 can mimic a smooth function. The first differencing might be used to allow higher local variation in the function.

Redefining the notation Δ for the whole vector

$$\Delta b \equiv \begin{pmatrix} b_2 - b_1 \\ b_3 - b_2 \\ \vdots \\ b_n - b_{n-1} \end{pmatrix}$$

and assuming that Δb is normal with mean zero and variance $\sigma_b^2 I_{n-1}$, we have the prior log-likelihood of b given by

$$\begin{aligned} \log p(b) &= -\frac{n-1}{2}\log\sigma_b^2 - \frac{1}{2\sigma_b^2} b'\Delta'\Delta b \\ &= -\frac{n-1}{2}\log\sigma_b^2 - \frac{1}{2\sigma_b^2} b' R^{-1} b \end{aligned} \qquad (18.4)$$

where

$$R^{-1} \equiv \Delta'\Delta = \begin{pmatrix} 1 & -1 & & & 0 \\ -1 & 2 & -1 & & \\ & \ddots & \ddots & \ddots & \\ & & -1 & 2 & -1 \\ 0 & & & -1 & 1 \end{pmatrix}.$$

Or, equivalently, we have assumed that b is normal with mean zero and inverse covariance matrix

$$D^{-1} \equiv \sigma_b^{-2} R^{-1}.$$

Note that $\log p(b)$ is a conditional log-likelihood given b_1; it is a convenient choice here, since b_1 does not have a stationary distribution. We may also view b as having a singular normal distribution, with D not of full rank; this is a consequence of specifying the distribution for only the set of differences. In contrast to the animal breeding application in Section 17.3, specifying R^{-1} here is more natural than specifying R (which is defined as the generalized inverse of R^{-1}; in practice we never need to compute it). In both applications R has a similar meaning as a scaled covariance matrix.

Using the second-order assumption that

$$\Delta^2 b \equiv \begin{pmatrix} b_3 - 2b_2 + b_1 \\ b_4 - 2b_3 + b_2 \\ \vdots \\ b_n - 2b_{n-1} + b_{n-2} \end{pmatrix}$$

is normal with mean zero and variance $\sigma_b^2 I_{n-2}$, the prior log-likelihood is the same as (18.4) with $(n-2)$ in the first term rather than $(n-1)$, and

$$R^{-1} \equiv (\Delta^2)'\Delta^2 = \begin{pmatrix} 1 & -2 & 1 & & & & 0 \\ -2 & 5 & -4 & 1 & & & \\ 1 & -4 & 6 & -4 & 1 & & \\ & \ddots & \ddots & \ddots & \ddots & \ddots & \\ & & 1 & -4 & 6 & -4 & 1 \\ & & & 1 & -4 & 5 & -2 \\ 0 & & & & 1 & -2 & 1 \end{pmatrix}.$$

18.4 Penalized likelihood approach

Combining the likelihood based on the observation vector y and the random effects b, and dropping terms not involving the mean parameters β and b, we obtain

$$\log L = -\frac{1}{2\sigma^2}\sum_{ij}(y_{ij} - \beta - b_i)^2 - \frac{1}{2\sigma_b^2} b' R^{-1} b.$$

The nonnegative quadratic form $b'R^{-1}b$ is large if b is rough, so it is common to call the term a roughness penalty and the joint likelihood a 'penalized likelihood' (e.g. Green and Silverman 1993). In the normal case, given σ^2 and σ_b^2, the estimates of β and b are the minimizers of a penalized sum of squares

$$\sum_{ij}(y_{ij} - \beta - b_i)^2 + \lambda b' R^{-1} b,$$

where $\lambda = \sigma^2/\sigma_b^2$.

There is a slight difference in the modelling philosophy between the roughness penalty and mixed model approaches. In the former the penalty term is usually chosen for computational convenience, and it is not open to model criticism. The mixed model approach treats the random effects b as parameters that require some model, and finding an appropriate model is part of the overall modelling of the data. It is understood that a model assumption may or may not be appropriate, and it should be checked with the data. There are two model assumptions associated with the penalty term:

- *The order of differencing.* The penalty approach usually assumes second-order differencing. Deciding what order of differencing to use in a particular situation is a similar problem to specifying the order of nonstationarity of an ARIMA model in time series analysis. It can be easily seen that under- or over-differencing can create a problem of error misspecification. For example, suppose the true model is a first-order random walk

$$\Delta b_j = e_j,$$

  where e_j's are an iid sequence. The second-order difference is

$$\Delta^2 b_j = \Delta e_j \equiv a_j,$$

  so a_j's are no longer an iid sequence, but a moving average (MA) of order one. This is a problem since, usually, the standard smoothing model would assume a_j's to be iid.
- *Normality.* A quadratic penalty term is equivalent to assuming normality. This is appropriate if $f(x)$ varies smoothly, but not if $f(x)$ has jump discontinuities as it would not allow a large change in $f(x)$, and it would force the estimate to be smooth. This is where the linear model setup is convenient, since it can be extended easily to deal with this case by using nonnormal mixed models.

18.5 Estimate of f given σ^2 and σ_b^2

The joint log-likelihood based on the observation vector y and the random effects b is

$$\begin{aligned}\log L &= -\frac{1}{2}\log|\Sigma| - \frac{1}{2}(y - X\beta - Zb)'\Sigma^{-1}(y - X\beta - Zb) \\ &\quad -\frac{n-d}{2}\log\sigma_b^2 - \frac{1}{2\sigma_b^2}b'R^{-1}b\end{aligned}$$

where d is the degree of differencing. Using the assumption $\Sigma = \sigma^2 I_N$, given σ^2 and σ_b^2, the estimates of β and b according to the general formula (17.12) are the solution of

$$\begin{pmatrix} X'X & X'Z \\ Z'X & Z'Z + \lambda R^{-1} \end{pmatrix} \begin{pmatrix} \beta \\ b \end{pmatrix} = \begin{pmatrix} X'y \\ Z'y \end{pmatrix}, \tag{18.5}$$

where $\lambda = \sigma^2/\sigma_b^2$. We can show that the combined matrix on the left-hand side is singular (Exercise 18.1); it is a consequence of specifying a model only on the set of differences of b. This implies we can set the level parameter β at an arbitrary value, but by analogy with the fixed effects model it is meaningful to set

$$\widehat{\beta} = \overline{y} = \sum_{ij} y_{ij}/N.$$

The estimate of b is the solution of

$$\{Z'Z + \lambda R^{-1}\}b = Z'(y - X\widehat{\beta}). \tag{18.6}$$

From the definition of Z in this problem, we can simplify (18.6) to

$$(W + \lambda R^{-1})b = W(\overline{y}^v - \overline{y}) \tag{18.7}$$

where $W = Z'Z = \text{diag}[n_i]$ is a diagonal matrix with n_i as the diagonal element, and

$$\overline{y}^v \equiv \begin{pmatrix} \overline{y}_1 \\ \vdots \\ \overline{y}_n \end{pmatrix}$$

is the 'raw' mean vector. If $n_i = 0$ the weight on $\overline{y}_i$ (which is not available) is zero, so it does not contribute in the computation; we can simply set $\overline{y}_i$ to zero. (The expression '$y - \overline{y}$' means that the scalar $\overline{y}$ is subtracted from every element of the vector y; this is a common syntax in array-processing computer languages.)

We can also write

$$\begin{aligned} \widehat{f} &= \overline{y} + (W + \lambda R^{-1})^{-1} W(\overline{y}^v - \overline{y}) \\ &= (W + \lambda R^{-1})^{-1} W\overline{y}^v + (W + \lambda R^{-1})^{-1}\{(W + \lambda R^{-1})\mathbf{1}_n\overline{y} - W\mathbf{1}_n\overline{y}\} \\ &= (W + \lambda R^{-1})^{-1} W\overline{y}^v + (W + \lambda R^{-1})^{-1}\lambda R^{-1}\mathbf{1}_n\overline{y} \\ &= (W + \lambda R^{-1})^{-1} W\overline{y}^v \end{aligned}$$

since $R^{-1}\mathbf{1}_n = 0$, where $\mathbf{1}_n$ is a vector of ones of length n.

For the purpose of interpretation, we define a *smoother matrix* S_λ as

$$S_\lambda = (W + \lambda R^{-1})^{-1} W \tag{18.8}$$

so that

$$\widehat{f} = S_\lambda \overline{y}^v.$$

We can see that

$$\begin{aligned} S_\lambda \mathbf{1}_n &= (W + \lambda R^{-1})^{-1} W \mathbf{1}_n \\ &= \mathbf{1}_n - (W + \lambda R^{-1})^{-1} \lambda R^{-1} \mathbf{1}_n \\ &= \mathbf{1}_n, \end{aligned}$$

meaning each row of the matrix adds up to one. So we can interpret each $\widehat{f}_i$ as a weighted average of the raw means $\overline{y}_i$'s, where the weights are determined by sample size n_i, the smoothing parameter λ and the choice of R^{-1}. If the smoothing parameter $\lambda = 0$ then there is no smoothing, and we are back to the previous estimate (18.1) based on assuming that b is fixed.

If the data are naturally in a regular grid format such that $n_i \equiv 1$ for all i, or we have pairs (x_i, y_i), for $i = 1, \ldots, n$, then $W = I_n$ and we get

$$\widehat{b} = (I_n + \lambda R^{-1})^{-1} (y - \overline{y}),$$

where $\overline{y} = \sum_i y_i / n$, and

$$\begin{aligned} \widehat{f} &= \overline{y} + (I_n + \lambda R^{-1})^{-1} (y - \overline{y}) \\ &= (I_n + \lambda R^{-1})^{-1} y. \end{aligned}$$

A particular $\widehat{f}_i$ is a weighted average of y_i's of the form

$$\widehat{f}_i = \sum_j k_{ij} y_j$$

where $\sum_j k_{ij} = 1$ for all i. Figure 18.5 shows the shape of the weights k_{ij}'s for $i = 1$, 10 and 20, and for $d = 1$ and 2.

A more 'physical' interpretation of the amount of smoothing can be given in terms of the model *degrees of freedom* or the number of parameters associated with the function estimate. This number of parameters is also useful to make a like-with-like comparison between different smoothers. By analogy with the parametric regression model the degrees of freedom are defined as

$$\text{df} = \text{trace } S_\lambda. \tag{18.9}$$

This is a measure of model complexity: as λ gets larger the estimate becomes more smooth, the degrees of freedom drop, and the estimate gets closer to a parametric estimate. If $\lambda \to 0$ we get the number of nonempty bins as the degrees of freedom.

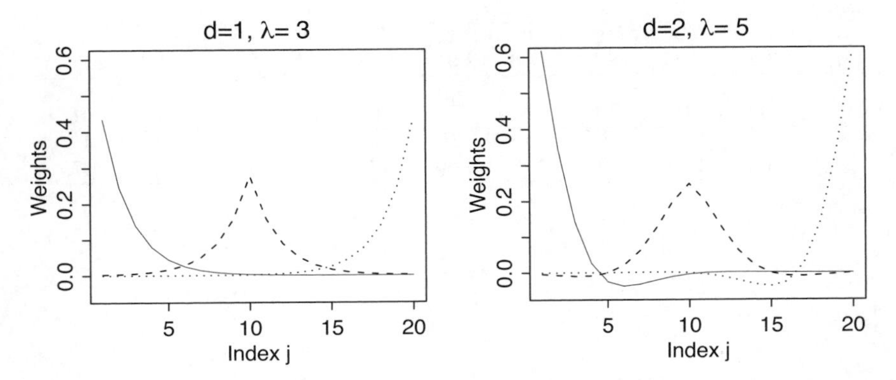

Figure 18.5: *The shape of the weights k_{ij}'s as a function of index j, for $i = 1$ (left edge, solid line), 10 (middle location, dashed line) and 20 (right edge, dotted line). The smoothing parameter λ is chosen so both smoothers for $d = 1$ and $d = 2$ have around 6 degrees of freedom.*

In principle we can compute $\widehat{b}$ by simply solving the linear equation (18.7), but in practice n may be large, so a simple-minded inversion of the matrix is not efficient. In all large-scale inversion problems we have to exploit the particular structure of the matrix:

- Note that R^{-1} is a band matrix (with one or two nonzero values on each side of the diagonal). Since W is a diagonal matrix, the matrix $(W+\lambda R^{-1})$ is also a band matrix. Finding a very fast solution for such a matrix is a well-solved problem in numerical analysis; see Dongarra *et al.* (1979, Chapter 2) for standard computer programs available in `Linpack` (a collection of programs for linear/matrix computations, such as finding solutions of linear equations).
- The Gauss–Seidel algorithm (Press *et al.* 1992, page 855) works well for this problem.
- If the weights n_i's are equal, so that W is a constant times the identity matrix, then we can use the Fourier transform method (Press *et al.* 1992, Chapters 12 and 13).

(The details of these algorithms are beyond the scope of this text, but serious students of statistics should at some point learn all of these methods.)

Example 18.2: We now apply the methodology to the SO_2 data given in Table 18.1 where $n = 20$ and $N = 41$. The bin statistics are given in Table 18.2, where 'NA' means 'not available'. Figure 18.6 shows the nonparametric smooth of $\overline{y}^v$ using smoothing parameters $\lambda = 5$ and $\lambda = 0.5$. □

18.6 Estimating the smoothing parameter

Estimating the smoothing parameter $\lambda = \sigma^2/\sigma_b^2$ is equivalent to estimating the variance components. We have described before the general problem

Bin i	1	2	3	4	5	6	7	8	9	10
n_i	1	2	0	1	3	3	0	5	1	3
$\overline{y}_i$	3.43	3.11	NA	3.58	2.94	2.68	NA	2.54	3.26	2.86

Bin i	11	12	13	14	15	16	17	18	19	20
n_i	8	4	3	3	2	0	0	1	0	1
$\overline{y}_i$	3.45	2.96	2.94	3.23	3.86	NA	NA	4.23	NA	4.7

Table 18.2: *Bin statistics for the SO_2 data*

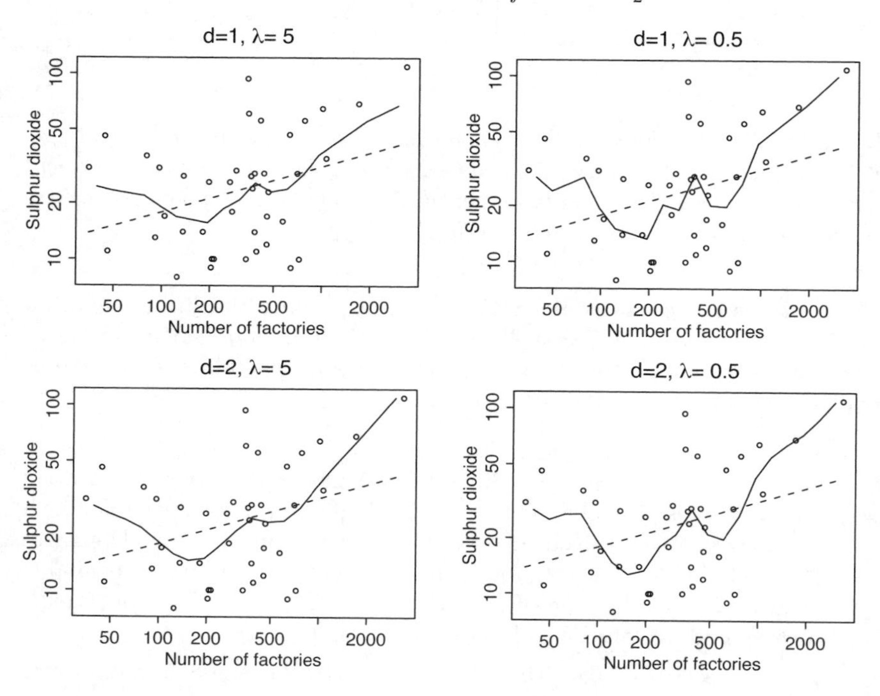

Figure 18.6: *Nonparametric smoothing of the SO_2 level against the industrial activity using the mixed model approach. The top row is based on the first-difference assumption and the bottom row on the second-difference. The dashed line on each plot is the linear fit.*

of estimating the variance components $\theta = (\sigma^2, \sigma_b^2)$ using the profile log-likelihood

$$\log L(\theta) = -\frac{1}{2}\log|V| - \frac{1}{2}(y - X\widehat{\beta})V^{-1}(y - X\widehat{\beta})$$

where $\widehat{\beta}$ is computed according to (17.7), and θ enters the function through

$$V = \sigma^2 I_N + \sigma_b^2 ZRZ',$$

or using the equivalent form described in Section 17.5. In this case we want to maximize

$$\begin{aligned} Q &= -\frac{N}{2}\log\sigma^2 - \frac{1}{2\sigma^2}(y - X\beta - Zb)'(y - X\beta - Zb) \\ &\quad -\frac{n-d}{2}\log\sigma_b^2 - \frac{1}{2\sigma_b^2}b'R^{-1}b \\ &\quad -\frac{1}{2}\log|\sigma^{-2}W + \sigma_b^{-2}R^{-1}|. \end{aligned}$$

with respect to all the parameters.

To apply the iterative algorithm in Section 17.5: start with an estimate of σ^2 and σ_b^2 (note that σ^2 is the error variance, so we can get a good starting value for it), then:

1. Compute $\widehat{\beta} = \overline{y}$, and $\widehat{b}$ according to (18.7), and the error $e = y - \widehat{\beta} - Z\widehat{b}$.
2. Compute the degrees of freedom of the model
$$\text{df} = \text{trace}\{(W + \lambda R^{-1})^{-1}W\},$$
and update θ using
$$\begin{aligned} \sigma^2 &= \frac{e'e}{N - \text{df}} \\ \sigma_b^2 &= \frac{1}{n-d}[b'R^{-1}b + \sigma^2\ \text{trace}\{(W + \lambda R^{-1})^{-1}R^{-1}\}], \end{aligned}$$
where all unknown parameters on the right-hand side are evaluated at the last available values during the iteration, and update $\lambda = \sigma^2/\sigma_b^2$. Recall that d is the degree of differencing used for the random effects.
3. Iterate 1 and 2 until convergence.

Example 18.3: To apply the algorithm to the SO_2 data we start with $\sigma^2 = 0.35$ (e.g. use a coarse partition on the data and obtain the error variance) and $\lambda = 5$ (or $\sigma_b^2 = 0.35/5$). For order of differencing $d = 1$ the algorithm converges to

$$\begin{aligned} \widehat{\sigma}^2 &= 0.3679 \\ \widehat{\sigma}_b^2 &= 0.0595 \end{aligned}$$

with the corresponding smoothing parameter $\widehat{\lambda} = 6.2$ and model degrees of freedom df $= 5.35$. The resulting estimate $\widehat{f}$ is plotted in Figure 18.7. Also shown is the quadratic fit of the data, which has 3 degrees of freedom for the model.

For $d = 2$, using the same starting values as above, the algorithm converges to

$$\begin{aligned} \widehat{\sigma}^2 &= 0.3775 \\ \widehat{\sigma}_b^2 &= 0.0038 \end{aligned}$$

with the corresponding smoothing parameter $\widehat{\lambda} = 99.2$ and model degrees of freedom df $= 3.56$, very close to the quadratic fit. The estimate using $d = 2$ is

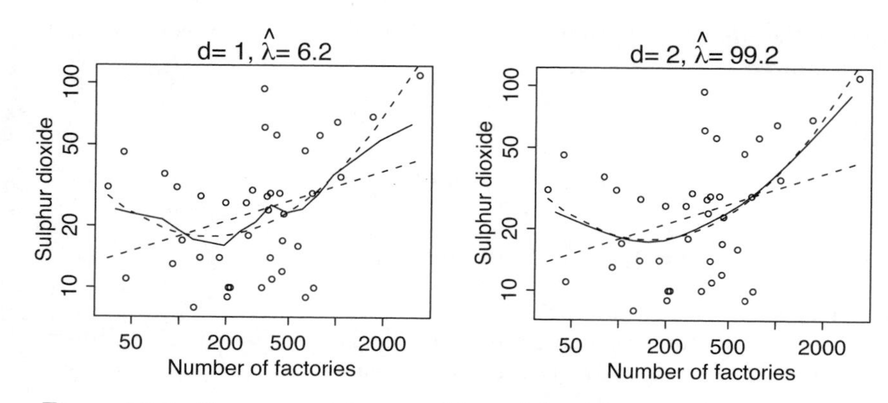

Figure 18.7: *Nonparametric smoothing of the SO_2 level against the industrial activity using the estimated smoothing parameter, with corresponding degrees of freedom df = 5.35 for d = 1, and df = 3.56 for d = 2. The dashed lines are linear and quadratic fits of the data.*

more 'pleasing', while using $d = 1$ the estimate appears to show some spurious local patterns. A formal comparison between the fits can be done using the AIC; for the current problem

$$\text{AIC} = N \log \widehat{\sigma}^2 + 2\ \text{df}.$$

We obtain AIC=−30.3 for $d = 1$, and a preferable AIC=−32.8 for $d = 2$. □

Generalized cross-validation

The generalized cross-validation (GCV) score was introduced by Craven and Wahba (1979) for estimation of the smoothing parameter λ in nonparametric regression. In our setting the score is of the form

$$\text{GCV}(\lambda) = \frac{e'e}{(N - \text{df})^2},$$

where the error $e = y - \widehat{\beta} - Z\widehat{b}$ and degrees of freedom df are computed at fixed λ. The estimate $\widehat{\lambda}$ is chosen as the minimizer of the GCV. The justification of the GCV (Wahba 1990, Chapter 4) is beyond the scope of our text.

In some sense GCV(λ) is a profile objective function for λ, which makes the estimation of λ a simple one-dimensional problem. Given $\widehat{\lambda}$ we can estimate the error variance as

$$\widehat{\sigma}^2 = \frac{e'e}{(N - \text{df})} \tag{18.10}$$

where e and df are computed at $\widehat{\lambda}$.

Figure 18.8 shows the GCV as a function of λ for the SO_2 data. The minimum is achieved at $\widehat{\lambda} \approx 135$, with a corresponding degrees of freedom df = 3.35, very close to the MLE given earlier.

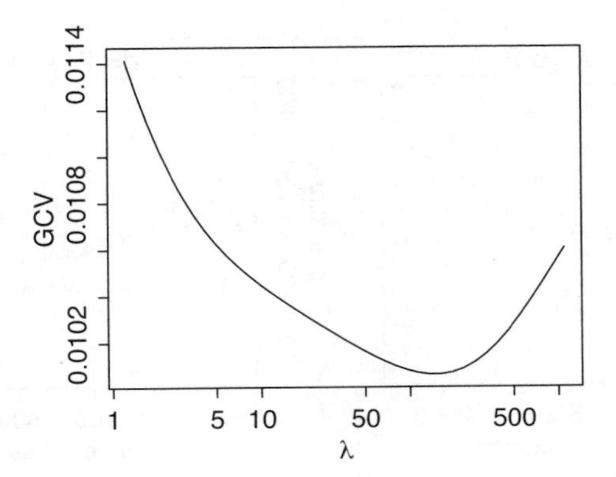

Figure 18.8: *The generalized cross-validation (GCV) score as a function of the smoothing parameter λ for the SO_2 data. The minimum is at $\widehat{\lambda} \approx 135$.*

18.7 Prediction intervals

Assuming θ is known the Fisher information for b based on the joint likelihood is given by (17.11). For the current problem

$$I(\widehat{b}) = (\sigma^{-2}Z'Z + \sigma_b^{-2}R^{-1}) = \sigma^{-2}(W + \lambda R^{-1}),$$

and the standard errors of the estimates are the square roots of the diagonal elements of

$$I(\widehat{b})^{-1} = \sigma^2(W + \lambda R^{-1})^{-1}.$$

In practice the unknown variance parameters are evaluated at the estimated values. Since $\widehat{\beta}$ is not estimated (recall that it is constrained to the mean value for identifiability reasons), $\text{se}(\widehat{f}_i) = \text{se}(\widehat{b}_i)$ and we can contruct the 95% prediction interval

$$\widehat{f}_i \pm 1.96\ \text{se}(\widehat{f}_i)$$

for each i.

Figure 18.9 shows the prediction band for $f(x)$ in the SO_2 data. We use the previously estimated values for σ^2 and λ. The upper limit is formed by joining the upper points of the prediction intervals, and similarly with the lower limit.

18.8 Partial linear models

Suppose we observe independent data (x_i, u_i, y_{ij}), for $i = 1, \ldots, n$, where x_i is a p-vector of predictors and u_i is a scalar predictor. A general model of the form

$$E(y_{ij}) = x_i'\beta + f(u_i)$$

and $\text{var}(y_{ij}) = \sigma^2$ is called a partial linear model (e.g. Speckman 1988). For example, this is used as a generalization of analysis of covariance, where β

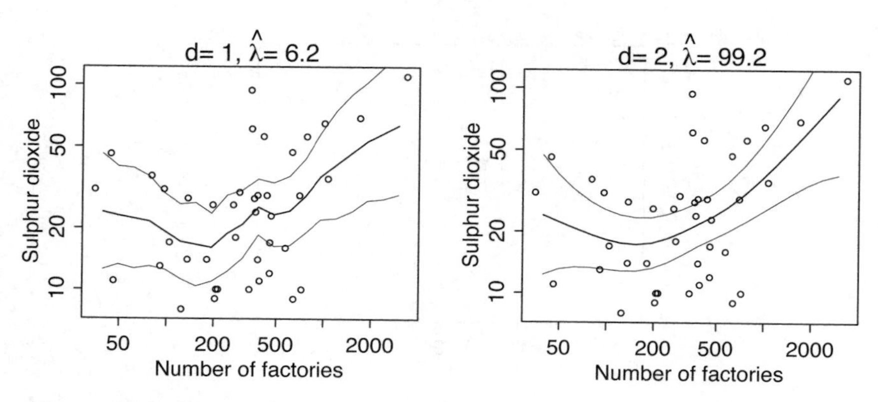

Figure 18.9: *The prediction band for the nonparametric function $f(x)$ based on pointwise 95% prediction intervals.*

is a measure of treatment effect, and $f(u_i)$ measures a nonlinear effect of the covariate u_i.

Assuming that u_i's are equispaced and following the previous development, we can write the partial linear model in the form

$$E(y|b) = X\beta + Zb,$$

where X is an $N \times p$ design matrix, β is a fixed effects parameter, and b is a random effects parameter satisfying a smoothness condition. The model is a linear mixed model, and the estimates of β and b are the solution of

$$\begin{aligned} (X'V^{-1}X)\beta &= X'V^{-1}y \\ (Z'\Sigma^{-1}Z + D^{-1})b &= Z'\Sigma^{-1}y, \end{aligned}$$

where $V = \Sigma + ZDZ'$. An iterative backfitting procedure may be used to avoid the computation of V^{-1}.

18.9 Smoothing nonequispaced data*

Occasionally we face an application where x_i's are not equispaced, and we are unwilling to prebin the data into equispaced intervals. The previous methodology still applies with little modification, but with more computations. The problem has a close connection with the general interpolation method in numerical analysis.

It is convenient to introduce the 'design points' $d_1, \ldots, d_p$, which do not have to coincide with the data points $x_1, \ldots, x_n$. These design points can be chosen for computational convenience, as with regular grids, or for better approximation, as with the so-called Chebyshev points for the Lagrange polynomial in Example 18.5. We will consider a class of functions defined by

$$f(x) = \sum_{j=1}^{p} b_j K_j(x),$$

where $K_j(x)$ is a known function of x and the design points, and b_j's are the parameter values determined by $f(x)$ at the design points. So, in effect, $f(x)$ is a linear model with $K_j(x)$'s as the predictor variables. In function estimation theory $K_j(x)$ is called *the basis function.* The nonparametric nature of $f(x)$ is achieved by allowing p to be large or, equivalently, by employing a rich set of basis functions.

Example 18.4: The simplest example is the power polynomial

$$f(x) = \sum_{j=1}^{p} b_j x^{j-1},$$

where we have used the basis function

$$K_i(x) = x^{j-1}.$$

Finding b_j's is exactly the problem of estimating the regression coefficients in a polynomial regression model; the design points do not play any role in this case. Extending the power polynomial to a high degree is inadvisable because of numerical problems. □

Example 18.5: Using the Lagrange polynomial

$$\begin{aligned} f(x) &= \sum_{j=1}^{p} f(d_j) \prod_{k \neq j} \frac{x - d_k}{d_j - d_k} \\ &= \sum_{j=1}^{p} b_j K_j(x) \end{aligned}$$

where $b_j \equiv f(d_j)$ and

$$K_j(x) \equiv \prod_{k \neq j} \frac{x - d_k}{d_j - d_k}.$$

Each $K_j(x)$ is a polynomial of degree $(p-1)$. The main advantage of the Lagrange polynomial is that the coefficients are trivially available. However, the choice of the design points can make a great difference; in particular, the uniform design is inferior to the Chebyshev design:

$$d_i = \frac{a+b}{2} + \frac{a-b}{2} \cos \frac{(2i-1)\pi}{2p}$$

for p points between a and b. □

Example 18.6: The *B-spline* basis (deBoor 1978) is widely used because of its local properties: $f(x)$ is determined only by values at neighbouring design points; in contrast, the polynomial schemes are global. The j'th B-spline basis function of degree m is a piecewise polynomial of degree m in the interval (d_j, d_{j+m+1}), and zero otherwise. The B-spline of 0 degree is simply the step function with jumps at points $(d_i, f(d_i))$. The B-spline of degree 1 is the polygon that connects $(d_i, f(d_i))$; higher-order splines are determined by assuming a smoothness/continuity condition on the derivatives. In practice it is common to use the cubic B-spline to approximate smooth functions (deBoor 1978; O'Sullivan

1987); this third-order spline has a continuous second derivative. The interpolating B-spline is

$$f(x) = \sum_{j=1}^{p-k-1} b_j K_j(x)$$

where $K_j(x)$'s are computed based on the design points or 'knots' $d_1, \ldots, d_p$. See deBoor (1978) for the cubic B-spline formulae. □

Methodology

Given observed data $(x_1, y_1), \ldots, (x_n, y_n)$, where $Ey_i = f(x_i)$, we can write a familiar regression model

$$\begin{aligned} y &= \sum_{j=1}^{p} b_j K_j(x) + e \\ &\equiv Zb + e \end{aligned}$$

where the elements of Z are

$$z_{ij} = K_j(x_i)$$

for some choice of basis function $K_j(x)$.

Since $f(x)$ is available as a continuous function consider a smoothness penalty of the form

$$\lambda \int |f^{(d)}(x)|^2 dx,$$

where $f^{(d)}(x)$ is the d'th derivative of $f(x)$. This is a continuous version of the penalty we use in Section 18.3. In view of $f(x) = \sum_{j=1}^{p} b_j K_j(x)$ the penalty can be simplified to a familiar form

$$\lambda b' P b,$$

where the (i, j) element of matrix P is

$$\int K_i^{(d)}(x) K_j^{(d)}(x) dx.$$

Hence the previous formulae apply, for example

$$\widehat{b} = (Z'Z + \lambda P)^{-1} Z'y.$$

18.10 Non-Gaussian smoothing

Using the GLMM theory in Section 17.8 we can extend nonparametric smoothing to non-Gaussian data.

Example 18.7: Suppose we want to describe surgical mortality rate p_i as a function of patient's age x_i. If we do not believe a linear model, or we are at an exploratory stage in the data analysis, we may consider a model where the outcome y_i is Bernoulli with probability p_i and

$$\text{logit } p_i = f(x_i),$$

for some function f. In this example there could be a temptation to fit a linear model, since it allows us to state something simple about the relationship between patient's age and surgical mortality. There are, however, many applications where such a statement may not be needed. For example, suppose we want to estimate the annual rainfall pattern in a region, and daily rainfall data are available for a 5-year period. Let y_i be the number of rainy days for the i'th day of the year; we can assume that y_i is binomial$(5, p_i)$, and

$$\text{logit } p_i = f(i),$$

where f is some smooth function. Rather than for analysing a relationship, the purpose of estimating $f(x)$ in this application is more for a description or a summary. □

As before, assume that we can arrange or pre-bin the data into regular grids, so our problem is as follows. Given the observations (x_i, y_{ij}) for $i = 1, \ldots, n$ and $j = 1, \ldots, n_i$, where x_i's are equispaced, y_{ij}'s are independent outcomes from the exponential family model (Section 6.5) of the form

$$\log p(y_{ij}) = \frac{y_{ij}\theta_i - A(\theta_i)}{\phi} + c(y_i, \phi).$$

Let $\mu_i \equiv Ey_{ij}$, and assume that for a known link function $h(\cdot)$ we have

$$h(\mu_i) = f(x_i) \equiv f_i,$$

for some unknown smooth function f.

To put this in the GLMM framework first vectorize the data y_{ij}'s into an N-vector y. Conditional on b, the outcome y has mean μ and

$$h(\mu) = X\beta + Zb, \tag{18.11}$$

and b satisfies some smoothness condition stated in Section 18.3. For the simple setup above

$$h(\mu) = f = \beta + b,$$

so X is a column of ones of length N, and Z is an $N \times n$ design matrix of zeros and ones; the row of Z associated with original data (x_i, y_{ij}) has value one at the i'th location and zero otherwise.

We will treat the general model (18.11) so that the inverse problems are covered, and all of our previous theories for smoothing and GLMM apply. The joint likelihood of β, θ and b is

$$\log L(\beta, \theta, b) = \log p(y|b) + \log p(b)$$

where $p(y|b)$ is in the exponential family given above, and $p(b)$ is the density of b. The parameter θ includes any other parameter in the model, usually the variance or dispersion parameters.

Estimating f given θ

We proceed as in Section 17.10, and some results are repeated here for completeness. Given a fixed value of θ we use a quadratic approximation of the likelihood to derive the IWLS algorithm; see Section 6.7. Starting with initial values for β^0 and b^0, the exponential family log-likelihood can be approximated by

$$-\frac{1}{2}\log|\Sigma| - \frac{1}{2}(Y - X\beta - Zb)'\Sigma^{-1}(Y - X\beta - Zb), \tag{18.12}$$

where Y is a working vector with elements

$$Y_i = x_i'\beta^0 + z_i'b^0 + \frac{\partial h}{\partial \mu_i}(y_i - \mu_i^0),$$

and Σ is a diagonal matrix of the variance of the working vector with diagonal elements

$$\Sigma_{ii} = \left(\frac{\partial h}{\partial \mu_i}\right)^2 \phi v_i(\mu_i^0),$$

where $\phi v_i(\mu_i^0)$ is the conditional variance of y_i given b. The derivative $\partial h/\partial \mu_i$ is evaluated at the current values of β and b. Alternatively we might use the term 'weight' $w_i = \Sigma_{ii}^{-1}$, and weight matrix $W = \Sigma^{-1}$.

If the random effects parameter b is assumed normal with mean zero and variance $\sigma_b^2 R$, where R is as described in Section 18.3, we have the familiar mixed model equation

$$\begin{pmatrix} X'\Sigma^{-1}X & X'\Sigma^{-1}Z \\ Z'\Sigma^{-1}X & Z'\Sigma^{-1}Z + \sigma_b^{-2}R^{-1} \end{pmatrix} \begin{pmatrix} \beta \\ b \end{pmatrix} = \begin{pmatrix} X'\Sigma^{-1}Y \\ Z'\Sigma^{-1}Y \end{pmatrix} \tag{18.13}$$

to update β and b. Or, using the iterative backfitting algorithm, we can solve

$$(Z'\Sigma^{-1}Z + \sigma_b^{-2}R^{-1})b = Z'\Sigma^{-1}(Y - Z\beta)$$

to update b, and similarly for β. By analogy with the standard regression model the quantity

$$\text{df} = \text{trace}\{(Z'\Sigma^{-1}Z + \sigma_b^{-2}R^{-1})^{-1}Z'\Sigma^{-1}Z\}$$

is called the degrees of freedom associated with b. The use of nonnormal random effects is described in Section 18.12.

Example 18.8: We will now analyse the surgical mortality data in Table 6.2, grouped into 20 bins given in Table 18.3. Let $y_i = \sum_j y_{ij}$ be the number of deaths in the i'th bin, and assume that y_i is binomial(n_i, p_i) with dispersion parameter $\phi = 1$. We want to estimate f such that

$$\text{logit } p_i = f_i = \beta + b_i.$$

To use the above methodology, start with β^0 and $b^0 = 0$ and compute the working vector Y with element

Bin i	1	2	3	4	5	6	7	8	9	10
Mid-x_i	50.5	51.6	52.6	53.7	54.7	55.8	56.8	57.9	58.9	60.0
n_i	3	0	1	3	2	2	4	1	1	2
$\sum_j y_{ij}$	0	NA	0	0	0	0	2	0	0	1

Bin i	11	12	13	14	15	16	17	18	19	20
Mid-x_i	61.0	62.1	63.1	64.2	65.2	66.3	67.3	68.4	69.4	70.5
n_i	4	4	3	2	1	0	2	2	1	2
$\sum_j y_{ij}$	2	3	1	1	0	NA	2	1	0	1

Table 18.3: *Bin statistics for the surgical mortality data in Table 6.2. 'NA' means 'not available'.*

$$Y_i = \beta^0 + b_i^0 + \frac{y_i - n_i p_i^0}{n_i p_i^0 (1 - p_i^0)}$$

and weight $w_i = \Sigma_{ii}^{-1} = n_i p_i^0 (1 - p_i^0)$. The matrix X is a column of ones and Z is an identity matrix I_{20}. We then compute the following updates:

$$\begin{aligned} \beta &= \frac{\sum_i w_i (Y - b)}{\sum_i w_i} \\ b &= (W + \sigma_b^{-2} R^{-1})^{-1} W (Y - \beta), \end{aligned}$$

where $W = \text{diag}[w_i]$. The iteration continues after recomputing Y and Σ. So the computation in non-Gaussian smoothing involves an iteration of the Gaussian formula. The model degrees of freedom associated with a choice of σ_b^2 are

$$\text{df} = \text{trace}\{(W + \sigma_b^{-2} R^{-1})^{-1} W\}.$$

Figure 18.10 shows the nonparametric smooth of p_i using smoothing parameter $\sigma_b^2 = 0.2$ and 2, with the corresponding 4.3 and 6.7 degrees of freedom. The matrix R used is associated with $d = 2$; see Section 18.3. For comparison the linear logistic regression fit is also shown. The result indicates some nonlinearity in the relationship between age and mortality, where the effect of age appears to flatten after age 62. □

Estimating the smoothing parameter

The discussion and method in Section 17.10 for estimating the variance components in GLMM apply here. In general we can choose θ to maximize

$$\log L(\theta) = \log L(\widehat{\beta}, \theta, \widehat{b}) - \frac{1}{2} \log |Z' \Sigma^{-1} Z + D^{-1}|, \qquad (18.14)$$

where θ enters through Σ and D^{-1}. This approximate profile likelihood can be maximized using any derivative-free optimization routine.

In the important special case of non-Gaussian outcomes involving a single function estimation, we typically assume $\phi = 1$, so $\theta = \sigma_b^2$. Since with smooth functions we do not expect σ_b^2 to be too large, we can use the following algorithm. Start with an initial estimate of σ_b^2, then:

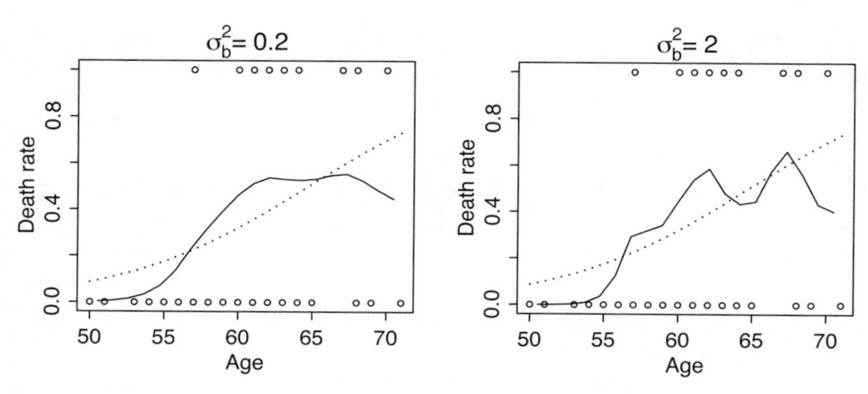

Figure 18.10: *Nonparametric smooth (solid) of mortality as a function of age compared with the linear logistic fit (dotted). The circles are the data points.*

1. Compute $\widehat{\beta}$ and $\widehat{b}$ given σ_b^2 according to the method in the previous section.
2. Fixing β and b at the values $\widehat{\beta}$ and $\widehat{b}$, update σ_b^2 using

$$\sigma_b^2 = \frac{1}{n-d}[b'R^{-1}b + \text{trace}\{(Z'\Sigma^{-1}Z + \sigma_b^{-2}R^{-1})^{-1}R^{-1}\}], \quad (18.15)$$

   where n is the length of b and d is the degree of differencing used to define R (so $n-d$ is the rank of R).
3. Iterate between 1 and 2 until convergence.

This procedure applies immediately to the mortality data example. Figure 18.11(a) shows the mortality rate as a function of age using the estimated $\widehat{\sigma}_b^2 = 0.017$, with corresponding df $= 2.9$.

Prediction intervals

From Section 18.7, assuming the fixed parameters are known at the estimated values, the Fisher information for b is

$$I(\widehat{b}) = (Z'\Sigma^{-1}Z + \sigma_b^{-2}R^{-1}).$$

We can obtain approximate prediction intervals for p_i as follows. First obtain the prediction interval for f_i in the logit scale

$$\widehat{f}_i \pm 1.96 \text{ se}(\widehat{b}_i),$$

where $\text{se}(\widehat{b}_i)$ is computed from $I(\widehat{b})$ above, then transform the end-points of the intervals to the original probability scale. A prediction band is obtained by joining the endpoints of the intervals. Figure 18.11(b) shows the prediction band for the mortality rate using the estimated $\widehat{\sigma}_b^2 = 0.017$.

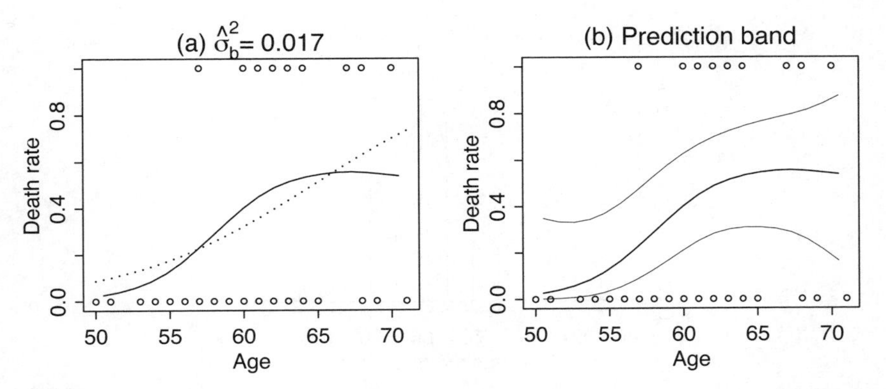

Figure 18.11: *(a) Nonparametric smooth (solid) of mortality rate using the estimated* $\widehat{\sigma}_b^2 = 0.017$*, compared with the linear logistic fit (dotted). (b) Prediction band for the nonparametric smooth.*

18.11 Nonparametric density estimation

The simplest probability density estimate is the histogram, a nonparametric estimate based on simple partitioning of the data. When there is enough data the histogram is useful to convey shapes of distributions. The weakness of the histogram is that either it has too much local variability (if the bins are too small), or it has low resolution (if the bins are too large).

The kernel density estimate is commonly used when a histogram is considered too crude. Given data $x_1, \ldots, x_N$, and kernel $K(\cdot)$, the estimate of the density $f(\cdot)$ at a particular point x is

$$f(x) = \frac{1}{N\sigma} \sum_i K\left(\frac{x_i - x}{\sigma}\right).$$

$K(\cdot)$ is typically a standard density such as the normal density function; the scale parameter σ, proportional to the 'bandwidth' of the kernel, controls the amount of smoothing. There is a large literature on the optimal choice of the bandwidth; see Jones *et al.* (1996) for a review.

Example 18.9: Table 12.1 shows the waiting time for $N = 299$ consecutive eruptions of the Old Faithful geyser in the Yellowstone National Park. The density estimate in Figure 18.12, computed using the Gaussian kernel with $\sigma = 2.2$, shows distinct bimodality, a significant feature that indicates nonlinear dynamics in the process that generates it. The choice $\sigma = 2.2$ is an optimal choice using the unbiased cross-validation score from Scott and Terrell (1987). □

There are several weaknesses of the kernel density estimate: (i) it is very inefficient computationally for large datasets, (ii) finding the optimal bandwidth (or σ in the above formula) requires special techniques, and (iii) there is an extra bias on the boundaries. These are overcome by the mixed model approach.

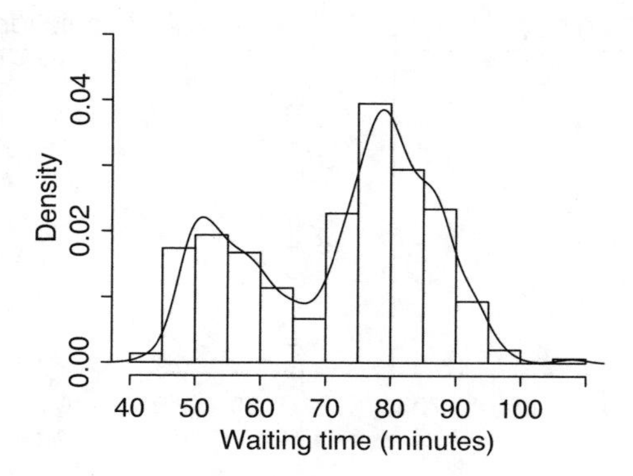

Figure 18.12: *The histogram and kernel density estimate (solid line) of the geyser data indicate strong bimodality.*

First we pre-bin the data, so we have equispaced midpoints $x_1, \ldots, x_n$ with corresponding counts $y_1, \ldots, y_n$; there is a total of $N = \sum_i y_i$ data points. The interval δ between points is assumed small enough such that the probability of an outcome in the i'th interval is $f_i\delta$; for convenience we set $\delta \equiv 1$. The likelihood of $f = (f_1, \ldots, f_n)$ is

$$\log L(f) = \sum_i y_i \log f_i,$$

where f satisfies $f_i \geq 0$ and $\sum_i f_i = 1$. Using the Lagrange multiplier technique we want an estimate f that maximizes

$$Q = \sum_i y_i \log f_i + \psi(\sum_i f_i - 1).$$

Taking the derivatives with respect to f_i we obtain

$$\frac{\partial Q}{\partial f_i} = y_i/f_i + \psi.$$

Setting $\frac{\partial Q}{\partial f_i} = 0$, so $\sum f_i(\partial Q/\partial f_i) = 0$, we find $\psi = -N$, hence f is the maximizer of

$$Q = \sum_i y_i \log f_i - N(\sum_i f_i - 1).$$

Defining $\lambda_i \equiv N f_i$, the expected number of points in the i'th interval, the estimate of $\lambda = (\lambda_1, \ldots, \lambda_N)$ is the maximizer of

$$\sum_i y_i \log \lambda_i - \sum_i \lambda_i,$$

exactly the log-likelihood from Poisson data. We no longer have to worry about the sum-to-one constraint. So, computationally, nonparametric density estimation is equivalent to nonparametric smoothing of Poisson data, and the general method in the previous section applies immediately.

To be specific, we estimate λ_i from, for example, the log-linear model

$$\log \lambda_i = \beta + b_i,$$

where b_i's are normal with mean zero and variance $\sigma_b^2 R$; the matrix R is described in Section 18.3. The density estimate $\widehat{f}_i$ is $\widehat{\lambda}_i/N$.

Computing the estimate

Given the smoothing parameter σ_b^2, start with β^0 and b^0, and compute the working vector Y with element

$$Y_i = \beta^0 + b_i^0 + \frac{y_i - \lambda_i^0}{\lambda_i^0}$$

and weight $w_i = \Sigma_{ii}^{-1} = \lambda_i^0$. Update these using

$$\begin{aligned} \beta &= \frac{\sum_i w_i(Y - b)}{\sum_i w_i} \\ b &= (W + \sigma_b^{-2} R^{-1})^{-1} W(Y - \beta), \end{aligned}$$

where $W = \text{diag}[w_i]$. In practice we can start with $b = 0$ and $\beta^0 = \log \overline{y}$.

Estimation of σ_b^2 is the same as in the previous binomial example; the iterative procedure and updating formula (18.15) for σ_b^2 also apply. As before, it is more intuitive to express the amount of smoothing by the model degrees of freedom associated with a choice of σ_b^2:

$$\text{df} = \text{trace}\{(W + \sigma_b^{-2} R^{-1})^{-1} W\}.$$

Example 18.10: For the geyser data, first partition the range of the data (from 43 to 108) into 40 intervals. The count data y_i's in these intervals are

```
 1  1  2 12 17  5 16  3 11  8  6  8  2  7  2  3  5 11  6 17
18 17 24 12 14 18  5 21  9  2 11  1  2  1  0  0  0  0  0  1
```

Figure 18.13 shows the density estimate of the waiting time using the above method (solid line) with $d = 2$ and an estimated smoothing parameter $\widehat{\sigma}_b^2 = 0.042$ (corresponding df = 11.1). The density estimate matches closely the kernel density estimate using the optimal choice $\sigma = 2.2$. □

Example 18.11: This is to illustrate the problem of the standard kernel estimate at the boundary. The data are simulated absolute values of the standard normal; the true density is twice the standard normal density on the positive side. The complete dataset is too long to list, but it can be reproduced reasonably using the following information. The values range from 0 to 3.17, and on partitioning them into 40 equispaced intervals, we obtain the following count data y_i:

```
17 14 15 20 17 15 16 17 19 14  7  9 14  7 10 11  5  8  5 10
10  4  6  7  4  2  5  3  1  1  2  0  1  0  0  1  1  0  0  1
```

Figure 18.14 shows the density estimates using the mixed model approach (solid line, based on $\widehat{\sigma}_b^2 = 0.0006$ or df = 4.3) and the kernel method (dotted line, with optimal choice $\sigma = 0.06$; and dashed line, with $\sigma = 0.175$). Using smaller σ the kernel estimate has less bias at the boundary, but the estimate is visibly too noisy, while larger σ has the opposite problem. □

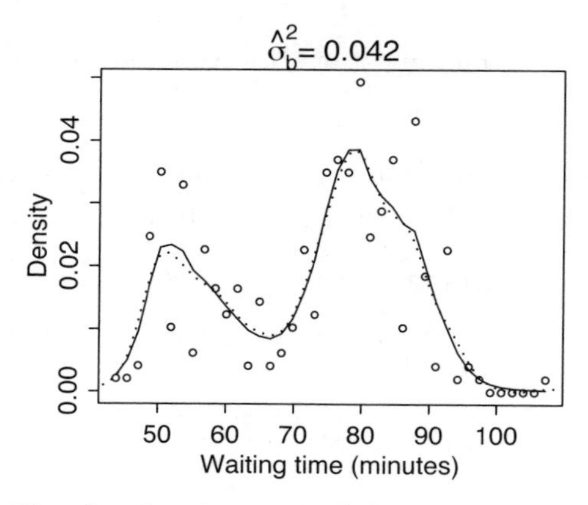

Figure 18.13: *The density estimate of the geyser waiting time using the mixed model approach (solid) and the kernel method (dotted). The smoothing parameters of both methods are estimated from the data. The scattered points are the counts y_i's scaled so that as a step function they integrate to one.*

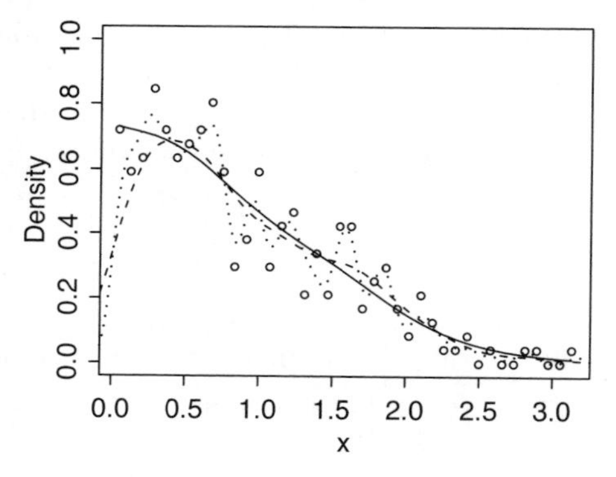

Figure 18.14: *The density estimates of a simulated dataset using the mixed model approach (solid) and the kernel method with* $\sigma = 0.06$ *(dotted) and* $\sigma = 0.175$ *(dashed). The scattered points are the scaled count data.*

18.12 Nonnormal smoothness condition*

The normal smoothness assumption is convenient computationally, and it is adequate in most applications. A nonnormal model may be required, for example, if we suspect the underlying function is discontinuous so its derivative might be heavy-tailed. Typically we still make an assumption that the d'th-order difference

$$\Delta^d b_i = a_i$$

is iid with some distribution with location zero and scale σ_b; these do not have to be mean and standard deviation, so, for example, the assumption covers the Cauchy or double-exponential models.

Let $\ell(a)$ be the log-likelihood contribution of a. Using starting value b^0, as in Section 17.10, we can first approximate $\ell(a)$ by

$$\ell(a) \approx \ell(a^c) - \frac{1}{2}(a - a^c)' D^{-1}(a - a^c),$$

where $D^{-1} = \text{diag}[-\ell''(a^0)]$, $a^0 = \Delta^d b^0$, and

$$a^c = a^0 + D\ell'(a^0).$$

Therefore,

$$\ell(a) \approx \ell(a^c) - \frac{1}{2}(\Delta^d b - a^c)' D^{-1}(\Delta^d b - a^c).$$

The derivative of $\ell(a)$ with respect to b is

$$(-\Delta^d)' D^{-1} \Delta^d b + (\Delta^d)' D^{-1} a^c.$$

Combining this with the quadratic approximation of $\log p(y|b)$, we obtain the updating equation

$$\begin{pmatrix} X'\Sigma^{-1}X & X'\Sigma^{-1}Z \\ Z'\Sigma^{-1}X & Z'\Sigma^{-1}Z + (\Delta^d)'D^{-1}\Delta^d \end{pmatrix} \begin{pmatrix} \beta \\ b \end{pmatrix} = \begin{pmatrix} X'\Sigma^{-1}Y \\ Z'\Sigma^{-1}Y + (\Delta^d)'D^{-1}a^c \end{pmatrix}.$$

In the normal case, $a^c = 0$, and the term $(\Delta^d)'D^{-1}\Delta^d$ reduces to $\sigma_b^{-2} R^{-1}$.

18.13 Exercises

Exercise 18.1: Show that the combined matrix on the left-hand side of (18.5) is singular.

Exercise 18.2: In Example 6.5 find the nonparametric smooth of the number of claims as a function of age. Compare it with the parametric fit. Find the confidence band for the nonparametric fit.

Exercise 18.3: Earthquake wave signals exhibit a changing variance, indicating the arrival of the different phases of the wave.

```
  -0.24  -0.19  -0.43  -1.30 -0.16  -1.15   1.42  -0.46   0.85  -0.62
   0.12   0.17  -0.32   0.48 -1.38   0.08  -0.22  -1.50  -0.27   2.38
  -1.72  -1.14  -0.47  -0.32  2.97  -1.76  -0.36   0.47  -0.89  -5.60
   9.30  -3.20   5.42  -7.51  3.44   0.02  -0.29  -9.37 -54.77   4.27
 -34.94  26.26  13.51 -87.68  1.85 -13.09 -26.86 -27.29   3.26 -13.75
  17.86 -11.87 -11.63   4.55  4.43  -2.22 -56.21 -32.45  12.96   9.80
  -6.35   1.17  -2.49  11.47 -7.25  -7.95  -8.03   7.64  25.63   9.12
  10.24 -19.08  -3.37 -13.86  7.60 -15.44   5.12   2.90   0.41  -4.92
  14.30   5.72 -10.87   1.86 -1.73  -2.53  -1.43  -2.93  -1.68  -0.87
   9.32   3.75   3.16  -6.34 -0.92   7.10   2.35   0.24   2.32  -2.72
  -2.95  -2.57  -1.63   2.06 -1.66   4.11   0.90  -2.21   2.71  -1.08
  -1.22  -0.68  -2.78  -1.91 -2.68  -0.95   1.17  -0.72
```

Assume the signal is observed at regular time points $i = 1, \ldots, N$, and y_i's are independent normal with mean zero and variance σ_i^2, where σ_i^2 changes smoothly over time.

(a) Develop a smoothing procedure, including the computational algorithm, to estimate the variance as a function of time.

(b) Apply it to the observed data.

(c) Find the prediction band for the variance function.

Exercise 18.4: Find the nonparametric estimate of the intensity function from the software failure data in Example 11.9, and compare it with the parametric estimate. Discuss the advantages and disadvantages of each estimate.

Exercise 18.5: For the epilepsy data in Section 11.9 find the nonparametric estimate of the baseline intensity function associated with the Cox model as described in Section 11.10.

Exercise 18.6: The following time series (read by row) is computed from the daily rainfall data in Valencia, southwest Ireland, from 1985 to 1994. There are 365 values in the series, each representing a calendar date with the 29th of February removed. Each value is the number of times during the ten year period the rain exceeded the average daily amount (4 mm); for example, on January 1st there were 5 times the rain exceeded 4 mm.

```
5 3 6 7 4 2 6 6 7 4 6 3 5 1 2 1 1 5 4 3 4 4 5 5 5 3 2 4 4
3 3 5 3 4 4 2 4 4 5 2 3 4 3 5 5 1 1 4 1 5 2 1 3 3 2 7 4 2
3 4 2 3 4 3 2 2 2 2 1 4 5 3 2 3 1 4 6 2 7 5 2 2 3 1 3 4 4
5 3 5 5 4 4 4 4 2 2 2 2 5 4 1 2 2 3 1 1 2 3 2 2 4 3 4 2 2
1 2 3 2 2 4 2 2 2 1 1 1 1 3 2 1 2 2 4 2 2 1 0 1 2 2 2 2 3
2 1 3 2 2 0 1 4 2 0 4 2 0 3 2 1 3 2 1 1 1 1 3 1 2 3 5 3 0
4 2 5 2 2 2 3 1 3 1 2 3 2 2 1 5 2 4 4 2 3 3 3 3 3 1 1 0 2
3 3 4 4 3 4 4 1 2 3 1 4 3 4 5 0 3 2 3 5 4 4 3 4 4 3 2 1 4
2 4 4 1 3 3 4 3 2 5 2 1 2 1 2 2 3 2 1 1 3 2 3 1 3 3 1 1 5
3 3 5 1 1 2 3 0 1 3 1 4 6 3 4 4 3 5 5 5 3 3 3 2 2 3 0 1 5
4 5 4 5 2 4 5 5 3 4 5 6 5 2 3 5 2 4 3 2 3 5 8 4 5 5 5 4 4
4 5 4 4 3 3 2 4 4 3 1 3 4 4 3 2 3 5 6 2 5 4 4 1 3 2 3 2 3
4 3 6 2 2 7 5 4 4 7 5 4 5 3 5 6 6
```

Assuming a binomial model for the observed data, compute the smoothed probability of exceeding the mean rainfall as a function of calendar time. Present also the prediction band around the smoothed estimate.

Bibliography

Agresti, A. (1996). *An introduction to categorical data analysis.* New York: Wiley.

Agresti, A. and Coull, B.A. (1998). Approximate is better than 'exact' for interval estimation of binomial proportions. *American Statistician*, **52**, 119–126.

Akaike, H. (1974). A new look at the statistical model identification. *IEEE Transaction on Automatic Control*, **AC-19**, 716–723.

Aldrich, J. (1997). R.A. Fisher and the making of maximum likelihood. *Statistical Science,* **12**, 162–176.

Anderson, T.W. (1984). *An introduction to multivariate statistical analysis.* New York: John Wiley & Sons.

Andersen, P.K., Borgan, O., Gill, R. and Keiding, N. (1993). *Statistical models based on counting processes.* Berlin: Springer-Verlag.

Armitage, P. (1961). Contribution to the discussion of 'Consistency in statistical inference and decision,' by C.A.B. Smith. *Journal of the Royal Statistical Society*, Series B, **23**, 1–37.

Apostol, T. (1974). *Mathematical analysis.* Reading, Massachusetts: Addison-Wesley.

Azzalini A. and Bowman A.W. (1990). A Look at some data on the Old Faithful geyser. *Applied Statistics*, **39**, 357–365.

Barnard, G.A. (1949). Statistical inference. *Journal of the Royal Statistical Society*, Series B, **11**, 115–149.

Barnard, G.A. (1963). Fisher's contribution to mathematical statistics. *Journal of the Royal Statistical Society*, Series A, **126**, 162–166.

Barnard, G.A., Jenkins, G.M. and Winsten, C.B. (1962). Likelihood inference and time series. *Journal of the Royal Statistical Society*, Series A, **125**, 321–372.

Barndorff-Nielsen, O.E. (1983). On a formula for a distribution of the maximum likelihood estimator. *Biometrika*, **70**, 343–365.

Barndorff-Nielsen, O.E. and Cox, D.R. (1984). Bartlett adjustment to the likelihood ratio statistic and the distribution of the maximum likelihood estimator. *Journal of the Royal Statistical Society*, Series B, **46**, 483–495.

Bartlett, M.S. (1953). Approximate confidence intervals. *Biometrika*, **40**, 12–19.

Bartlett, M.S. (1965). R.A. Fisher and the fifty years of statistical methodology. *Journal of the American Statistical Association*, **60**, 395–409.

Bartlett, M.S. (1969). When is inference *statistical* inference. In *Probability, Statistics and Time*, 98–110, by M.S. Bartlett (1975). London: Chapman and Hall.

Bartlett, M.S. (1971). Two-dimensional nearest neighbour systems and their ecological applications. In *Statistical Ecology*, Vol. I, 179–194. University Park: Pennsylvania State University Press.

Bayarri, M.J., DeGroot, M.H., and Kadane, J.B. (1987). What is the likelihood function? In *Statistical Decision Theory and Related Topics IV*, Vol. 1, S.S. Gupta and J. Berger (Eds.). New York: Springer Verlag.

Bennett, J.H. and Cornish, E.A. (1974). *Collected papers of R.A. Fisher,* 5 Vols. University of Adelaide.

Berger, J.O. (2000). Bayesian analysis: a look at today and thoughts of tomorrow. *Journal of the American Statistical Association*, **95**, 1269–1276.

Berger, J.O. and Berry, D. (1987). The relevance of stopping rules in statistical inference. In *Statistical Decision Theory and Related Topics IV*, Vol. 1, S.S. Gupta and J. Berger (Eds.). New York: Springer Verlag.

Berger, J.O. and Wolpert, R.L. (1988). *The likelihood principle.* Hayward: Institute of Mathematical Statistics.

Besag, J.E. (1974). Spatial interaction and the statistical analysis of lattice systems (with discussion). *Journal of the Royal Statistical Society*, Series B, **36**, 192–236.

Besag, J.E. (1975). Statistical analysis of non-lattice data. *The Statistician*, **24**, 179–195.

Besag, J.E. and Clifford, P. (1989). Generalized Monte Carlo significant tests. *Biometrika*, **76**, 633–642.

Besag, J.E., Yorke, J. and Mollié, A. (1991). Bayesian image restoration with two applications in spatial statistics. *Annals of the Institute of Statistics and Mathematics*, **43**, 1–59.

Bickel, P.J. and Doksum, K.A. (1981). An analysis of transformations revisited. *Journal of the American Statistical Association*, **76**, 296–311.

Birnbaum, A. (1962). On the foundation of statistical inference. *Journal of the American Statistical Association*, **57**, 269–326.

Birnbaum, A. (1970). More on concepts of statistical evidence. *Journal of the American Statistical Association*, **67**, 858–861.

Bjørnstad, J.F. (1996). On the generalization of the likelihood function and likelihood principle. *Journal of the American Statistical Association*, **91**, 791–806.

Boole, G. (1854). *The Laws of thought.* Reissued as Vol. II of the *Collected Logical Works.* (1952). Illinois: La Salle.

Box, G.E.P. and Cox, D.R. (1964). An analysis of transformation (with discussion). *Journal of the Royal Statistical Society*, Series B, **26**, 211–252.

Box, G.E.P. and Cox, D.R. (1982). An analysis of transformations revisited, rebutted. *Journal of the American Statistical Association*, **77**, 209–210.

Box, G.E.P., Jenkins G.M. and Reinsel, G.C. (1994). *Time series analysis: forecasting and control.* Englewood Cliffs, N.J.: Prentice-Hall.

Box, J.F. (1978). *R.A. Fisher, the life of a scientist.* New York: Wiley.

Breslow, N.E. (1981). Odds ratio estimators when the data are sparse. *Biometrika*, **68**, 73–84.

Breslow, N.E. and Clayton, D. G. (1993). Approximate inference in generalized linear mixed models. *Journal of the American Statistical Association*, **88**, 9–25.

Breslow, N.E. and Day, N. (1980). *Statistical methods in cancer research.* Lyon: IARC.

Brownlee, K.A. (1965). *Statistical theory and methodology in science and engineering*, 2nd edn. New York: Wiley.

Buehler, R.J. and Fedderson, A.P. (1963). Note on a conditional property of Student's *t*. *Annals of Mathematical Statistics*, **34**, 1098–1100.

Burnham, K.P. and Anderson, D.R. (1998). *Model Selection and inference: a practical information-theoretic approach.* New York: Springer Verlag.

Butler, R.W. (1986). Predictive likelihood inference with applications (with discussion). *Journal of the Royal Statistical Society*, Series B, **48**, 1–38.

Butler, R.W. (1987). A likely answer to 'What is the likelihood function?'. In *Statistical decision theory and related topics IV*, Vol. 1, eds. S.S. Gupta and J.O. Berger (Eds.). New York: Springer Verlag.

Campbell, R. and Sowden, L. (1985). *Paradoxes of rationality and cooperation: prisoner's dilemma and Newcomb's problem.* Vancouver: University of British Columbia Press.

Christensen, R. and Utts, J. (1992). Bayesian resolution of the 'exchange paradox'. *American Statistician*, **46**, 274–276. Correction in the same journal in 1996, page 98.

Chung, K.L. (1974). *A course in probability theory*, 2nd edn. New York: Academic Press.

Clopper, C.J. and Pearson, E.S. (1934). The use of confidence or fiducial limits illustrated in the vase of the binomial. *Biometrika*, **26**, 404–413.

Cox, D.R. (1958). Some problems connected with statistical inference. *Annals of Mathematical Statistics*, **29**, 357–372.

Cox, D.R. (1972). Regression models and life tables (with discussion). *Journal of the Royal Statistical Society*, Series B, **34**, 187–220.

Cox, D.R. (1975). Partial likelihood. *Biometrika*, **62**, 269–276.

Cox, D.R. (1978). Foundations of statistical inference: the case for ecclectism. *Australian Journal of Statistics*, **20**, 43–59.

Cox, D.R. (1990). The role of models in statistical analysis. *Statistical Science*, **5**, 169–174.

Cox, D.R. and Reid, N. (1987). Parameter orthogonality and approximate conditional inference (with discussion). *Journal of the Royal Statistical Society*, Series B, **49**, 1–39.

Cox, D.R. and Snell, E.J. (1981). *Applied statistics: principles and examples.* London: Chapman and Hall.

Cramér, H. (1955). *The elements of probability theory and some of its applications.* New York: Wiley.

Craven, P. and Wahba, G. (1979). Smoothing noisy data with spline functions. *Numerische Mathematik*, **31**, 377–403.

Crowder, M.J. (1978). Beta-binomial Anova for proportions. *Applied Statistics*, **27**, 34–37.

Daniel, C. and Wood, F.S. (1971). *Fitting equations to data.* New York: Wiley.

Darwin, C. (1876). *The effect of cross- and self-fertilization in the vegetable kingdom*, 2nd edn. London: John Murray.

Davison, A.C., Hinkley, D.V. and Worton, B.J. (1992). Bootstrap likelihoods. *Biometrika*, **79**, 113–30.

Davison, A.C., Hinkley, D.V. and Worton, B.J. (1995). Accurate and efficient construction of bootstrap likelihoods. *Statistical Computing*, **5**, 257–64.

deBoor, C. (1978). *A practical guide to splines.* New York: Springer Verlag.

Dempster, A.P., Laird, N.M. and Rubin, D.B. (1977). Maximum likelihood from incomplete data via the EM algorithm (with discussion). *Journal of the Royal Statistical Society*, Series B, **39**, 1–38.

Denby, L. and Mallows, C.L. (1977). Two diagnotic displays for robust regression analysis. *Technometrics*, **19**, 1-13.

Diaconis, P. (1985). Theories of data analysis: from magical thinking through classical statistics. In *Exploring data, tables, trends and shapes*, D.C. Hoaglin, F. Mosteller and J.W. Tukey (Eds.). New York: Wiley.

DiCiccio, T.J., Hall, P. and Romano, J.P. (1989). Comparison of parametric and empirical likelihood functions. *Biometrika*, **76**, 465–476.

Diggle, P. (1983). *Statistical analysis of spatial point patterns.* London: Academic Press.

Dongarra, J.J., Bunch, J.R., Moler, C.B. and Stewart, G.W. (1979). *LINPACK Users' Guide.* Philadelphia: SIAM.

Draper, N.R. and Smith, H. (1981). *Applied regression analysis*, 2nd edn. New York: Wiley.

Durbin, J. (1960). Estimation of parameters in time-series regression models. *Biometrika*, **47**, 139–153.

Durbin, J. (1980). Approximations for densities of sufficient estimators. *Biometrika*, **67**, 311–333.

Edwards, A.W.F. (1974). The history of likelihood. *International Statistical Review*, **49**, 9–15.

Edwards, A.W.F. (1992). *Likelihood*, expanded edition. Baltimore: Johns Hopkins University Press.

Edwards, W., Lindman, H. and Savage, L.J. (1963). Bayesian statistical inference for psychological research. *Psychological Review*, **70**, 193–242.

Efron, B. (1971). Does an observed sequence of numbers follow a simple rule? (Another look at Bode's law). *Journal of the American Statistical Association*, **66**, 552–559.

Efron, B. (1977). The efficiency of Cox's likelihood function for censored data. *Journal of the American Statistical Association*, **72**, 557–565.

Efron, B. (1979). Bootstrap methods: another look at the jackknife. *Annals of Statistics*, **7**, 1–26.

Efron, B. (1982). *The Jackknife, the bootstrap and other resampling plan.* Vol. 38 of *CBMS-NSF Regional Conference Series in Applied Mathematics.* Philadelphia: SIAM.

Efron, B. (1986a). Why isn't everybody a Bayesian? *American Statistician*, 40, 1–11.

Efron, B. (1986b). Double exponential families and their use in generalized linear regression. *Journal of the American Statistical Association*, **81**, 709–721.

Efron, B. (1987). Better bootstrap confidence intervals (with discussion). *Journal of the American Statistical Association*, **82**, 171-200.

Efron, B. (1993). Bayes and likelihood calculations from confidence intervals. *Biometrika* **80**, 3–26.

Efron, B. (1998). R.A. Fisher in the 21st century. *Statistical Science*, **13**, 95–122.

Efron, B. and Hinkley, D.V. (1978). Assessing the accuracy of the maximum likelihood estimator: observed versus expected Fisher information. *Biometrika*, 65, 457–482.

Efron, B. and Tibshirani, R. J. (1993). *An Introduction to the Bootstrap.* New York: Chapman and Hall.

Elston, R.C. and Grizzle, J.E. (1962). Estimation of time-response curves and their confidence bands. *Biometrics*, **18**, 148–159.

Evans, D. (1953). Experimental evidence concerning contagious distributions in ecology. *Biometrika*, **40**, 186–211.

Fadeley, R.C. (1965). Oregon malignancy pattern physiographically related to Hanford, Washington, radioisotope storage. *Journal of Environmental Health*, **27**, 883–897.

Fairley, W.B. (1977). Accidents on Route 2: two-way structure for data. In *Statistics and Public Policy*, W.B. Fairley and F. Mosteller (Eds). Reading, Massachusetts: Addison-Wesley.

Fears, T.R., Benichou, J. and Gail, M.H. (1996). A reminder of the fallibility of the Wald statistic. *American Statistician*, **50**, 226–227.

Feller, W. (1968). *An introduction to probability theory and its applications*, 3rd edn. Vol. 1. New York: Wiley.

Fienberg, S.E. and Hinkley, D.V. (1980). *R.A. Fisher: An appreciation.* New York: Springer Verlag

Fisher, R.A. (1912). On an absolute criterion for fitting frequency curves. *Messenger of Mathematics*, **41**, 155-160.

Fisher, R.A. (1921). On the 'probable error' of a coefficient of correlation deduced from a small sample. *Metron*, **1**, 2–32.

Fisher, R.A. (1922). On the mathematical foundations of theoretical statistics. *Philosophical Transactions of the Royal Society of London*, Series A, **222**, 309–368.

Fisher, R.A. (1925). Theory of statistical estimation. *Proceedings of the Cambridge Philosophical Society*, **22**, 700–725.

Fisher, R.A. (1930). Inverse probability. *Proceedings of the Cambridge Philosophical Society*, **26**, 528–535.

Fisher, R.A. (1933). The concepts of inverse probability and fiducial probability referring to unknown parameters. *Proceedings of the Royal Society A*, **139**, 343–348.

Fisher, R.A. (1934). Two new properties of mathematical likelihood. *Proceedings of the Royal Society A*, **144**, 285–307.

Fisher, R.A. (1936). Uncertain inference. *Proceedings of the American Academy of Arts and Sciences*, **71**, 245–258.

Fisher, R.A. (1973). *Statistical methods and scientific inference*, 3rd edn. New York: Hafner.

Fraser, D.A.S., Monette, G. and Ng, K.W. (1984). Marginalization, likelihood and structural models. In *Multivariate Analysis VI,* P.R. Krisnaiah (Ed.). Amsterdam: North-Holland.

Gatsonis, C., Hodges, J.S., Kass, R.E. and McCulloch, R.E. (1997). *Case studies in Bayesian statistics*, Vol. 3. New York: Springer Verlag.

Gilks, W.R., Spiegelhalter, D.J. and Richardson, S. (1995). *Markov Chain Monte Carlo in practice.* London: Chapman and Hall.

Gill, P.E., Murray, W., Saunders, M.A. and Wright, M.H. (1986). User's guide to NPSOL (Version 4.0): a FORTRAN package for nonlinear programming. Technical Report SOL 86-2, Department of Operations Research, Stanford University.

Godambe, V.P. (1960). An optimum property of a regular maximum likelihood estimation. *Annals of Mathematical Statistics*, **31**, 1208–1212.

Godambe, V.P. and Thompson, M.E. (1976). Philosophy of survey sampling practice. In *Foundations of Probability Theory, Statistical Inference and Statistical Theories of Science, Vol. II,* W.L. Harper and C.A. Hooker (Eds.). Dordrecht, The Netherlands: Reidel.

Goldstein, M. and Howard, J.V. (1991). A likelihood paradox (with discussion). *Journal of the Royal Statistical Society*, Series B, **53**, 619–628.

Gong, G. and Samaniego, F.J. (1981). Pseudo maximum likelihood estimation: theory and applications. *Annals of Statistics*, **9**, 861–869.

Good, I.J. (1969). A subjective evaluation of Bode's law and an objective test for approximate numerical rationality. *Journal of the American Statistical Association*, **64**, 23-66.

Grambsch, P. and Therneau, T. (1994). Proportional hazards tests and diagnostics based on weighted residuals. *Biometrika*, **81**, 515–526.

Green, P.J. and Silverman, B.W. (1993). *Nonparametric regression and generalized linear models: a roughness penalty approach.* London: Chapman and Hall.

Greenwood, M. and Yule, G.U. (1920). An inquiry into the nature of frequency distributions representative of multiple happenings with particular reference to the occurrence of multiple attacks of disease or of repeated accidents. *Journal of the Royal Statistical Society*, **83**, 255–279.

Hacking, I. (1965). *Logic of Statistical Inference.* Cambridge: Cambridge University Press.

Hald, A. (1999). On the history of maximum likelihood in relation to inverse probability and least squares. *Statistical Science*, **14**, 214–222.

Harville, D. (1974). Bayesian inference for variance components using only error contrasts. *Biometrika*, **61**, 383–385.

Harville, D. (1977). Maximum likelihood approaches to variance component estimation. *Journal of the American Statistical Association*, **72**, 320–340.

Helland, I.S. (1995). Simple counterexamples against the conditionality principle. *American Statistician,* **49**, 351–356.

Hinkley, D.V. (1979). Predictive likelihood. *Annals of Statistics*, **7**, 718–728.

Hoaglin, D.C. (1985). Using quantiles to study shape. In *Exploring data, tables, trends and shapes,* D.C. Hoaglin, F. Mosteller and J.W. Tukey (Eds.). New York: Wiley.

Hoaglin, D.C. and Tukey, J.W. (1985). Checking the shape of discrete distributions. In *Exploring data, tables, trends and shapes,* D.C. Hoaglin, F. Mosteller and J.W. Tukey (Eds.). New York: Wiley.

Hochberg, Y. and Tamhane, A.C. (1987). *Multiple comparison procedures.* New York: Wiley.

Hotelling, H. (1951). The impact of R.A. Fisher on statistics. *Journal of the American Statistical Association*, **46**, 35–46.

Ihaka, R. and Gentleman, R. (1996). R: a language for data analysis and graphics. *Journal of Computational and Graphical Statistics*, **5**, 299-314.

Jeffreys, H. (1961). *Theory of probability.* 3rd ed. Oxford: Clarendon Press.

Jenkins, B.M. and Johnson, J. (1975). *International terrorism: a chronology, 1968–1974.* Report R-1597-DOS/ARPA, Rand Corporation.

Johnson, R. and Wichern, D. (1992): *Applied Multivariate Statistical Analysis,* 3rd ed. Englewood Cliffs, N.Y.: Prentice-Hall.

Jones, M.C., Marron, J.S. and Sheather, S.J. (1996). Progress in data-based bandwidth selection for kernel density estimation. *Computational Statistics*, **11**, 337–381.

Jørgensen, B. (1987). Exponential dispersion models (with discussion). *Journal of the Royal Statistical Society*, Series B, **49**, 127–162.

Kalbfleisch, J.D. and Prentice, R.L. (1980). *Statistical analysis of failure time data*, New York: Wiley.

Kalbfleisch, J.D. and Sprott, D.A. (1969). Application of likelihood methods to models involving large number parameters. *Journal of the Royal Statistical Society*, Series B, **32**, 125–208.

Kass, R. and Raftery, A. (1995). Bayes factors and model uncertainty. *Journal of the American Statistical Association*, **91**, 1343–1370.

Kendall, M.G. (1963). Ronald Aylmer Fisher, 1890–1962. *Biometrika*, **50**, 1–15.

Kendall, M.G., Stuart, A. and Ord, J.K. (1977). *Advanced theory of statistics*, 4th edn. London: Griffin.

Kiefer, J. and Wolfowitz, J. (1956). Consistency of the maximum likelihood estimator in the presence of infinitely many incidental parameters. *Annals of Mathematical Statistics*, **27**, 887–906.

Kolaczyk, E.D. (1994). Empirical likelihood for generalized linear models. *Statistica Sinica*, **4**, 199–218.

Kullback, S. and Leibler, R.A. (1951). On information and sufficiency. *Annals of Mathematical Statistics*, **22**, 79–86.

Lancaster, H.O. (1961). Significance tests in discrete distributions. *Journal of the American Statistical Association*, **56**, 223–234.

Lange, K.L. (1995). A quasi-Newton acceleration of the EM algorithm. *Statistica Sinica,* **5**, 1–18.

Lange, K.L., Little, R.J.A. and Taylor, J.M.G. (1989). Robust statistical modelling using the t distribution. *Journal of the American Statistical Association*, **84**, 881–896.

Lawless, J.F. (1987). Regression methods for Poisson process data. *Journal of the American Statistical Association*, **82**, 808–815.

Lee, Y. and Nelder, J. A. (1996). Hierarchical generalized linear models (with discussion). *Journal of the Royal Statistical Society*, Series B, **58**, 619–678.

Lee, Y. and Nelder, J. A. (2000). Two ways of modelling overdispersion in non-normal data. *Journal of the Royal Statistical Society*, Series C, **58**, 591–599.

Lehmann, E.L. (1983). *Theory of point estimation.* New York: Wiley.

Lehmann, E.L. (1986). *Testing statistical hypotheses.* New York: Wiley.

Lehmann, E.L. (1990). Model specification: the views of Fisher and Neyman, and later developments. *Statistical Science*, **5**, 160–168.

Lehmann, E.L. (1993). The Fisher, Neyman-Pearson theories of testing hypotheses: one theory or two? *Journal of the American Statistical Association*, **88**, 1242–1249.

Lehmann, E.L. and Scheffé, H. (1950). Completeness, similar regions, and unbiased estimation. *Sankhya*, **10**, 305–340.

Li, G. (1985). Robust regression. In *Exploring data, tables, trends and shapes,* D.C. Hoaglin, F. Mosteller and J.W. Tukey (Eds.). New York: Wiley.

Liang, K-Y. and Zeger, S.L. (1986). Longitudinal data analysis using generalized linear models. *Biometrika*, **73**, 13–22.

Lindsey, J.K. (1995). Fitting parametric counting processes by using log-linear models. *Applied Statistics*, **44**, 201–212.

Lindsey, J.K. (1996). *Parametric statistical inference.* Oxford: Oxford University Press.

Lindsey, J.K. (1999a). Some statistical heresies (with discussion). *The Statistician*, **48**, 1–40.

Lindsey, J.K. (1999b) Relationships among model selection, likelihood regions, sample size, and scientifically important differences. *The Statistician*, **48**, 401–411.

Lindsey, J.K. (2000). A family of models for uniform and serial dependence in repeated measures studies. *Applied Statistics*, **49**, 305–320.

Lindsey, J.K. and Altham, P.M.E. (1998). Analysis of human sex ratio by using overdispersed models. *Applied Statistics,* **47**, 149–157.

Lipton, P. (1993). *Inference to the best explanation.* London: Routledge.

Louis, T.A. (1982). Finding the observed information matrix when using the EM algorithm. *Journal of the Royal Statistical Society*, Series B, **44**, 226–233.

McCullagh, P. and Nelder, J.A. (1989). *Generalized linear models*, 2nd edn. London: Chapman and Hall.

McLachlan, G.J. and Basford, K.E. (1988). *Mixture models: inference and applications to clustering.* New York: Marcel Dekker.

Mehta, C.R. and Walsh, S.J. (1992). Comparison of exact, mid-p and Mantel-Haenszel confidence intervals for the common odds ratio across several 2×2 contigency tables. *American Statistician,* **46**, 146–150.

Mélard, G. (1984). A fast algorithm for the exact likelihood of autoregressive-moving average models. *Applied Statistics*, **33**, 104–114.

Meng, X.L. and Rubin, D.B. (1991). Using the EM algorithm to obtain asymptotic variance-covariance matrices: the SEM algorithm. *Journal of the American Statistical Association*, **86**, 899–909.

Meng, X.L. and van Dyk, D. (1997). The EM algorithm: an old folk song sung to a fast new tune (with discussion). *Journal of the Royal Statistical Society*, Series B, **59**, 511–540.

Miller, R.G. (1964). A trustworthy jackknife. *Annals of Mathematical Statistics*, **35**, 1594–1605.

Miller, R.G. (1981). *Simultaneous statistical inference.* New York: Springer Verlag.

Miralles, F.S., Olaso, M.J., Fuentes, T., Lopez, F., Laorden, M.L. and Puig, M.M. (1983). Presurgical stress and plasma endorphin level. *Anesthesiology*, **59**, 366–367.

Monti, A.C. and Ronchetti, E. (1993). On the relationship between the empirical likelihood and empirical saddlepoint approximation for multivariate M-estimators. *Biometrika*, **80**, 329–338.

Mosteller, F. and Wallace, D.L. (1964). *Inference and disputed authoship: The Federalist.* Reading, MA: Addison-Wesley.

Musa, J.D., Iannino, A. and Okumoto, K. (1987). *Software reliability: measurement, prediction, application.* New York: McGraw-Hill.

Navidi, W. (1997). A graphical illustration of the EM algorithm. *American Statistician*, **51**, 29–31.

Nelder, J.A. and Pregibon, D. (1987). An extended quasi-likelihood function. *Biometrika*, **74**, 221–232.

Neyman, J. (1935). On the problem of confidence intervals. *Annals of Mathematical Statistics*, **6**, 111-116.

Neyman, J. (1961). Silver jubilee of my dispute with Fisher. *Journal of the Operation Research Society of Japan*, **3**, 145–154.

Neyman, J. (1967). R.A. Fisher (1890-1962): an appreciation. *Science*, **156**, 1456–1460.

Neyman, J. (1977). Frequentist probability and frequentist statistics. *Synthese*, **36**, 97–131.

Neyman, J. and Pearson, E.S. (1933). On the problem of the most efficient tests of statistical hypotheses. *Philosophical Transactions of the Royal Society of London*, Series A, **231**, 289–337.

Neyman J. and Scott E.L. (1948). Consistent estimates based on partially consistent observations. *Econometrika*, **16**, 1–32.

Oakes, D. (1999). Direct calculation of the information matrix via the EM algorithm. *Journal of the Royal Statistical Society*, Series B, **61**, 479–482.

Orchard, T. and Woodbury, M.A. (1972). A missing information principle: theory and applications. Proceedings of the 6th Berkeley Symposium on Mathematical Statistics and Probability, **1**, 697–715.

O'Sullivan, F. (1986). A statistical perspective on ill-posed inverse problems *Statistical Science*, **1**, 502–518.

O'Sullivan, F. (1987). Fast computation of fully automated log-density and log-hazard estimators. *SIAM Journal of Scientific and Statistical Computing*, **9**, 363–379.

Owen, A.B. (1988). Empirical likelihood ratio confidence intervals for a single functional. *Biometrika*, **75**, 237–249.

Owen, A.B. (1990). Empirical likelihood ratio confidence regions. *Annals of Statistics*, **18**, 90–120.

Owen, A.B. (1991). Empirical likelihood for linear models. *Annals of Statistics*, **19**, 1725–1747.

Patterson, H.D. and Thompson, R. (1971). Recovery of inter-block information when block sizes are unequal. *Biometrika*, **58**, 545–554.

Pawitan, Y. (2000). Computing empirical likelihood using the bootstrap. *Statistics and Probability Letters*, **47**, 337–345.

Pearson, K. (1920). The fundamental problems of practical statistics. *Biometrika*, **13**, 1–16.

Pearson, E.S. (1974). Memories of the impact of Fisher's work in the 1920s. *International Statistical Reviews*, **42**, 5–8.

Pocock, S.J. (1977). Group sequential methods in the design and analysis of clinical trials. *Biometrika*, **64**, 191–199.

Press, W.H., Teukolsky, S.A., Vetterling, W.T. and Flannery, B.P. (1992). *Numerical recipes.* Cambridge: Cambridge University Press.

Ramsey, F.P. (1931). *The foundation of mathematics and other logical essays.* London: Routledge and Kegan Paul.

Rao, C.R. (1973). *Linear Statistical Inference and Its Applications*, 2nd edn. New York: Wiley.

Redelmeier, D.A. and Tibshirani, R.J. (1997). Association between cellular-telephone calls and motor vehicle collisions. *New England Journal of Medicine*, **336**, 453-458.

Reid, N. (1988). Saddlepoint methods and statistical inference. *Statistical Science*, **3**, 213–227.

Reid, N. (1995). The roles of conditioning in inference. *Statistical Science*, **10**, 138–199.

Reilly, M. and Lawlor, E. (1999). A likelihood-based method of identifying contaminated lots of blood product. *International Journal of Epidemiology*, **28**, 787-792.

Ripley, B.D. (1988). *Statistical inference for spatial processes.* Cambridge: Cambridge University Press.

Rosner, G.L. and Tsiatis, A.A. (1988). Exact confidence intervals following a group sequential trial: a comparison of methods. *Biometrika*, **75**, 723–729.

Royall, R.M. (1997). *Statistical evidence.* London: Chapman and Hall.

Savage, L.J. (1954). *Foundation of statistics.* New York: Wiley.

Savage, L.J. (1976). On rereading Fisher (with discussion). *Annals of Statistics*, **4**, 441–500.

Scott, D.W. and Terrell, G.R. (1987). Biased and unbiased cross-validation in density estimation. *Journal of the American Statistical Association*, **82**, 1131–1146.

Serfling, R.J. (1980). Approximation theorems of mathematical statistics. New York: Wiley.

Shafer, G. (1990). The unity and diversity of probability (with discussion). *Statistical Science*, **5**, 435–444.

Silverman, B.W. (1986). *Density estimation for statistics and data analysis.* London: Chapman and Hall.

Simon, H.A. (1985). The sizes of things. In *Statistics: a guide to the unknown*, J.M. Tanur *et al.* (Eds). Monterey, California: Wadsworth and Brooks/Cole.

Singh, M., Kanji, G.K. and El-Bizri, K.S. (1992). A note on inverse estimation in nonlinear models. *Journal of Applied Statistics*, **19**, 473–477.

Sokal, R.R. and Rohlf, F.J. (1981). *Biometry*, 2nd edn. San Francisco: W.H. Freeman.

Speckman, P. (1988). Kernel smoothing in partial linear models. *Journal of the Royal Statistical Society*, Series B, **50**, 413–436.

Sprott, D.A. (1975). Application of maximum likelihood methods to finite samples. *Sankhya*, **37**, 259–270.

Sprott, D.A. (2000). *Statistical inference in science.* New York: Springer Verlag.

Sprott, D.A. and Kalbfleisch, J.D. (1969). Examples of likelihoods and comparisons with point estimates and large sample approximations. *Journal of the American Statistical Association*, **64**, 468–484.

Steering Committee of the Physicians' Health Study Research Group (1989). Final report on the aspirin component of the ongoing Physicians' Health Study. *New England Journal of Medicine*, **321**(3): 129–35.

Stigler, S.M. (1977). Do robust estimators work with real data? *Annals of Statistics*, **5**, 1055–1098.

Stigler, S.M. (1982). Thomas Bayes's Bayesian inference. *Journal of the Royal Statistical Society*, Series A, **145**, 250–258.

Stone, M. (1974). Cross-validatory choice and assessment of statistical predictions. *Journal of the Royal Statistical Society*, Series B, **36**, 111–147.

Stone, M. (1977). An asymptotic equivalence of choice of model by cross-validation and Akaike's criterion. *Journal of the Royal Statistical Society*, Series B, **39**, 44–47.

Strauss, D. (1992). The many faces of logistic regression. *American Statistician*, **46**, 321–327.

Student (W.S. Gosset) (1908). The probable error of a mean. *Biometrika*, **6**, 1–25.

Thall, P.F. and Vail, S.C. (1990). Some covariance models for longitudinal count data with overdispersion. *Biometrics*, **46**, 657–671.

Venn, J. (1876). *The logic of chance*, 2nd edn. London: Macmillan.

Wahba, G. (1990). *Spline models for observational data.* Philadelphia: SIAM.

Wald, A. (1943). Tests of statistical hypotheses concerning several parameters when the number of observations is large. *Transaction of the American Mathematical Society*, **54**, 426–482.

Wald, A. (1947). *Sequential Analysis.* New York: Wiley.

Wald, A. (1949). Note on the consistency of the maximum likelihood estimate. *Annals pf Mathematical Statistics*, **20**, 493–497.

Weinberg, C.R. and Gladen, B.C. (1986). The beta-geometric distribution applied to comparative fecundibility studies. *Biometrics*, **42**, 547–560.

Weisberg, S. (1985). *Applied linear regression*, 2nd edn. New York: Wiley.

Whitehead, J. (1980). Fitting Cox's regression model to survival data using GLIM. *Applied Statistics*, **29**, 268–275.

Whitehead, J. (1986). On the bias of maximum likelihood estimation following a sequential test. *Biometrika*, **73**, 573–581.

Whitehead, J. Todd, S. and Hall, W.J. (2000). Confidence intervals for secondary parameters following a sequential test. *Journal of the Royal Statistical Society*, Series B, **62**, 731–745.

Wilks, S.S. (1938). The large sample distribution of the likelihood ratio for testing composite hypotheses. *Annals Mathematical Statistics*, **9**, 60–62.

Wu, C.F.J. (1983). On the convergence properties of the EM algorithm. *Annals of Statistics*, **11**, 95–103.

Yates, F. (1990). Foreword to *Statistical methods, experimental design and statistics inference,* by R. A. Fisher. Oxford: Oxford University Press.

Yates, F. and Mather, K. (1963). Ronald Aylmer Fisher. In *Collected Papers of R.A. Fisher,* J.H. Bennett and E.A. Cornish (Eds.). University of Adelaide.

Zeger, S.L. (1988). A regression model for time series of counts. *Biometrika*, **75**, 621–629.

Zeger, S.L. and Liang, K-Y. (1986). Longitudinal data analysis for discrete and continuous outcomes. *Biometrics*, **42**, 121–130.

Index